# *Dictionary of Literary Biography*

1 *The American Renaissance in New England,* edited by Joel Myerson (1978)

2 *American Novelists Since World War II,* edited by Jeffrey Helterman and Richard Layman (1978)

3 *Antebellum Writers in New York and the South,* edited by Joel Myerson (1979)

4 *American Writers in Paris, 1920-1939,* edited by Karen Lane Rood (1980)

5 *American Poets Since World War II,* 2 parts, edited by Donald J. Greiner (1980)

6 *American Novelists Since World War II, Second Series,* edited by James E. Kibler Jr. (1980)

7 *Twentieth-Century American Dramatists,* 2 parts, edited by John MacNicholas (1981)

8 *Twentieth-Century American Science-Fiction Writers,* 2 parts, edited by David Cowart and Thomas L. Wymer (1981)

9 *American Novelists, 1910-1945,* 3 parts, edited by James J. Martine (1981)

10 *Modern British Dramatists, 1900-1945,* 2 parts, edited by Stanley Weintraub (1982)

11 *American Humorists, 1800-1950,* 2 parts, edited by Stanley Trachtenberg (1982)

12 *American Realists and Naturalists,* edited by Donald Pizer and Earl N. Harbert (1982)

13 *British Dramatists Since World War II,* 2 parts, edited by Stanley Weintraub (1982)

14 *British Novelists Since 1960,* 2 parts, edited by Jay L. Halio (1983)

15 *British Novelists, 1930-1959,* 2 parts, edited by Bernard Oldsey (1983)

16 *The Beats: Literary Bohemians in Postwar America,* 2 parts, edited by Ann Charters (1983)

17 *Twentieth-Century American Historians,* edited by Clyde N. Wilson (1983)

18 *Victorian Novelists After 1885,* edited by Ira B. Nadel and William E. Fredeman (1983)

19 *British Poets, 1880-1914,* edited by Donald E. Stanford (1983)

20 *British Poets, 1914-1945,* edited by Donald E. Stanford (1983)

21 *Victorian Novelists Before 1885,* edited by Ira B. Nadel and William E. Fredeman (1983)

22 *American Writers for Children, 1900-1960,* edited by John Cech (1983)

23 *American Newspaper Journalists, 1873-1900,* edited by Perry J. Ashley (1983)

24 *American Colonial Writers, 1606-1734,* edited by Emory Elliott (1984)

25 *American Newspaper Journalists, 1901-1925,* edited by Perry J. Ashley (1984)

26 *American Screenwriters,* edited by Robert E. Morsberger, Stephen O. Lesser, and Randall Clark (1984)

27 *Poets of Great Britain and Ireland, 1945-1960,* edited by Vincent B. Sherry Jr. (1984)

28 *Twentieth-Century American-Jewish Fiction Writers,* edited by Daniel Walden (1984)

29 *American Newspaper Journalists, 1926-1950,* edited by Perry J. Ashley (1984)

30 *American Historians, 1607-1865,* edited by Clyde N. Wilson (1984)

31 *American Colonial Writers, 1735-1781,* edited by Emory Elliott (1984)

32 *Victorian Poets Before 1850,* edited by William E. Fredeman and Ira B. Nadel (1984)

33 *Afro-American Fiction Writers After 1955,* edited by Thadious M. Davis and Trudier Harris (1984)

34 *British Novelists, 1890-1929: Traditionalists,* edited by Thomas F. Staley (1985)

35 *Victorian Poets After 1850,* edited by William E. Fredeman and Ira B. Nadel (1985)

36 *British Novelists, 1890-1929: Modernists,* edited by Thomas F. Staley (1985)

37 *American Writers of the Early Republic,* edited by Emory Elliott (1985)

38 *Afro-American Writers After 1955: Dramatists and Prose Writers,* edited by Thadious M. Davis and Trudier Harris (1985)

39 *British Novelists, 1660-1800,* 2 parts, edited by Martin C. Battestin (1985)

40 *Poets of Great Britain and Ireland Since 1960,* 2 parts, edited by Vincent B. Sherry Jr. (1985)

41 *Afro-American Poets Since 1955,* edited by Trudier Harris and Thadious M. Davis (1985)

42 *American Writers for Children Before 1900,* edited by Glenn E. Estes (1985)

43 *American Newspaper Journalists, 1690-1872,* edited by Perry J. Ashley (1986)

44 *American Screenwriters, Second Series,* edited by Randall Clark, Robert E. Morsberger, and Stephen O. Lesser (1986)

45 *American Poets, 1880-1945, First Series,* edited by Peter Quartermain (1986)

46 *American Literary Publishing Houses, 1900-1980: Trade and Paperback,* edited by Peter Dzwonkoski (1986)

47 *American Historians, 1866-1912,* edited by Clyde N. Wilson (1986)

48 *American Poets, 1880-1945, Second Series,* edited by Peter Quartermain (1986)

49 *American Literary Publishing Houses, 1638-1899,* 2 parts, edited by Peter Dzwonkoski (1986)

50 *Afro-American Writers Before the Harlem Renaissance,* edited by Trudier Harris (1986)

51 *Afro-American Writers from the Harlem Renaissance to 1940,* edited by Trudier Harris (1987)

52 *American Writers for Children Since 1960: Fiction,* edited by Glenn E. Estes (1986)

53 *Canadian Writers Since 1960, First Series,* edited by W. H. New (1986)

54 *American Poets, 1880-1945, Third Series,* 2 parts, edited by Peter Quartermain (1987)

55 *Victorian Prose Writers Before 1867,* edited by William B. Thesing (1987)

56 *German Fiction Writers, 1914-1945,* edited by James Hardin (1987)

57 *Victorian Prose Writers After 1867,* edited by William B. Thesing (1987)

58 *Jacobean and Caroline Dramatists,* edited by Fredson Bowers (1987)

59 *American Literary Critics and Scholars, 1800-1850,* edited by John W. Rathbun and Monica M. Grecu (1987)

60 *Canadian Writers Since 1960, Second Series,* edited by W. H. New (1987)

61 *American Writers for Children Since 1960: Poets, Illustrators, and Nonfiction Authors,* edited by Glenn E. Estes (1987)

62 *Elizabethan Dramatists,* edited by Fredson Bowers (1987)

63 *Modern American Critics, 1920-1955,* edited by Gregory S. Jay (1988)

64 *American Literary Critics and Scholars, 1850-1880,* edited by John W. Rathbun and Monica M. Grecu (1988)

65 *French Novelists, 1900-1930,* edited by Catharine Savage Brosman (1988)

66 *German Fiction Writers, 1885-1913,* 2 parts, edited by James Hardin (1988)

67 *Modern American Critics Since 1955,* edited by Gregory S. Jay (1988)

68 *Canadian Writers, 1920-1959, First Series,* edited by W. H. New (1988)

69 *Contemporary German Fiction Writers, First Series,* edited by Wolfgang D. Elfe and James Hardin (1988)

70 *British Mystery Writers, 1860-1919,* edited by Bernard Benstock and Thomas F. Staley (1988)

71 *American Literary Critics and Scholars, 1880-1900,* edited by John W. Rathbun and Monica M. Grecu (1988)

72 *French Novelists, 1930-1960,* edited by Catharine Savage Brosman (1988)

73 *American Magazine Journalists, 1741-1850,* edited by Sam G. Riley (1988)

74 *American Short-Story Writers Before 1880,* edited by Bobby Ellen Kimbel, with the assistance of William E. Grant (1988)

75 *Contemporary German Fiction Writers, Second Series,* edited by Wolfgang D. Elfe and James Hardin (1988)

76 *Afro-American Writers, 1940-1955,* edited by Trudier Harris (1988)

77 *British Mystery Writers, 1920-1939,* edited by Bernard Benstock and Thomas F. Staley (1988)

78 *American Short-Story Writers, 1880-1910,* edited by Bobby Ellen Kimbel, with the assistance of William E. Grant (1988)

79 *American Magazine Journalists, 1850-1900,* edited by Sam G. Riley (1988)

80 *Restoration and Eighteenth-Century Dramatists, First Series,* edited by Paula R. Backscheider (1989)

81 *Austrian Fiction Writers, 1875-1913,* edited by James Hardin and Donald G. Daviau (1989)

82 *Chicano Writers, First Series,* edited by Francisco A. Lomelí and Carl R. Shirley (1989)

83 *French Novelists Since 1960,* edited by Catharine Savage Brosman (1989)

84 *Restoration and Eighteenth-Century Dramatists, Second Series,* edited by Paula R. Backscheider (1989)

85 *Austrian Fiction Writers After 1914,* edited by James Hardin and Donald G. Daviau (1989)

86 *American Short-Story Writers, 1910-1945, First Series,* edited by Bobby Ellen Kimbel (1989)

87 *British Mystery and Thriller Writers Since 1940, First Series,* edited by Bernard Benstock and Thomas F. Staley (1989)

88 *Canadian Writers, 1920-1959, Second Series,* edited by W. H. New (1989)

89 *Restoration and Eighteenth-Century Dramatists, Third Series,* edited by Paula R. Backscheider (1989)

90 *German Writers in the Age of Goethe, 1789-1832,* edited by James Hardin and Christoph E. Schweitzer (1989)

91 *American Magazine Journalists, 1900-1960, First Series,* edited by Sam G. Riley (1990)

92 *Canadian Writers, 1890-1920,* edited by W. H. New (1990)

93 *British Romantic Poets, 1789-1832, First Series,* edited by John R. Greenfield (1990)

94 *German Writers in the Age of Goethe: Sturm und Drang to Classicism,* edited by James Hardin and Christoph E. Schweitzer (1990)

95 *Eighteenth-Century British Poets, First Series,* edited by John Sitter (1990)

96 *British Romantic Poets, 1789-1832, Second Series,* edited by John R. Greenfield (1990)

97 *German Writers from the Enlightenment to Sturm und Drang, 1720-1764,* edited by James Hardin and Christoph E. Schweitzer (1990)

98 *Modern British Essayists, First Series,* edited by Robert Beum (1990)

99 *Canadian Writers Before 1890,* edited by W. H. New (1990)

100 *Modern British Essayists, Second Series,* edited by Robert Beum (1990)

101 *British Prose Writers, 1660-1800, First Series,* edited by Donald T. Siebert (1991)

102 *American Short-Story Writers, 1910-1945, Second Series,* edited by Bobby Ellen Kimbel (1991)

103 *American Literary Biographers, First Series,* edited by Steven Serafin (1991)

104 *British Prose Writers, 1660-1800, Second Series,* edited by Donald T. Siebert (1991)

105 *American Poets Since World War II, Second Series,* edited by R. S. Gwynn (1991)

106 *British Literary Publishing Houses, 1820-1880,* edited by Patricia J. Anderson and Jonathan Rose (1991)

107 *British Romantic Prose Writers, 1789-1832, First Series,* edited by John R. Greenfield (1991)

108 *Twentieth-Century Spanish Poets, First Series,* edited by Michael L. Perna (1991)

109 *Eighteenth-Century British Poets, Second Series,* edited by John Sitter (1991)

110 *British Romantic Prose Writers, 1789-1832, Second Series,* edited by John R. Greenfield (1991)

111 *American Literary Biographers, Second Series,* edited by Steven Serafin (1991)

112 *British Literary Publishing Houses, 1881-1965,* edited by Jonathan Rose and Patricia J. Anderson (1991)

113 *Modern Latin-American Fiction Writers, First Series,* edited by William Luis (1992)

114 *Twentieth-Century Italian Poets, First Series,* edited by Giovanna Wedel De Stasio, Glauco Cambon, and Antonio Illiano (1992)

115 *Medieval Philosophers,* edited by Jeremiah Hackett (1992)

116 *British Romantic Novelists, 1789-1832,* edited by Bradford K. Mudge (1992)

**117** *Twentieth-Century Caribbean and Black African Writers, First Series,* edited by Bernth Lindfors and Reinhard Sander (1992)

**118** *Twentieth-Century German Dramatists, 1889-1918,* edited by Wolfgang D. Elfe and James Hardin (1992)

**119** *Nineteenth-Century French Fiction Writers: Romanticism and Realism, 1800-1860,* edited by Catharine Savage Brosman (1992)

**120** *American Poets Since World War II, Third Series,* edited by R. S. Gwynn (1992)

**121** *Seventeenth-Century British Nondramatic Poets, First Series,* edited by M. Thomas Hester (1992)

**122** *Chicano Writers, Second Series,* edited by Francisco A. Lomelí and Carl R. Shirley (1992)

**123** *Nineteenth-Century French Fiction Writers: Naturalism and Beyond, 1860-1900,* edited by Catharine Savage Brosman (1992)

**124** *Twentieth-Century German Dramatists, 1919-1992,* edited by Wolfgang D. Elfe and James Hardin (1992)

**125** *Twentieth-Century Caribbean and Black African Writers, Second Series,* edited by Bernth Lindfors and Reinhard Sander (1993)

**126** *Seventeenth-Century British Nondramatic Poets, Second Series,* edited by M. Thomas Hester (1993)

**127** *American Newspaper Publishers, 1950-1990,* edited by Perry J. Ashley (1993)

**128** *Twentieth-Century Italian Poets, Second Series,* edited by Giovanna Wedel De Stasio, Glauco Cambon, and Antonio Illiano (1993)

**129** *Nineteenth-Century German Writers, 1841-1900,* edited by James Hardin and Siegfried Mews (1993)

**130** *American Short-Story Writers Since World War II,* edited by Patrick Meanor (1993)

**131** *Seventeenth-Century British Nondramatic Poets, Third Series,* edited by M. Thomas Hester (1993)

**132** *Sixteenth-Century British Nondramatic Writers, First Series,* edited by David A. Richardson (1993)

**133** *Nineteenth-Century German Writers to 1840,* edited by James Hardin and Siegfried Mews (1993)

**134** *Twentieth-Century Spanish Poets, Second Series,* edited by Jerry Phillips Winfield (1994)

**135** *British Short-Fiction Writers, 1880-1914: The Realist Tradition,* edited by William B. Thesing (1994)

**136** *Sixteenth-Century British Nondramatic Writers, Second Series,* edited by David A. Richardson (1994)

**137** *American Magazine Journalists, 1900-1960, Second Series,* edited by Sam G. Riley (1994)

**138** *German Writers and Works of the High Middle Ages: 1170-1280,* edited by James Hardin and Will Hasty (1994)

**139** *British Short-Fiction Writers, 1945-1980,* edited by Dean Baldwin (1994)

**140** *American Book-Collectors and Bibliographers, First Series,* edited by Joseph Rosenblum (1994)

**141** *British Children's Writers, 1880-1914,* edited by Laura M. Zaidman (1994)

**142** *Eighteenth-Century British Literary Biographers,* edited by Steven Serafin (1994)

**143** *American Novelists Since World War II, Third Series,* edited by James R. Giles and Wanda H. Giles (1994)

**144** *Nineteenth-Century British Literary Biographers,* edited by Steven Serafin (1994)

**145** *Modern Latin-American Fiction Writers, Second Series,* edited by William Luis and Ann González (1994)

**146** *Old and Middle English Literature,* edited by Jeffrey Helterman and Jerome Mitchell (1994)

**147** *South Slavic Writers Before World War II,* edited by Vasa D. Mihailovich (1994)

**148** *German Writers and Works of the Early Middle Ages: 800-1170,* edited by Will Hasty and James Hardin (1994)

**149** *Late Nineteenth- and Early Twentieth-Century British Literary Biographers,* edited by Steven Serafin (1995)

**150** *Early Modern Russian Writers, Late Seventeenth and Eighteenth Centuries,* edited by Marcus C. Levitt (1995)

**151** *British Prose Writers of the Early Seventeenth Century,* edited by Clayton D. Lein (1995)

**152** *American Novelists Since World War II, Fourth Series,* edited by James and Wanda Giles (1995)

**153** *Late-Victorian and Edwardian British Novelists, First Series,* edited by George M. Johnson (1995)

**154** *The British Literary Book Trade, 1700-1820,* edited by James K. Bracken and Joel Silver (1995)

**155** *Twentieth-Century British Literary Biographers,* edited by Steven Serafin (1995)

**156** *British Short-Fiction Writers, 1880-1914: The Romantic Tradition,* edited by William F. Naufftus (1995)

**157** *Twentieth-Century Caribbean and Black African Writers, Third Series,* edited by Bernth Lindfors and Reinhard Sander (1995)

**158** *British Reform Writers, 1789-1832,* edited by Gary Kelly and Edd Applegate (1995)

**159** *British Short Fiction Writers, 1800-1880,* edited by John R. Greenfield (1996)

**160** *British Children's Writers, 1914-1960,* edited by Donald R. Hettinga and Gary D. Schmidt (1996)

**161** *British Children's Writers Since 1960, First Series,* edited by Caroline Hunt (1996)

**162** *British Short-Fiction Writers, 1915-1945,* edited by John H. Rogers (1996)

**163** *British Children's Writers, 1800-1880,* edited by Meena Khorana (1996)

**164** *German Baroque Writers, 1580-1660,* edited by James Hardin (1996)

**165** *American Poets Since World War II, Fourth Series,* edited by Joseph Conte (1996)

**166** *British Travel Writers, 1837-1875,* edited by Barbara Brothers and Julia Gergits (1996)

**167** *Sixteenth-Century British Nondramatic Writers, Third Series,* edited by David A. Richardson (1996)

**168** *German Baroque Writers, 1661-1730,* edited by James Hardin (1996)

**169** *American Poets Since World War II, Fifth Series,* edited by Joseph Conte (1996)

**170** *The British Literary Book Trade, 1475-1700,* edited by James K. Bracken and Joel Silver (1996)

**171** *Twentieth-Century American Sportswriters,* edited by Richard Orodenker (1996)

## *Documentary Series*

**1** *Sherwood Anderson, Willa Cather, John Dos Passos, Theodore Dreiser, F. Scott Fitzgerald, Ernest Hemingway, Sinclair Lewis,* edited by Margaret A. Van Antwerp (1982)

**2** *James Gould Cozzens, James T. Farrell, William Faulkner, John O'Hara, John Steinbeck, Thomas Wolfe, Richard Wright,* edited by Margaret A. Van Antwerp (1982)

**3** *Saul Bellow, Jack Kerouac, Norman Mailer, Vladimir Nabokov, John Updike, Kurt Vonnegut,* edited by Mary Bruccoli (1983)

**4** *Tennessee Williams,* edited by Margaret A. Van Antwerp and Sally Johns (1984)

**5** *American Transcendentalists,* edited by Joel Myerson (1988)

**6** *Hardboiled Mystery Writers: Raymond Chandler, Dashiell Hammett, Ross Macdonald,* edited by Matthew J. Bruccoli and Richard Layman (1989)

**7** *Modern American Poets: James Dickey, Robert Frost, Marianne Moore,* edited by Karen L. Rood (1989)

**8** *The Black Aesthetic Movement,* edited by Jeffrey Louis Decker (1991)

**9** *American Writers of the Vietnam War: W. D. Ehrhart, Larry Heinemann, Tim O'Brien, Walter McDonald, John M. Del Vecchio,* edited by Ronald Baughman (1991)

**10** *The Bloomsbury Group,* edited by Edward L. Bishop (1992)

**11** *American Proletarian Culture: The Twenties and The Thirties,* edited by Jon Christian Suggs (1993)

**12** *Southern Women Writers: Flannery O'Connor, Katherine Anne Porter, Eudora Welty,* edited by Mary Ann Wimsatt and Karen L. Rood (1994)

**13** *The House of Scribner, 1846–1904,* edited by John Delaney (1996)

**14** *Four Women Writers for Children, 1868–1918,* edited by Caroline C. Hunt (1996)

## *Yearbooks*

**1980** edited by Karen L. Rood, Jean W. Ross, and Richard Ziegfeld (1981)

**1981** edited by Karen L. Rood, Jean W. Ross, and Richard Ziegfeld (1982)

**1982** edited by Richard Ziegfeld; associate editors: Jean W. Ross and Lynne C. Zeigler (1983)

**1983** edited by Mary Bruccoli and Jean W. Ross; associate editor: Richard Ziegfeld (1984)

**1984** edited by Jean W. Ross (1985)

**1985** edited by Jean W. Ross (1986)

**1986** edited by J. M. Brook (1987)

**1987** edited by J. M. Brook (1988)

**1988** edited by J. M. Brook (1989)

**1989** edited by J. M. Brook (1990)

**1990** edited by James W. Hipp (1991)

**1991** edited by James W. Hipp (1992)

**1992** edited by James W. Hipp (1993)

**1993** edited by James W. Hipp, contributing editor George Garrett (1994)

**1994** edited by James W. Hipp, contributing editor George Garrett (1995)

**1995** edited by James W. Hipp, contributing editor George Garrett (1996)

## *Concise Series*

**Concise Dictionary of American Literary Biography,** 6 volumes (1988-1989): *The New Consciousness, 1941-1968; Colonization to the American Renaissance, 1640-1865; Realism, Naturalism, and Local Color, 1865-1917; The Twenties, 1917-1929; The Age of Maturity, 1929-1941; Broadening Views, 1968-1988.*

**Concise Dictionary of British Literary Biography,** 8 volumes (1991-1992): *Writers of the Middle Ages and Renaissance Before 1660; Writers of the Restoration and Eighteenth Century, 1660-1789; Writers of the Romantic Period, 1789-1832; Victorian Writers, 1832-1890; Late Victorian and Edwardian Writers, 1890-1914; Modern Writers, 1914-1945; Writers After World War II, 1945-1960; Contemporary Writers, 1960 to Present.*

Dictionary of Literary Biography® • Volume One Hundred Seventy-One

# Twentieth-Century American Sportswriters

Dictionary of Literary Biography® • Volume One Hundred Seventy-One

# Twentieth-Century American Sportswriters

Edited by
Richard Orodenker
*Pennsylvania State University, Ogontz*

A Bruccoli Clark Layman Book
Gale Research
Detroit, Washington, D.C., London

Printed in the United States of America

The paper used in this publication meets the minimum requirements of American National Standard for Information Sciences–Permanence Paper for Printed Library Materials, ANSI Z39.48-1984. ♾™



**Library of Congress Cataloging-in-Publication Data**

Twentieth-century American sportswriters / edited by Richard Orodenker.
p. cm. – (Dictionary of literary biography; v. 171)
"A Bruccoli Clark Layman book."
Includes bibliographical references and index.
ISBN 0-8103-9934-2 (alk. paper)
1. Sportswriters – United States – Biography – Dictionaries. I. Orodenker, Richard. II. Series.
GV742.4.T94 1996
070.4'49'7960922 – dc20 96-38561
CIP

10 9 8 7 6 5 4 3 2 1

# Contents

# Plan of the Series

*. . . Almost the most prodigious asset of a country, and perhaps its most precious possession, is its native literary product – when that product is fine and noble and enduring.*

Mark Twain*

The advisory board, the editors, and the publisher of the *Dictionary of Literary Biography* are joined in endorsing Mark Twain's declaration. The literature of a nation provides an inexhaustible resource of permanent worth. We intend to make literature and its creators better understood and more accessible to students and the reading public, while satisfying the standards of teachers and scholars.

To meet these requirements, *literary biography* has been construed in terms of the author's achievement. The most important thing about a writer is his writing. Accordingly, the entries in *DLB* are career biographies, tracing the development of the author's canon and the evolution of his reputation.

The purpose of *DLB* is not only to provide reliable information in a convenient format but also to place the figures in the larger perspective of literary history and to offer appraisals of their accomplishments by qualified scholars.

The publication plan for *DLB* resulted from two years of preparation. The project was proposed to Bruccoli Clark by Frederick C. Ruffner, president of the Gale Research Company, in November 1975. After specimen entries were prepared and typeset, an advisory board was formed to refine the entry format and develop the series rationale. In meetings held during 1976, the publisher, series editors, and advisory board approved the scheme for a comprehensive biographical dictionary of persons who contributed to North American literature. Editorial work on the first volume began in January 1977, and it was published in 1978. In order to make *DLB* more than a reference tool and to compile volumes that individually have claim to status as literary history, it was decided to organize volumes by topic, period, or genre. Each of these freestanding volumes provides a biographical-bibliographical guide and overview for a particular area of literature. We are convinced that this organization – as opposed to a single alphabet method – constitutes a valuable innovation in the presentation of reference material. The volume plan necessarily requires many decisions for the placement and treatment of authors who might properly be included in two or three volumes. In some instances a major figure will be included in separate volumes, but with different entries emphasizing the aspect of his career appropriate to each volume. Ernest Hemingway, for example, is represented in *American Writers in Paris, 1920–1939* by an entry focusing on his expatriate apprenticeship; he is also in *American Novelists, 1910–1945* with an entry surveying his entire career. Each volume includes a cumulative index of the subject authors and articles. Comprehensive indexes to the entire series are planned.

With volume ten in 1982 it was decided to enlarge the scope of *DLB*. By the end of 1986 twenty-one volumes treating British literature had been published, and volumes for Commonwealth and Modern European literature were in progress. The series has been further augmented by the *DLB Yearbooks* (since 1981) which update published entries and add new entries to keep the *DLB* current with contemporary activity. There have also been *DLB Documentary Series* volumes which provide biographical and critical source materials for figures whose work is judged to have particular interest for students. One of these companion volumes is entirely devoted to Tennessee Williams.

We define literature as the *intellectual commerce of a nation:* not merely as belles lettres but as that ample and complex process by which ideas are generated, shaped, and transmitted. *DLB* entries are not limited to "creative writers" but extend to other figures who in their time and in their way influenced the mind of a people. Thus the series encompasses historians, journalists, publishers, and screenwriters. By this means readers of *DLB* may be aided to perceive literature not as cult scripture in the keeping of intellectual high priests but firmly positioned at the center of a nation's life.

**From an unpublished section of Mark Twain's autobiography, copyright by the Mark Twain Company*

*DLB* includes the major writers appropriate to each volume and those standing in the ranks immediately behind them. Scholarly and critical counsel has been sought in deciding which minor figures to include and how full their entries should be. Wherever possible, useful references are made to figures who do not warrant separate entries.

Each *DLB* volume has a volume editor responsible for planning the volume, selecting the figures for inclusion, and assigning the entries. Volume editors are also responsible for preparing, where appropriate, appendices surveying the major periodicals and literary and intellectual movements for their volumes, as well as lists of further readings. Work on the series as a whole is coordinated at the Bruccoli Clark Layman editorial center in Columbia, South Carolina, where the editorial staff is responsible for accuracy of the published volumes.

One feature that distinguishes *DLB* is the illustration policy – its concern with the iconography of literature. Just as an author is influenced by his surroundings, so is the reader's understanding of the author enhanced by a knowledge of his environment. Therefore *DLB* volumes include not only drawings, paintings, and photographs of authors, often depicting them at various stages in their careers, but also illustrations of their families and places where they lived. Title pages are regularly reproduced in facsimile along with dust jackets for modern authors. The dust jackets are a special feature of *DLB* because they often document better than anything else the way in which an author's work was perceived in its own time. Specimens of the writers' manuscripts are included when feasible.

Samuel Johnson rightly decreed that "The chief glory of every people arises from its authors." The purpose of the *Dictionary of Literary Biography* is to compile literary history in the surest way available to us – by accurate and comprehensive treatment of the lives and work of those who contributed to it.

The *DLB* Advisory Board

# Introduction

While there have been memoirs and autobiographies of individual sportswriters, chapters on sportswriters in works on newspaper journalists, and "how-to" books on sportswriting, this volume of the *Dictionary of Literary Biography* represents the first full-length study of American sportswriters. For some of the writers, the entries published here are the first studies to appear anywhere.

Writing about sports should not be relegated, as it often has been, to the lower rungs of the literary ladder. Sportswriting is rich in language and treats important and timeless themes, such as the value of heroism and the consequences of failure. In addition, sportswriters have had to deal with social issues such as the commercialization of sport and the conflict between amateurism and professionalism; race relations, from the early days of black-white boxing matches and the myth of the "Great White Hope" to the legacy of Jackie Robinson and the current age of black dominance in many sports; gambling, both as a menace to the integrity of sports and as part of the lore and low comedy of baseball, boxing, and thoroughbred racing; and the effects of technology on the ways sports are played and watched. Sportswriters continue to debate who the true heroes are and which sporting events will stand as keystone cultural or historical markers. In these respects, sportswriters have produced a literature whose value often transcends that of "mere" sport.

Sports have been treated in literature since ancient times. Odysseus wrestles with Ajax in the *Iliad* and throws the discus in the *Odyssey;* Virgil gives an account of a boxing match in the *Aeneid.* There are hundreds of other literary references to sports, but the first significant publication devoted to a sport is Izaak Walton's *The Compleat Angler* (1653). British writers of sports literature included Joseph Strutt, author of *Sports and Pastimes of the People of England* (1801); Pierce Egan, whom the American sportswriter A. J. Liebling called the "great historian of Boxiana"; William Hazlitt, author of the well-known pieces and "Cavanagh the Fives-Player" (1819) and "The Fight" (1822); and the noted golf writer Sir Bernard Darwin, who in 1907 began a regular Saturday column for the *Times* of London.

Some scholars consider William Trotter Porter's *Spirit of the Times* (1831) the first important American sports publication. The first American newspaper sports editor is believed to have been Charles R. Wright of the *Syracuse Courier,* beginning in 1878; at about the same time, Francis C. Richter of the *Philadelphia Public Ledger* is credited with being the first person to run a full-fledged sports department. Other newspapers in the vanguard of sports coverage in America included the New York dailies the *Clipper, Mercury, Sun,* and *Mail.* The *New York Post* sportswriter Stanley Frank says in his *Sports Extra* (1944) that Arthur Brisbane's 10 March 1888 account in the *New York Sun* of the John L. Sullivan–Charley Mitchell fight in Chantilly, France, "marked the first full-dress coverage of an international sporting event." According to the New York journalist Jack Lang, a long-time executive of the Baseball Writers Association of America, "By the 1890s, most newspapers in America had created a sports staff. It was in the final decade of the nineteenth century that sports writing began to develop as a full time job on the nation's newspapers." Among magazines, *Harper's Weekly, Century, Frank Leslie's Illustrated Weekly, Outing, Collier's, The Police Gazette, Sporting Life, The Sporting News,* and *Field & Stream* provided a market for freelance sportswriters and would continue to do so into the next century.

The styles of nineteenth-century American sportswriters ranged from the stately Victorian cadences of Henry Chadwick, the "Father of Baseball," to the comical quips of Leonard Dana Washburn of the *Chicago Inter-Ocean.* Chadwick wrote the first hardbound book on baseball, *The Game of Base Ball* (1868), in addition to many guides and pamphlets, such as *Beadle's Dime Baseball Player* (1860), and he edited the influential *Spalding Guide* (1877). Washburn covered "Grandpa" Cap Anson's colorful Chicago White Stockings; brilliant, witty similes punctuate virtually every line of his prose: "Willie Forget-Me-Not Hutchinson hit the ball a lick that brought tears to its eyes. Kittridge, who was just due, got strong reverse English on the leather and started an artesian well in far-away left field." As the sportswriter Hugh Fullerton wrote in *The Saturday Evening Post* (21 April 1928), "Washburn was permitted to lift baseball out of the literary class of the market reports." O. P. Caylor, sports editor of the *Cincinnati Enquirer* and author of yearly baseball previews for *Harper's Weekly,* was a better stylist than Chadwick.

Besides Washburn, members of the humorous and metaphorically rich "Chicago School" of sportswriting included Charlie Seymour and Finley Peter Dunne (later Dunne, as editor of *Collier's,* would solicited many sports articles). Walter Camp was to football what Chadwick was to baseball: his *American Football* (1891) was the first of his four major volumes on the sport, and it was he who came up with the idea of All-American college squads. According to sportswriter and editor Tom Meany in *Collier's Great Sports Stories* (1955), Camp was "the first to break into *Collier's* with a sports story." Among the fathers of American sportswriting, Fullerton cites a possible matriarch, as well: the wife of *Chicago Daily News* editor Charles Chapin "occasionally, without signing a name . . . wrote some of the keenest and most searching analyses of teams and players ever printed."

Significant "literary" sports publications of the late nineteenth century included Ernest L. Thayer's poem "Casey at the Bat" in the *San Francisco Examiner* (3 June 1888) and the nearly as popular ditty "Slide, Kelly, Slide!" (1889), by John W. Kelly. The Frank Merriwell series of novels, written by Gilbert Patten under various pseudonyms, began to appear in 1896. Like Hazlitt before him, the journalist and short-story writer Richard Harding Davis is an example of a "literary" author who occasionally indulges in sportswriting. In a 24 November 1895 front-page *New York Journal* account of a Princeton-Yale football game Davis wrote: "It is not possible to describe [Yale captain Thorne's] run. It would be as easy to explain how a snake disappears through the grass, or an eel slips from your fingers, or to say how a flash of linked lightning wriggles across the sky." His article in *Collier's* (June 1904) on Japanese sumo wrestlers includes the sentence "The great hands slap the bare flesh with the crack of a whip, the men snort and puff like elephants, the rolls of muscle and fat rise under the clean brown skin in knots as round as hawsers."

Twentieth-century authors who crossed over from "serious" literature to sportswriting include Roger Angell and A. J. Liebling. Conversely, Westbrook Pegler, Heywood Broun, Paul Gallico, Damon Runyon, and Charles E. Van Loan left sportswriting to pursue successful careers as fiction writers or political and social commentators, although they sometimes turned back to sports: Pegler covered the 1936 Olympics in Berlin and championship boxing bouts; Broun continued to write on sports in his home-published paper, *Broun's Nutmeg;* and Gallico treated sports in many short stories. W. C. Heinz and Ring Lardner wrote fiction, some of it related to sports in addition to their newspaper and magazine prose. Sportswriters, especially in the early years of the century, also wrote poetry as a regular fixture of their columns. Lardner dabbled in doggerel; Hugh Edmund Keough and John Kieran wrote accomplished verse, as did Grantland Rice, whose best-known lines come from his poem "Alumnus Football" (1908): "For when the One Great Scorer comes to write against your name / He marks – not that you won or lost – but how you played the Game." Hugh Fullerton, Fred Lieb, and Allison Danzig were not great prose stylists but were simply writers who knew their subjects well and wrote about them competently.

Washburn and Davis laid the foundation for a more consciously literary sportswriting that was directed toward the general reader rather than the sports fanatic. Sports, the public learned as the nineteenth century was coming to a close, could be big news. As Michael Oriard observes in *Reading Football: How the Popular Press Created an American Spectacle* (1993): "By the mid-1890s both the quantity and the quality of the football coverage in the daily papers in New York, Philadelphia, and Boston were staggering: front-page, full-page accounts of the big games, accompanied by sometimes dozens of sensationalistic illustrations." The full flowering of nineteenth-century styles, best represented here by Charles Dryden and Bozeman Bulger, occurred in the years immediately after the turn of the century. In 1908 the formation of the Baseball Writers Association of America at a meeting of twenty-two writers at the Hotel Pontchartrain in Detroit was a sign that attitudes toward sports and the profession of sportswriting were beginning to change.

Two distinct styles of sportswriting emerged in the early twentieth century: the "Gee Whiz" and "Aw, Nuts" schools. Though Stanley Walker of the *New York Herald Tribune* may have been the first to use these terms in print (in his book *City Editor,* 1934), he did not coin them; the expressions had long been accepted in journalism circles. The Gee Whiz group, according to Irving T. Marsh and Edward Ehre, practiced "the hero-worship type of writing," while the Aw, Nuts school favored "the cynical and even critical type." Gee Whiz writers included Broun, Gallico, and Rice; among the Aw, Nuts writers were Lardner, Pegler, and W. O. McGeehan (although Pegler had been a Gee Whiz writer early in his career). These styles tended to predominate before 1930. As Roger Kahn pointed out in a 1974 *Esquire* piece, "Where Have All Our Heroes Gone?," "The best writers composed with wit, but the stories, even the finest, lack proximity.

They were researched from where they were written: the press box. . . . Newspaper and magazine sports journalism have come about since 1940. Some old idolatry persists, but it is modified by a new muckraking. Editors want athletes quoted as they speak, portrayed almost as they are. Reading sports pages and articles, one has little sense of strolling through a pantheon." In *Sports Page* (1949) Woodward added a third classification, naming a school that came to prominence after World War II: "the On-the-Button school which attempts straight reporting. . . . It is a return to the old Who, What, Where, When, Why school of journalism without its unadorned baldness."

Just at the time editors began toning down the epigones of Washburn and Davis, writers with original voices of their own, such as Keough, Lardner, McGeehan, Rice, Bozeman Bulger, Charles Dryden, and John Kieran were making their presence felt on the sports pages. These writers tended to become columnists rather than straight reporters; "In the Wake of the News" in the *Chicago Tribune* and "Sports of the Times" in *The New York Times* would become the two most prominent sports columns in America. "In the Wake of the News" was written first by Keough, then briefly by Fullerton, and then by Lardner; Kieran's was the first byline on "Sports of the Times." Readers throughout the country would get to know many sportswriters through their syndicated columns, such as Rice's "Sportlight" and Red Smith's "Views of Sport." Sportswriters such as Fullerton considered themselves reporters, albeit colorful ones. Unhappy as a columnist – the newspapers' "writer's" position – Fullerton turned "In the Wake of the News" over to Lardner.

Not all sportswriters are great stylists. Walker's 1934 quip – "At its best, sports writing is all that its admirers claim for it, but even when it is considered pretty good, it may be unnecessarily bad" – still holds true. It is probably more accurate to say that sportswriting tends to be uneven. Few sportswriters have been as heavily criticized, for instance, as Arthur Daley of *The New York Times;* yet Daley also won a Pulitzer Prize.

Much of the history of sportswriting has been documented by sportswriters themselves, including Frank, Fullerton, Kahn, Rice, and Fred Russell, usually in the form of memoirs, autobiographies, and anthologies. Walker's *City Editor* and Woodward's *Sports Page* tell the story of American sportswriting to the midpoint of the twentieth century. Since then sportswriting has received little attention. The single most valuable source is a collection of interviews with sportswriters, *No Cheering in the Press Box* (1974; revised edition, 1995), by Jerome Holtzman of the *Chicago Sun-Times.* Nine of Holtzman's subjects are featured in *DLB 171.* Other important recent works relevant to sportswriting include Charles Fountain's well-received biography of Grantland Rice, *Sportswriter* (1993), and the scholarly works of Oriard, a former professional football player and current professor of English at Oregon State University.

The most important source of information on sportswriting in the twentieth century is, of course, the writing itself. While publishers had long brought out books for the juvenile market – Fullerton's three books about the fictional baseball player Jimmy Kirkland, for example, had been published in 1915 – they began to show greater interest in the adult sports market as the twentieth century progressed. Harper and Brothers published *The Omnibus of Sport,* edited by Rice and Harford Powel, in 1932. Rice and Powel gathered, as they say in the preface, "the best material available, long and short, in the way of fiction and articles, verse and comment that will help to tell the long and thrilling story of all the better known games and to present the characteristics of those who play these games." In more than eight hundred pages Rice and Powel cover "The Ancient Pageant" from Homer to Hazlitt, with chapters on football, baseball, golf, tennis, combat, racing, and hunting and fishing. As other editors have done after them, Rice and Powel include "literary" sportswriting – Mark Twain's "The Celebrated Jumping Frog of Calaveras County," Ernest Hemingway's "The Undefeated," and Lew Wallace's "The Chariot Race" from his *Ben-Hur* (1880) – along with short stories and articles by Davis, Broun, Darwin, Gallico, Kieran, Lardner, Pegler, Van Loan, George Ade, Bill Corum, O. B. Keeler, Jack London, Irving Cobb, and Rice himself.

A. S. Barnes of New York became the most prolific company to publish sports books. Among its most important titles were Frank G. Menke's *Encyclopedia of Sports* (1939); *Sports Extra* (1944), edited by Stanley Frank; *Sport's Golden Age: A Close-Up of the Fabulous Twenties* (1948) and *The Greatest Sports Stories from the New York Times* (1951), both edited by Allison Danzig and Peter Brandwein; *The Greatest Sports Stories from the Chicago Tribune* (1953), edited by Arch Ward; and *Collier's Greatest Sports Stories* (1954), edited by Tom Meany. Meany also edited several volumes of the Barnes All-Star Library. The company also published Rice's best-selling *The Tumult and the Shouting* (1954); James T. Farrell's *My*

*Baseball Diary* (1957); and *Nashville Banner* sportswriter Fred Russell's autobiography *Bury Me in an Old Press Box* (1957).

G. P. Putnam's Sons got into the business of sports publishing, most significantly with its Sports Series, which included team histories of all sixteen major league ball clubs, most of them written by Lieb. Graham's biography of Lou Gehrig, *A Quiet Hero* (1942), went through ten printings. Simon and Schuster published Herbert Warren Wind's anthology *The Complete Golfer* (1954) and Charles Einstein's *The Fireside Book of Baseball* (1956); the latter was followed by two additional volumes in 1958 and 1968. Danzig and Peter Schwed's comprehensive reader *The Fireside Book of Tennis* (1972) covers nearly the entire history of the sport in contemporary primary writings, including many by professional players such as Bill Tilden and Helen Wills Moody. Cheap paperbacks also contributed to the popularity of sports literature. *New York Herald Tribune* boxing writer Caswell Adams edited a small but valuable paperback anthology, *Fifty Famous Sports Stories* (1946).

The most comprehensive anthologies are the *Best Sports Stories* volumes published by the E. P. Dutton firm: the series was edited by Marsh and Ehre from 1944 to 1980; it suspended in 1980 and resumed in 1982, edited by Ehre and *The Sporting News* and published through the rest of the decade. The publisher awarded money prizes in three categories – Best Magazine Story, Best News-Feature Story, and Best News-Coverage Story – which were judged by such writers as Caswell Adams, John Chamberlain, Bob Considine, Frank Graham, Quentin Reynolds, and Red Smith. The judges read the entries without knowing who the authors were or where they had been published. In the early 1990s Houghton Mifflin began publishing *Best American Sports Writing,* edited by Glenn Stout; these volumes emphasize feature-length magazine articles on a wider variety of sports than those covered in *Best Sports Stories.*

Sportswriters have often served as ghostwriters for books ostensibly written by athletes. For nearly half of the century the usual practice was not to list the collaborator's name. Sometimes the athletes were substantial contributors to the books, as in the case of Christy Mathewson's classic *Pitching in a Pinch* (1912), ghostwritten by John N. Wheeler, but in other cases the writer did most of the work for none of the credit. Later bylines read "as told to" or "with" the ghostwriter; today the athlete and the writer are frequently listed as coauthors.

Sportswriters also wrote for magazines. *American Golfer,* founded in 1908, was edited by Rice in the 1920s. *The Ring,* founded by Nat Fleischer in 1922, became America's most renowned boxing authority. *Football Digest,* founded in 1939; *Baseball Digest,* started in 1941; and *Golf Digest,* begun in 1950, followed the format popularized by *Readers' Digest,* reprinting and condensing articles originally published elsewhere, though they eventually began printing original articles as well. The digests allowed readers across the country to read the work of prominent sportswriters from many different cities, such as Shirley Povich of the *Washington Post* and Gordon Cobbledick of the *Cleveland Plain Dealer,* as well as the well-known New York writers. Beginning in 1946 *Sport* magazine attempted, as Al Silverman recalls its manifesto, to "stress the human interest side of people in athletics." It published freelance articles by well-known sportswriters from around the country. Two important anthologies of articles from the magazine are *The Best of Sport, 1946–1971* (1972), edited by Silverman, and *Sport Magazine's All-Time All-Stars* (1977), edited by Tom Murray.

It was, however, *Sports Illustrated,* first published on 16 August 1954, that changed the face of sports journalism in America. There had been a *Sports Illustrated* in the 1930s – a slick, country-club type sporting magazine. The name had been picked up after World War II by Dell Publishing, which employed Stanley Woodward as editor of a new version of the magazine. It ceased publication after only a few issues. The title was purchased by Henry R. Luce in 1954. Initially a more literary magazine that published guest essays by such authors as Robert Frost and William Saroyan, *Sports Illustrated* soon began to cover mainstream sports for those who, like Robert W. Creamer, one of the magazine's first senior editors, preferred "watching and reading about sports to participating in them." In time *Sports Illustrated* moved to investigative sports journalism. Sportswriters also found a market in men's magazines of the 1950s such as *True, Argosy,* and *Real;* the last was edited by journeyman sportswriter Ray Robinson, who published the first sports piece by the young Jimmy Breslin.

Before the 1920s sportswriters had greater stylistic freedom than they ever would again, as well as more space in which to develop their prose. They were quick to pick up on conflicting class attitudes about sports: tennis and golf, played by the well-to-do, were seen as sissified sports by the "bugs" in the baseball bleachers, while members of the "smart set" looked down even on Ivy League football players. Sports heroes were often considered palookas, ne'er-do-wells, and illiterates, and amateurs were more highly regarded than professionals.

Sportswriters, who often traveled with athletes on Pullman cars, depicted them in stylized fashion. The writers could be opinionated or satiric, but they were more often than not respectful of sporting institutions and deferential to the individuals they were covering. Frequently subsidized by the professional teams or promoters, writers were careful not to bite the hands that fed them. When news broke of the "Black Sox" scandal – the fixing of the 1919 World Series by Chicago White Sox ballplayers and known gamblers – some writers, like many fans, became disenchanted, bitter, and cynical.

The 1920s are considered the Golden Age of Sports. In this era of superstars, writers did not have to create athletic heroes; they merely recorded the athletes' exploits, albeit usually in glowing terms. Writers still tended to gloss over or repress negative qualities belonging to national heroes, though writers such as Pegler and McGeehan sometimes wrote with little apparent restraint. Nonetheless, the Gee Whiz writers outnumbered the Aw, Nuts writers as each sport was dominated by a single personality: Jack Dempsey in boxing, Babe Ruth in baseball, Bill Tilden in tennis, Bobby Jones in golf, Red Grange in football, and Man o' War in horse racing. The 1920s also saw the dawn of the age of "hype" as newspapers devoted enormous amounts of coverage to the promotion of upcoming events. Sportswriters, wrote Frank in *Sports Extra,* "sat down at their typewriters prepared to invest the event they were covering with the epic grandeur of the Creation." In *City Editor,* Walker titled his chapter on sportswriting "Valhalla's Bull-Pen."

The profession of sportswriting changed during the Great Depression of the 1930s. While sports could still be front-page news, the sports section was coming into its own, running from four to five pages daily and six to eight pages on Sunday. Teams stopped the practice of covering hometown reporters' expenses on road trips. Newspapers cut back on staff and column space, and some sportswriters (especially on smaller newspapers) had to learn quickly to be experts on many sports rather than specialists on one. Writers were less interested in the myth and lore of the game or buffoonish players or in putting athletes up on pedestals than in what these players might actually have to say as professional athletes or, in the case of human-interest stories, as ordinary people. The decade saw the emergence of a tougher breed of sportswriters, including Meany, Woodward, Cannon, John Lardner, and Red Smith.

The 1930s saw the rise of the "pro" in sports, both on the field and in the newspaper office. In 1935 William H. Taylor, yachting reporter for the *New York Herald Tribune,* won the first Pulitzer Prize for a sports story, though the prize was not specifically awarded for sportswriting: his coverage of the America's Cup touched not only on the races but also on the difficult diplomatic relations between England and the United States. With the victories of heavyweight champion Joe Louis and Olympic gold medal runner Jesse Owens in the 1930s, sportswriters suddenly had a new issue with which to contend. The careers of Louis and Owens did much to raise the consciousness of many of these writers about race relations and set the stage for wider – if belated – support of Jackie Robinson and the integration of major league baseball in the 1940s. In *Champion* (1985), his biography of Louis, Chris Mead notes that in time sportswriters "came out unabashedly for Louis and assumed the rest of the country was rooting for him, too."

The years immediately after World War II were the heyday of the *New York Herald Tribune* sports page under Woodward's editorship. The *Tribune* had a long history of premier sportswriting since the days of McGeehan, Hanna, Rice, Kieran, and Broun, but Woodward's sports department is considered by most scholars to be the finest of the twentieth century. Woodward left the *Herald Tribune* in 1948 (he would return as sports editor from 1959 to 1962), but its writers continued to thrive either at the paper or elsewhere.

In the 1950s and early 1960s the most dynamic sportswriting was coming from magazine journalists and freelance authors, including Kahn, John Lardner at *Newsweek* and Creamer at *Sports Illustrated.* Arthur John Daley won the first Pulitzer Prize specifically awarded for sports reporting in 1956. (Sportswriters Red Smith [1976], Dave Anderson [1981], and Jim Murray [1990] would win Pulitzers for commentary.) While there are several sportswriting awards of note, including the Red Smith Award and the J. G. Taylor Spink Prize, two of the most prestigious sportswriting prizes date from this period: the Sportswriter of the Year award has been presented annually since 1959, when it was won by Red Smith, and the National Sportscasters and Sportswriters Hall of Fame prize was first won posthumously by Grantland Rice in 1962.

This volume does not claim to be a comprehensive history of American sports journalism. Instead, *DLB 171* presents an overview of sportswriting in the first seven decades of the twentieth century by focusing on some of its finest practitioners. The volume does not concentrate exclusively on the sports and journalism centers of New York and Chicago but includes regional sportswriters such as Cobbledick, Furman Bisher, Si Burick, and Chet Smith. African American sportswriters, excluded from mainstream positions during the early part of the century because of discrimination, wrote for the thriving black press; they are represented in this volume by the entries on Sam Lacy and Wendell Smith. Chronologically, the volume ranges from sportswriters who spent the most substantial portions of their careers in the early part of the twentieth century to those who were plying their trade in the late 1940s and the 1950s, even though they may not have come to national prominence until the 1960s or 1970s. The cutoff point for *DLB 171* is 1970. If there was anything left of the Gee Whiz school, it virtually disappeared with the publication that year of former New York Yankee pitcher Jim Bouton's baseball exposé, *Ball Four*. That book started a trend, which continues to this day, of unexpurgated "tell-all" sports biographies and autobiographies, many of them far more revealing and explicit than Bouton's. The new era brought the dominance of the New Journalism and of "chipmunk" sportswriting – the latter term coined by sportswriter Jimmy Cannon, who disliked the incessant chattering of these writers in the press box. The New Journalism sportswriters, led by Jimmy Breslin and Gay Talese, brought a greater degree of subjectivity and a more novelistic feel to the medium, while the chipmunks, Maury Allen and Leonard Shecter among them, tended toward the iconoclastic, refusing to take the "ideal" of sports seriously and not hesitating to be aggressive in their interviewing techniques. The New Journalists emphasized unconventional narrative styles and deliberate ambiguity, while the chipmunks emphasized humor, farce, and, most of all, controversy.

Many of the books by the sportswriters in this volume are long out of print and difficult to find. Yet there is every reason to hope that students and scholars of sport will take a renewed interest in sportswriting and sports literature and, particularly, in the works of American sportswriters whose best writings languish on microfilm. Sportswriting was once America's great literature of the breakfast table. It has come a long way since and deserves a second look.

– *Richard Orodenker*


## Acknowledgments

This book was produced by Bruccoli Clark Layman, Inc. Karen L. Rood is senior editor for the *Dictionary of Literary Biography* series. George P. Anderson and Philip B. Dematteis were the in-house editors.

Production manager is Samuel W. Bruce. Photography editors are Julie E. Frick and Margaret Meriwether. Photographic copy work was performed by Joseph M. Bruccoli. Layout and graphics supervisor is Emily Ruth Sharpe. Copyediting supervisor is Laurel M. Gladden. Typesetting supervisor is Kathleen M. Flanagan. Systems manager is Chris Elmore. Laura Pleicones and L. Kay Webster are editorial associates. The production staff includes Phyllis A. Avant, Ann M. Cheschi, Melody W. Clegg, Patricia Coate, Joyce Fowler, Brenda A. Gillie, Stephanie C. Hatchell, Penelope M. Hope, Kathy Lawler Merlette, Pamela D. Norton, Delores Plastow, William L. Thomas Jr., and Allison Trussell.

Walter W. Ross, Steven Gross, and Mark McEwan did library research. They were assisted by the following librarians at the Thomas Cooper Library of the University of South Carolina: Linda Holderfield and the interlibrary-loan staff; reference-department head Virginia Weathers; reference librarians Marilee Birchfield, Stefanie Buck, Stefanie DuBose, Rebecca Feind, Karen Joseph, Donna Lehman, Charlene Loope, Anthony McKissick, Jean Rhyne, Kwamine Simpson, and Virginia Weathers; circulation-department head Caroline Taylor; and acquisitions-searching supervisor David Haggard.

Dictionary of Literary Biography® • Volume One Hundred Seventy-One

# Twentieth-Century American Sportswriters

# Dictionary of Literary Biography

## Roger Angell

*(19 September 1920 – )*

Steven P. Gietschier
The Sporting News

MAJOR POSITIONS HELD: Editor and writer, *Magazine X* (1946–1947); senior editor, *Holiday* (1947–1956); fiction editor and general contributor, *The New Yorker* (1956– ).

BOOKS: *The Stone Arbor and Other Stories* (Boston: Little, Brown, 1960; London: Gollancz, 1961);
*A Day in the Life of Roger Angell* (New York: Viking, 1970; revised edition, New York: Penguin, 1990);
*The Summer Game* (New York: Viking, 1972);
*Five Seasons: A Baseball Companion* (New York: Simon & Schuster, 1977);
*Late Innings: A Baseball Companion* (New York: Simon & Schuster, 1982);
*Season Ticket: A Baseball Companion* (Boston: Houghton Mifflin, 1988);
*Once More Around the Park: A Baseball Reader* (New York: Ballantine, 1991).

OTHER: "Baseball in the Mind," in *This Great Game,* edited by Doris Townsend (Englewood Cliffs, N.J.: Prentice-Hall, 1971), pp. 25–31;
*A Baseball Century: The First 100 Years of the National League,* edited by Jeanne McClow, with an introductory essay by Angell (New York: Macmillan, 1976; London: Collier-Macmillan, 1976), pp. 18–21;
*Baseball,* photographs by Walter Iooss Jr., with text by Angell (New York: Abrams, 1984);
"Early Innings," in *Birth of a Fan,* edited by Ron Fimrite (New York: Macmillan, 1993; Oxford: Macmillan, 1993), pp. 1–25.

*Roger Angell (photograph by Judy Tompkins)*

Roger Angell has combined his lifelong enthusiasm for baseball with a talent for composing stylish, intelligent prose to craft masterful essays admired by sports fans and general readers alike. In 1962 he accepted an invitation from the editor of *The New Yorker* to travel to spring training in Florida. The pieces he wrote during the next decade attracted a fairly select audience that multiplied when

*Dust jacket for the 1972 collection of baseball articles that Angell wrote for* The New Yorker

his work was collected in a book. During the course of more than thirty years he has been chronicling the sport he loves as a fan, writing with a rare and caring eloquence while taking his readers deep into baseball's elegant complexity.

Angell was born in New York City on 19 September 1920 to Ernest and Katharine Shepley Sergeant Angell. He attended a progressive grammar school, Lincoln School of Teacher's College, and lived with his father after his parents divorced in 1929. "I saw a lot of my mother," Angell told interviewer Jared Haynes for *Writing on the Edge* in 1992, "but she was away. I was young and I yearned for her, so what she did, working for *The New Yorker,* was of great significance to me." Katharine Angell, in fact, was one of the first people hired by Harold Ross as an editor at his new magazine, founded in 1925, and she is credited with encouraging the early fiction of such writers as John O'Hara, Vladimir Nabokov, Mary McCarthy, John Cheever, Irwin Shaw, and John Updike.

Angell's father was a Harvard-educated attorney who moved from Cleveland to New York after World War I to practice corporate law. In 1936 Ernest Angell became New York regional director of the Securities and Exchange Commission, then headed by Joseph P. Kennedy. Angell became a leader in the American Civil Liberties Union, eventually becoming the long-term chairman of its national board. In "Early Innings," an autobiographical essay published in Ron Fimrite's *Birth of a Fan* (1993) and in *The New Yorker,* Angell remembered his father as "a lawyer of liberal proclivities who voted for Norman Thomas, the Socialist candidate, in the presidential election of 1932 and again in 1936." Ernest Angell also sparked his son's interest in baseball. "My father began taking me and my four-years-older sister to games at some point in the latter twenties," Angell recalled. "He had grown up in Cleveland in the Nap Lajoie–Addie Joss era" and was a big fan of the Indians.

Katharine Angell remained an important influence in her son's life. "My mother always supported my wishes to be a writer, my baby efforts," Angell told Haynes. Angell soon found another role model in E. B. White, whom his mother had urged Ross to hire. Soon after White and Katharine Angell were married in 1929, the young Angell saw his written work published for the first time in "The Conning Tower," the newspaper column written by Franklin P. Adams that was open to contributions by its readers. White became a permanent influence on Angell. "He was there at hand, and he wrote so well. I learned things from him," he explained to Haynes, "the main one being to try to write simply and directly and to try to make it sound easy."

In "Early Innings" Angell remembered the early to mid 1930s as "palmy days" for a boy who rooted for the Yankees, the Giants, and the New Deal. He recalled Carl Hubbell's "magical screwball" and Joe DiMaggio, "the first celebrated rookie to come up to either team after I had attained full baseball awareness: *my* Joe DiMaggio." Other sports interested him as well, including tennis, Ivy League football, and New York Rangers hockey at Madison Square Garden, where he went on weekends with his mother and stepfather, "who were red-hot hockey fans." As editor of his high school newspaper, Angell interviewed New York mayor Fiorello La Guardia. "I can't remember anything he said, but I can still see his feet, under the mayoral swivel chair, not quite touching the floor."

Angell kept abreast of baseball by reading *The New York Times* and the *New York Herald Tribune,* delivered each morning, and the *Sun* and the *World-Telegram,* brought home by his father at night, and he became familiar with the celebrated sports bylines of that era. Angell wrote that the *New York Herald Tribune* writer Richards Vidmer, "if memory is to be trusted, was my favorite scribe, but before that, back when I was nine or ten years old, what I loved best in the sports pages were box scores and, above all, names." In *The Summer Game* (1972) he recalled those "Dickensian names . . . that have stuck with me ever since I deciphered my first box scores and began peopling the lively landscape of baseball in my mind."

Encouraged by his father, who played amateur baseball well into maturity, Angell attempted to become a pitcher. "I used to have dreams," he admitted to Michael Mok in 1972. "When I was nine years old, I would wait for the bus on Fifth Avenue on my way to school and would sometimes pass Jake Ruppert, owner of the New York Yankees. He had red German cheeks, snappy eyes and pointed mustaches. Every time I saw him I would slam the ball into my mitt with real force, hoping one day he'd say, 'Lad, take this note up to Yankee Stadium.' He never did."

After graduating from Pomfret School, in Connecticut, in 1938 Angell attended Harvard University. He remembered some extraordinary composition teachers in several top-notch freshman classes. "But there isn't much to say in a classroom about writing," he told Haynes. He graduated in 1942, joined the army air force, and married Evelyn Ames Baker. They had two daughters, Caroline and Alice, and were divorced in 1963. Angell then married Carol Rogge in October 1963. They had one son, John Henry, who has figured into some early essays. In the service Angell was managing editor of *Brief,* a weekly magazine published for enlisted men in the Pacific theater of World War II. After the war he worked for *Magazine X* and *Holiday* as senior editor before accepting a position as fiction editor and general contributor with *The New Yorker* in 1956. That same year his own *New Yorker* short story, "The Killing" (1949), was reprinted in Charles Einstein's *The Fireside Book of Baseball* (1956). In the story Linwood P. Schumacher, a door-to-door salesman, enters the apartment of two intoxicated ladies, one of whom thinks that he is Hal Schumacher (her own "Prince Hal"), a real-life pitcher with the New York Giants. The woman, who has been infatuated with the ballplayer since the mid 1930s, buys the brushman's entire line (hence the "killing"), hoping that he will use his big commission to "have that chipped elbow operated on." Angell also published *The Stone Arbor and Other Stories* (1960), a warmly reviewed collection of thirteen stories that originally appeared in *The New Yorker,* and the satires and humorous pieces gathered under the title *A Day in the Life of Roger Angell* (1970).

The idea to ask Angell to write about baseball came in 1962 from William Shawn, Harold Ross's successor as editor of *The New Yorker*. "Our magazine is in the enviable position of 'covering' only those things which appeal to specific writers and editors," Angell told Mok, and Shawn wanted to do more pieces on sports. Shawn assigned Angell to report about spring training. The result was "The Old Folks Behind Home" (7 April 1962), a string of four vignettes written from the viewpoint of the senior citizens, who outnumbered other spring training fans at that time.

This first foray was happily published by Shawn, but Angell recalled no plan to have him pursue this work regularly. "I just kept going," he told Haynes. "I had no idea it would go on this long. . . . I just went on from year to year because I always found something else I wanted to write about. It seemed to be a good fit."

Angell wrote about the New York Mets in "The 'Go!' Shouters" in the 16 June 1962 *New Yorker,* and in October he chronicled the National League play-off matching the Los Angeles Dodgers and the San Francisco Giants and the World Series pitting the Giants against the New York Yankees in "A Tale of Three Cities." Quite deliberately he maintained the fan's perspective in these essays and others composed during the following few years. Knowing that major league baseball was routinely covered in the newspapers, Angell wanted to present the game from a different angle. He "decided to sit in the stands . . . to pick up the feel of the game as it happened to the people around me."

Soon, though, he opted for an occasional seat in the press box, especially during the World Series. Reflecting upon a rhythm of attendances that produced two or three pieces a year for "The Sporting Scene" department of *The New Yorker,* Angell told Haynes that he "eventually came to learn more about baseball." But he admitted to Haynes that:

> I was still faking it as a news reporter. Writing at length for a leisurely and most generous weekly magazine, I could sum things up, to be sure, and fill in a few gaps that the newspapermen were too hurried or too cramped for space to explore, but my main job, as I conceived it, was to continue to try to give the feel of things – to explain the baseball as it happened to me, at a distance and in retrospect.

Angell saw the first decade of his baseball writings collected in *The Summer Game,* a book dedicated to his father and published to critical acclaim. In the 11 June 1972 *New York Times Book Review,* literary editor Ted Solotaroff wrote: "Page for page, *The Summer Game* contains not only the classiest but also the most resourceful baseball writing I have ever read." Mike Shannon, editor of *Spitball: The Baseball Literary Magazine,* included the book in *Diamond Classics: Essays on 100 of the Best Baseball Books Ever Published* (1989). The Society for American Baseball Research put *The Summer Game* and its two successor volumes in "The Essential Baseball Library," a 1987 list of fifty-seven books published in the *SABR Review of Books: A Forum of Baseball Literary Opinion* (volume two).

Angell's prose style and powers of observation also received praise, often in baseball metaphors. One reviewer called him "a Koufax of the keyboard"; another said his commentary "is as sharp as a curve that breaks perfectly as the letters." Shannon, citing Angell's precision, vivid images, and delightful comparisons, asserted that *The Summer Game* "demonstrated a new style of literary baseball writing, resonant and eloquent, that enormously pleased jaded readers and jolted ordinary sportswriters into a recognition, not that the rules of writing had changed, but that within the same old rules Angell was making wonderful plays that nobody had ever seen before."

Some observers of American sports in the early 1970s believed that football was eclipsing baseball as the new national pastime. Solotaroff made this point explicitly in his review, and Angell faced it repeatedly as he toured to promote his book. "[M]ost of the TV and radio people I encountered seemed amused and a little scornful that anyone would still want to write about baseball," he told Dick Johnson for the *SABR Review of Books* in 1988. "Well, I thought all along that baseball was different from other sports, and that some of its special qualities were conducive to good writing," he added.

What sets baseball apart, Angell wrote in "The Interior Stadium," the final and most memorable essay in *The Summer Game,* is

> one saving grace that distinguishes it – for me, at any rate – from every other sport. Because of its pace, . . . its clean lines can be restored in retrospect. . . . [I]t is a game of recollections, recapturings, and visions. . . . Baseball, I must conclude, is intensely remembered because only baseball is so intensely watched. . . . Within the ballpark, time moves differently, marked by no clock except the events of the game. This is the unique, unchangeable feature of baseball, and perhaps explains why this sport . . . remains somehow rustic, unviolent, and introspective.

Baseball's freedom from the constraints of time has become an oft-repeated point of analysis, but the scholar A. James Memmott lauded Angell for this insight, which reminded Memmott of a famous insight from the poet William Wordsworth. Memmott noted that Wordsworth, in his "Preface to the Second Edition of Lyrical Ballads" (1800), characterized poetry as "the spontaneous overflow of powerful emotions: it takes its origin from emotion recollected in tranquillity." Memmott saw this same sequence in Angell's prose, "getting away in time and space from the experience in order to understand it."

Following publication of *The Summer Game* Angell continued his bi- or triannual baseball "reports," as he called them, for *The New Yorker.* The essays he wrote between 1972 and 1976 were collected in *Five Seasons: A Baseball Companion* (1977), the book that found him a wider audience and advanced him to the first rank of baseball writers. Edward Hoagland, in the July 1977 *Harper's* magazine, called *Five Seasons* "a zinging new book jammed with glee and eagerness and lore and exact fact." Both Donald Hall in the 14 October *National Review* and Jack Kroll in the 30 May *Newsweek* declared it better than its predecessor. In the August 1977 issue of *The Atlantic,* C. Michael Curtis said Angell occupied "ground shared with practically no one else." Peter

*Dust jacket for the 1982 collection of Angell's essays and reflections on baseball*

Andrews in the 14 May *Saturday Review* called him "the best baseball writer ever." Hall wrote simply, "Angell can *pick it*."

Excluding the first chapter, the essays in *Five Seasons* are arranged chronologically to form more of an integral history than *The Summer Game*. In his foreword Angell called these years "the most significant half-decade in the history of the game," singling out "remarkable individual feats – by Hank Aaron, Lou Brock, and Nolan Ryan, among others; . . . two vivid and absolutely different champion clubs – the Oakland A's and the Cincinnati Reds; and . . . a World Series [in 1975] . . . of unmatched intensity, brilliance, and pleasure." Still, he complained that the game on the field was "almost obscured by . . . the strikes and lockouts and other labor skirmishings of the players and owners, the bartering of franchises, the adulteration of the game by sudden gate-enhancing innovations, the deadening influence of network television, the arrival of player free-agency, the inflation of player salaries, and the purchased loyalties of most of the principals in the game."

Reviewers observed that this book was anchored by the five essays summarizing each regular season and analyzing each year's World Series. "The writing in these chapters," commented Shannon, is "polished, precise, and original. . . . While they do unquestionably fulfill the obligation implied . . . they command our attention and admiration long after we have become well acquainted with the facts they impart." Most critics were even more taken by the chapters in which Angell struck out in new directions, including the opening essay on the ball itself and the mysteries of pitching, a string of humorous tales woven together into the one piece ("Stories for a Rainy Afternoon") not previously published in *The New Yorker,* and a 1976 essay, "In the Counting House," on the impact of free agency.

*Five Seasons* received a front-page review in the 15 May 1977 *New York Times Book Review,* which also published an interview with Angell conducted by the sports aficionado and author George Plimpton. In this talk Angell said he continued to write about baseball because he never stopped learning more and more about the mysteries of the game:

> In the past few years I've become aware of the whole idea of giving yourself up in order to move men around

> the bases. . . . I've begun to appreciate the patterns of pitching. . . . I'm beginning to have more feeling about the difficulty of hitting, and to understand the accomplishment of someone like George Brett, for example.

Angell also told Plimpton that:

> With baseball it's helpful to notice little things. Perhaps that's why I take so many notes. . . . And of course there's so much to put down. In a sense, baseball is the perfect writer's sport because it is linear: You can feel that things are going to happen.

Still, the essays in *Five Seasons* that garnered the most acclaim were a quartet in which Angell gave up his seat in the press box to study the game from yet new perspectives. Similar to other profiles often published in *The New Yorker,* these chapters painted distinctive word portraits of Horace Stoneham, seventy-two-year-old owner of the Giants; a trio of longtime Detroit Tigers fans; the California Angels scout Ray Scarborough; and the Pittsburgh Pirates pitcher Steve Blass, a man who suddenly and inexplicably lost the ability to throw the ball over the plate.

Almost unanimously *Five Seasons* was also lauded for its language. Hoagland gently scolded Angell for "pleasuring himself with the traditional chestnuts of baseball scribes: 'the starting hurler,' 'the starboard garden,' 'the visiting nine.' " But others were lavishly kind. "Angell's prose is graceful and pleasant," wrote Hall, "with never a misstep, never cliché or corn or overstatement or pomposity." Andrews added, "Angell ranges over his territory like Marty Marion. . . . He has crafted two flawless baseball masterpieces in a row, and that hasn't been done since Johnny Vander Meer pitched back-to-back no-hitters in 1938."

By the time Angell assembled his third collection of essays, *Late Innings: A Baseball Companion* (1982), his faithful readers were quite accustomed to his publication schedule. "Angell's appearances in *The New Yorker,*" wrote Michael Oriard, "have themselves become seasonal, a complement to the rhythms of the sport he covers." Reflecting upon Angell's habit of writing one piece around the opening of each season, Oriard called this practice "a rite of spring rebirth nearly as necessary to reader-fans as the rites he describes." Then, "without fixed regularity, he has also written a series of profiles and reflections." And, Oriard added, "no baseball season has been complete until Angell's final recap has appeared in print: a summary from his jottings and scorecards through the summer, followed by highlights of the World Series."

*Late Innings,* covering 1977 through 1981, both confirmed this pattern and departed from it. The book opened with a single non-*New Yorker* essay, an expanded version of the introduction Angell wrote for *A Baseball Century: The First 100 Years of the National League* (1976). There were five so-called spring training pieces and a handful of profiles. But Angell's succession of annual postseason surveys and analyses came to a halt after a strike by the Major League Baseball Players Association cut a fifty-day hole in the 1981 season. Angell reported on the strike in "The Silence" (30 November 1981) and stuffed his coverage of the special elimination play-offs, the League Championship Series, and the World Series into one disheartened and dismissive paragraph.

Still, *Late Innings* appeared to excellent reviews, and Angell received the customary plaudits. The reviewer for the 17 May 1982 *Sports Illustrated* said, "the stuff is as good as ever," and in the 10 May *Time* the book was called "another quinquennial classic." The reviewer in the 10 May *Newsweek,* borrowing baseball's argot, said, "Ted Williams can *hit.* Bob Gibson can *throw.* Roger Angell can *write.*" The novelist Mark Harris, in the 23 May *New York Times Book Review,* summarized his opinion of Angell's fresh prose: "he marvelously avoids the clichés and weary formulations that characterize baseball writing. He never draws upon the stock of jargon that has been gathering for a century. He sees things by looking at them, not by remembering what has been so often said about them." George Hunt, in the 4 September *America,* agreed: "If even *some* writers could write as well as Angell writes about *anything,* here specifically baseball, the electronic revolution would become a dismaying memory."

Harris and other reviewers identified "Distance" (22 September 1980), a penetrating portrait of former Saint Louis Cardinals pitcher Bob Gibson, as one of the best chapters in the book. Others favored "In the Country" (17 August 1981), a piece written during the strike about baseball fans in the hinterlands. The essay that has endured, though, and been anthologized more than once, is "The Web of the Game" (20 July 1981), a remarkable but informal look at baseball's capacity for surprises and interconnections. Taking in an NCAA regional tournament game, St. John's at Yale, Angell put himself in the company of ninety-one-year-old former Yale coach "Smokey" Joe Wood, who had been the pitching hero of the 1912 World Series. As St. John's Frank Viola defeated Ron Darling, 1–0 in thirteen innings (Darling threw eleven innings of no-hit ball), Angell merged his recollection of a game in which Wood had defeated Walter Johnson 1–0 and the final game of the 1912 Series that Angell's father had attended.

In retrospect, scholarly critics have seen Angell's vision darkening in *Late Innings* as he was forced to consider "the substrata of the sport – money, celebrity, power, traditions, and social change – that lie beneath its green and ordered fields," according to the foreword. James Gindin argued that "the treatment of money, franchises, and owners, what he [Angell] once calls 'corporate masochism,' obscures baseball in some of the . . . essays . . . collected into *Late Innings*." But Oriard pointed out that "distasteful events are always followed by a newly determined but increasingly tenuous affirmation." Angell himself tended to agree with this latter view:

> I am still a baseball fan as well as a baseball writer. . . . [A]lthough big-league baseball is long past its innocent youth and well into what appears to be a disordered and self-destructive sort of middle age, it is a flushed and vigorous reprobate, with astounding recuperative powers. . . . More than half of my own baseball-watching is behind me, but I have no intention of leaving the park.

In 1984 Angell provided an extended essay for Walter Iooss Jr.'s collection of 133 color photographs titled simply *Baseball*. Reviewers of this book praised Iooss, who took the photos while working for *Sports Illustrated*, but recognized, too, that Angell's contribution was more than formulaic. The reviewer in the 5 November 1984 *New Yorker* commented:

> Mr. Angell discusses some of the more arresting photographs, which are otherwise given only perfunctory captions, and digresses on the aesthetics and subtleties of baseball, on the inability of television to capture the feel and space of a game that is essentially three-dimensional (and the regrettable tendency of announcers to fill every dramatic pause and hush in the stands with chatter), and on the poignancy inherent in albums like this, since, given the nature of this young man's game, they inevitably remind us of the passage of time.

Slowly Angell's style of reporting changed. "I came to know some players and I began to feel free about going onto the field and into the clubhouse," he told Haynes. "One Hard Way to Make a Living" (4 May 1981), reprinted in *Late Innings*, is an investigation of hitting. But Angell called "In the Fire" (12 March 1984), an essay on catching that focused on Ted Simmons, "the first 'What do you do?' piece I ever wrote." This new emphasis on the skills needed to play the game became a distinctive theme in Angell's fourth collection, *Season Ticket: A Baseball Companion* (1988), published with a Ted Williams quotation as an epigram: "Don't you know how hard this all is?"

In the preface to this book Angell explained his new interest:

> I have become engrossed in the craft and techniques of the game: not just how the runs were scored but why. I have talked for hours on end with catchers and pitchers . . . and infielders, hitters, coaches, managers, scouts, executives, and owners, and most of the time I started with the same question: "How do you do what you do?"

In addition to essays on catching, *Season Ticket* includes essays on pitching and playing the infield, as well as profiles of Roy Eisenhardt, an owner of the Oakland A's, and Dan Quisenberry, a relief pitcher.

In "The Arms Talks" (4 May 1987) Angell wrote, "Trying to learn the game . . . protects us from its overattachments and repeated buffetings, and for me, as the years go by, this has become the best part of baseball." To do these pieces, Angell told Haynes, he sought out the "great talkers. Of course any reporter knows enough to go to a good talker. You remember who talks well – who talks in sentences and now and then even in paragraphs." And he took lots of notes:

> People who have known me in various pressboxes around the league know that I write a lot during the games. It's a kind of joke – all the notes I take. But the reason is that I'm going to write much later than anybody else. . . . You don't know when you're watching a game if this is going to fit into something else that happens in September and something else that happens in October, or some recurrent theme I want to pick up on, or something about this particular player that I'm going to see later on.

The result of this research was a volume that garnered the usual fine reviews for both its content and its author. "The essays have an unhurried pace appropriate to the game they annotate," wrote Joel Conarroe in the 20 March 1988 *New York Times Book Review*, "and a reader does well to take them in one at a time, with an occasional doubleheader to vary the rhythm." David Lehman, in the 11 April *Newsweek*, remarked, "As a stylist, Angell resembles a pitcher with pinpoint control. As a chronicler of the game, he's in a class with Ring Lardner and Red Smith."

Baseball between 1983 and mid 1987 – the essays on the 1982 season and the 1987 World Series did not make *Season Ticket* – severely tested Angell's

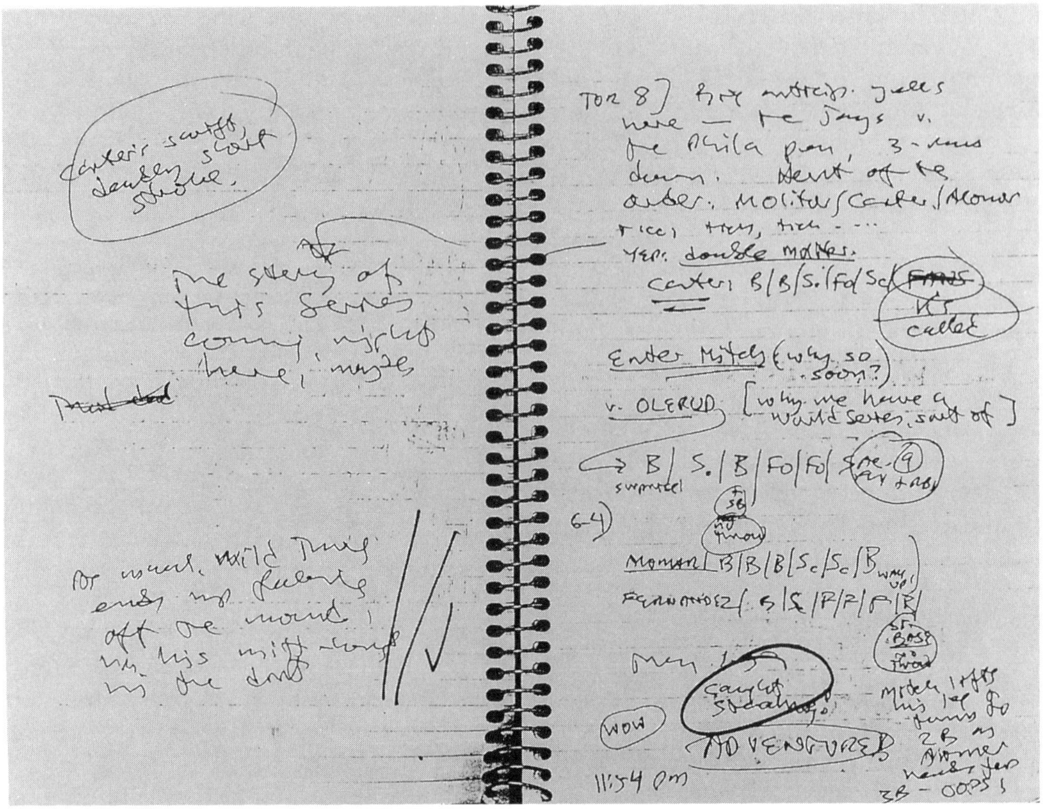

*Pages from Angell's notes on the 1993 World Series (courtesy of Roger Angell)*

devotion. In particular, he felt wrenched in two directions simultaneously as he contemplated composing an essay about the problem of drugs in professional baseball. "As a baseball writer," he said in "The Cheers for Keith" (5 May 1986), "I dutifully did my work and tried to pay attention. . . . But at the same time another part of me – the fan side – still felt aggrieved and bitter about the whole thing. None of this," he said, "had anything much to do with *baseball* – the delectable and marvelously difficult old game that has brought me almost undiminished pleasure since I first began to take notice of it, more than fifty years ago." He worked out his solution to this dilemma by writing about first baseman Keith Hernandez, an admitted cocaine user but one of the players whose skills Angell greatly admired.

The severest test for Angell the fan, however, did not come from the drug problem or any other off-the-field distraction, but from the 1986 World Series in which his two favorite teams, the Boston Red Sox and the Mets, opposed each other after surviving tense League Championship Series. Angell reported on this conflict in "Not So, Boston" (8 December 1986), an essay with a palindrome for a title. As the Mets eked out their slim, seven-game victory, Angell resolved his personal crisis and discovered where his heart truly lay: "I am a Mets fan. I had no idea how this private Series would come out, but when the Mets almost lost the next-to-last game of the Series I suddenly realized that my pain and foreboding were even deeper than what I had felt when the Red Sox came to the brink out in Anaheim [playing the Angels in the American League Championship Series]."

*Season Ticket* stands as the fourth installment in an informal tetralogy, a quartet of books capturing all but a few of Angell's essays written over a quarter-century. After its publication Angell continued to pen his essays on about the same schedule through the end of the 1993 season. He published another book, *Once More Around the Park: A Baseball Reader* (1991), but it was primarily, in his own words, "a re-collection," with nineteen of the twenty-one essays reprinted from earlier books. Only "Celebration" (22 August 1988), a profile of A. Bartlett Giamatti, then president of the National League, and "No, But I Saw the Game" (31 July 1989), a study of baseball movies, were appearing in book form for the first time.

When the 1994 baseball season was interrupted by a players' strike, Angell used the editorial platform of *The New Yorker* to express his disgust for the emerging situation. In a "Comment" published 15 August 1994, he wrote:

> Caring about the issues . . . may be too much to ask of an adult, particularly one who retains a modicum of fanly interest in the old game. We've been here too often before – seven times, counting lockouts and some very brief cessations, since 1972 – and in those years the parties to the disputes, once partners in a sentimental little mini-business, have flourished and fattened, particularly around the wallet.

After the balance of the season was canceled by decision of the club owners, he wrote "Hardball" (17 October 1994), an evenhanded but passionate analysis of the sport's seemingly intractable labor-management war. Angell did not expect either side in the dispute to care much about what he had to offer in the way of suggestions to resolve their differences, but he did hope they would understand his concerns as one of the millions of fans who would have to decide whether or not to go back to the ballparks when the strike was settled: "I'll probably want to go," he concluded, "but right now – and here's another baseball first – I'm not too sure about that."

Angell is that rare baseball fan who has been able to make a career out of his pleasure. Without abandoning the wonder, the affection, and the detachment that characterize a fan's kinship to baseball, Angell has fashioned a string of remarkable essays that explore the sport in consistently new ways. His work possesses a grace and elegance previously unknown in sports journalism and has earned a lasting place in the literature spawned by the national pastime.

**Interviews:**

Michael Mok, "Roger Angell," *Publishers Weekly,* 202 (10 July 1972): 22;

George Plimpton, "Talk with Roger Angell," *New York Times Book Review,* 15 May 1977, pp. 1, 32–34;

Dolly Langdon, "Is Winning the Only Thing? Not in His Book, Says the Keeper of Baseball's Flame, Roger Angell," *People,* 18 (26 July 1982): 81–83;

Dick Johnson, "A Conversation with Roger Angell," *SABR Review of Books,* 3 (1988): 43–52;

Jared Haynes, "An Interview with Roger Angell: 'They Look Easy, But They're Hard,'" *Writing on the Edge,* 4 (Fall 1992): 133–150.

**References:**

James Gindin, "Roger Angell and the Annals of Baseball," *Michigan Quarterly Review,* 25 (Summer 1986): 568–581;

A. James Memmott, "Wordsworth in the Bleachers: The Baseball Essays of Roger Angell," *Journal of American Culture,* 5 (Winter 1982): 52–56;

Michael Oriard, "An Appreciation on Roger Angell's Silver Anniversary (Notes on a Fan's Notes)," *Aethlon,* 5 (Fall 1987): 35–54;

William Palmer, "History, Tradition, and Hubris: The Baseball Universe of Roger Angell," *Journal of Popular Culture,* 20 (Fall 1986): 17–27;

Mike Shannon, *Diamond Classics: Essays on 100 of the Best Baseball Books Ever Published* (Jefferson, N.C.: McFarland, 1989).

# Furman Bisher

*(4 November 1918 –   )*

Josh Riley

MAJOR POSITIONS HELD: Editor, *Lumberton Voice* (1938–1939); wire service and sports editor, *High Point Enterprise* (1939–1940); state editor (1940–1942), sports editor (1946–1950), *Charlotte News;* sports editor, *Atlanta Constitution* (1950–1957), *Atlanta Journal and Sunday Journal-Constitution* (1957–   ); columnist, *Sporting News.*

BOOKS: *With a Southern Exposure* (New York: Nelson, 1962);

*Miracle in Atlanta: The Atlanta Braves Story* (Cleveland: World, 1966);

*Strange but True Baseball Stories* (New York: Random House, 1966);

*Aaron, r.f.,* by Bisher and Henry Aaron (Cleveland: World, 1968); revised as *Aaron* (New York: Crowell, 1974);

*Arnold Palmer's Golden Year* (Englewood Cliffs, N.J.: Prentice-Hall, 1971);

*The Atlanta Falcons: Violence and History* (Englewood Cliffs, N.J.: Prentice-Hall, 1973);

*The Masters: Augusta Revisited. An Intimate View* (Birmingham, Ala.: Oxmoor House, 1976);

*The Furman Bisher Collection* (Dallas: Taylor, 1989);

*A Series for the World: Baseball's First International Fall Classic,* by Bisher, W. P. Kinsella, and Dave Perkins (San Francisco: Woodford Press, 1992).

OTHER: "Shinola's Game (World Series Game IV)," in *Best Sports Stories: 1958 Edition,* edited by Irving T. Marsh and Edward Ehre (New York: Dutton, 1958), pp. 52–54;

"Greed, Inc.," in *Best Sports Stories: 1960 Edition,* edited by Marsh and Ehre (New York: Dutton, 1960), pp. 62–64;

"The Many Moods of Mauch," in *Best Sports Stories: 1961 Edition,* edited by Marsh and Ehre (New York: Dutton, 1961), pp. 104–106;

"Memories of a National Resource," in *Best Sports Stories: 1982 Edition,* edited by Ehre and *Sporting News* (New York: Dutton, 1982), pp. 94–96;

"The South's Sporting Times," *Southern Living,* 25 (June 1990): 136, 138, 140–141.

*Furman Bisher (*Atlanta Journal-Constitution*)*

Furman Bisher's move to Atlanta, Georgia, could not have come at a better time for an up-and-coming sportswriter. Soon after he joined the *Atlanta Constitution* in 1950, the city began the rapid growth that would result in its becoming the "hub of the South." As the city developed into a major

*Dust jacket for the revised edition of Henry Aaron's "autobiography," written by Bisher after extensive interviews with Aaron. The book was originally published in 1968 as* Aaron, r.f. *and was republished in 1974, the year Aaron broke Babe Ruth's career home-run record.*

metropolis, it was inevitable that professional sports teams would find their way there. Bisher was uniquely qualified to cover this emerging force in American athletics. He is now one of the country's premier sportswriters; his columns appear regularly in the pages of the *Atlanta Journal and Sunday Journal-Constitution,* where he serves as sports editor. Additionally, he has written several books and contributed to many periodicals. Much of his writing centers around the South and chronicles the changes that have occurred in that section of the nation in the second half of the twentieth century.

James Furman Bisher was born on 4 November 1918 in Denton, North Carolina, to Chisholm Bisher, a textile manufacturer, and Mamie Morris Bisher. Growing up in a small town hundreds of miles from the nearest major-league city, Bisher developed his love of baseball by watching, and then becoming batboy and later statistician for, his hometown team – a perennial winner in the Davidson County League; as he recalled later, it played "the most excellent baseball a kid could see." As a journalist he would write with special affection about the handful of players who came from the ranks of farmers and shoe salesmen and made it to the majors.

As a boy Bisher read everything he could, especially the works of Horatio Alger Jr. Of the sportswriters of the period he most liked Joe Williams of the *New York World-Telegram,* a paper that, in his opinion, had the best sports page until the *New York Herald Tribune* came along in the late 1930s.

Bisher attended Furman University in Greenville, South Carolina, for two years, transferring to the University of North Carolina when Furman discontinued its journalism program. Graduating in 1938, he served as an editor on the *Lumberton Voice* before joining the *High Point Enterprise* in 1939. A year later he left to become state editor of the *Charlotte News.* He was a naval aviator from 1942 to 1946, reaching the rank of lieutenant, senior grade. In 1946 he returned to the *Charlotte News* as sports editor. In 1950 he moved to Atlanta and became sports editor for the *Constitution.* In 1954 Bisher married Montyne Harrell; they subsequently had four children: Clark, Roger, James Furman Jr., and Monte. They later divorced, and Bisher married Lynda Landon.

In 1957 he joined the *Atlanta Journal* as sports editor and columnist, retaining these positions when the two newspapers merged in 1957 to form the *Journal and Sunday Journal-Constitution.*

In the 1950s Bisher's newspaper and magazine pieces began to demonstrate his versatile narrative voice. Dizzy Dean's career, from the pitcher's mound to radio "commertator" and television, provoked a *Constitution* article written entirely in "Deanese": "Jay Hanna Dean came out of Chickalah, Arkansaw, more years ago than a coon hound's got fleas." Bisher was impressed that a man who "never did know no English . . . can start out broadcasting manure and wind up broadcasting television." For *The Saturday Evening Post* article "Father Is a Football Coach" (20 November 1954) he interviewed the daughter of University of Georgia football coach Wally Butts, providing insight into a local sports figure from a rarely heard point of view (though at the stadium, Jean Butts Jones admited, "our view of him is no better than any spectator's"). Bisher gets his subject to move between sports-related topics and personal ones: she had recently married a successful high-school football coach, and "my father was flattered that after observing a coach's life from the inside for the greater part of my twenty-one years, I should still be brave enough to marry one."

In the preface to *Best Sports Stories: 1958 Edition* Bob Considine calls Bisher's piece on the fourth game of the 1957 World Series a "modern retelling of the classic about 'for the want of a nail the shoe was lost.' A delightful account of an engrossing ball game." In the top of the ninth inning the Yankees had tied the score, with two outs and two men on base, when Warren Spahn of the Milwaukee Braves tried to keep the ball low and outside to Elston Howard (in the next sentence Bisher refers, albeit good-naturedly, to "the dark brown Howard"; such racial references were still common, especially among southern sportswriters). The pitch "turned out to be delinquent, low and inside." In the top of the tenth inning, Tony Kubek, "a willful and resourceful traitor who was born here [Milwaukee] and went to New York to make his living," put the Yankees ahead by one run. In the bottom of the tenth, pinch hitter Vernal Jones reached first base by arguing successfully that he had been hit on the foot by a pitch, showing the umpire the shoe polish mark on the ball. The Braves then won the game on Eddie Matthews's "manfully expressive" home run, but Jones was the hero of the day for Bisher in a tale that brought "rare glory" to "the department of nondistinction entitled 'hit by pitcher' " as well as demonstrating an Aesopian moral: "always keep your shoes shined."

At the end of the 1950s Bisher, disgusted at "the deteriorating state of the game today," railed against the proposed second All-Star Game of the 1959 season, which was scheduled to be played at the ninety-three-thousand-seat Los Angeles Coliseum, "home of the charitable Walter O'Malley," rather than in San Francisco, which "lacked the eleemosynary attractiveness of a $500,000 gate." Bisher disliked the idea of prostituting the national pastime for any reason. While he commended the fact that some of the money would go to the newly established players' pension fund, he wondered about those players who had been unable to get into the plan a year or two before. He saw the episode as symptomatic of what was becoming "Greed, Inc.," the title of the essay, faulting both the owners and the players and foreshadowing the state of the game some thirty years later. For Bisher, baseball, which he had learned to love "in its rawest, crudest backwoods form," will always be, in the pure dictionary definition, a "pastime." Admiring the article's "indignation quotient," John Hutchens observed in the preface to *Best Sports Stories: 1960 Edition,* where it was reprinted, "that this blast at the acquisitive instinct strikes me as particularly welcome."

Bisher's first book, *With a Southern Exposure,* was published in 1962. A collection of columns he had written in the 1950s and early 1960s on a wide variety of sporting subjects, including baseball, football, tennis, boxing, golf, and horse racing, the book offers the best overview of his early work. The singer Bing Crosby noted in the introduction that "the good ones, the outstanding writers, are those who can see the color, the personalities, and their relationships, the amusing and interesting sidelights, and can tell about them in the good earthy prose the sports follower understands and appreciates." Included in the volume is "The Destitute Slugger," which describes the decline of former baseball great Jimmy Foxx. While other ballplayers had invested their money wisely and had maintained lives beyond the baseball diamond, Foxx had done neither. Bisher is unsympathetic:

> What seemed to have happened to Foxx was a failure to readjust to life off the stage, out of the range of the cheers and the adulation. This is an occupational hazard. Some survive it and some don't. But this kind of thing comes up frequently, the old hero down and out and the worshipful public rallying to the wall. Somehow, I fail to respond to the cry that baseball and the world have a debt to Jimmy Foxx. It paid him hand-

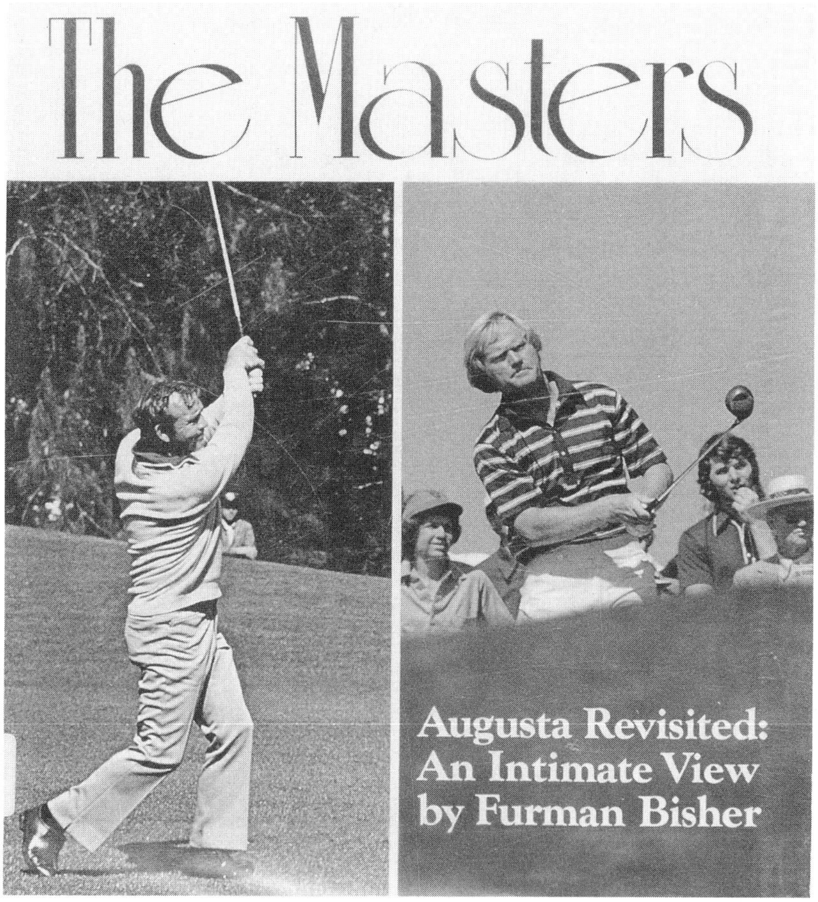


*Dust jacket for Bisher's coffee-table book about one of the nation's most prestigious golf tournaments*

somely while he performed. It offered him opportunities after he could play no longer.

Also included in *With a Southern Exposure* is "The Many Moods of Mauch." Bisher had met Gene Mauch when the latter was the twenty-seven-year-old manager of the Atlanta Crackers. Mauch, Bisher judged, unquestionably had the analytical skills for the job but was lacking in the area of temperament. Years later Bisher watched Mauch, then the manager of the Philadelphia Phillies, transform himself from "a patient, temperate, reserved, mild, objective, humanitarian" and start a bench-clearing brawl after three of his batters were hit by the pitcher. Before it was over, "Billy Martin, former welterweight champion of the American League, was led off the field bloody and torn. Robin Roberts, customarily a disciple of peace, and Frank Robinson, the Cincinnati first baseman, had squared off like Peter Jackson and Bob Fitzsimmons."

Another column collected in the book is "Ty Cobb . . . the Motivating Force," which was written in 1961 just after Cobb's death. Bisher reminisces about a visit with Cobb at the baseball legend's home in Banks County in northern Georgia. The two men looked out at the valley called the Narrows, where Cobb said that he was born:

> The smoke curled up from the chimney of the house in view and flattened out over the valley. You could imagine the baby child in the arms of his mother, remembered in those parts as a raven-haired beauty, and you looked at this soul-weary man beside you searching for something out of reach. A baby had gone out from that wilderness to the greatest fame an athlete could know, and you wondered what it could possibly be that he still sought.

At the time *With a Southern Exposure* was published, Atlanta was on its way toward becoming a major city, but it still lacked a professional sports team. The country's growing population and the beginning of the massive migration to the Sunbelt increased the likelihood that new cities, especially a few in the South, would be allowed to join the pres-

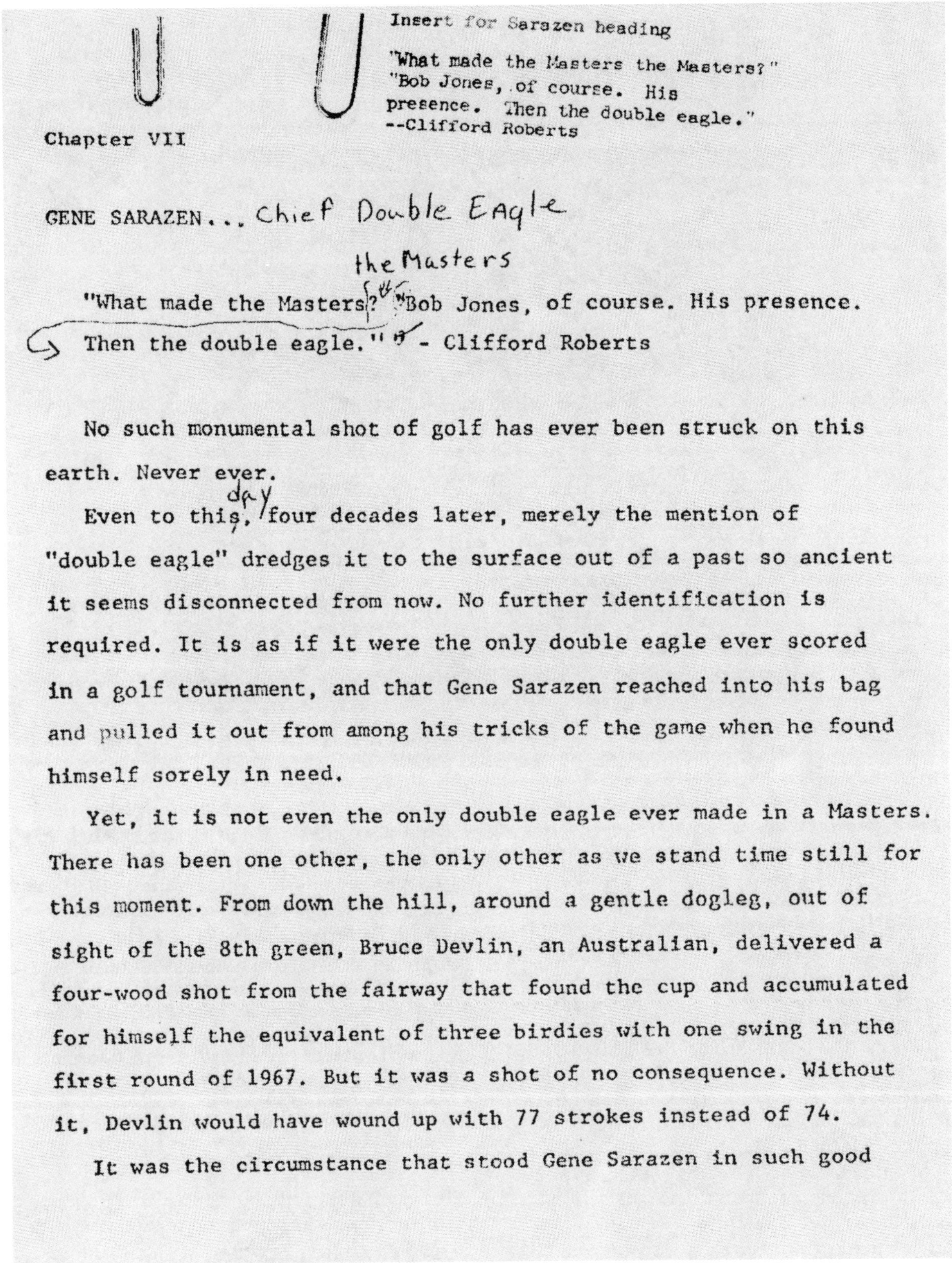

Insert for Sarazen heading

"What made the Masters the Masters?"
"Bob Jones, of course. His
presence. Then the double eagle."
--Clifford Roberts

Chapter VII

GENE SARAZEN... Chief Double Eagle

the Masters

"What made the Masters?" "Bob Jones, of course. His presence.

Then the double eagle." - Clifford Roberts

No such monumental shot of golf has ever been struck on this earth. Never ever.

Even to this day, four decades later, merely the mention of "double eagle" dredges it to the surface out of a past so ancient it seems disconnected from now. No further identification is required. It is as if it were the only double eagle ever scored in a golf tournament, and that Gene Sarazen reached into his bag and pulled it out from among his tricks of the game when he found himself sorely in need.

Yet, it is not even the only double eagle ever made in a Masters. There has been one other, the only other as we stand time still for this moment. From down the hill, around a gentle dogleg, out of sight of the 8th green, Bruce Devlin, an Australian, delivered a four-wood shot from the fairway that found the cup and accumulated for himself the equivalent of three birdies with one swing in the first round of 1967. But it was a shot of no consequence. Without it, Devlin would have wound up with 77 strokes instead of 74.

It was the circumstance that stood Gene Sarazen in such good

*Page from the typescript for* The Masters *(courtesy of Furman Bisher)*

tigious fraternity of cities with big-league teams. Residents of Atlanta were clamoring to have a team of their own, but no team wanted to move to a city that had no stadium. Soon the business and political leaders of Atlanta decided to do whatever was necessary to bring a team to the city. The luring of the Milwaukee Braves to Atlanta became the subject of Bisher's second book, *Miracle in Atlanta* (1966).

With his intimate firsthand knowledge of the personalities involved, Bisher provides an insightful look at the secret meetings between business leaders, politicians, and the baseball community where the move was negotiated. Bisher begins by describing the attempts of Charles Finley, owner of the Athletics, to move his franchise out of Kansas City. Bisher writes that Finley, who had a reputation as a deal maker, was supremely confident that he had the votes he needed from the other owners to permit the move – even Calvin Griffith, owner of the Minnesota Twins, who had previously not been agreeable to the idea:

> There was some little inside deal between the two of them . . . something Calvin wanted and which Finley would support him on. It didn't come through clearly in our conversation, but Finley was all confidence. Suddenly, he had the American League by the short hair, and to put him into Atlanta his compatriots would have to rush to his support in the meeting room.

Although the Athletics ultimately moved to Oakland, the stage was set for baseball to head south. Atlanta mayor Ivan Allen, whose "knowledge of baseball extended barely beyond the ability to identify Ty Cobb and to recall that nine men make up a team, nine innings make up a game and there shall be no gambling in the left field bleachers," had nevertheless run for election promising to bring a team to the city. Working with the business community, Allen constructed an $18 million stadium in less than a year. Later, Bisher says, Allen would brag that "We built a stadium on land we didn't have with money we didn't have for a team we didn't have."

Bisher recounts the negotiations that preceded the Braves' move to Atlanta with a clarity and realism that makes the reader feel as though he or she is in the room as the talks are going on, and he provides background information so that the reader can understand the viewpoints of the participants. Of the cloak-and-dagger circumstances surrounding the meetings he says:

> Ordinarily, a board of directors meeting of a major league baseball club is about as well attended by news hounds as a weekly meeting of the temperance society. But this one in Chicago drew quite a following, and it came off in the surreptitious way of the high councils of old gang lords, that is, if I can assume that the movies I've seen on television are accurate portrayals of gang lords arriving at a high council.

Bisher also discusses the groundbreaking impact the move had in legal circles, chronicling the lawsuits and injunctions and the resulting controversial legislation granting immunity to Major League Baseball from federal antitrust laws. The book includes a history of the Braves franchise, tracing the team to its roots in Boston in 1871.

*Miracle in Atlanta* foreshadows the present situation in which baseball finds itself – as more a business than a sport – as well as the many "cynic projects," as Bisher calls them, in which writers engage when writing about baseball. In *The New York Times Book Review* (12 June 1966) Rex Lardner called Bisher's book "a complicated tale of temporary injunctions, permanent injunctions, rancor, secret conferences, rumor-mongering, rumor-scotching, secret deals, catcalling, acrimony, subtle suasion, intrigue, accusations, veiled and open threats and bad will." After reading the book one is forced to conclude that the "miracle" is that the deal was ever concluded and that the politicians and businessmen involved actually lived up to their promises.

Also in 1966 Bisher published a work geared toward younger readers. *Strange but True Baseball Stories* is a collection of twenty-six anecdotes drawn from baseball lore. Bisher writes in the introduction: "There is a popular misconception that baseball is a game played nine men to the side. This is for narrow people, those who believe that horseshoes can be played only by horses." Baseball, according to Bisher, can be played in the imagination anywhere and any time that "two people get together who like to tell stories." Baseball stories are often filled with exaggeration and embellishment, he says, but the

> stories in this book are different. Some are strange, some are uncommon, some are almost unbelievable. Some are old, some have never been told before. Not all of the people in this book are famous. Some are like actors who have only one short line in a drama, then disappear and never appear again. These stories have only one thing in common: they actually did happen.

Among the tales Bisher relates are "The Catcher Who Pitched All Night" and "The Dog That Made a Box Score." "The Midget of St. Louis" is a comedic tale of how, in 1951, Browns owner Bill Veeck

Jr. sent the three-foot-seven-inch-tall, sixty-seven-pound Eddie Gaedel to the plate with instructions not to swing. Of course, with such a small strike zone, Gaedel walked and was promptly replaced at first base with a pinch runner. Dramatic finishes, such as Bobby Thompson's homer in the 1951 World Series, are highlighted, as are stories about players such as Johnny Vander Meer, who once pitched back-to-back no-hitters.

By the mid 1960s Bisher's reputation extended well beyond Atlanta. His "World Series Wrap-up" for 1966, reprinted in *Best Sports Stories: 1967 Edition,* shows Bisher at the top of his form. The stunning sweep of Los Angeles by the upstart Baltimore Orioles "made the Dodgers look as though they didn't belong here, and," Bisher admits, "I'd have sworn that the Dodgers were the best team in baseball when it began." He describes Baltimore as "a town that God had forgot": "The Chesapeake Bay looked like a good place to dump the garbage, so they built Baltimore here." Baltimore had been the home of a venerable nineteenth-century ball club; then "somebody moved the St. Louis Browns here," which "wouldn't make any history books as a sign of progress." Having known the Browns in their inglorious years in the 1940s, Bisher finds it all the more amazing that these "new St. Louis Browns, with the birds on the shirts and the Orioles for a name, not only just won a World Series, they won one like nobody ever won one before." Bringing to his stories a knowledge of the past is essential for Bisher as he writes for a generation of readers who have no idea who pitchers "like Kretlow, Chakales, Shallock, Schmitz, and even Saul Rogovin" were. The article shows Bisher's knack for streamlining his stories, never burdening them with unnecessary facts that the reader could easily infer; analogy serves a better purpose than detailed exposition and also lends itself to humor: "The fact is, the Orioles won it looking like the Dodgers figured to look." The games are spared analysis: "How do you analyze a 1–0 ball game in which only two runners get as far as second base? And the one run isn't scored with a rally?" Talking to losing manager Walter Alston is "a lot like interviewing a Japanese general in 1945." The piece, which began with memories of the old Orioles of "Wilbert Robinson, John McGraw and Willie Keeler," ends with the Orioles' Frank Robinson "telling everybody about the kind of pitch he hit" and the sudden realization that "the baseball season was over."

In 1968 Bisher conducted extensive interviews with Henry Aaron that appeared the same year as the book *Aaron, r.f.* Bisher writes in the introduction that Aaron's principal "characteristic is stability, making the right play in the most direct, unflamboyant manner"; he "is a kind of hitter all to himself."

> He is not the greatest player who ever lived. He may not be the greatest player in the game today, though I am inclined to think he is. But for the long haul, for giving a performance that pleases the critics on a consistent basis over the years, there is no one around like him, and there hasn't been since Stan Musial. In fact, I like to think of him as a right-handed Stan Musial in color.

Aaron recalls for Bisher his childhood, his father's stress on the importance of education, his discovery as a softball player, and his years with the Indianapolis Clowns of the old Negro League, when black players touring the South were required to stay in old, run-down hotels above pool halls. "Everytime I see a kid come into spring training and start complaining about something that's not just right, I feel like going over to him and saying to him, 'Hey, kid, if you think this is bad, you should have been with me when I was with the Indianapolis Clowns in Winston-Salem, North Carolina.' " Bisher recounts how Aaron was thrust into the starting lineup of the Milwaukee Braves in 1954 when Bobby Thompson suffered a broken ankle during spring training in Florida and the drastic change that this event produced for Aaron, who, by his own admission, was naive and easily taken advantage of. The book follows Aaron's career through the Braves' move to Atlanta.

Particularly interesting in retrospect is Aaron's perspective on Babe Ruth's all-time home-run record. He recognizes that he has a chance of breaking the record (which he eventually did) but says:

> I want to make certain that I don't get set in my mind to stay around until I wear out my welcome, until the fans look upon me as a welfare case. I want to bow out gracefully with the club, with the other players and with the fans on my side. I don't want to go . . . like a drowning man clutching for a straw. When I get ready to go, my home-run standing is not going to have anything to do with it. I'm not going to accept bench-warming, waiting for a chance to pinch-hit in hopes that I have another home run or two in my bat.

Aaron tells Bisher that he believes that the Minnesota Twins' Harmon Killebrew will be the one to break Ruth's record. In 1974, the year Aaron broke the record, the book was revised and republished as *Aaron,* "by Henry Aaron with Furman Bisher."

*Dust jacket for the collection of eighty-four of Bisher's best columns*

While studying in North Carolina Bisher had formed a friendship with a Wake Forest College student golfer named Arnold Palmer. As Americans obtained more leisure time in the 1950s and 1960s the golf craze began, and Palmer was at the forefront of the game. In 1971 Palmer became the subject of Bisher's next book, *Arnold Palmer's Golden Year.* Bisher allows the reader to accompany Palmer from his early tournaments, when he "suffered through the trials of a young man attempting to weed out the real from the visionary, through the uncertainties and the strange new valleys of a boy growing into manhood and making decisions that turn him into a course of life." Bisher describes the crucial year of 1960, when Palmer rebounded from losing the 1959 Masters Tournament and established himself by winning the Masters and U.S. Open and finishing one shot short of taking the British Open. He traces Palmer's subsequent highly successful career and the enthusiastic following ("Arnie's Army") the golfer picked up along the way:

> Palmer was becoming the god of a personal religion that moved among the idolatrous of the country, later the world. What no other golfer had been to the public, Palmer was becoming, a man to whom the average man, woman, child, squealing emotionaries could relate, and who related back through some gesture, some characteristic move that was to become a trademark of his.

Bisher ends the book in the early 1970s with Palmer's winning of the newly established Heritage Classic, played on a course in Hilton Head Island, South Carolina, that

> had been freshly hacked out of the marshland, the palmetto scrub, the bayous and the thick pine and oak forests on the southwestern side of the island . . . played in an almost haunting atmosphere of dripping Spanish moss, gnarled, almost grotesquely shaped oak and pine trees, alligator-infested lagoons, sand traps that were more like open pits, wiry seaside grass, and came to a quite logical conclusion on an eighteenth hole whose fairway was bordered by an ancient cemetery.

In 1976 Bisher returned his attention to golf with a coffee-table book, *The Masters: Augusta Revisited.* This colorful work includes photographs and diagrams of the Masters' course in Augusta, Georgia, along with Bisher's descriptions:

> Standing on this vast terrace, the lawn not up to spring standard, bare in some places beneath the huge water oak trees that steal the sunlight, you feel a vast incompetence in your effort to appreciate the history, the grandeur, the greatness of those who have stood on this

> ground before you, and to absorb what all of this has come to mean in terms of golf. Across the vast expanse, rising still elegantly above the pines, is the upper thrust of the one-time Forest Hills Hotel to remind you of the past.

Bisher describes the terrain on which the course is laid out as

> a gentle, rolling spread of hill and valley with a spring-fed stream that only a near century of preparation could have put in such perfect condition to receive a golf course . . . while America was still a wilderness, Indians had camped upon this land, attracted by the spring that is the course's source of water, that feeds the lakes and Rae's creek. De Soto and his marchers are said to have passed through it in their search for the Mississippi River.

Bisher writes with firsthand knowledge about activities on the course and in the clubhouse and includes profiles of and interviews with golfing greats of the 1970s such as Palmer, Ben Hogan, Byron Nelson, Sam Snead, and Jack Nicklaus. A piece on Bobby Jones, the founder of the Masters, was excerpted for *This Is My Best,* a collection of sports stories edited by Howard Siner.

As sports editor of the recently merged *Atlanta Journal and Sunday Journal-Constitution* Bisher hired a young graduate of the University of Georgia who quickly became his executive sports editor at age twenty-three. The late Lewis Grizzard, who a decade later would become a nationally known humorist and syndicated columnist, had been a fan of Bisher's since, as a ten-year-old, he had eagerly awaited the sports page each morning to read Bisher's column. Bisher and Grizzard clashed in the beginning; in a column in the *Journal and Sunday Journal-Constitution* on 19 October 1990 Grizzard wrote:

> Bisher made my life miserable for three years. One comma out of place, he'd raise hell. The attendance line left off the second game box score . . . and he'd chew on me. Bisher was a tyrant. He once fired a staffer, back when he was writing his column and running the department, too, by saying, "I've got a great new assignment for you. Find another job."

The two became close friends, however, and Bisher asked Grizzard to write the foreword for *The Furman Bisher Collection* (1989). There Grizzard describes Bisher's writing:

> He was profoundly different from most of the other sportswriters I read. He had a way. A style. He stood back from the tumult and the shouting and found truths and real meanings. The words from other writers marched drone-like. Furman's danced. . . . His passion for and understanding of life, its participants, and the games they play are evident in each word.

In 1990 Grizzard dedicated his book *If I Ever Get Back to Georgia I'm Gonna Nail My Feet to the Ground* "To Bisher, who knew what he was talking about after all."

*The Furman Bisher Collection* comprises eighty-four of Bisher's best columns. The first section of the book is devoted to baseball and includes pieces on Denny McLain ("It didn't matter that his fastball was gone, he was ingloriously out of condition, his earned-run average was about the size of his waistline, and he was unemployed"), Pete Rose ("He arrived in a dirty uniform and ten minutes after it was cleaned, it was dirty again. He played for the love of it"), Al Campanis ("a man so devoid of bias that he once traded his own son for a player to be named later"), and the folly of the designated-hitter rule ("Surely in these times of advancing idiocy we haven't lost appreciation for the classic pitching duel. Nor for the pitcher who can break up his own game with the bat"). The section on football features columns on Woody Hayes, Bill Curry, and Super Bowl IX, and the stirring "Damning Indictment of Garbagemen," on college athletes who fail to receive an education:

> Persons who can't read or write have no earthly reason for taking up space in college. How do you educate those who don't recognize words and can't make letters and sign their report card with an X? We are coming to the point where we must decide what college is for – education or a refuge for illiterates hired to put on shows for television, for which the school is paid a handsome fee.

The next section delves into Bisher's beloved world of golf, with pieces on Palmer, Nicklaus, Hogan, and Tom Watson. It is followed by sections on horse racing and tennis (in "Battle of Brats – Past and Present" he says that John McEnroe "is an unfortunate youth who appears to be quite disappointed with this planet"). The section on boxing includes columns on Muhammad Ali and Joe Frazier, Ken Norton, and Marvin Hagler and Sugar Ray Leonard; of the Hagler-Leonard fight he says,

> Prizefighters don't win fights with a scowl and a sneer. The guy you wouldn't want to meet in a dark alley isn't always the one who wins under bright lights. This was little Lord Fauntleroy taking The Hulk. Popeye over Bluto again.

*Bisher at his desk* (Atlanta Journal-Constitution)

Noteworthy is a piece written at the time of Joe Louis's death, in which Bisher says: "He was solid American gold. It may read like solid American corn, but here was a man who loved his country and put it where his heart was." Before appearing in Bisher's own collection, the Louis piece had been selected for *Best Sports Stories: 1982 Edition;* the "Judge's Comments" on the article noted that "Bisher tries to forget all the sad moments of Louis' life. . . . His is critical of the 'carnival' atmosphere that surrounded Louis' first funeral – when a group of celebrities paid their last 'respects' to the Brown Bomber by giving the corpse a standing ovation. . . . But his admiration for Louis is best seen when he calls the man from eastern Alabama a 'national resource.' " The following section gathers columns about personalities and events in a variety of fields, including Cobb, Shoeless Joe Jackson, Pete Maravich, Ted Williams, and the U.S. hockey team's victory over the Soviet Union in the 1980 Olympics. Columns on athletes who had recently died are particularly poignant; Bisher brings out the lesser-known private side of these public figures, reminding the reader that the humanity of the athlete is more important than his or her sport.

Rounding out the book are a few of Bisher's own favorite columns. Particularly insightful is "Ah, the Enlightened Age," in which Bisher presents his opinions on several issues in American sports. For example, he writes of "the specialist":

> We have the baseball player who never goes to bat, the placekicker who never plays football, the relief pitcher, the career pinch-runner and pinch hitter, the kick returner, the pass rusher, the third down player, the tight end, split end, holder, long snapper, the hatchet man in basketball and the brawler in hockey. Laugh, if you will, but you never knew a real hero unless you saw a four-letter man, the athlete who could play any game – and make it in class, too.

Bisher has written articles for such magazines as *Sports Illustrated, True, Collier's, Golf, Saga, American Weekly,* and *Esquire;* a 1990 piece for *Southern Living* discusses the contributions southerners have made to the sporting world. He has also been a columnist for *The Sporting News.* He has contributed to collections and anthologies, including many volumes of *Best Sports Stories of the Year, Grantland Rice Prize Sports Stories, The Fireside Book of Baseball, Sportswriters' Choice, Treasury of Sports Stories, Best of Sport,* and *Miracle Season,* a colorful hardcover history of the 1991 Atlanta Braves. His many awards include the Associated Press Sportswriting Award, the United Press International Sportswriting Award, the Turf Writing Award, the Georgia Sports Writer of the Year Award, the North Carolina Governor's Award, and the Bobby Jones Sportsman of the Year Award. He has been inducted into the National Sportscasters and Sportswriters Hall of Fame, the International Golf Writer's Hall of Fame, the Georgia Sports Hall of Fame, and the North Carolina Sports Hall of Fame. In 1961 *Time* magazine called him one of the nation's five best columnists. He is a member of the Football Writers Association, of which he was president in 1959–1960, and the National Sportscasters and Sportswriters Association, serving as its president from 1974 to 1976.

Bisher remains popular, particularly among southern readers and golf lovers, as he continues his no-nonsense approach to collegiate and professional sports. Ironically, he never realized his lifelong dream of becoming a "sportswriter" – that is, a beat reporter who covers a particular team on a regular basis – because he began his career in sports journalism as an editor. Nonetheless, his stories demonstrate his credo that a writer should first of all be a good reporter and, second, should write entertainingly. As Charles Einstein observed in volume four of *The Fireside Book of Baseball* (1987), "two generations of sports fans had been reading stories by Furman Bisher. . . . Reading the stories and remembering them. Especially the way they ended." At an *Atlanta Journal and Constitution* reunion in 1990, Grizzard summed up the feelings he and his colleagues had for Bisher:

> We didn't have the enormous space and resources sports departments have today, but we had our moments. And always there to guide us and show us daily what quality journalism was all about was Bisher. Always, just "Bisher." I wouldn't have missed it for the world.

Primarily a regional writer, Bisher has never had the national audiences enjoyed by Joe Williams or Red Smith; but he continues to leave his mark on what he calls "the lost art" of sportswriting, an art that should provide entertainment, a sense of history, and a sense of fascination for America's games and athletic heroes.

**Reference:**

Lewis Grizzard, "Toast to Bisher," *Atlanta Journal and Sunday Journal-Constitution,* 19 October 1990, p. E1.

# Bob Broeg

*(18 March 1918 – )*

James A. Vlasich
*Southern Utah University*

MAJOR POSITIONS HELD: Reporter, Associated Press, Jefferson City, Missouri (1941), Boston (1942); sportswriter, *Saint Louis Star-Times* (1942); sportswriter (1945–1958), sports editor (1958–1977), assistant to the publisher (1977–1984), contributing editor, *Saint Louis Post-Dispatch* (1984– ); columnist, *Sporting News* (1967–1977).

Bob Broeg in 1993

BOOKS: *Don't Bring That Up! Skeletons in the Sports Closet,* by Broeg and Bob Burrill (New York: Barnes, 1946);

*Stan Musial: "The Man's" Own Story,* by Broeg and Stanley Frank Musial (Garden City, N.Y.: Doubleday, 1964); revised as *The Man, Stan: Musial Then and Now* (Saint Louis: Bethany Press, 1977);

*Super Stars of Baseball* (Saint Louis: Spink, 1971); revised and expanded as *Superstars of Baseball: Their Lives, Their Loves, Their Laughs, Their Laments* (South Bend, Ind.: Diamond Communications, 1994);

*Ol' Mizzou: A Story of Missouri Football* (Huntsville, Ala.: Stroud, 1974); revised as *Ol' Mizzou: A Century of Tiger Football* (Marceline, Mo.: Walsworth, 1990);

*We Saw Stars,* by Broeg, Musial, and Jack Buck (Saint Louis: Bethany Press, 1976);

*Football Greats,* by Broeg, Buck, and Weeb Ewbank (Saint Louis: Bethany Press, 1977);

*St. Louis Is for Families,* by Broeg, Paula Knoderer, and Carole Hollander, edited by Guin Tuckett (Saint Louis: Bethany Press, 1979);

*The Pilot Light and the Gas House Gang* (Saint Louis: Bethany Press, 1980);

*Bob Broeg's Redbirds, a Century of Cardinals' Baseball* (Saint Louis: River City Publishers, 1981; revised, 1987; revised again, Marceline, Mo.: Walsworth, 1992);

*My Baseball Scrapbook* (Saint Louis: River City Publishers, 1983);

*Baseball from a Different Angle,* by Broeg and William J. Miller Jr. (South Bend, Ind.: Diamond Communications, 1988);

*Baseball's Barnum: Ray "Hap" Dumont, Founder of the National Baseball Congress* (Wichita, Kans.: Center for Entrepreneurship, W. Frank Barton School of Business Administration, Wichita State University, 1989);

*Bob Broeg: Memories of a Hall of Fame Sportswriter* (Champaign, Ill.: Sagamore Publishing, 1995).

OTHER: *Front Page: A Century of Post-Dispatch News and Sports,* edited by Broeg (Saint Louis: River City Publishers, 1982);

"Remembrance of Summers Past," *National Pastime: A Review of Baseball History,* 1 (Fall 1982): 22–24.

In the age of political correctness, Bob Broeg's iconoclastic nature stands out. Broeg's personality fused with his talent to produce one of the giants in sportswriting in the generation following World War II. Indeed, Broeg helped to define American sportswriting in that era. Those who know him only through his books or the many articles he has written for magazines and newspapers would find, on meeting him, a certain familiarity: Broeg writes as he speaks – entertainingly and accurately. Unlike many current writers who focus on statistical measurements of a player's career, Broeg hones in on the person inside the uniform or behind the team owner's desk. His emphasis is on the anecdotal, and his stories are heavily laden with his own opinions; his writing is crisp and sprightly, with a frequent use of alliteration.

Robert William Broeg was born in Saint Louis on 18 March 1918 to Robert M. Broeg, a bread-truck driver, and Alice Wiley Broeg. His delivery was difficult, and, as a result, Broeg was left almost blind in his left eye. Although this disability would hinder his performance as an athlete, it would not lessen his interest in sports: Broeg would simply redirect his energies from playing sports to writing about them. An old family story has it that a granduncle once felt the infant Broeg's head and predicted phrenologically that Broeg would become a writer.

Germans and Irish were dominant ethnic groups in Saint Louis during Broeg's early years, and his own heritage combined both strains. His surname, in its original German form, would be unpronounceable for many Americans; the family Americanized the pronunciation, but the spelling remained confusing. Broeg would use the rhyme "*Broeg* as in *plague*" to help out in this regard. His mother's father had come to America from Ireland. Neither of Broeg's parents had gone beyond the sixth grade. Broeg's father had been an amateur boxer; his interest in baseball was limited to the lowly Saint Louis Browns, but Broeg's mother became an avid baseball fan after she followed her son's suggestion and started listening to Cardinals games on the radio for relaxation in the evenings.

On Memorial Day 1927 Broeg's uncle took him to his first game at Sportsman's Park. "I've never seen blue that was bluer, green that was greener, or white that was whiter than the sky, the grass and the players' dry-cleaned home uniforms," of a major league outfield," Broeg recalled. It was not long before Frankie Frisch became his boyhood idol, much as Stan Musial would be his favorite as an adult. Broeg's only two player biographies would focus on these Cardinal greats. A neighbor, Louis Sika, accompanied him to most games and once arranged for Broeg to meet Babe Ruth under the stands behind the Yankee dugout. The autograph Broeg received that day was the biggest thrill of his childhood. Sika, a Hungarian, typified what Broeg would discover over the years – that some of the greatest followers of the game were immigrants.

As Broeg's interest in the game expanded during these ballpark sojourns, so did his fascination with baseball literature. While Broeg was in the hospital recovering from an appendectomy in 1927, Sika brought him the first book in the Baseball Joe series. These were corny books but great reading for a boy, and Broeg still mentions them when he talks to children about writing. He also read adventure stories and was especially fascinated with Robin Hood, Jerry Todd, Don Sturdy, and Tom Swift, but he devoured any baseball book he could find.

By his early teens Broeg had decided to become a baseball writer. Mary Culver, who taught him for three years at Mount Pleasant grade school, recognized a diamond in the rough and encouraged him to write about his favorite subject. This suggestion led to a mimeographed newspaper and a meeting with a man who would later become his mentor, J. Roy Stockton, sports editor of the *Saint Louis Post-Dispatch;* Stockton encouraged Broeg to get as much education as possible before beginning a career.

Broeg began to write sports at Cleveland High School; he was sports editor of the student newspaper by his sophomore year and general editor the next year. In the fall of 1935 his high school baseball coach, Bert Fenenga, arranged for him to interview Branch Rickey, the innovative general manager of the Cardinals. The teenage writer asked Rickey what players Rickey would place on an all-star team. The following year Broeg got a job with the club as a ticket taker. Four years later he began the summer doing some writing for the *Sporting News,* but the Cardinals offered him more money to join their publicity department. He apologized as gracefully as he could to Taylor Spink, the editor of the *Sporting News,* but Spink sent a caustic reply: "When you grow older, you'll learn to appreciate the feelings of others." It was an incident that neither man would forget. (In 1977 Spink's son, Johnson,

*Broeg (second from left) at the 1957 World Series in Yankee Stadium with Fred Russell, sports editor of the* Nashville Banner; *Dave Grote; and Tom Siler, sports editor of the* Knoxville News-Sentinel

would fire Broeg, who had been writing a column for the "Bible of Baseball" for ten years.)

In 1936 Broeg followed Stockton's advice and enrolled at the University of Missouri in Columbia. Journalism was a relatively new subject, having been introduced in the undergraduate program in 1908. Under the direction of Walter Williams the school developed an outstanding reputation that virtually assured job placement for its graduates. As a sophomore Broeg was employed in the university publishing department and wrote for the school newspaper. During his last two years in college he served as Associated Press correspondent in Columbia, covering general news. For Broeg, the job was like castor oil: "I didn't like the taste, but it was good for me."

The summer before he graduated from college Broeg went to work for Ray "Hap" Dumont, who ran the National Baseball Congress tournament in Wichita, Kansas. As Dumont's public-relations man, Broeg got a lesson in business as well as writing. Almost a half century later Broeg was still writing about Dumont: *Baseball's Barnum* (1989) is, in many ways, his most endearing work. Broeg gives a vivid account of Dumont's innovative measures to fill the stands, such as a pneumatic plate duster that the umpire could operate with a foot pedal; a jack-in-the-box microphone that would pop out of the ground for announcements; and an electronic timer to speed up games. More than Broeg's reminiscences of an old friend, the book is an account of an aspect of the game that does not often appear in print. Long before television, Dumont's tournament brought major league baseball to new regions of the country.

Broeg's Associated Press job delayed completion of his studies by half a year, but he graduated with honors in 1941. His work in Columbia served as a springboard to a job with the Associated Press's Jefferson City bureau, where AP executive Paul Mickelson, a former sports writer and columnist, assigned him to help cover the state legislature. When the legislative session ended in the summer of 1941 Broeg returned to Wichita; but a short time later Mickelson found him another Associated Press job in Boston that included some sports reporting. Always diffident about his talent, Broeg became apprehensive about his depth as a writer. In addition, in-

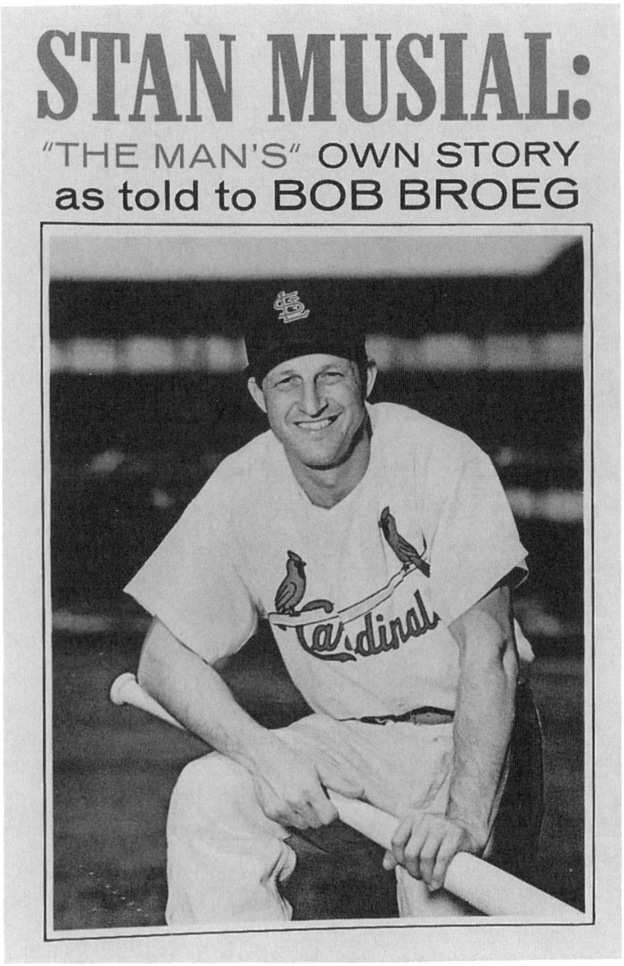


*Dust jacket for Musial's autobiography, as told to Broeg*

flation caused by expanding defense manufacturing after the country's entry into World War II made it difficult for Broeg to make ends meet. Worse yet, as the third man in the sports department, he suffered abuse from an alcoholic sports editor. Eager to return home, he wrote to Stockton, who wrote back that Broeg had gone a long way before he met his "first son-of-a-bitch" in journalism. He encouraged Broeg to contact Stockton's opposition, the *Saint Louis Star-Times,* and do such a good job that the *Post-Dispatch* would be forced to hire him. After thirteen months in Boston, Broeg returned to Saint Louis.

Broeg joined the U.S. Marines in December 1942 after only three months with the *Star-Times* as a sportswriter. He had met Dorothy Carr of Oak Park, Illinois, at college; they were married on 19 June 1943. A fellow graduate of the school of journalism, she went to work for *Better Homes and Gardens* in Des Moines, Iowa. When Broeg was transferred to Washington, D.C., where he wrote for a marine bulletin, she was able to obtain a job as news editor for the Red Cross in the same city. He covered the World Series of 1944 that featured the Saint Louis Browns of the National League and the Saint Louis Cardinals of the American League for *Leatherneck Magazine.*

While he was still in the service Broeg coauthored a book with Bob Burrill. The idea came from a conversation he had with a friend following a one-sided Notre Dame loss to Army in 1944. Broeg decided that there were enough of these embarrassing incidents in sports to warrant a book, so, doing most of the research at the Library of Congress himself, Broeg wrote, *Don't Bring That Up! Skeletons in the Sports Closet* (1946). According to Broeg, two things hurt sales of the book, the title and the timing: the former was too cute, and the latter coincided with the end of the war.

Following his discharge in the summer of 1945 Broeg turned down several offers, including one from the Associated Press in New York, in order to work as a sportswriter for the last newspaper that would ever employ him; in the minds of Saint Louis sports fans, Broeg and the *Post-Dispatch* would become synonymous. Mickelson, who was then the general news editor in the paper's New York bureau, gave Broeg a piece of advice that the younger man would never forget: "Always be neat, never lose your sense of humor, and don't take yourself too seriously." The new position would also lead to Broeg's regular radio show, which was carried on the powerful KMOX from 1963 until 1990, when it was reduced to three weekly sports vignettes.

Though his first experience with Taylor Spink had ended on a sour note, when the *Sporting News* started publishing a weekly piece on each baseball team in 1946 Broeg wrote the one on the Cardinals. Spink liked Broeg's work so much that he asked him to produce some feature articles. In addition, Broeg was assigned by Al Perkins, the first editor of *Sport Magazine,* to write some pieces for that popular publication. Broeg's work for the *Sporting News* eventually led to a column, "Broeg on Baseball" (1967–1977).

During the immediate postwar period Broeg developed a new approach to sportswriting: rather than relating a play-by-play description of a game, he would provide interesting sidelights and background information. For example, in August 1945 during a Yankees-Browns series he interviewed manager Joe McCarthy about the Yankees' glut of second basemen; the article certainly made for more interesting reading than the typical "Bombers Bash

Browns" story. He applied the same philosophy to a piece for *The Saturday Evening Post* on 24 November 1945. Stockton had originally been asked to write about the Tulsa University football program, but he declined and recommended Broeg, who was more knowledgeable about the sport. Broeg contacted the coach, Henry Frnka, but Frnka would not talk about football because he believed that doing so would jinx the team. Therefore, Broeg gleaned some information from other sports editors and wrote a human-interest piece about Frnka. The article pleased both the coach and the publisher, and Broeg would do many more pieces for *The Saturday Evening Post*. Frnka said that Broeg had been fair in his writing, which led the writer to propose as his own epitaph "He was fair."

One of the more bizarre stories Broeg covered involved Saint Louis Browns owner Bill Veeck and Eddie Gaedel, the only midget player in baseball history. The night before a Browns–Detroit Tigers game on 19 August 1951 Veeck revealed to Broeg, after a few beers, that as a stunt he would use Gaedel the next day. Usually the newspaper would not have a photographer stay for all eight innings at a Sunday doubleheader; but, having the advantage of inside information, Broeg tipped Jack January to stay long enough to take the famous picture of the umpire looking over Bob Swift's shoulder as Swift knelt to catch the ball with the midget at bat.

Broeg's own choice for the most significant story he has covered concerns the beginning of the players' pension system shortly after World War II. In June 1946, when club owners asked players for recommendations to ease their financial burdens, Broeg stumbled on the Cardinals' proposal for a pension plan. He was directed to the room of shortstop Marty Marion and trainer Dr. Harrison J. "Doc" Weaver, who had begun to draw up a list of measures to finance the program: player contributions of two dollars per day, to be matched by the clubs; the use of All-Star Game receipts and World Series radio money; and the possibility of raising new revenue by matching natural rivals (such as the Cardinals versus the Browns or the Yankees versus the Giants) in exhibition contests at midseason. Broeg's story on the plan for the next day's *Post-Dispatch* was carried at considerable length by the Associated Press. The owners recognized the practicality of the idea and adopted it – except for the exhibition games – with little protest. It was a step in the direction of management concessions to players, but no one could have predicted that this relatively modest plan would mushroom into the gargantuan salary packages of modern-day athletes.

A sore point with Broeg to this day is the alleged threat by Cardinal players to stage a protest strike against the Brooklyn Dodgers' use of Jackie Robinson in his inaugural season of 1947. While Cardinal players denied that there had been any vote on such an action, it became part of the Robinson legend and helped to establish support for the long-overdue integration of the sport. Unfortunately for Saint Louis supporters, it also branded their team as racist. Broeg does not deny that a certain segment of the team voiced opposition to admitting a black player to the major leagues, but he points out that members of other organizations, including the Dodgers themselves, had done the same. In fact, he says, Brooklyn traded Dixie Walker the next year because of his opposition. When the story reappeared in Roger Kahn's *The Era* (1993), Broeg expressed outrage. "I said, Roger, you're Jewish and you know about conclusion jumping, and you know what you guys have gone through, and yet you automatically want to go black and white when there were a lot of shades of grey." Broeg's resentment stems from the fact that the reputations of some good men were hurt by the story.

Broeg also defends Cardinals outfielder Enos Slaughter, a hot-tempered southerner who was accused by Red Barber and others of spiking Robinson during the 1947 season. Broeg acknowledges Slaughter's reputation as an aggressive player, but he believes, as did sportswriters Bill Corum and Robert Burnes, that there was no premeditation on Slaughter's part. Others point out that the location of the wound – eight inches above the left ankle and on the outside of the leg – indicates that it probably was a deliberate act. Broeg also contends that the real problem might have been that Robinson was playing out of position, as he was not a natural first baseman.

Broeg admits to having had major difficulties with former Cardinals (now Chicago Cubs) announcer Harry Caray, whose second-guessing of plays and patronizing of sportswriters were constant irritants to Broeg. Broeg found it almost impossible to converse with Caray without getting into an argument; on separate occasions Gus Mancuso and then Dixie Walker had to pull him off Caray. Especially upsetting to Broeg were Caray's frequent broadcast comments that began with such phrases as "even though you haven't read this in the newspaper . . ." or "in spite of what you read in the newspaper. . . ."

In addition to the Cardinals, Broeg covered University of Missouri football and college and pro-

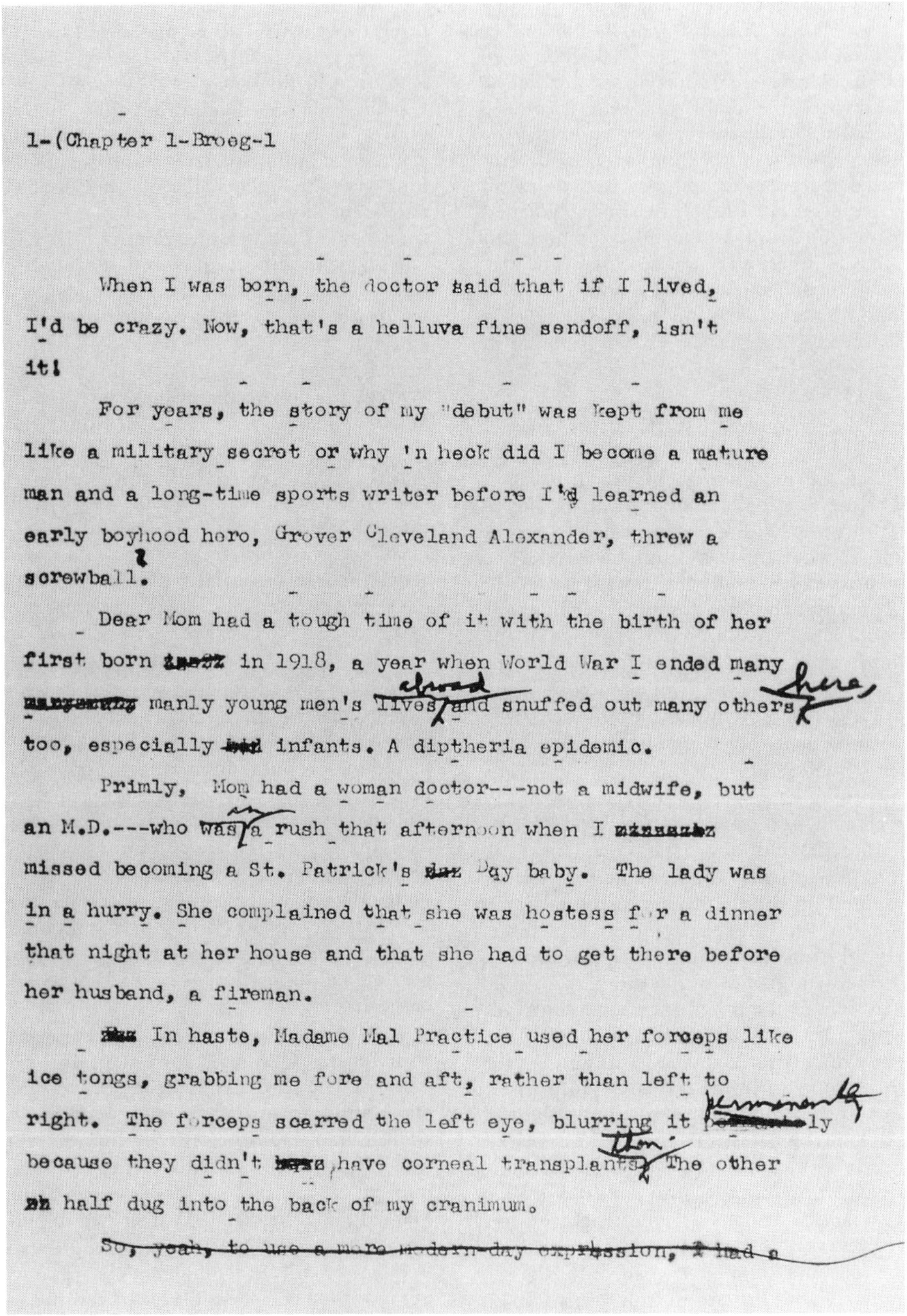

1-(Chapter 1-Broeg-1

When I was born, the doctor said that if I lived, I'd be crazy. Now, that's a helluva fine sendoff, isn't it!

For years, the story of my "debut" was kept from me like a military secret or why 'n heck did I become a mature man and a long-time sports writer before I'd learned an early boyhood hero, Grover Cleveland Alexander, threw a screwball.

Dear Mom had a tough time of it with the birth of her first born in 1918, a year when World War I ended many manly young men's ~~lives and~~ abroad snuffed out many others here too, especially infants. A diptheria epidemic.

Primly, Mom had a woman doctor---not a midwife, but an M.D.---who ~~was~~ in a rush that afternoon when I missed becoming a St. Patrick's Day baby. The lady was in a hurry. She complained that she was hostess for a dinner that night at her house and that she had to get there before her husband, a fireman.

In haste, Madame Mal Practice used her forceps like ice tongs, grabbing me fore and aft, rather than left to right. The forceps scarred the left eye, blurring it permanently because they didn't have corneal transplants then. The other half dug into the back of my cranimum.

~~So, yeah, to use a more modern-day expression, I had a~~

*First page of the typescript for Broeg's autobiography (courtesy of Bob Broeg)* fessional basketball until 1958, when he replaced his longtime mentor Stockton by becoming only the third *Post-Dispatch* sports editor in the twentieth century. That same year Broeg decided to initiate a banquet for the Saint Louis chapter of the Baseball Writers' Association; as national president, he was embarrassed that Saint Louis was one of the few branches of the organization not to have such a function. (It is now one of the few that do.) The first banquet made a profit, and the tradition has continued ever since. Today the chapter provides between $7,500 and $10,000 to a foundation for journalism scholarships.

After Broeg's first book in 1946, eighteen years passed before the next appeared. Stan "the Man," as Broeg was the first to label him, was the subject of *Stan Musial: "The Man's" Own Story* (1964); an updated version, *The Man, Stan: Musial Then and Now,* would appear in 1977. In 1964 Broeg won the Rockne Club's award as Sports Writer of the Year. The same year he was cited by the Elks Club for meritorious service to the sporting community. In 1971 he became the first sportswriter to receive the gold medal award from the respected University of Missouri School of Journalism.

In 1958 Dorothy Broeg helped Ken Smith, a former writer with the *New York Mirror* and a director of the Baseball Hall of Fame, rewrite the organization's constitution. She also used her considerable expertise as an editor to cut the manuscript for her husband's *Ol' Mizzou: A Story of Missouri Football* (1974) down to size by "squeezing out a lot of hot air," as Broeg would put it later. (An updated version, *Ol' Mizzou: A Century of Tiger Football,* would appear sixteen years later.) Dorothy died in 1975, following a yearlong battle with cancer. Two years later Broeg married Lynette Anton Emmenegger of Saint Louis; Broeg has a stepson and a stepdaughter from this marriage.

In 1977 *Post-Dispatch* publisher Joseph Pulitzer Jr. made Broeg his assistant. In this position Broeg represented the paper at community functions while continuing to write sports columns. The same year he received the Sportswriter of the Year Award from the Eagles fraternal organization. Seven years later he was elevated to contributing editor. He could vote for the induction of players to the Baseball Hall of Fame and was also the only sportswriter allowed to cast his ballot in the college and pro football Hall of Fame elections. Another unique aspect of Broeg's resume is his double position as a member of both the Board of Directors and the Veterans' Committee of Coopertown's museum. In 1979 he was elected to the writer's wing of the Baseball Hall of Fame.

Since his boyhood, Broeg had had great respect for the scholar-athlete. He especially admired Barry Wood, an all-American quarterback at Harvard who later became president of the Washington University School of Medicine and of Johns Hopkins University. Wood had been the subject of one of Broeg's stories for the *Post-Dispatch* in 1945; colleagues had told Broeg that Wood generally refused to talk to the media, but, undaunted, Broeg pressed on. Stories on University of Colorado football star Byron "Whizzer" White, who became an associate justice of the U.S. Supreme Court, and Pete Dawkins, a famous halfback for Army who later became a general and served in Vietnam, further exemplify Broeg's idea of honoring those who are successful both in sports and in other fields. In 1967 Broeg suggested to Frank Leeming, the newspaper's public affairs director, that the *Post-Dispatch* start an annual dinner to honor the state's top high-school senior scholar-athletes. Broeg estimates that the *Post-Dispatch* has spent more than a half million dollars on the event, in spite of the fact that Pulitzer was not a sports buff. Current publisher Nick Penniman declared at the twenty-fifth anniversary dinner in 1992 that the group of annual winners would be known as the "Bob Broeg Top Ten." Broeg is also credited with saving the Missouri Sports Hall of Fame when the State Fair decided to abandon it in 1963. It is now located in Springfield, Missouri.

The book for which Broeg is best known is *Super Stars of Baseball* (1971), expanded and revised in 1994 as *Superstars of Baseball: Their Lives, Their Loves, Their Laughs, Their Laments.* The book deals with fifty-five stars, including Hank Aaron, Sandy Koufax, and Honus Wagner, detailing both the strengths and the weaknesses of the former greats. Of his boyhood idol Frisch, Broeg writes:

> Frisch is recalled as the pilot light of the Old Gas Gang and as an umpire baiter, summa cum laude. Later, he was regarded as a hairshirt because of his speeches and writings in which, for instance, he referred to present-day spring training as "a country club without dues" and the modern plastic batting helmet as "an inverted garbage can."

In a 28 August 1971 review of the book for the *Sporting News,* Milton Richman praised the "crisp, simple language, the kind that everyone understands and enjoys; [Broeg] comes as humanly close to bridging baseball's generation gap as anyone I know."

*The Pilot Light and the Gas House Gang* (1980) follows Frisch's exploits from his early days with the Giants to his years as a player and manager in Saint Louis. *Bob Broeg's Redbirds, a Century of Cardinals' Baseball* appeared in 1981, with revised editions in 1987 and 1992. In 1988 Broeg co-authored *Baseball from a Different Angle* with William J. Miller Jr., a professor of history at Saint Louis University. Miller did the research and Broeg the writing, and the result is a perfect combination of scholarship and light prose. While the book was designed as a text for a sports-history class, it is, as Joseph Overfield observed in a 1988 review for *The SABR Review of Books,* not a standard history but "a delightful mixture of baseball history, humor, and opinion" with writing that "is sprightly, informal . . . and anecdotal, wherein lies its greatest charm." Each chapter focuses on a particular aspect of the game, such as equipment, announcers, salaries, and rule changes.

*Baseball from a Different Angle* provides excellent examples of Broeg's fondness for alliteration. The Baltimore Orioles "outfoxed and outfought the foe in the Naughty Nineties of nifty baseball"; old-time sportswriters were "purple-prosed-and-poetic-put-'em-on-pedestal" types; Ray Schalk was "a pint-sized, pepperpot catcher"; and cheating players were "cunning cutthroat characters." The technique can be overdone.

Broeg can be as critical as he is entertaining, but he is always fair. His comments on modern sportswriters are a case in point. On the one hand, he is quick to point out that journalists today are better educated than those of his generation and that many of them are talented writers. He also thinks, however, that they should rely more on their own judgment, ask more pointed questions, and produce well-written narratives rather than depend on locker-room quotes. He also contends that writers of his generation were too ornate in style, while modern ones are far too caustic; a balance between these extremes should be sought, he says.

Broeg does not shy away from self-criticism. He confesses that he has probably worn out as many commas as he has the bow ties for which he is renowned. He aims for complex sentences that flow as easily as casual conversation; for example, discussing in *Superstars of Baseball* the controversy over the Frisch–Rogers Hornsby trade before the 1927 season, Broeg notes that the new Cardinal second baseman had much to prove when the season started, and the comparisons "moved from the white paper of winter conjecture to the green grass of direct summer competition."

Broeg has never been satisfied that he has achieved his goals as a sportswriter. His readers would disagree, and his many awards testify that other sportswriters concur with the fans on this point. More important, Broeg will always be embraced by the average fan. He came from their ranks, and his writing shows that he has never forgotten them.

# Heywood Broun

*(7 December 1888 – 18 December 1939)*

Bill Knight
*Western Illinois University*

See also the Broun entry in *DLB 29: American Newspaper Journalists, 1926–1950.*

MAJOR POSITIONS HELD: Reporter, *New York Morning Telegraph* (1908, 1910–1911), *New York Evening Sun* (1909); copyreader, rewrite man, sportswriter, sports editor, war correspondent, critic, and columnist, *New York Tribune* (1911–1921); columnist, *New York World* (1921–1928), *New York Telegram* (1928–1931), *New York World-Telegram* (1931–1939), *New York Post* (1939).

BOOKS: *The A.E.F.: With General Pershing and the American Forces* (New York & London: Appleton, 1918);

*Our Army at the Front* (New York: Scribners, 1918);

*Seeing Things At Night* (New York: Harcourt, Brace, 1921);

*Pieces of Hate and Other Enthusiasms* (New York: Doran, 1921);

*The Boy Grew Older* (New York & London: Putnam, 1922);

*The Sun Field* (New York & London: Putnam, 1923);

*Sitting on the World* (New York & London: Putnam, 1924);

*A Shepherd* (New York: Privately printed, 1926);

*Gandle Follows His Nose* (New York: Boni & Liveright, 1926);

*Anthony Comstock: Roundsman of the Lord,* by Broun and Margaret Leech (New York: Boni, 1927);

*Christians Only: A Study in Prejudice,* by Broun and George Britt (New York: Vanguard, 1931);

*It Seems To Me: 1925–1935* (New York: Harcourt, Brace, 1935);

*A Tale of Toyland,* edited by Charles C. Fleming (Richmond, Va.: Press of the Hiccupy Herring, 1957).

**Collection:** *Collected Edition of Heywood Broun,* edited by Heywood Hale Broun (New York: Harcourt, Brace, 1941).

*Heywood Broun (illustration by Seymour; The Newspaper Guild Archives)*

PLAY PRODUCTIONS: *No, Sirree,* New York, 49th Street Theater, 30 April 1920;

*Shoot The Works,* New York, George M. Cohan Theater, 21 July 1931.

OTHER: "Nonsense-orship," in *Nonsense-orship: Sundry Observations Concerning Prohibitions, Inhibitions and Illegalities,* edited by G.P.P. (New York: Putnam, 1922);

J. B. Mussey, ed., *The Cream of the Jesters,* includes a contribution by Broun (New York: Boni, 1931).

*Broun, circa 1913, when he began to cover the New York Giants for the* New York Tribune *(The Newspaper Guild Archives)*

Heywood Broun was a man of contrasts – gregarious and introspective, brave and fearful, direct and evasive, a celebrity and a champion of the underdog. His fellow newspaperman Franklin P. Adams called Broun "a lion in print but a lamb in his personal relationships" and added that Broun "will be a legend, and a quoted writer, when the Pulitzers, the Danas and Greeleys are only names."

Broun was a big man. At six feet three inches and between 250 and 300 pounds, he wore size thirteen shoes (frequently without shoelaces) and baggy, shapeless suits that some said made him look like an unmade bed. His hair was tousled, his ties askew, his hats slouched. He favored ice cream with gin poured over it. Raised in Manhattan, he spoke with an accent that "always sounded to me suspiciously like a Southern drawl," observed Joseph F. Collis, who added, "His drawl was as characteristic of him as his politeness."

Broun may be remembered most often for his role as a founder and first president of the Newspaper Guild (1933–1939), the AFL-CIO–affiliated trade union that represents reporters and other media employees in the United States and Canada. "It is the only important work I ever did in my life," he said in *The New Republic* in 1938. "The Guild did for me what Harvard could not do."

Others may remember Broun almost as much for his newspaper column, "It Seems To Me," which he wrote from 1921 until shortly before his death in 1939. There for more than eighteen years he fiercely fought for social justice and also quietly reflected on ordinary life, the deaths of friends, and changes in society. People still mention him as one of the colorful members of the famous Algonquin Round Table group or the infamous "Thanatopsis Literary and Inside Straight Club," which included some of the same individuals who met regularly to play poker. Some may remember him as only a journalistic anachronism or a frustrated novelist.

He was all these things – and much more. He enjoyed athletics and competition; he appreciated participants and spectators alike; and he loved his experience as a sports journalist, bringing enthusiasm and action to a form of reporting that was often staid.

Born on 7 December 1888 in Brooklyn, New York, Broun was the third of four children born to Heywood Cox Broun and Henriette Brose Broun. His Scottish immigrant father was a successful printer and stationer; his German mother was a devout Episcopalian who spoke French as well as German fluently.

Heywood Broun attended the private Horace Mann School in New York City, where his activities included football. He was named editor of the school newspaper and voted best-all-around man in his class. At Harvard University (1906–1910) Broun played basketball until a knee injury ended his career, and he studied the Bible as literature. Part of the Harvard class of 1910 – which also included future journalists John Reed and Walter Lippmann and poets T. S. Eliot and Alan Seeger – Broun had a short story published in the *Harvard Advocate.* Yet he failed in repeated attempts to make the staff of the *Harvard Crimson,* the student newspaper, even though he was hired for summer work at the *New York Morning Telegraph* after his sophomore year and the *New York Evening Sun* at the end of his junior year.

After failing French prevented him from earning a degree, he applied for a full-time job at the *New York Morning Telegraph,* described by the *American Newspaper Directory* of that time as "the brightest and breeziest published this side of Paris." According to Robert K. DeArment, the *Telegraph* specialized in "theatricals, racing, automobiling and the higher branches of sport." Its staff included the young gossip columnist Louella Parsons, as well as the old, former gunfighter Bat Masterson, who was sports editor.

In Broun's tryout with the *Telegraph* he was assigned to write an article on golf, which he did confidently and quickly. An editor read it twice and exclaimed, "This is all right! Give us some more like that!" At the *Telegraph* Broun did a bit of everything, from working at the copy desk and writing editorials to contributing items to a column called "'Round The Town." Before long his baseball stories attracted attention. More than a year later, Broun asked for a two-dollar raise, and the paper responded by firing him.

During the next six months he freelanced for the *New York Sun* and applied for work at every New York daily, finally landing a job as a copyreader for the conservative *New York Tribune.* By 1912 Broun's copy-desk skills had been noticed. Editor George Burdick promoted Broun to rewrite, where he next gained the attention of fellow rewrite man Ogden Mills Reid (future editor of the *Tribune* and son of the owner, Whitelaw Reid). Despite the progress he had made at the paper, Broun was tempted by an offer from theatrical producers to go to China to research background for a proposed play, and he took a leave of absence to travel to Shanghai and Beijing and soak up local color.

After Broun's return to New York, *Tribune* sports editor George Daley moved him to sports. There Broun helped to create some respectability for *Tribune* coverage of sports in general and baseball in particular. In 1913 the legendary Cubs first baseman Frank Chance, then managing the New York Yankees, complained to Broun and sportswriter Fred Lieb of the *New York Press* that Hal Chase – the talented but notoriously crooked Yankee first baseman – was throwing games. The editor at Lieb's paper advised against using the remark, but Broun published it. Yankees owner Frank Farrell criticized Broun, but within days the Yankees had traded Chase to the White Sox for two run-of-the-mill players.

Others criticized Broun simply because he wrote about sports. During a recess in a 1915 meeting between the major leagues, Lieb accompanied Broun to dine at the Harvard Club, where one of Broun's former professors paused at their table. Learning Broun's current vocation, the professor exclaimed, "Bah! Have you fallen to that?" Yet, according to Adolph Gillis, acceptance from the intelligentsia did not matter to Broun, who thought "the national interest in sports was altogether creditable." Broun's sports articles, said Gillis, are "steeped in a robust understanding of the folk."

Moreover, Broun was able to covey his unbridled enthusiasm. According to Broun biographer Dale Kramer, "Broun's greatest asset was his ability to impart a feeling of the game's excitement. He integrated the facts with colorful allusions, and the discovery was soon made that a story could stand on action and characterization by themselves." In an unpublished interview Broun's son, Heywood Hale Broun, recalled that his father "was notorious for rooting in the press box. There was a strong, unspoken rule that you *never* showed any enthusiasm in the press box, but he'd be carrying on and yelling." Broun himself said:

> These peculiar performances embittered my colleagues who had not slapped one palm against another for 50 years. I could not help taking sides. I was wholly unable to attain objectivity. . . . Sid Mercer and Damon Runyon were shocked each afternoon as I broke all traditions of the craft by rooting violently in the press box.

Broun also enjoyed the behind the scenes of sportswriting and socializing with athletes he respected, including legendary baseball pitcher Christy Mathewson. Broun's son recalled that "one of the proudest moments of his life was becoming a regular checkers players with Mathewson." The elder Broun also was not above hero worship, as is apparent in this comment about New York Giants manager John McGraw: "An important part of McGraw's capacity for leadership [was] that he could take kids out of coal mines and wheat fields and make them walk and talk and chatter and play ball with the look of eagles."

Broun and many of his contemporaries helped to make sports figures heroic. With uncompromising zeal, they adored and attacked their subjects and their games. Confessing that "I belong to the 'Ataboy!' school of criticism," Broun also modestly conceded in a September 1932 Scripps-Howard newsletter that he was "a good baseball reporter." Charles Fountain agreed, calling Broun – along with Runyon and Grantland Rice – a "Gee Whiz" sportswriter, as opposed to an "Aw, Nuts" type like Ring Lardner. Others were more laudatory. "With the *Tribune,* Broun turned in some of the greatest baseball and football stories that have ever been written," recalled Bennett Cerf. Biographer Richard O'Connor observed that Broun "loved the sport and writing about it so much that he soon ranked with Damon Runyon, Bozeman Bulger of the *Sun,* and Grantland Rice of the *Mail* as a fabricator of prose glorifying the accomplishments of Christy Mathewson and Rube Marquard and their teammates."

The younger Broun added that his father "thought he flirted a little with immortality when he successfully talked McGraw out of making Mathewson a first baseman." Broun indeed showed reverence for the Giants pitcher:

> Robbed of his catcher, a wrecked machine tottering to ruin at his back, and the greatest sluggers in baseball poised for the onslaught, old Chris Mathewson – master, king, emperor, and ruler of all baseball pitchers at home and in the dominions beyond the seas – annihilated the attack of the Philadelphia Athletics by the might of his wizardry.

On another occasion he wrote:

> Oeschger was in the box for Philadelphia, but his delivery is not nearly as difficult as his name. He was hit so hard that [Phillies manager] Charlie Dooin was driven at last to send in another pitcher of such obscurity that he was not even indicated on the scorecard. Matteson was the newcomer. He lacks more than a few letters of being a Mathewson.

Broun covered sports for the *Tribune* for about three years, before becoming a drama critic for the paper in 1915. "The day of the transfer, he acted as official scorer at a Giant-Cub ball game in the afternoon, and covered Ethel Barrymore's opening in an Edna Ferber play called *Our Mrs. McChesney* in the evening," Cerf wrote. "At the ball game, he scored a close play as an error for the visiting shortstop, thereby depriving the Giant batter of a base hit in the records. That evening, he roasted Miss Barrymore's performance to a fare-thee-well. The next day, the *Tribune*'s managing editor received two indignant communications. One, from the Giant batsman, read, 'What's the big idea of sending a lousy dramatic critic up here to be official scorer?' The other, signed by Miss Barrymore, concluded, 'How dare you assign a cheap baseball reporter to cover the opening of a Barrymore play?'" Broun's attitude bridged the disciplines:

> We like the drama so far. In the office of the National League . . . there hangs a framed motto. "Take nothing for granted in baseball," it says. This is not the motto of the theater.
>
> The drama has succeeded in eliminating many of the uncertainties that harass baseball. Nobody can tell how a game will end until the ninth inning, but the least experienced theater-goer can predict the end of a play pretty accurately just as soon as the hero has taken one long lingering glance at the heroine.

In October 1915 Broun met Russian ballet dancer Lydia Lopokova, after writing a rave review of her performance on the New York stage. They became engaged, but she soon spurned him for another dancer and later married economist John Maynard Keynes. While Broun was still covering sports, *Tribune* Sunday feature writer Alice Duer Miller had introduced him to journalist and feminist Ruth Hale in the Polo Grounds press box. Hale thought some of Broun's baseball stories were inaccurate – a challenging attitude that he found stimulating. Broun and Hale were married on 7 June 1917, and their only child, Heywood Hale Broun, was born the following March.

Named *Tribune* sports editor in 1916, Broun brought an unorthodox approach to the job. In 1928 he confided to H. Le B. Bercovici: "I'm not interested in the usual accounts that so-and-so socked Tunney in the jaw at the training camp yesterday afternoon. I am interested in Tunney's opinion of Goethe, and I believe my readers would be inter-

*Broun at work on his column "It Seems To Me" (The Newspaper Guild Archives)*

ested in that, too. If Tunney quotes Shakespeare while sparring with one of his partners, I want that news."

Broun also wanted sportswriters to find out if athletes used the products they endorsed and to avoid the appearance of conflicts by paying their own way. "I'd concentrate on personalities in sport," Broun said. "Just the usual stuff about a great baseball player having started his career with a sand-lot team is not news. But if my reporter finds out that a baseball player struck out with the bases full because he was out on a beer party the night before the game, that's the story I want."

Soon after becoming sports editor, Broun hired W. O. McGeehan from William Randolph Hearst's *New York Journal,* where he had written a column signed "Right Cross." Broun offered the newcomer a byline and a raise. The *Tribune* "sports page, carrying the witty Broun, the romantic Grantland Rice, and caustic William McGeehan, became a fan's delight," noted Richard Kluger in *The Paper: The Life and Death of the New York Herald Tribune* (1986). Broun's lead in a story about one lackluster stretch of games demonstrates this wit – and silliness: "An eye for an eye and a tooth for a tooth may have been a good policy in Mosaic days, but it represents only an average of .500 when applied to baseball. Just now it is the doctrine which the Giants seem bent on upholding." Another comment about a new Phillies shortstop is a good example of Broun's specialty, the critical put-down: "He does not look like a shortstop; he does not act like a shortstop; and only by grossly circumstantial evidence will he ever be convicted of shortstopping. Now he may drop his disguise and go back to the outfield."

When the United States entered World War I, Broun became a war correspondent for six months. (Hale also went to Europe, working for the Red Cross and the *Chicago Tribune* until her pregnancy forced her to return to New York.) Broun's enthusiastic and irreverent reports, revealing poor conditions and preparation among U.S. troops, antagonized Gen. John J. Pershing.

After returning to New York and the *Tribune* in 1918, Broun wrote his first two books. In *The A.E.F.: With General Pershing and the American Forces* (1918) he tried to personalize an impersonal war, much as journalist Ernie Pyle did during World War II. "We can't do less for our war heroes,"

Broun wrote, "than we have done for Ty Cobb and Christy Mathewson and Smokey Joe Wood." Broun's *Our Army at the Front* (1918) follows the advance of the American military, praising the acts of thousands of brave individuals.

During this postwar period Broun also spent about three years covering the literary beat, writing a column called "Books and Things," which became less frequently about books than things. "His was one of the very first 'personalized columns,'" said his son. "He wrote about the World Series and Congress, fishing or someone's dog dying."

In 1919 the Algonquin Round Table (which took its name from the shape of the table where the group sat in the dining room of the Algonquin Hotel) included Broun, Hale, Adams, Alexander Woollcott, George S. Kaufman, Robert Benchley, Dorothy Parker, Harold Ross, and Ferber. The following spring Broun organized a stage revue, *No, Sirree,* featuring as many of the Algonquin group (also known as "the Vicious Circle") as he could persuade to appear for the single-night run of the show on 30 April 1920.

In summer 1921 Herbert Bayard Swope, editor of the *New York World,* who was assembling his "op-ed" page (opposite the editorial page), hired Broun to write a column for the Pulitzer flagship paper. Within the next several years the paper was also featuring Adams's column, drama criticism by Woollcott, book reviews by Laurence Stallings, music criticism by Deems Taylor, and editorials by James M. Cain and Walter Lippmann. Broun was more comfortable at the *World* than at the conservative *Tribune,* and he became one of the most widely read and respected columnists in the country, earning $25,000 a year in the 1920s – a high salary for that time.

Broun was also a regular contributor to *The Nation* and *The New Republic,* for which he began writing a column, "Shoot The Works," in 1935, as well as writing occasionally for *Collier's, Harper's, McCall's,* the *American Mercury, Vanity Fair, Bookman,* and *Judge,* where in 1922 he served as editor of the sports department. He also became an advisory editor for *The New Yorker,* founded in 1925 by his friend Harold Ross.

Broun's *World* column, "It Seems To Me," always reflected his diverse interests. It touched on "yacht races, baseball, football, boxing, books, plays, politics and, most often, his heirs and assigns," recalled critic Charles Baldwin. Broun covered the 1923 Jack Dempsey–Tom Gibbons boxing match in Montana; mused about horse racing at Saratoga and the blessing and curse of "Southpaws"; wrote about Joe Louis and tennis and "immortalized" Babe Ruth. "We know that traditionally the Jacks and the Davids bring down their giants, but in those days there was no infighting. David kept out of the clinches," Broun wrote in "Carpentier Was Too Eager When Victory Seemed Near" for the *Tribune* in 1921.

Sometimes Broun's prose was overshadowed by that of colleagues. Broun covered the same 1924 Army–Notre Dame football game for which Rice wrote his legendary lead, "Outlined against a blue-gray October sky, the Four Horsemen rode again. . . ." Broun's column also had an arresting lead: "The Army thought the attack would come from the sky, but Notre Dame switched tactics and defeated the West Pointers with sweeping cavalry charges around the ends." Yet his lead is largely forgotten. (Covering an Army-Navy game a few years earlier, Broun used the presence of a World War I general as an angle for his story: "In a field box sat the French leader who helped to establish at Verdun the slogan 'They shall not pass,' and both Army and Navy tried it continuously.")

Broun collected some of his articles in his third book, *Seeing Things At Night* (1921), and the subjects range from sports and the press to romance and religion. In "Southpaws" Broun relies on a biblical reference (a device he used frequently); betraying himself as a learned writer while his tongue-in-cheek approach shows himself to be a man of the masses. Linking Aod from the Old Testament to Leonardo da Vinci, Charlie Chaplin, Babe Ruth, and finally all southpaws, Broun explains that this unusual class – including mostly athletes from Ty Cobb to polo player Watson Webb – has evolved a defensive disposition:

> On account of the extent and duration of the ill will to left-handers there has come to be definitely such a thing as a left-handed temperament. This is no more than natural. The left-hander is a rebel. He is the descendant of staunch ancestors who refused to conform to the pressing demands of the church, the army, and the business world.

By treating a light subject with such eloquence and heavy sobriety, Broun adds to the reader's enjoyment. Broun again used Scripture in a column lead the day after the second game of the 1923 World Series, won because of two home runs hit by the Yankees' Babe Ruth: "The Ruth is mighty and shall prevail."

Broun's years with the *World* were mostly tranquil. In fact, Broun once reminisced, "It may be that newspapers which die in a state of grace go to

*Broun with wife Connie Madison and political reporter Emmet Crozier on a guild picket line at the 1934 strike on the* Newark Ledger *(State Historical Society of Wisconsin)*

heaven, and if my luck holds and the time come, St. Peter will say to me, 'Heywood, here's your old job back on the old *Morning World*.' And that would be Heaven." Broun's happiness in his work during the 1920s shows in the quantity and quality of his writing. He published three collections of articles, three novels, and a nonfiction collaboration.

The topics covered in the articles collected in *Pieces of Hate and Other Enthusiasms* (1921) vary from the literary habits of heavyweight champ Jack Dempsey to a frustrated interview with G. K. Chesterton. Two of the best pieces concern sports. "Sport for Art's Sake" traces the theatrical merit of Dempsey's defeat of challenger Georges Carpentier, and "I'd Die for Dear Old Rutgers" describes a Forest Hills tennis match: "To a true American sportsman a set of tennis is held in about the same esteem as a popular playwright holds a woman's honor," Broun wrote. "There is no point at which 'I give up' can be sanctioned."

Broun's writing is admirable for its effortlessness, as well as its warm wit and whimsy. "Carpentier went down with his head up" is a good lead. "Harvard locked the stable door just as the horse was going and caught him by the tail" is more forced, but funny. To start a 1920 football game story, Broun alluded to the title of a novel by Princeton alumni F. Scott Fitzgerald: "Princeton, hitherto believed to be this side of paradise, sent a line smash through the pearly gates of this afternoon and defeated Yale by 20–0."

Broun wrote as an innocent, empathetic onlooker who appreciated heroes and delighted in amusements. In "Ruth vs. Roth" he passes on his shock at finding a Professor Filibert Roth but not George Herman Ruth in *Who's Who in America,* and in "Popup John" he recalls his own old line: "Jack Kleinow goes down to first as if he were pulling Big Bill Taft in a ricksha." In "Saratoga Fades" he criticizes even as he reminisces:

> The wolves of the gambling casinos howl dismally at night in Saratoga because they are forbidden to fleece their victims. But the sheep themselves are even more mournful and morose because they may not bare their throats to the friendly fangs of the predatory. Hell hath no fury like that of an unshorn lamb.

Broun's largely autobiographical first novel, *The Boy Grew Older* (1922), has a poker-playing protagonist, Peter Neale, who writes a column called "Looking Them Over" for the *Bulletin,* edited by

*Broun as depicted by Carl Rose (from Bennett Cerf,* Try and Stop Me, *1945)*

Rufus Twice (either *Tribune* editor Reid or Swope of the *World*). The novel celebrates life and sport: "When you say newspaper work's stupid, you're practically saying that life's stupid," Neale tells his son. Elsewhere he thinks that his son could fill his shoes on the sportswriting beat, "could put in it this thing that was sport, and beauty, and life and fighting and everything else worth while in life."

A second novel, *The Sun Field* (1923), is also loosely autobiographical, featuring a sportswriter narrator with a heart condition (a long-running Broun fear) and a journalist not unlike Lippmann. There are also characters based on Babe Ruth (home-run king Tiny Tyler) and Ruth Hale (*To-Morrow* magazine writer and feminist Judith Winthrop). According to Charles Baldwin in his *The Men Who Make Our Novels* (1924), "Judith falls in love with Tiny because of the beauty of his throwing arm after he sends the ball home from a catch in center field," while Broun "treats the whole thing with just the right mixture of humor and restraint."

In another collection of articles, *Sitting on the World* (1924), Broun successfully avoids some frivolity without sacrificing whimsy. Whether he writes about politics or God, a stuntman diving into water or a catfish struggling out of water, Broun's prose is leaner and cleaner than in earlier pieces.

In "A Teeth-Gritter Fails," which covers tennis player Suzanne Lenglen's defeat of Molla Bjurstedt Mallory, he writes, "In the modern world a man without will or character travels naked through a jungle. In our own case, for instance, we live in constant danger of being carried off by insurance agents, stock promoters and men with genuine stray bits of authentically smuggled English woolens." Although Lenglen was acknowledged as a better player, Mallory seemingly had an edge in the will-to-win area. Broun openly admired such strength of character and courage throughout his own career. (Writing about boxer Joe Louis in 1939, Broun concluded, "In order to amount to anything a man must learn how to shiver and shake and still keep coming in.")

Yet in "A Teeth-Gritter Fails" Broun repeatedly recognizes the limits of determination: "Battling Nelson, for instance, was almost without boxing skill," he writes. "He had nothing but perseverance. In him character was developed to such a remarkable extent that he was able to wear down opponents by hitting their fists with his face."

In a reflection on McGraw's *My Thirty Years in Baseball,* Broun has some fun with the Giants manager, likening him to the classicists: "Beginning with the intention of doing no more than a book about thirty years of baseball, McGraw suddenly thrusts himself into a position where he must make some attempt to interpret the whole history of civilization."

Broun's "Convention Studies" includes a great line: "In Madison Square Saturday night William Jennings Bryan testified his love of Christ and voted for the Ku Klux Klan."

The love of Christ and Christmas were familiar themes for Broun, who seemed to specialize in Christmas columns. President Franklin D. Roosevelt read one, "Even To Judas," on his 1938 Christmas Eve radio broadcast. Broun's *A Shepherd* (1926) – a privately printed, sixteen-page, boxed pamphlet published in a limited edition of 250 copies – is a sentimental tale about some of the first witnesses to the Nativity. Broun later used the story as a column, and his column "That Child in Bethlehem" was retitled *A Tale of Toyland* (1957), a sixteen-page booklet published in an edition of 250 copies.

*Broun at bat for the Nutmegger Club, with Connie Madison preparing to run the bases for him*

In 1926 Broun's third and last novel came out. *Gandle Follows His Nose* is an adult fairy tale in which Bunny Gandle confronts giants and evil spirits and battles a sorcerer and an army in an allegory that seems to frolic at the expense of logic. Broun's friend and colleague Deems Taylor considered adapting the novel as an opera. *Gandle Follows His Nose* was a critical and popular disappointment, but Broun's next book was his most successful. Broun wrote *Anthony Comstock: Roundsman of the Lord* (1927), a biography of a well-known antipornography crusader, with Margaret Leech, wife of Ralph Pulitzer, the publisher of the *New York World.*

By the time this book was published, however, the comfortable common ground of liberal political beliefs that Broun shared with the *World* had begun to erode. Broun drifted left as Pulitzer became a political centrist. In May 1928 Broun was fired from his job at the *World,* after a disagreement with the management that stemmed from his ardent protests in his column of the scheduled executions of Italian American anarchists Nicola Sacco and Bartolomeo Vanzetti, who had been convicted of armed robbery and murder on evidence that many people, liberals like Broun and his friends in particular, considered inconclusive.

After Broun's 5 August 1927 column about the case, Swope and Pulitzer asked Broun to stop writing about it. Nevertheless – angered by the newspaper's editorial position, which accepted the sentence of execution – Broun published another column on the subject and wrote two more that the *World* refused to run. In response Broun went on sabbatical for more than four months, while continuing to write for magazines. He returned to the paper in January 1928, apparently reconciled with *World* management, but after he wrote a piece calling for a genuinely liberal newspaper in New York and published it in the issue of *The Nation* dated 4 May 1928, the *World* fired him on 3 May 1928.

Broun was hired by Roy Howard of Scripps-Howard as local columnist for the *New York Telegram.* His syndicated column was read by as many as one million readers coast to coast, and during the

Great Depression Broun's salary started at $30,000 and went up to $49,000. He worked for Scripps-Howard for more than a decade, a time of furious activity, productivity, and change for Broun.

In 1930 Broun ran for Congress on the Socialist ticket, finishing third in a field of three. Early in 1931, the *World* folded, was bought by Scripps-Howard, and merged with the *Telegram.* That same year Broun and *Telegram* reporter George Britt wrote *Christians Only: A Study in Prejudice,* a 333-page study on discrimination in general and anti-Semitism in particular. Examining jobs, schools, neighborhoods, and society as a whole, the book was well received. (Broun acknowledged that he had dictated most of his contribution to the volume.)

In July 1931 Broun produced *Shoot The Works,* a revue in which he was featured along with Connie Madison, the chorus girl who would become his second wife, and occasional unpaid cameos by celebrities such as Al Jolson, Bill "Bojangles" Robinson, Helen Morgan, and the Marx Brothers. "His friends agreed," wrote Margaret Case Harriman in *The Vicious Circle* (1951), "that his chief motive in staging this production was to help the starving young actors. But a few skeptics have pointed out that Broun himself was stage-struck." The show ran for eighty-seven performances.

Broun and Hale had maintained separate homes for years, and although they were on civil terms, she had sought a divorce for a long time. He finally agreed, and she was granted a divorce in Mexico on 17 November 1933. In less than a year she became gravely ill, lapsed into a coma, and died on 18 September 1934. "My best friend died yesterday," Broun wrote in his column. "Ruth Hale was a valiant fighter in an important cause. 'Come on and be a radical,' I used to say, but Miss Hale insisted on being a militant feminist – all that and nothing more and nothing less."

Broun grieved but went on with his life, continuing to care for his teenage son. On 5 January 1935 he married Connie Madison (born Maria Incoronata Fruscella), the widow of Johnny Dooley. He adopted her daughter, Patricia. After ten weeks of instruction by Monsignor Fulton J. Sheen, he converted to her faith, Roman Catholicism; his confirmation was presided over by Archbishop Francis Spellman.

By that time Broun had formed the American Newspaper Guild (later renamed the Newspaper Guild), which became affiliated with the AFL in 1936 and the more militant CIO in 1937. Serving as guild president for the rest of his life, Broun threw himself into its needs. "The Newspaper Guild will, with time, probably prove to be Heywood Broun's most permanent monument," said Roy Howard of Scripps-Howard.

The busy Broun also was a driven man of leisure. People recalled his being able to churn out a solid column in a half hour, in the midst of a card game or a Newspaper Guild convention. A claustrophobic hypochondriac who usually worked at home on his messy bed in an untidy bedroom, Broun also sought escape by painting watercolors; playing softball, golf, or billiards; or fishing or bowling. In the late 1930s he was a member of a softball team called the Nutmegger Club – which included at various times Gene Tunney, Gene Sarazen, Grantland Rice, Harold Ross, Deems Taylor, Quentin Reynolds, and other celebrity neighbors in Connecticut. One of their opponents was the Nine Old Men, a team that included Lowell Thomas and Theodore Roosevelt Jr. "Whenever Broun, who weighed about 265 pounds, was lucky enough to get on base – which was seldom – his wife, Connie, ran for him," recalled newsman John Wheeler.

*It Seems To Me: 1925–1935* (1935) collects many of Broun's columns but omits most of those about athletic competition. Instead readers were treated to "classic Broun": Sacco and Vanzetti columns (including an exchange of letters between Broun and Pulitzer and *The Nation* piece that got him fired), the first guild column, and militant musings about Socialist Eugene V. Debs, faceless agitators, and soulless strikebreakers. Though Broun's later writing tends have a darker tone suited to serious looks at difficult times, two sentimental gems about sports shine through: "Popup John" (1936) and "The Happy Days of Baseball" (1939).

In "Popup John" the longtime Giants beat writer and fan confesses his youthful devotion to the pre-Yankee Highlanders, whose first baseman, John Ganzel, seemed to be a specialist in reverse clutch hitting, popping up towering vertical drives to end rallies. In "The Happy Days of Baseball" nostalgia for old ballplayers is balanced with a begrudging respect for rowdy Dodger fans: "Brooklyn is one of the few bright spots in the entire circuit. There is no sportsmanship among the Flatbush fans. They do not cheer even the most spectacular play by any invader and they do throw pop bottles. Of such stuff is the greatness of baseball made."

In an August 1936 issue of *Collier's* Broun published one of his last short stories, "The Last Signal," which includes some memories from his press-box days covering the Yankees back in the early teens. The story concerns an upper-class fellow named Chauncey Frothingham Wigglesworth, who

*Broun's son, Heywood Hale Broun (right), as he accepts the announcement of his father's induction into Baseball's Hall of Fame (*The Guild Reporter; *The Newspaper Guild Archives)*

made the Yankees' pitching squad after flunking out of Harvard, and Hardcoal Harry Haggerty, an illiterate, aging catcher-coach who says he "don't know one letter from another – not even those on my baseball shirt."

Chauncey and Hardcoal devise a system of hand signals, when Chauncey is pitching, that is also designed to teach Hardcoal how to read and spell ("c-a-t for a curve, f-a-t is the fastball, and so on"). Chauncey, who is "a good deal less than bookish" himself and spells almost as poorly as Ring Lardner's fictional bush-league player Jack Keefe, failed his Shakespeare exam at Harvard because he took the question "Was Hamlet truly mad?" to mean "Was Hamlet really good and sore?" He barely gets the last batter to strike out in the final game of the World Series after Hardcoal misspells "bug" – their pitch for "emergencies."

In 1938 Broun and several of his literary and celebrity friends who had also moved to Connecticut started a weekly tabloid. First called *Connecticut Nutmeg,* it featured an eclectic blend of writing – Broun's "Nature Notes" on gardening, athlete Gene Tunney on Neville Chamberlain and the Munich Pact with Nazi Germany, and novelist Ursula Parrott's tips on beauty, diet, and etiquette. The endeavor was a disorganized drain on energy and finances, and within a year Broun accepted individual responsibility for the paper, retitling it *Broun's Nutmeg.*

In 1939 the *World-Telegram* owners decided that paying Broun $49,000 had become too expensive. Howard told Broun that his annual contract would not be renewed, because the board of directors thought streamlining should begin with high-priced specialists to protect rank-and-file reporters. Broun replied, "I've made that speech to Guildsmen a thousand times. I won't take it back because I'm the one to be chopped off." Broun accepted a *New York Post* offer of about $12,000 a year and wrote one column, an analysis of the possibility that President Franklin D. Roosevelt would seek a third term in 1940.

A case of the grippe quickly developed into pneumonia. During the night of 17 December 1939, his close friend Morris Ernst heard Broun say, "If I pull through, I'll remember that Ring Lardner would have lived longer had he only written what he really wanted to write. And I will write only about horse racing, night clubs, baseball, gambling and life." Administered the last rites, Broun slipped into a coma about four the next morning and died at 9:50 A.M. on 18 December 1939.

On hearing of the death of their president, guild activists from around the country sent condolences. "The doctors say that he died of a weak heart," wired the Cleveland Local, "but no one who knew him will ever believe it. No heart was stronger in the service of just causes." Not long before his death, commenting in his *Nutmeg* on mortality and on his recent birthday, Broun seemed peculiarly prophetic:

> When a man is halfway home, he begins to fret.... But at 51 I have more faith than I ever had before....
>
> It may be that in small towns where farms begin around the bend in the road there is more familiarity with birth and death. There are things which can be learned from the fields and the barnyards where animals come into life and go out of it with slight articulate protest.

Two months after Broun's death, the Newspaper Guild of New York held a memorial meeting featuring tributes from Swope, Reynolds, John L. Lewis, Edward G. Robinson, and many other friends and colleagues. More than twelve thousand people attended. "Sports fans read him when he wrote about sports," Morris Watson said in his introduction. "He was an equal authority on subjects foreign or remote to sports fans."

Broun always made writing look fun, and easy. But he worked hard at it. "Heywood loved to play with words, find fun in them," recalled Fred Lieb in his autobiography, citing Broun's lead about Giants manager John McGraw's habit of starting pitcher Rube Schauer and relieving him with his only other decent pitcher, Ferdie Schupps: "It never Schauers, but it Schupps."

Yet as Saul Schraga once recalled, "There were many more Brouns than just Heywood the phrase-maker. There were Broun the humanitarian, Broun the angry foe of special privilege, Broun the devoted father, Broun the thoughtful son and husband, and, of course, Heywood the soft touch, who sprinkled dimes and $20 bills with the same abandon to men standing in bread lines." Heywood Hale Broun adds that his father "was a driven perfectionist. He'd have much of a piece composed in his head, then sit down and type as fast as he could, pausing occasionally to murmur aloud to get the rhythms right."

Some critics, however, have said Broun could have been even better. "Broun had spent years writing usually interestingly if not always well about baseball, the theater and political conventions," wrote John Moreau in *Journalism Quarterly* (Autumn 1967)."By his own admission much of his writing was dashed off to meet a deadline, and often when he had nothing to say. In [his] virtuous spreading of himself in so many directions, he as an artist spread himself thin."

Lewis Gannett, an old friend of Broun, scoffed at such notions: "It was a silly thing to say that his 'extracurricular' activities interfered. It was precisely because Heywood Broun played the races, visited the bread lines, produced a Broadway play, ran for Congress, walked picket lines, organized the Guild and joined a church, that he and his column stayed young." *St. Louis Post-Dispatch* arts editor Thomas Sherman commented in 1957 that "I followed Broun as a sports reporter, a drama critic and a columnist. Everything he wrote struck me as being a remarkable combination of lightness and depth."

In 1976 Broun was inducted into the International Labor Hall of Fame in Detroit, where he is honored with such labor luminaries as John L. Lewis, Samuel Gompers, and Walter Reuther. Five years earlier the Baseball Writers Association of America made Broun the eighth sportswriter to be inducted into the Baseball Hall of Fame in Cooperstown, New York, where he joined Rice, Runyon, Lieb, and Lardner.

In *Whose Little Boy Are You? A Memoir of the Broun Family* (1983), Heywood Hale Broun wrote, "Almost from the beginning Heywood was a vast success in the newspaper business. In his day he was straight reporter, war correspondent, sports writer, drama critic, book critic, and general columnist. If some people did one of those things better than he did, no one could match him at all of them." Broun, however, "was never more at home than when writing about baseball," his son recalled.

**Biographies:**

Dale Kramer, *Heywood Broun: A Biographical Portrait* (New York: Wyn, 1949);

Richard O'Connor, *Heywood Broun: A Biography* (New York: Putnam, 1975);

Heywood Hale Broun, *Whose Little Boy Are You? A Memoir of the Broun Family* (New York: St. Martin's Press/Marek, 1983).

**References:**

Charles Alexander, *John McGraw* (New York: Viking, 1988);

Charles Baldwin, *The Men Who Make Our Novels,* revised edition (New York: Dodd, Mead, 1924);

H. Le B. Bercovici, "Here's Paper Broun Would Edit," *American Press* (August 1928);

Bennett Cerf, *Try and Stop Me* (New York: Simon & Schuster, 1945);

Jim Cesnick, "Baseball Writers Honor Broun," *Guild Reporter,* 13 August 1971;

Negley D. Cochran, ed., "Heywood Broun, As He Seems to One Who Knows Him Best," *Scripps-Howard News,* September 1932;

Joseph F. Collis and others, *Biographies in Sound* (New York: NBC Radio, 1956);

Bob Cooke, ed., *Wake Up The Echoes, from the Sports Pages of the New York Herald Tribune* (Garden City, N.Y.: Hanover House, 1956);

Robert K. DeArment, *Bat Masterson: The Man and the Legend* (Norman: University of Oklahoma Press, 1979);

Morris Ernst, *The Best Is Yet . . .* (New York: Harper, 1945);

Charles Fountain, *Sportswriter: The Life and Times of Grantland Rice* (New York: Oxford University Press, 1993);

Adolph Gillis and Roland Ketchum, *Our America: A Survey of Contemporary America As Exemplified in the Lives and Achievements of Twenty-Four Men and Women Drawn from Representative Fields* (Boston: Little, Brown, 1936);

Margaret Case Harriman, *The Vicious Circle: The Story of The Algonquin Round Table* (New York: Rinehart, 1951);

Richard Kluger, with the assistance of Phyllis Kluger, *The Paper: The Life and Death of The New York Herald Tribune* (New York: Knopf, 1986);

Dale E. Larson, "Advocate Journalist, Even in Sports," *Guild Reporter,* 30 April 1971;

Larson, "The Newspaperman Who Founded A Union," *Guild Reporter,* special supplement, 25 December 1964;

John L. Lewis, Franklin P. Adams, and others, *Heywood Broun As He Seemed To Us* (New York: Published for the Newspaper Guild of New York by Random House, 1940);

Fred Lieb, *Baseball As I Have Known It* (New York: Coward, McCann & Geoghegan, 1977);

John Adam Moreau, "The Often Enraged Heywood Broun: His Career and Thought Revisited," *Journalism Quarterly,* 44 (Autumn 1967): 497–507;

David Nemec, *The Ultimate Baseball Book,* edited by Daniel Okrent and Harris Lewine (New York: Houghton Mifflin, 1979): 104, 162, 184;

M. B. Schnapper, ed., *Heywood Broun* (Washington, D.C.: American Council on Public Affairs, 1940);

Saul Schraga, "Knight in a Rumpled Suit," in *Page One Year Book* (Philadelphia: Newspaper Guild of Greater Philadelphia, 1958);

Fulton J. Sheen, *The Conversion of Heywood Broun* (New York: Saint Paul Guild, 1940);

Thomas Sherman, *Page One* (Saint Louis: St. Louis Newspaper Guild, 1957);

Louis Leo Snyder and Richard Morris, eds., *Treasury of Great Reporting* (New York: Simon & Schuster, 1949);

John Wheeler, *I've Got News for You* (New York: Dutton, 1961).

**Papers:**

Some of Heywood Broun's papers and correspondence are at the Newspaper Guild international headquarters in Silver Spring, Maryland; other papers are at the Library of Congress.

# Bozeman Bulger

*(22 November 1877 – 23 May 1932)*

William J. Plott
The Birmingham News

MAJOR POSITIONS HELD: Night city editor and sports editor, *Birmingham Age-Herald* (1899–1905); sports reporter and editor, *New York Evening World* (1906–1931); staff writer, *Saturday Evening Post* (1931–1932).

BOOK: *Thirty Years in Baseball,* by Bulger and John J. McGraw (New York: Boni & Liveright, 1923).

SELECTED PERIODICAL PUBLICATIONS – UNCOLLECTED:

FICTION

"Yas-Suh, 'Ats Er Dog!," *Saturday Evening Post,* 192 (10 April 1920): 20–22, 189–190;

"High Hats," *Collier's,* 83 (25 May 1929): 10–11.

NONFICTION

"Fainting Finish," *Everybody's,* 32 (April 1915): 512–522;

"Big League Promise," *McClure's,* 47 (May 1916): 30–32;

"Dry Manhattans," *Saturday Evening Post,* 192 (21 February 1920): 14, 124, 127;

"The Great White Wall," *Saturday Evening Post,* 192 (24 April 1920): 38–42;

"What to Cook When the Wife's Away," *Ladies' Home Journal,* 38 (July 1921): 75;

"Real Thrill in the Circus," *Collier's,* 69 (April 1922): 13;

"The Psychology of the Sucker," *Saturday Evening Post,* 194 (15 April 1922): 10–11, 107, 109–110, 113;

"Growing Golf Germ," *Saturday Evening Post,* 195 (24 February 1923): 20;

"Making Owners Like Their Cars," *Saturday Evening Post,* 195 (5 May 1923): 32–34, 173–174;

"25 Years in Sport," *Saturday Evening Post,* 200 (28 April 1928): 8–9; 200 (12 May 1928): 36–38; 200 (26 May 1928): 37;

"Warm Feet and Hot Hockey," *Saturday Evening Post,* 203 (9 May 1931): 33, 117;

"Drive Thrills," *Saturday Evening Post,* 203 (23 May 1931): 24, 35;

*Bozeman Bulger (right) with Damon Runyon in 1912*

"The Polo Pony Grows Up," *Saturday Evening Post,* 203 (30 May 1931): 31, 42;

"Night Baseball Versus the Mosquito," *Saturday Evening Post,* 203 (6 June 1931): 35;

"Don't Monkey With a Boat," *Saturday Evening Post,* 204 (11 July 1931): 24, 47, 50;

"Auto Follows the Food," *Saturday Evening Post,* 203 (20 June 1931): 35, 101;
"How the Umpire Sees 'Em," *Saturday Evening Post,* 204 (8 August 1931): 29, 47, 50;
"Tennis Seeks a Slugger," *Saturday Evening Post,* 204 (22 August 1931): 31, 93;
"Harness Horse Sniffs," *Saturday Evening Post,* 204 (12 September 1931): 28, 97;
"America's Fall Tonic," *Saturday Evening Post,* 204 (3 October 1931): 10–11, 34, 38;
"Sheepskins for Thoroughbreds," *Saturday Evening Post,* 204 (21 November 1931): 29, 93–94;
"And Along Came Ruth," *Saturday Evening Post,* 204 (28 November 1931): 6–7, 36–38; 204 (5 December 1931): 16–17, 48, 50; 204 (12 December 1931): 28, 30, 86, 88–89; 204 (19 December 1931): 16–17, 36, 38;
"Army of Basket Ball," *Saturday Evening Post,* 204 (9 January 1932): 43, 142–143;
"Pitch Horseshoes and Live Long," *Saturday Evening Post,* 204 (20 February 1932): 34, 108;
"Catching the Coyote," *Saturday Evening Post,* 206 (23 September 1933): 84–85.

PLAY PRODUCTIONS: *The Battle Cry of Freedom,* San Francisco, Orpheum Theatre;
*Curves,* Hammerstein's Victoria Theatre, 1910;
*The $11,000 Beauty,* New York, circa 1907;
*The Silver Bottle*;
*Swat Milligan,* circa 1908;
*Throwing the Bluff,* New York, Keith's Union Square Theater, 18 November 1912;
*Us Marshalls,* by Bulger and Irvin S. Cobb.

Millions of newspaper and *Saturday Evening Post* readers came to consider Bozeman Bulger a great sportswriter whose work showed keen wit and insight into America's games. But Bulger found sportswriting a calling that was difficult to accept. Born to a distinguished southern family, Bulger at first was ashamed of sportswriting as a profession. Writing in a three-part 1928 series for *The Saturday Evening Post* titled "25 Years in Sports," Bulger admitted, "For years, I actually felt embarrassed at telling anybody I was a sporting editor. My aim, influenced perhaps by my father's friends, was what I then considered higher." Bulger's desire, perhaps inspired by the four years he spent as secretary to Congressman (later Senator) Oscar W. Underwood and his service as a clerk to the Alabama state legislature, was to be a political writer. His interest in and knowledge of sports, though, determined his career.

When he died in 1932, Bulger was the dean of American sportswriters, a mentor to Damon Runyon, a raconteur and toastmaster without equal. In a 1910 photograph taken at the Polo Grounds in New York, Bulger, a walking cane leaning across one leg, is at the center of a gathering of baseball scribes. Seated and standing around him are such notables as Sam Crane, Fred Lieb, Grantland Rice, and Runyon. But while the fame of such writers as Rice and Runyon has survived because of a relatively large body of work that remains accessible to readers, Bulger has been forgotten because nothing of his work exists in book form under his name. There are no collections of his columns and magazine pieces, no acknowledgments of his contributions in the books that he is known or believed to have ghostwritten. Even the titles of most of the plays and sketches he wrote for Broadway, Off Broadway, and vaudeville have been lost; only one manuscript of a theatrical piece is known to exist.

Bulger's importance in his own time is revealed by the notice given his death. Among those expressing regrets were Rice, Ring Lardner, Babe Ruth, author Albert Payson Terhune, sportscaster Graham McNamee, and baseball executive Charles Stoneham. His sudden death at age fifty-four was the lead item on *The New York Times* obituary page, and the subsequent obsequy reports in that and other newspapers reveal the esteem with which he was held by his contemporaries. The Associated Press obituary said Bulger "numbered his friends by thousands, both in the baseball world and out of it. He was well known for his endless humorous stories of the diamond and the Southern Negro, ranking with Irvin Cobb as a teller of tales in the drawling Southern dialect." Dan Daniel, writing in the *New York World-Telegram,* credited Bulger with revolutionizing baseball writing "by the use of funny incidents and humorous delineation." In his column, "Daniel's Dope," he wrote that Bulger

> brought baseball closer to the people by writing into the game a more intimate link between the fans and players. He gave to the devotees of the game a complete and veracious portrait of the player, with his virtues and vices, his fine points and his drawbacks. . . . He found our diamond Boswells devoted to a rather dull school of dreary actualities, allegedly tempered here and there by such efforts as "Jones died on first" and "Smith got a life." . . . He appreciated that baseball shouted for dramatization – that the game had its daily thrill behind the dry-as-dust facts.

Daniel, like many of his peers, found it difficult to believe "that laughing Boze" was dead. Runyon wrote that the sports world was "fairly stunned" by the news of Bulger's death: "It was difficult to believe that its most beloved sports writer, jovial, whole-souled 'Boze' Bulger, had passed from this earth."

In almost all of the tributes written after his death and in the biographies and autobiographies of such contemporaries as Irvin S. Cobb and Runyon, Bulger is noted for his sense of humor. He apparently could see the absurd, or the potential for it, in every aspect of his life. His sense of humor is evident in what survives of his theatrical efforts and in his baseball columns. It is also evident in his dialect stories of Negro life, which, though filled with stereotypes, appear to have been told with gentle paternalism and affection, not mean-spiritedness. An unidentified writer in the *New York Herald-Tribune* wrote: "He was always the toastmaster at baseball dinners and was as good a story teller as Irvin S. Cobb, who was one of his closest friends. They were known to their intimates as 'Big Irv' and 'Little Irv.' Mr. Bulger shortened many a train journey for the sportswriters with his Negro dialect tales and was a master of dialect." Grantland Rice wrote in "The Sportlight," his nationally syndicated column, that "The gayest and most gallant note in sports literature is still. There will be mourning in the major and the minor leagues. Magnates, players and fans will miss the music of his laughter, for Colonel Bulger was one of those who did more for putting drama and humor into the game of baseball than any of the magnates or any of the players."

Born 22 November 1877 in Dadeville, Alabama, Bulger came from a family steeped in journalism, the military, and the law – all of which would figure in his own life. Although his grandfather Michael Jefferson Bulger was among the delegates opposed to the secession ordinance in Alabama, he quickly enlisted once the die was cast and rose to be a brigadier general in the Confederate army. His heroism under fire was extraordinary. During an early battle he bound a wounded arm to his chest and continued to command his troops. Receiving an artery-severing wound to the leg, he stopped the flow of blood by placing corncobs over the entrance and exit wounds and holding them in place with a suspender. Later, he was so severely wounded at Gettysburg that he was left on the field for dead. When discovered alive by a Union officer of lesser rank, he refused to surrender his sword until an officer of equal or greater rank could be located. Michael Jefferson Bulger also served as a state senator and lived to be one hundred years old. Bulger's father, William Douglass Bulger, was also a Confederate officer, fighting in several major campaigns and suffering wounds at Chickamauga. After the war he became an attorney, practicing in Dadeville and eventually Birmingham. Both Bulger's father and grandfather had worked for the *Dadeville Record* before the war.

Bulger spent his first ten or twelve years in Dadeville, then moved with his family to Birmingham. After finishing his secondary education, he attended the University of Alabama in Tuscaloosa and became an attorney, though it is not clear if this was by academic degree or by reading with another lawyer. In a brief memoir, written about 1916 and found in one of his scrapbooks, he described himself as having early ambitions to be a ballplayer: "I progressed far enough for them to force me to become a lawyer." In this piece he advocates women's suffrage and notes Oscar W. Underwood's suggestion that he "become a newspaper man or fiction writer." Bulger spent about four years with Underwood, serving the lawmaker until the Spanish-American War broke out, and Bulger carried on the family military tradition by enlisting in the army, rising to the rank of sergeant in Cuba.

During his Washington years with Underwood he witnessed many outstanding baseball and boxing events involving the more celebrated practitioners of the late nineteenth century. One of his experiences in Washington became an anecdote he recalled in the reminiscence for *The Saturday Evening Post.* On 17 March 1897 he participated in the fake Washington "broadcast" of the heavyweight championship fight between James J. Corett and Bob Fitzsimmons that was taking place in Carson City, Nevada. An itinerant sportswriter and an old telegrapher decided to boost their sagging bankrolls by hiring a hall, selling tickets, and doing a blow-by-blow broadcast of the fight. At the last minute, however, the telegraph company refused to provide the wire without advance payment. With the help of Bulger and others, the entrepreneurs set the receiver on a table in the hall and ran the wire down to a transmitter in the basement. Bulger and friends monitored bulletins being posted outside the *Washington Star* building across the street. They relayed these to the man in the basement, and he quickly sent them upstairs:

> There was never a more lifelike description given of a distant prize fight. In fact, they were so intent on their work that when we got the flash . . . of the knock-out, the old telegrapher upstairs strung out a description of five minutes' duration, bringing it to a real fitting climax. The crowd went away delighted, and so did this pair of ingenious old boys.

It was the kind of harmless foolery and subsequent great story that Bulger so loved and made a trademark of his life and his writing.

When he left Washington, Bulger became a legislative clerk in Montgomery, Alabama. He received his baptism as a sportswriter in that city during the fall of 1898 or 1899. Football was beginning to attract attention in the South, and a game between Auburn (then known as Alabama Polytechnic Institute) and the University of the South at Sewanee, Tennessee, was scheduled to be played in Montgomery. The *Atlanta Constitution* assigned its Montgomery correspondent, Robert McDavid, to cover the game. "Incredible as it may sound now, there wasn't a newspaperman in the state capital – with the legislature in session, too – who knew how to write a football game," Bulger recalled in *The Saturday Evening Post.* Someone told McDavid that Bulger had been a scrub on a football team. The correspondent collared Bulger, saying, "You don't have to know how to write. If you understand the game and will merely put down the plays on a pad of paper, we will make a story out of it and I will see that you get ten dollars."

Possibly taking advantage of his having a brother on the Auburn team, Bulger persuaded the coach, the legendary John Heisman, to allow him to stay on the Auburn sideline throughout the game. The young reporter-to-be so diligently recorded the details of each play that Heisman was later able to use the notes to figure out rushing statistics. Bulger recalled that McDavid wrote "a nice flowing lead which I admired greatly and the next day I saw my first sporting story – as dreary as it was – in print. I wonder if any newspaperman ever forgot that thrill of seeing his effort in cold type. No, he couldn't." While Bulger would not be the first journalist inspired by seeing his prose in print, he was already somewhat inclined toward the profession from his Washington experiences with Underwood. "All of my life," he wrote in *The Saturday Evening Post,* "I have had a love for association with newspaper men. I had a habit of trailing around with them long before I was actually initiated as a member. I suppose the calling gets in one's blood that way."

When the legislative session ended in Montgomery, Bulger got a job with the *Birmingham Age-Herald,* in 1899. He was hired as night city editor, and his sporting editor role was seen as secondary in importance. Putting the situation in the perspective of the early 1900s, Bulger wrote that sports "was a side line. . . . [ I ] wrote the baseball story in the afternoon just because I liked it." With his title of sporting editor came the duties of dealing with

*John McGraw (left) with Walter Johnson, George Sisler, and Christy Walsh at the 1924 World Series between the New York Giants and Washington Senators (photograph © Kadel & Herbert)*

anything remotely of a sports nature and also of settling all bets, regardless of their subject matter. "The public, for some reason, had the firm belief that a sporting editor knew everything or could find it on a moment's notice." One night he was called by two men in dispute over the question of whether circus contortionists used snake oil for suppleness. But Bulger would see remarkable development in American sportswriting during the course of his career. Coverage grew from "a bare two columns in some newspapers to eight pages" by the 1920s. By 1928 the average metropolitan newspaper's sports staff would grow to a size twice as large as the entire staff of the *Birmingham Age-Herald.* He noted in his *Saturday Evening Post* reminiscence that the sporting editor of his youth had become in many cases a nonwriting executive called a "sports editor."

There was no baseball his first year in Birmingham, the old Southern League having broken up during the Spanish-American War. Bulger and other newspaper, baseball, and sporting men began trying to organize a new league. The result was the Southern Association, one of baseball's venerable minor leagues, which survived from 1901 to 1963, when some owners preferred to shut down the league rather than allow black participation. Bulger discussed his role in the beginning of the league in *The Saturday Evening Post*:

> The old trick in organizing a minor league was, in case of financial difficulties, to keep it going until the Fourth of July receipts could be taken in and then quit. . . . The pioneer manager often had to take it on the run, with sheriff, the unpaid players and indignant fans in hot pursuit. That is how I came to own the Birmingham franchise for a few days. Sam Mills, unable or unwilling to finance the operations any further, walked out of town, leaving us flat. He even took the home plate with him, claiming that to be a part of his personal property.

With the club $500 in debt, the league president turned the franchise over to Bulger, who was approached by four businessmen. He considered them "good fellows and real sportsmen" and gave them the franchise. By 1928 the club was worth an estimated $200,000. "That will give the present-day fan an idea of how unimportant sports were regarded twenty-five years ago," he wrote in *The Saturday Evening Post.* However, he did buy two shares from his friends, which he sold later to finance his "assault on the newspapers of New York."

Although difficult to attribute for certain to Bulger, the baseball columns in the *Birmingham Age-Herald* began to show a sparkle and wit in the early 1900s that fit later descriptions of Bulger's personality and parallel the work he did later for the *New York Evening World.* The following lines appeared at the beginning of an account of Birmingham's 3–2 win over Atlanta on 21 August 1903:

> Away down yonder in old West End
> Occasional plays have jostled great men.
> But big Pat Millerick has skinned them all
> By scoring from second on an infield ball.
> And I guess that's running some.

Over the course of several seasons, much such verse appeared, all concluding with a "some" line apropos to the event and the descriptive verse.

In 1906 or 1907 Bulger, with his wife, Louise Strane Bulger, and young daughter, Gene, in tow, launched his New York adventure. He described himself as "waiting at the editorial door of the old *Evening World* office . . . hankering to be a great political writer." A week earlier he had secured an interview with William Jennings Bryan, who had just returned from Russia and was full of comments on the revolution. That scoop had earned Bulger a five-dollar bonus and, he figured, a good shot at a political writing job. "I was determined not to disclose the fact that I had been a sporting editor down South. It seemed so terribly inconsequential," he wrote.

One of Bulger's baseball writing colleagues, possibly Sam Crane, apparently had already dropped a note to Vincent Treanor, the editor in charge of the sporting desk, alerting him to Bulger's possible arrival. Buoyed by Treanor's mention of a Fourth of July speech given by Bulger's father at Tammany Hall, Bulger related his political reporting ambitions. Then, abruptly, Treanor asked Bulger if he knew anything about the new spitball that pitcher Jack Chesbro was throwing for the New York team in the American League. "I admitted that I did, that I had really experimented with it myself. It developed in a moment that my political writing career had suffered a setback in advance."

Treanor asked Bulger to go talk with manager Clark Griffith about the pitch and arrange for photos showing the pitcher's hands when throwing the ball. When Bulger hesitated, still hoping for a political writing job, Treaner told him "Everybody knows about politics, but doggoned few know about the spitball. In this town Jack Chesbro is a much more important man than William J. Bryan. You can make more money writing about him." The *Evening World,* which for two years had been seeking the right man to cover baseball regularly, had found its writer. Bulger knew Griffith and was quickly able to get the story and photos Treanor wanted. He was hired at thirty-five dollars a week:.

> With a bankroll worn down to twenty dollars and a wife and baby waiting uptown to hear the news of my New York triumph, I broke the running-broad-jump records for the telephone booth, there to be initiated in the most humiliating manner. The office boys, according to years-old usage, had filled the concave top of the stool with ice water, which I promptly sat in. All new men were initiated that way.

Bulger would get another initiation when he prepared to leave for his first out-of-town trip. The office boys covered the doorknob with mustard. The ritual had "a chastening effect" on the young man embarking on his adventure.

The Alabama native hit New York at a time when baseball writers were an intellectual fraternity as likely to be familiar with the latest novel as with the latest pitcher from the minor leagues. Bulger wrote that the press box at the Polo Grounds "could easily have been mistaken for an outdoor literary club. Many authors sat with us daily. . . . Baseball writers as a rule were extensive readers and sharp critics of current literature. . . . Any author could find out the moment he shoved through the little iron gate just what was the frank verdict of his latest work. . . . Any time a new figure bobbed up in literature, Charley Van Loan and Damon Runyon would make a point of bringing him around for appraisal."

Among the literary friends Bulger made was William Sydney Porter, the popular short story writer under the pseudonym O. Henry. One night in the Metropole bar, Porter was commenting on a piece of Bulger's that he had enjoyed. "'You'll get along fine,'" he told me, "'if you'll just keep on trying to be yourself. Don't try to get fancy, but say right out what you think. A little later on you ought to try fiction.' Nothing ever stirred my ambition as that advice from this little insignificant-looking man who was then rapidly rising to great popularity. . . . His was a gentle soul. His greatest efforts were spent in avoiding parties where he would be lionized. I cherished for years O. Henry's coral scarfpin that was left to me when he died."

In New York Bulger got a chance to hone the blend of humor and reportage he had brought to the baseball writing in Birmingham. An example of Bulger's humor in a baseball column is the following anecdote related in a 25 February 1911 report on doings in the New York Giants training camp in Texas. In it, Bulger describes a disagreement between umpire Charles Hansell and the Giants' Beals Becker:

> "You are as crazy as [Bugs] Raymond!" yelled Becker. "That wasn't a strike!"
>
> "I'll call them when they are over!" replied the umpire, with great dignity.
>
> "Yes," retorted Becker, "but what in Halifax do you call them when they ain't over?"
>
> "Go on and play ball!" sternly ordered Mr. Hansell, thereby dodging the question at issue.
>
> "You ought to have come out here two months ago," snorted Becker as he menacingly approached the arbiter.
>
> "What for?" inquired Hansell, who thought he referred to his need of practice.
>
> "Why to help Murphy fix up the grounds!" answered Becker. "And partly earn your money!"
>
> That was too strong and Hansell ordered Becker out of the game and called upon the high sheriff, who was present, to give him protection.

Bulger wrote for the *Evening World,* except for a break during World War I, until the paper was sold in 1931. Although as a forty-year-old family man he was exempt from military service, Bulger quickly enlisted when the United States entered the war. During his second stint of military service, he was named the chief press liaison for American war correspondents, received a citation for bravery under fire in the Argonne Forest offensive, and was promoted to lieutenant colonel shortly before his discharge. After World War I he resumed his coverage of baseball and other sports for the *Evening World* but also began freelancing pieces to the *The Saturday Evening Post, Collier's,* and other magazines. He wrote sketches and plays for vaudeville, Broadway and Off-Broadway shows, and dabbled occasionally in short fiction.

Bulger was one of the earliest ghostwriters, a practice he began in a series of articles on Edward Payson Weston, an elderly man who became a celebrity in the early 1900s for his walking adventures, particularly a 105-day, 3,805-mile transcontinental trek from New York City to San Francisco: "I had to walk with him for four days around New York, a mob of kids following. . . . My job was to walk with the old gentleman and prepare daily articles – signed by him – on feet and how to take care of them." He recalled how ghostwriting became a major part of his career, "Later I was assigned to similar work for Young Corbett, Battling Nelson, John L. Sullivan and many others. As a final test of my ability as a ghost writer they even had me impersonate a race horse and tell his story." Bulger's ghostwriting, however, is the one of the most elusive aspects of his career. With the exception of John J. McGraw's autobiography, *Thirty Years in Baseball* (1923), no other titles can be certainly attributed to him. However, in his *Satuday Evening Post* reminiscence he wrote at some length about his ghostwriting experiences with Ty Cobb, Honus Wagner, and Sullivan, all of which appear to have been book-length manuscripts.

In his ghostwriting Bulger insisted that he was merely a facilitator, that Cobb and Wagner essentially wrote their own memoirs:

> The impression that they merely consented to the use of their names and had nothing to do with what went under them is downright unfair to them. They furnished all the detailed information, worked out their ideas with me and were responsible for everything printed. My main task was to keep them on the line that they were hewing and then assemble the whole into shape for the printer. Naturally, men like that are not familiar with the technical forms of writing, of continuity or of assembly.

To support his contention, Bulger recalled when he had asked Wagner to jot down on a pad some of the greatest plays he had ever seen in his career and also dig up some photographs from his playing days. Bulger wanted only memory-joggers that he

*Baseball writers Fred Lieb, Bulger, Jim Harrison, Bob Ripley, and George Daley at Lake Oscawana, New York, in August 1927*

would flesh out on the typewriter. "Four days later he showed up with a whole suitcase full of the funniest photographs I have ever seen.... After exhibiting these he calmly handed me eighty-five pages from a writing pad on which he had written down the plays with a stub of a pencil. He had headed them Great Plays I Saw. This, I think, ought to convince the skeptical that some great athletes do prepare their stuff."

Bulger described a similar experience with boxer John L. Sullivan, after the pugilist "had reached the days of his mellowness." Bulger traveled a vaudeville circuit with Sullivan one winter, ostensibly to help with the preparation of the fighter's autobiography. The plan was for them to work on the book in the mornings and for Bulger to be Sullivan's companion at evening social events. On their first day together Sullivan tossed about six hundred pages of manuscript material on the table and said, "There, young fellow, is what a half dozen different fellows have got together at one time and another. Now you go through it and if there's anything else you want to know, just ask me questions. I'll stand for anything that's in there if it ain't balled up." Sullivan apparently enjoyed Bulger's work and company because he gave the writer a gold penknife, "his well-known method of bestowing the decoration of friendship," wrote Bulger. "That quaint custom, kept up all his life, was well known in sports circles. My knife, which bears his name, is set on one side with a small diamond, a sapphire and a ruby, the reverse being reserved for the engraved inscription. . . . To possess one of John L.'s friendship knives was quite a distinction twenty years ago."

Bulger's ghostwriting John J. McGraw's autobiography, *Thirty Years in Baseball,* resulted from their more than twenty-year friendship. Bulger had a singular relationship with McGraw, which he attributed in part to an incident in a billiard parlor frequented by writers and baseball players. McGraw bet quietly 15–1 odds against fifty cents that Bulger had laid on the table. Bulger gained McGraw's respect when he won the bet and collected McGraw's $7.50. The New York Giants manager enjoyed watching "a dark horse come through the gate." Thereafter, Bulger said he had McGraw's ear for stories and "a foundation for big-league baseball understanding." Bulger's understanding was evident in his assessment of McGraw's overpowering bench during the National League pennant race of 1908. McGraw, Bulger understood, had seen the necessity of a "reserve force" and had gotten Giants' owner John Brush to "thr[o]w open the bankroll." The Giants were able to make a run for the pennant and ultimately tie for second-place behind Chicago, largely because, as Bulger had written on 21 September, "McGraw is prepared for any emergency."

Despite Bulger's tribute to his celebrity collaborators, one suspects that he undersold the gifts he brought to his ghostwriting. The McGraw book shows both an erudition and a storytelling ability that likely came from Bulger. McGraw's engaging use of interesting details, especially in his discussion of strategy in the 1923 World Series defeat of his New York Giants by the New York Yankees, and his skill in telling many humorous stories seems to fit in with Bulger's style. According to his obituaries Bulger had completed a second McGraw book just before his untimely death. This was probably a reference to "Genius of the Game," a three-part series on McGraw that *The Saturday Evening Post* published posthumously, the first installment running just a few days after Bulger's death.

Bulger's theatrical writing is as elusive as is his ghostwriting. Although he apparently wrote humorous vaudeville skits as well as several short plays, few titles can be definitely attributed to him and almost none of his work in this vein survives. Harold

Seymour, in the second volume of his baseball history, mentions *Curves,* a play featuring ballplayers Christy Mathewson and Chief Bender as well as actress May Tully. Bulger in *The Saturday Evening Post* reminiscence recalls *The $11,000 Beauty,* a Broadway musical comedy centered around Rube Marquard, whose minor league contract with Indianapolis had been sold to Brooklyn for that then-extraordinary sum.

On one occasion Bulger helped to resuscitate a failing baseball play, title unknown, that had been written for vaudeville around 1910. Responding to the director's request for "a real comedy character," Bulger and two unidentified cowriters introduced Van Cortlandt Parks, a Briton who did little more than walk across the stage wearing knickers and carrying a golf bag. At the time, he wrote, golf was viewed with much disdain by Americans. Baseball players referred to the game as "mountain billiards and cow pasture pool." Recalled Bulger, "The mere sight of that golf bag was a big laugh the moment he emerged from the wings. . . . The comedy of the golfer and the golf bag – just the mere sight of him – put our act over so well that it ran three years."

Bulger's sportswriting and playwriting as well as his winning personality brought him his own measure of celebrity and fame. He created several fictional characters – baseball players Jerry Hopps and Swat Milligan and scout "Spotty" Malloy – for humorous tall tales that ran in the *World.* Milligan, a slugger for the Poison Oak Nine, eventually became the central character of a play titled *Swat Milligan* (1908). As Edward O'Connor asserts in *Heywood Broun: A Biography* (1975), Bulger, along with Runyon, Rice, and Broun, were the sportswriters who ranked as the best stylists of their time. Each year on his birthday Bulger organized a golf tournament at the Cold Spring Country Club. According to an unidentified obituary from Bulger's scrapbooks, the trophy was "an aluminum cup costing about the tenth part of a dollar. The engraving on the trophy cost many times that amount."

Bulger was so well established that when the *Evening World* was sold in 1931, he was quickly picked up by *The Saturday Evening Post* and given charge of its "Out-of-Doors" column. He subsequently wrote on such diverse sporting topics as baseball, polo, and motoring. Bulger also wrote skillfully, in *The Saturday Evening Post* and other magazines, on topics other than sports. The 15 April 1922 *Saturday Evening Post* published a Bulger article on "The Psychology of the Sucker," but the man always getting taken was a stock buyer, not a sports fan. He also wrote such articles as "What to Cook When the Wife's Away" for the July 1921 *Ladies' Home Journal* and "Dry Manhattans" for the 21 February 1920 *Saturday Evening Post,* a look at what the bartenders were doing since the arrival of Prohibition. Bulger, certainly an imbiber himself, discretely avoided any mention of speakeasies. (Tom Clark in his 1978 book, *The World of Damon Runyon,* relates an afternoon at the Polo Grounds when Irvin S. Cobb and Bulger punctuated their anecdotes by "each tossing off a stiff one from the half-pint medicine bottle Bulger kept tucked in the lining of his broad-brim black hat.") The casual tone of Bulger's late *Saturday Evening Post* pieces suggests he was enjoying life as a magazine writer. But it was to be a short career. In the early morning hours of 23 May 1932, he died of a heart attack at his home at Lynbrook, Long Island. He had played golf the day before.

Although Bulger's once-prominent position among American sportswriters has been obscured by time and the unavailability of his writing, his career and his influence on peers was significant. During his life he saw an unparalleled growth in sports and noted the construction of concrete and steel stadiums to accomodate the nation's increasing interest in sports in the years before and after World War I. While Bozeman witnessed baseball's relatively unchallenged mastery of that interest as the sport became America's game in the first quarter of the twentieth century, he also wrote about golf, tennis, soccer, basketball, and professional ice hockey.

**References:**

Tom Clark, *The World of Damon Runyon* (New York: Harper & Row, 1978);

Irvin S. Cobb, *Exit Laughing* (Indianapolis & New York: Bobbs-Merrill, 1941);

Clifton Daniel, ed., *Chronicle of the 20th Century* (Mount Kisco, N.Y.: Chronicle Publications, 1987);

Edward O'Connor, *Heywood Broun: A Biography* (New York: Putnam, 1975);

Grantland Rice, *The Tumult and the Shouting* (New York: A. S. Barnes, 1954);

Harold Seymour, *Baseball, Vol. II: The Golden Age* (New York: Oxford University Press, 1971);

Ed Weiner, *The Damon Runyon Story* (New York: Longmans, Green, 1948).

# Si Burick

*(14 June 1909 – 10 December 1986)*

Mike Lackey
The Lima News

MAJOR POSITIONS HELD: Sports editor and columnist, *Dayton Daily News* (1928–1986).

BOOKS: *Alston and the Dodgers*, by Burick and Walter Alston (Garden City, N.Y.: Doubleday, 1966);

*The Main Spark: Sparky Anderson and the Cincinnati Reds,* by Burick and Sparky Anderson (Garden City, N.Y.: Doubleday, 1978);

*Byline: Si Burick – A Half Century in the Press Box* (Dayton, Ohio: Dayton Daily News, 1982).

Si Burick's career as a sports editor and columnist for *The Dayton Daily News* spanned nearly sixty years. Burick covered major national and international sports events from the 1920s to the 1980s and gained a national reputation without ever working for a major metropolitan newspaper.

Born 14 June 1909 in Dayton, Ohio, Simon Burick was the son of Rabbi Samuel F. and Lillian Burick. Burick's father, a Polish Russian immigrant, was the spiritual leader of Dayton's Beth Abraham Synagogue for more than fifty years. On 28 June 1935 Burick married Rachel "Rae" Siegal. They had two daughters, Lenore and Marcia.

After graduating from Dayton's Stivers High School in 1926, Burick entered the University of Dayton, intending to study medicine. But by then he was already launched on his journalistic career; offered a full-time job at *The Dayton Daily News* as he prepared for his second year at the university, he readily gave up his aspirations in medicine. "It was a boon to humanity," he wrote years later in the preface to a collection of his columns. "I would have been a terrible doctor."

Burick first worked for *The Dayton Daily News* in 1925, covering Stivers as a two-dollar-a-week high-school correspondent. After high school he worked at the newspaper as office boy, telephone operator, and eventually sportswriter. He was appointed sports editor in November 1928, at age nineteen. His first byline – "By S. Burick" – appeared 26 August 1925 over a six-paragraph preview of the Stivers football season. His first column appeared 16 November 1928. A half-century later he made fun of the jargon and clichés that riddled his early efforts. Though the lead paragraph of his inaugural column included the words *grid* and *clash,* he pointed out, the newly minted sports editor could not quite manage the phrase *football game.*

But the newspaper's top executive quickly recognized Burick's promise. Burick was an immediate favorite of the publisher James Middleton Cox, former governor of Ohio and 1920 Democratic nominee for president of the United States. Cox encouraged Burick to expand his horizons beyond Dayton. Soon Burick was traveling far and wide in search of stories. One milestone was his trip to Saint Louis for the 1930 World Series. If some of his reportage sounded a bit awestruck (6 October 1930: "Your correspondent was waiting for him at the Forest Park Hotel, the St. Louis club's headquarters . . . "), Burick brought home a local angle of particular interest to his readers – a profile of Jesse Haines, one of Saint Louis's heroes of the series and a onetime automobile mechanic in Phillipsburg, Ohio, just north of Dayton.

Before long it seemed the world was Burick's beat. He covered his first Kentucky Derby in 1929, his first heavyweight championship boxing match (Max Schmeling versus Young Stribling in Cleveland) in 1931. He later covered all but two Kentucky Derbys from 1929 to 1986; major bouts of heavyweight champions from Schmeling to Muhammad Ali; all but one World Series from 1934 to 1984; opening day of the baseball season in Cincinnati annually from 1929 through 1984; the first twenty Super Bowl football games; and the Olympic Games in Rome (1960), Mexico City (1968), Munich (1972), Montreal (1976), and Los Angeles (1984).

His travels took him to the capitals of the world and to places far off the beaten path. On 10–11 July 1958 Burick gave his readers back-to-back

*Si Burick (photograph by Sol Smith, courtesy of* The Dayton Daily News*)*

columns datelined Washington, D.C., and Cut and Shoot, Texas. The first described New York Yankees' manager Casey Stengel testifying before a Senate committee; the second recounted a visit to the hometown of little-known Roy Harris as he trained for his heavyweight title fight against Floyd Patterson.

No matter how far Burick traveled, he never forgot his audience. He always covered the whole story but never missed an observation or viewpoint that would especially interest the Miami Valley of Ohio. When Daytonian Sam Hall won a silver medal in the Olympic diving competition in 1960, Burick watched the finals with Hall's father. When Burick saw his first bullfight, he described it in terms the folks back home would understand: "Man invariably wins," he reported on 22 October 1968 from Mexico City, "but from what I saw for my 165 pesos Sunday at the Plaza Mexico, the matadors and their aides do not always pitch shutouts."

In that column, written as a letter reassuring his "bosses" at *The Dayton Daily News* that he was not abusing their largesse, Burick cheerfully cast himself as the midwestern bumpkin abroad. That light, self-deprecating tone was part of Burick's stock-in-trade. Until the mid 1970s his column was called "Si-ings" – a title suggesting the ephemeral ruminations of a literary lightweight and at the same time indulging Burick's lifelong addiction to puns. Inevitably, the column reflected the personality of the man behind it. One of Burick's favorite paragraphs was written in tribute to Secretariat, winner of thoroughbred racing's Triple Crown:

> This horse is the opposite of everything I am. . . . He is young, he is beautiful, he is fast, he has hair, he has strong legs, he is durable, he has a large bank account. And his entire sex life is in front of him.

When that line appeared in his column on 12 June 1973, Burick did not even take credit for it; he reported it as a comment overheard on the banquet circuit. But the observation was surely his own, and pure Burick. For years he kept it on his desk encased in plastic, and his admirers quoted it repeatedly in various forms.

Colleagues suggested that Burick's sense of humor was the secret of his longevity. Arnold Rosenfeld, who worked with Burick for fifteen years as managing editor and later editor of *The Dayton Daily News,* admired him as a man who "has stood above the easy putdown and cheap, synthetic anger. When he speaks in rare outrage, the reader can know he means it, and something's wrong."

Over time Burick made many personal friends among the sports figures he covered. He collaborated on the autobiographies of two friends – *Alston and the Dodgers* (1966) and *The Main Spark: Sparky Anderson and the Cincinnati Reds* (1978). Both subjects were successful major league baseball managers with southern Ohio connections. Walter Alston, manager of the Brooklyn and Los Angeles Dodgers from 1954 through 1976, was born near Cincinnati; Burick referred to him as the "Squire of Darrtown," for the southwestern Ohio hamlet where Alston lived most of his life. George "Sparky" Anderson managed the Cincinnati Reds from 1970 through 1978, during the period of the team's greatest success. Both books were gentle, affectionate portraits of men Burick liked and respected. Written in the first person, each book allowed its subject to tell his own story the way he wanted it told. Rex Lardner, writing on 16 October 1966 in *The New York Times Book Review,* dismissed *Alston and the Dodgers* as "somewhat bland and sugary." *The Library Journal* of 15 November 1966 found it a pleasant self-portrait of "an uncomplicated country boy . . . (who) has no ax to grind."

But Burick did not allow his subjects to sidestep controversy entirely. Alston gave his explanations of his supposed feud with Dodger infielder Jackie Robinson, modern major league baseball's first black player, and his differences with the outspoken Dodger coach Leo Durocher. Anderson devoted an entire chapter, titled "My Biggest Mistake," to the series of events leading to the acrimonious departure of outfielder Bobby Tolan from the Reds and another chapter to criticism of pitcher Don Gullett, who left the Reds to sign a free-agent contract with the New York Yankees.

Though each book was told from its subject's point of view, Burick found ways to introduce other voices and other viewpoints. *Alston and the Dodgers* incorporated Burick's interview with the longtime Dodger executive Emil J. "Buzzy" Bavasi, delving into Alston's relationship with his boss and exploring the question of whether Alston – who signed a tenuous string of one-year contracts with the ball club – had ever been in danger of being fired. Four chapters of *The Main Spark* were turned over to four of the Reds' star players, allowing Pete Rose, Johnny Bench, Joe Morgan, and Tom Seaver to reflect on Anderson as a man and as a manager.

In both instances the result was pretty much a mutual admiration society; Bavasi generally praised Alston, and the Cincinnati players mostly expressed affection for Anderson. But Burick the newsman did not shy away from criticizing sports figures – even friends – when he thought they deserved it. Good examples were his warts-and-all portraits of Woody Hayes, the tempestuous football coach at Ohio State University. Burick followed Hayes's twenty-eight years as coach of the Buckeyes from start to finish. For years Burick exchanged cards with Hayes and his wife on Valentine's Day, which coincided with Hayes's birthday. But after Hayes refused to talk with reporters following a defeat in the Rose Bowl game, Burick devoted his column of 3 January 1976 to "an open letter to Wayne Woodrow Hayes," lecturing his old friend in print about his responsibilities as a coach and role model. The letter began:

> Dear Woody,
> This is man-to-man, not boy to boy. You and I have known each other for almost 30 years; through good times and bad, fat years and lean. You'll be 63, come Feb. 14. That's a grown-up age, Woody.

Burick continued in that vein for 750 words before concluding:

> You are a person, Woody, not a robot. You are entitled to your emotions. But it is time you learn you are a public person. You owe us, the agents of the public, some answers. Sorry, but that's how it is. You can't win them all. No one ever has.

Hayes did not speak to Burick for several months. After Hayes was fired for throwing a punch at an opposing player late in the 1978 Gator Bowl game, Burick unflinchingly catalogued Hayes's tantrums and indiscretions. But this time Burick was more sad than critical. "Ah, if Woody had only remembered that football was a game," Burick reflected on 31 December 1978. "The tragedy is this is what he'll be remembered for: The Gator Bowl incident, and the yardmarker incident at Michigan, and shoving a camera into the face of a photographer[.]"

Burick's final book was a collection of columns, *Byline: Si Burick – A Half Century in the Press Box* (1982), published by *The Dayton Daily News.* At that time Burick calculated that he had churned out thirteen thousand columns – an average of five a week, sometimes seven, for nearly fifty-four years. Accepting his estimate of at least one thousand words per column, Burick probably wrote fifteen million words for *The Dayton Daily News* in his career.

*Burick with baseball manager Walter Alston (photograph by Ryan Sanders, courtesy of* The Dayton Daily News*)*

He also did a fifteen-minute show five days a week on Dayton radio station WHIO from 1935 to 1961; for a time he also did a show on television. His schedule did not allow for much other writing. Burick's newspaper work, plus *Alston and the Dodgers* and *The Main Spark,* encompassed virtually his entire literary output.

*Byline* forms a representative collection but includes barely an average of one column for every year of his long career. Burick, selecting his favorites, included the obligatory "big" events – both Joe Louis–Max Schmeling boxing matches, Don Larsen's perfect game in the 1956 World Series, the first Super Bowl. Looking back, he concluded that the events he covered usually overshadowed his prose. But he also made room in the book for some strictly local pieces, such as his coverage of a high school–college football doubleheader at the University of Dayton on Thanksgiving Day in 1930.

The collection also reflected Burick's vision, growing over time, of sport as a mirror of the world around it. He dealt thoughtfully with black protest at the 1968 Olympics and terrorism at the 1972 Olympics. Earlier, he took a deep interest in the breaking of baseball's color barrier. After Jackie Robinson's first season with the Brooklyn Dodgers, Burick scored a major "scoop" on the subject when a bombshell was dropped practically in his backyard. Branch Rickey, president of the Dodgers and one of baseball's wiliest operators, chose to reveal the story six hundred miles from New York. Speaking at a football banquet at predominantly black Wilberforce State College near Dayton, Rickey disclosed the previously untold story of a meeting at which baseball's team owners had formally voted to keep black players out of the major leagues.

Burick called the speech "Rickey's Baseball Emancipation Address." Burick's account of it, based partly on notes Rickey turned over to Burick after the speech, captured the appropriate tone of moral outrage.

Burick's column hit the street on 17 February 1948. By the time the story went national the following day, baseball executives were howling with indignation and piling denials on top of denials. Rickey, returning to New York, expressed surprise at the reaction and even suggested that some of the team officials might have forgotten taking the vote.

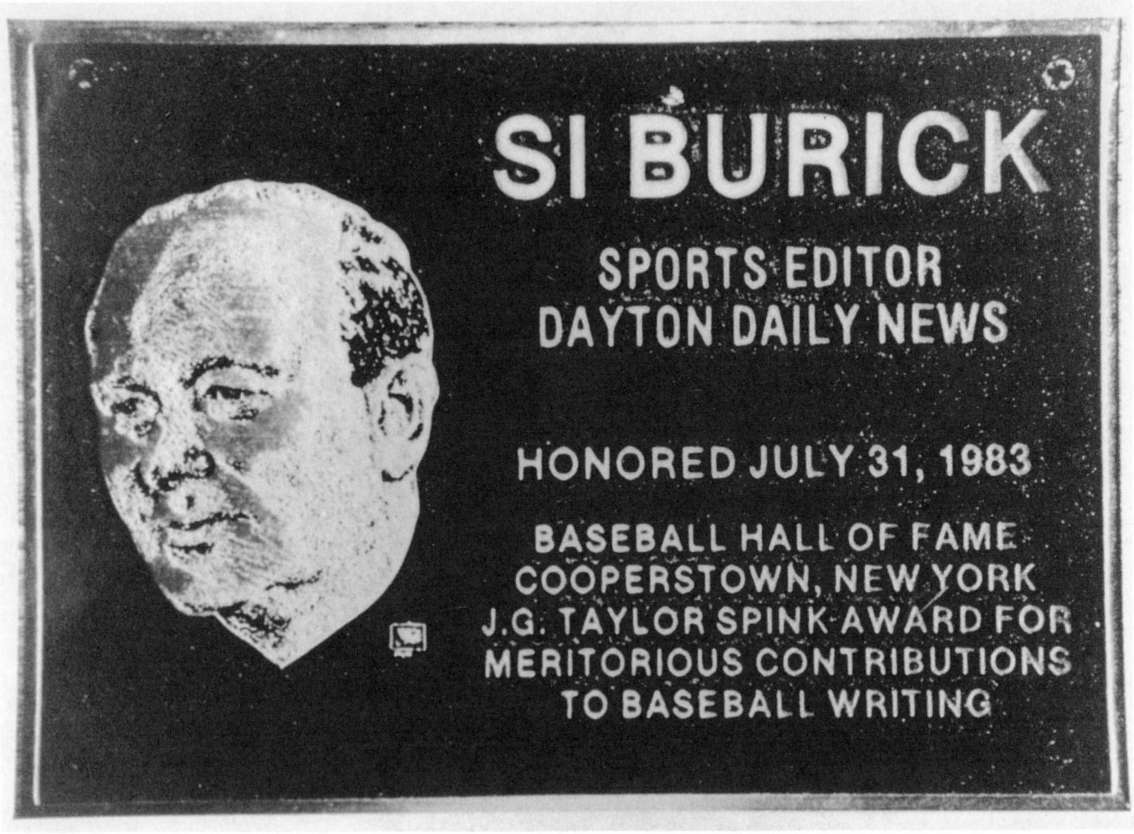


*Plaque honoring Burick at the National Baseball Hall of Fame (courtesy of* The Dayton Daily News*)*

But he stood by his story and said specifically that his remarks at Wilberforce had been accurately reported.

Burick was convinced, disclaimers to the contrary notwithstanding, that Rickey had made the speech with calculated intent and had gotten precisely the response he expected. "Later, Rickey would say in New York, with obvious tongue in cheek, that he did not think the speech would create such a furor," Burick wrote in an introductory note in *Byline.* "Yet, during his remarks, he turned to me saying: 'What I am saying, Si, is on the record. If your pencil is not sharp, I will sharpen it for you.'"

Burick also made infrequent but memorable forays outside the world of sports. His 1963 tribute to John F. Kennedy was reprinted in 1964 in Irving T. Marsh and Edward Ehre's annual *Best Sports Stories* anthology. When *The Dayton Daily News* launched a series of stories on Dayton's changing neighborhoods, Burick volunteered a bittersweet, two-thousand-word reminiscence about the near east side, the ethnic melting pot in which he was born, survived the flood of 1913, and lived into manhood. Going back, he found "a holocaust." His father's synagogue was gone, as were the grade school Burick attended as a boy and all the houses in which his family had lived. Whole streets had disappeared, swept away by construction of new highways. He concluded:

> I have no quarrel with progress, but I do wish a few landmarks had been left behind. My grandchildren, and maybe theirs, might have been a little interested in the old places. Somebody said, when I mentioned this, "Why not a plaque?" but where could you put it? On an expressway ramp?

The story appeared on 10 November 1971. In its wake Burick received mail from transplanted Daytonians all over the country. He later said nothing else he ever wrote prompted as much reader response.

Burick's writing style was never ornate or elaborate. His mature work was distinguished by its sound reporting, gentle wit, and polished storytelling. The generations of sportswriters who learned their trade under Burick's tutelage knew him as an exacting grammarian and painstaking writer who admonished them never to use *win* as a noun. Such niceties of style were acquired through years of refinement. In his fiftieth anniversary column of 16 November 1978 he reflected on how far his profession – and he – had come:

> When I broke in, sports writing was a mass (or mess) of clichés. Literacy did not become a virtue in this field until much later. One bit of idiocy required that you did not repeat the same word even if it properly told the story.

By the time Burick wrote these words, he was an institution in Dayton and a nationally respected figure in sports journalism. His protégés, traveling the country on their own assignments for *The Dayton Daily News,* found themselves instantly recognized in the sportswriting fraternity as representatives of "Si's paper." And as the years accumulated, so did the awards.

Sixteen times between 1954 and 1979 Burick's columns were reprinted in *Best Sports Stories.* He was voted Ohio Sportswriter of the Year by the National Sportswriters and Sportscasters Association nineteen times, including fourteen times in succession between 1962 and 1975. He was inducted into the association's hall of fame in 1985; to attend induction ceremonies in Salisbury, North Carolina, he missed the Reds' opening game for the first time since 1929. In 1986 he became the sixth recipient of the Red Smith Award from the Associated Press Sports Editors. The University of Dayton presented him a doctor of humane letters degree in 1977 and its distinguished alumnus award in 1980.

His proudest honor came on 31 July 1983, when he accepted the J. G. Taylor Spink Award and was enshrined in the writers' wing of the National Baseball Hall of Fame in Cooperstown, New York. At the time of his death he remained the only writer from a city without a major league baseball team to be so honored. At the presentation Burick was praised for his dedication to "common sense, moderation and fundamental decency."

"Mine has been a career of joy," Burick told the gathering. "Joy has been the keynote of my life as a sportswriter and as a baseball writer. I never considered it hard work. I never considered it work. Else why would I have chosen to stay with it all these years?"

Burick never retired. He remained both sports editor and a featured columnist at *The Dayton Daily News,* one of the last sportswriters in the country to perform both jobs at a metropolitan daily newspaper. He never became comfortable with computers; when he was on the road he banged out his columns on a portable typewriter and called them in by telephone.

He filed his fiftieth anniversary column from Hiroshima, where he was covering the Cincinnati Reds on a tour of Japan. Despite undergoing open-heart surgery in March 1986 and being diagnosed with leukemia, he continued to write as long as he lived. The last event he covered was the football game between Ohio State and the University of Michigan on 22 November 1986; he watched the game with Woody Hayes and wrote about the old coach's reactions. Burick's last column appeared on 7 December 1986, three days before his death.

**Interviews:**

Joe Falls, "Burick is Dayton," *Dayton Daily News Magazine,* 24 July 1983, pp. 13–18;

Ralph Morrow, "This and That With Si Burick," *Dayton Daily News Magazine,* 24 July 1983, pp. 19–20.

**Papers:**

*The Dayton Daily News* maintains an extensive file of articles and photographs relating to Burick's life, career, and achievements. Among them are two special retrospective sections, "Sixty Years in Sports" (eight pages), published on 25 August 1985; and "Si Burick: 1909–1986" (ten pages), published at the time of Burick's death and reprinted on 19 January 1987. The National Baseball Hall of Fame's National Baseball Library and Archive maintains a file on Burick, including articles from *The Dayton Daily News;* his obituary, dated 22 December 1986, from *The Sporting News;* and a transcript of acceptance speeches from the Hall of Fame induction ceremony on 31 July 1983.

# Jimmy Cannon

*(10 April 1910 – 5 December 1973)*

Robert Klein
*Editorial Board, NINE*

MAJOR POSITIONS HELD: Reporter, *New York Daily News* (1927–1930); columnist, *New York World-Telegram* (1930–1934); feature writer, International News Service (1935–1936); sportswriter, *New York American* (later *Journal and American*) (1936–1939); correspondent, *PM* (1941–1942), *Stars and Stripes* (1942–1945); sports columnist, *New York Post* (1945–1959, 1972–1973), *New York Journal-American* (later *World Journal Tribune*) (1959–1967), Hearst Headline Service (1967–1972).

BOOKS: *The Sergeant Says,* edited by Frank Sullivan (New York: Knopf, 1943);

*Nobody Asked Me* (New York: Dial, 1951);

*Who Struck John?* (New York: Dial, 1956);

*Nobody Asked Me, But . . . : The World of Jimmy Cannon,* edited by Jack Cannon and Tom Cannon (New York: Holt, Rinehart & Winston, 1978).

SELECTED PERIODICAL PUBLICATIONS – UNCOLLECTED: "Club Fighter," *True,* 20 (January 1948);

"Beautiful Hot Dog," *McCall's,* 82 (July 1955): 46–47;

"This Prize Fight Racket," *Esquire,* 82 (October 1974): 279–392.

*Jimmy Cannon*

A no-nonsense, sentimental newspaperman who wore his heart on his rolled-up sleeve, Jimmy Cannon was, along with Red Smith, one of the two most influential sportswriters of his generation. He also became, in 1959, its highest paid, with a salary of $1,000 a week. At the height of his career, during the twenty years from the mid 1940s to the mid 1960s, Cannon's syndicated columns defined the genre of the personal profile. He was the first sportswriter to examine in a daily newspaper the character and innermost motivations of such legendary sports figures as Joe Louis, Sugar Ray Robinson, and Joe DiMaggio, whom he lyrically described in *Who Struck John?* (1956) as "a man who was meant to play ball on hot afternoons on the grass of big cities. He doesn't belong in the rain."

Cannon's stylistic innovations and keenly self-conscious observations established a new approach to reporting. Thoughtful and perceptive, he tried to understand completely who and what he wrote about. "He revolutionized the sports column," author and journalist Jack Newfield asserted in an unpublished 1994 interview. "He told the story of the person like no one did before him." In the process, Jerome Holtzman wrote, Cannon "gained a genuine cult of readers." Equally at ease with track touts, bookmakers, and assorted Broadway types – he called them "a kind of delicatessen nobility" in a 17 November 1952 *Time* article – Cannon was widely acknowledged as heir to the mantle of Broadway's preeminent chronicler, Damon Runyon. But Cannon went beyond the uncritical colorful features that placed his mentor's name in the American lexicon.

Cannon's eye and ear were finely tuned to the rhythms and inflections of the habitués of the cafés, cafeterias, and street corners of Manhattan, where he lived all his life, as well as the baseball dugouts and boxing rings. His phrasemaking and pungent prose draw the reader intimately close to his subject matter, broaching a familiarity that allows one to imagine being in the same room with a dying show-business legend or listening in on prerace paddock jabber. He offered opinions on everything from dog shows to rowing machines to the pitfalls of marriage, all delivered in a tone suggesting the accumulation of hard-won information and resultant wisdom. Under Cannon's byline, the world was organized into brilliantly shaped images and delivered in a vernacular simultaneously poignant and tough.

Called "an artisan of the language" by Dave Anderson in the 6 December 1973 *New York Times* and "an excellent sportswriter and . . . a very good writer aside from sports" by his friend Ernest Hemingway, Cannon was a three-time Dutton *Best Sport Stories* Award recipient. "He brought literature to the newspaper," according to Newfield. "It was the first time there was a stylist who was able to meet a deadline. His writing was like poetry." Cannon became a pioneer in the art of forcing daily journalism, with its space and time constraints, into a signature narrative form that often approached the impassioned viewpoint of a crusading but empathic urban preacher. Paul Gallico wrote in his foreword to Cannon's collected *PM* pieces that Cannon "reproduces not only the sounds of the little men, but their dreams and aspirations."

Although his byline was ultimately found on the sports pages of three of New York City's major afternoon dailies, his subject matter ranged far and wide. During the course of his forty-six-year journalism career, Cannon covered murders and street crime, presidential inaugurations, and the trial of the kidnapper of Charles Lindbergh's child. (He also briefly worked in the mid 1930s as a copywriter for an advertising agency.) Before and during the early years of U.S. involvement in World War II (1941–1942), he banged out stories about the barracks and KP lives of soldiers at Fort Dix for *PM* under the heading, "The Sergeant Says. . . ." He then moved overseas, where he filed combat dispatches for the European edition of *Stars and Stripes* while attached to the third army of Gen. George Patton Jr. Less than a decade later, in the early 1950s, Cannon was back at the war front, reporting for *New York Post* readers his behind-the-lines observations from the Korean peninsula.

An odd combination of a late-night, coffee-guzzling socializer who no longer drank and a lifetime bachelor who resided for much of his professional life in the Edison, a Broadway-area hotel, Cannon did not hesitate to bring his personal values and convictions into his writing. Invariably, the soul-bending reflections energized his output, often turning the humdrum into celebration. War experiences in particular tapped some measure of personal responsibility that permeated his writing. Cannon's most compelling works essentially eschewed the profession's dictum that a detached objectivity is the journalist's highest calling; he found it almost constitutionally impossible to distance himself emotionally from his subjects, whether it was a bloodied preliminary fighter struggling to get up from the canvas, a cancer-stricken Babe Ruth, or an exhausted army private behind a mud-encrusted machine gun.

Though he periodically returned to the theme of the hard-pressed GI serving in a foreign land, over the years Cannon remained most attracted and sympathetic to those athletes who struggled to make a living. The level of skill was secondary, whether it was a crippled star such as Mickey Mantle or the ordinary college gridiron player participating in "the great fraud of university football," which Cannon in the 18 November 1951 *New York Post* called "a commemoration of hypocrisy and swindle which fleeces spectators of their senses and transforms them into fools."

He was proud of his eclectic output, and he was not shy about revealing his formula. "A guy shouldn't hang out strictly with baseball writers," he told Holtzman late in his career. "It gets incestuous. Baseball, that's all you talk about. You close the world out, and the world must be let in. The thing I try to do is get the world in over the bleacher walls. That was my whole idea in sportswriting." Cannon, A. J. Liebling noted, "is interested more in motivations than in box scores." Not only did Cannon explore and excavate the soul of his subjects, but he invariably appropriated a piece of their psyche, often as not kneading it into the more polished clay of his columns. His empathy was genuine and extraordinarily personalized; he suffered along with the sufferer. In an anecdote reported in the 17 November 1952 issue of *Time,* Cannon once was asked to accompany a friend to a fancy social gathering. He declined. "As a sportswriter," he said, "I'm only interested in people who sweat."

James Joseph Victor Cannon, the oldest of three sons of Thomas J. and Loretta Monahan Cannon, was born on 10 April 1910 and grew up "down in the unfreaky part of Greenwich Village" in New York City. He attended St. Anthony's school on MacDougal Street, next door to the church in which he was baptized and where family members had been christened and married for generations. The Cannons lived in a cold-water flat several blocks from the Hudson River docks, alongside of which young Cannon learned to swim "because I was shoved off the pier," he recalled in a piece included in *Nobody Asked Me, But . . . : The World of Jimmy Cannon* (1978), "which was the standard procedure used by the coach, who was any bigger kid you turned your back on." Cannon's father, a minor Tammany Hall politician, worked as a clerk for the city. "He believed in what he did," Cannon told Holtzman. "He thought it was disgraceful to take money from a neighbor for doing them a favor. And he spent his whole life doing small favors for people in the neighborhood. . . . He liked all of his kids. He was a very decent man."

The neighborhood and its comfortable quotidian occurrences served as a journalist's primer and springboard for Cannon, teaching him to attune his sensibilities to the nuances and graces, to the behaviors and impulses, of the people who often carried no spare change in the repatched pockets of their work clothes. The resilience and essential goodness of the underdog was a theme that readily took anchor. In "The Not-So-Good Old Days," a bittersweet reminiscence of his youth included in the 1978 collection, he wrote about the bone-honest sturdiness of the elderly coal and ice deliverymen of a time gone by. His recollected portrait of obscure but heroic figures who "spent their old age laboring up tenement stairs with their burdens on their backs or slung over their shoulders" is as vivid and compelling as are his descriptions of the "eccentric supervision" and "unique greatness" of a Yogi Berra or the "swift and awkward" fierceness of a Kid Gavilan in the 16 December 1951 *New York Post.*

An attention to the familiar lives of others was just one source of information. "As long as I can remember I was reading books," he told Holtzman. "My family didn't encourage me. It was my own idea. Some of the people in the family were worried I would ruin my eyes. I used to go to the public library on Leroy Street, in Manhattan. I read the usual things that everybody read in those days – Frank Merriwell, Horatio Alger, *Boys' Life.*" At age fourteen, upon leaving Regis High School after his freshman year, Cannon was hired to work in the classified advertising department at *The New York Evening World.* Less than two years later he landed a job as copyboy at *The New York Daily News* and became hooked on the life of the journalist. Cannon never returned to the classroom. "I worked from midnight to 8 A.M.," he recalled of the undesirable eight-hour slot dubbed the "lobster shift." "I was sorry for anybody else who didn't work these hours because I thought it was the best of all possible lives." Not long after his arrival at the *News,* he encountered an uninvited citizen in the newsroom who insisted he was God. "Pleased to meetcha," Cannon was reported to have replied in the 17 November 1951 article in *Time.* "Heard a lot aboutcha."

With the support of the first of his journalistic mentors, Mark Hellinger, the paper's important Broadway columnist, Cannon was promoted to general assignment reporter. "Hellinger was Jimmy's rabbi," Cannon's youngest brother, Jack, remembered. In what was to prove a typical pattern, except for his uncharacteristically lengthy stay at *The New York Post,* Cannon remained at the *News* for three years, until 1930, when he moved to the *New York World-Telegram,* where he wrote a radio column. In 1935 he was hired to write features for the International News Service, leaving in 1936. Cannon's coverage of the Lindbergh kidnapping trial during this period brought him to the attention of the second of the influential men in his career, Damon Runyon. At the time a columnist for William Randolph Hearst's *New York Journal,* Runyon, impressed by the young writer's flair and "natural style," urged Cannon to write sports and helped him get a sportswriting job at Hearst's *New York American.* "The best way to be a bum and make a living is to be a sportswriter," Runyon advised Cannon in a remark reported in the 12 November 1959 *Newsweek.* Runyon also saw something of himself in the city-spawned journalist who was able to extract unflinchingly a certain color and vivacity from the shadowed and heretofore unreported recesses of society's less influential citizens, as well as its more

*Cannon with Joe DiMaggio*

flamboyant denizens. Before Runyon died he anointed the by-now-successful Cannon "the custodian of my reputation when I'm gone." At the time of Runyon's passing in 1946, Cannon's column "Jimmy Cannon Says:" had begun its nearly fourteen-year run as the feature attraction in the *Post*'s widely read sports section.

He remained grateful to Runyon, whom he classed with Westbrook Pegler and Ben Hecht as "the three greatest writers I ever read in newspapers." But Cannon also insisted on maintaining his own identity. "Maybe subconsciously you could say Damon influenced my writing. But I'm not sure if that's true. I'm not being pompous when I say this, but my style just developed. It's the way I write. Damon never acted as an editor with me. He treated me like a kid brother."

As did Runyon, Cannon liked being around the nightlife and goings-on of the Broadway set. Comedian Joe E. Lewis was a good friend, as were nightclub owner Toots Shor and movie actresses Nina Foch and Joan Blondell, whom he dated for many years. Cannon even moved into Leo Lindemann's old apartment at Sixty-third and Central Park West after the death in 1957 of the famed proprietor of Lindy's, the quintessential Broadway hangout. Cannon also became close to Frank Sinatra. Along with lifelong friendships with such sports figures as Joe Louis and Joe DiMaggio ("My two favorite people in sports"), these show-business connections solidified Cannon's range of source material for the columns, color-packed stories, and memorable phrases to come.

Cannon's time at the *Journal and American,* as it came to be called, furnished fertile territory for the increasingly noticed sportswriter. Writing on 5 December 1937 of the pummeling administered by heavyweight boxer Harry Thomas to his sparring partners, Cannon reported: "So hard were the sock stooges hit and so hard did they fall, that this bystander was certain an earthquake was rocking the hills as the ring shook and creaked and clouds of resin drifted about the dazed marionettes of mace the way dust arises from the wreckage of a razed building." The deftness of one-line summary dismissals, eventually the hallmark of his columns, is evident in this handling of a particularly poor managerial performance by Brooklyn Dodgers manager Burleigh Grimes on 5 May 1938: "Master minds and trapeze artists," Cannon observed almost in passing, "only realize they're in the wrong business when they miss."

World War II proved to be another opportunity for Cannon to sharpen his skills. It also revealed his knack for capturing the limits of human endurance, regardless of the arena of battle. In a 2 January 1945 dispatch for *Stars and Stripes,* datelined "With American Forces," Cannon captures in a few sentences the color, emotions, and facts of awful personal tragedy. The rolling beat and connecting quotations, harbingers of newspaper columns to come, embrace the universality of American soldiers in their struggle and sacrifice.

> Detroit shouted, but all you could hear was a smeared whisper through his bandages.
>
> "I'll never forget that town," he muttered. "I left a piece of my nose in it."
>
> The wounded sat stiffly in the trucks, and they rose tautly when they came to a rut in the frozen road. The dust of the road had made their hair gray, but it did not look strange because their faces were old with suffering and fatigue.
>
> "Give me Paris!" Brooklyn shouted. "Give me Paris and a nice Momma who'll wash and shave me and tuck me to sleep for 9,000,000 years."
>
> They all laughed, but you couldn't tell they were laughing by looking at their faces.

Upon his return from Europe and World War II, Cannon settled into his *New York Post* column and soon gained a loyal following. But he found it difficult to shake the memory of the experiences he had had under severe battle conditions. He especially could not rid himself of the images of American men under the most horrible conditions imaginable. And perhaps he did not want to. "Combat soldiers are the loneliest people in the world," he later wrote in *Nobody Asked Me* (1951). "What a man does in a period of war, he carries around inside him forever." Cannon transferred that verity to his sportswriting.

But Cannon could also be trenchantly frivolous. His post–Korean War pieces in the *Post,* varied both in structure and in subject matter, became famous as much for their muscular insights as for the acerbic wit of a man intent on proclaiming his views to the world. Secure in his columnist's job, Cannon settled into the routine and rhythm of the style of writing that was to become identified with him. During the next decade and a half, his output far outpaced the production and eclecticism of his earlier years. Sometimes, however, he overreached. This is especially apparent in those increasingly rare times when it seemed Cannon needed to pay homage to his now-dead mentor, Runyon. More than just Runyonesque, however, many of Cannon's characters evoked in their chronicler a harsh passion that occasionally missed its mark. When Cannon portrayed a member of the mob who liked to sit at ringside and burn the thighs of nightclub girls with a lighted cigar, Liebling observed that "none of the Cannon gangsters has a heart of gold. And unless a gangster has that, he is a valueless literary property."

Likewise, Cannon's portrayal of women was not infrequently condescending. The hint of the judgmental Puritan, always ready to punch through to the surface in Cannon's published observations on life's vicissitudes, shone most clearly in some of his comments on women, which lack the humanizing ironies, the telling idiosyncrasies, of the men and their world with whom Cannon felt more comfortable. Many of these opinions threaded their way through the columns headed "Nobody Asked Me, But . . .":

> The toughest of women theatrical agents have a tendency to be kittenish (28 November 1951).
>
> Any time I hear a girl described as dreamy-eyed, I figure she needs specs (28 November 1951).
>
> The longer the false eye lash the dumber the girl (5 December 1951).

Another sexist remark is included in the 17 November 1952 article in *Time:*

> Any man is in difficulty if he falls in love with a woman he can't knock down with the first punch.

Cannon also had his alter ego, Two Head Charlie, the bookmaker column character who, along with pals such as The Blotter and Attention Please, made pronouncements on all manner of human endeavor, sports-related and otherwise. Gamblers and gambling were an integral part of Cannon's sports scene, and Two Head Charlie's musings on betting, especially at the racetrack, served both as a vehicle for Cannon to pay homage to Hellinger and Runyon as well as an outlet allowing for some self-indulgent creative writing not otherwise chanced. It was often cutting, though Cannon leavened the effect with typical Broadway humor. In the 25 November 1951 column, Two Head Charlie has gathered with some betting buddies at a Manhattan goulash joint after midnight . A challenge is issued and money is put down – the action is on Two Head Charlie's homely looks. Stitch, his opponent in this makeshift beauty contest, is a "handsome waiter who knits to appease his horse player's nervous system." Two Head Charlie immediately seeks to sway the odds. "He looks like a kangaroo who owes a lot of money," Cannon has him say.

Always in control of the well-trimmed phrase, Cannon could just as readily turn the grace note of a simple sentence into a glistening image. "The hooves of the horses were painted black and shone like a book-maker's shoes on Saturday night," he remarked in a 4 November 1951 column on a New York society horse show, which he called "a snob's rodeo." He also was capable of composing such prose lyrics as "Baseball is music played on an adding machine." Reading Cannon often turned into a multidimensional exercise, wherein he colorized and gave shape to the words on the page. Nor was he averse to writing about himself, typically using smart-alecky tough-guy one-liners, at which he became an acknowledged master.

To get material for his celebrated and much-imitated columns, many of them collected in *Nobody Asked Me, But . . .*, Cannon, an inveterate walker and observer of the New York scene, spent hours perambulating about the familiar neighborhoods and side streets and eateries of his beloved Manhattan. His travels on foot resulted in some of the best one-liners in the newspaper business. They also revealed an incipient cantankerousness in a man who eventually decried the decline of civility in the approach of his younger contemporaries. An increasing melancholia began to creep into more of his columns. The authoritative tone could be querulous: "Why do drug store chemists become infuriated if you mention you're in a hurry?"; amusingly opinionated: "Examine the guys who drink alone at the bar the next time you go into a saloon and I'll bet you'll find the majority are round-shouldered"; or, like the homegrown New York Irishman he was, rhetorically self-effacing: "I don't like Boston because all the men look like me." "If Howard Cosell was a sport," he once wrote, "it would be Roller Derby." The signature series of columns, "Guaranteed to Happen At . . . ," provided still another setting for his sometimes barbed, sometimes sympathetic epigrams. For example, in a column included in *Who Struck John?,* Cannon promised that the following were "Guaranteed to Happen At A Training Camp":

> One of the ball writers will fall in love with a waitress.
>
> The old-timers will tell the young fellows how tough it was years ago when they tried to break in.
>
> There will be endless arguments about how to pitch to Willie Mays, Stan Musial, and Ted Williams.

Cannon's importance, along with that of the two other nationally syndicated sports columnists who dominated the trade during the 1950s and 1960s, Arthur Daley of *The New York Times* and Red Smith of *The New York Herald Tribune,* was both immediate and lasting. "They did not form new schools of sportswriting so much as they inspired the next generation by extending the possibilities of the craft," Robert Lipsyte observed. "Cannon's passion and Smith's wit and Daley's prestige helped make America's second major postwar sports boom seem not only significant and interesting, but somehow psychohistorically inevitable, like manifest destiny." It was a delicious time for New York sports fans, particularly those of baseball. For all the settled diversity of literary choices on the sports page, spread generously among three of the city's major dailies for a range of reading appetites, there was simultaneously through much of the 1950s an uncanny personification of the three major journalists' styles in the center fielders of the city's three major league teams: the Dodgers' Duke Snyder seemed as passionate as Cannon; the Giants' Willie Mays as clever as Smith; and the Yankees' Mickey Mantle as worthy of respect as Daley. The reporter and the reported-upon came to be made for each other.

But Cannon took the parallel a step further. In what has come to be a column hallmark, he neatly and subtly wrapped the reporter and subject into a tight psychological bundle that obscured the reportorial distance between the two. He became a part of the subject but stayed comfortably and safely in the background. He bore into the flesh of the story while appearing to be aloof from the action. It was a defining creation. For millions of readers, the second-person singular was never the same.

> You're Eddie Stanky, a big league second baseman who played on teams that won pennants in three cities. Now

> you're the manager of the St. Louis Cardinals. You always regarded yourself as a manager. Vanity didn't produce this image of yourself. You know what you are. They'll never ship your glove to Cooperstown. Your number won't be retired when you quit. You're a baseball player of queer and inconsequential skills. You're not a Hall of Fame guy. But your capacity for enthusiasm is unlimited. Your ambition is boundless.
>
> You ran slower than the other guy. Your arm was weaker. You were clumsier. You were less agile. But you studied batters and played them right. . . . Behind you, in center field, will be Stan Musial, the greatest ball player in either league. You must envy him because baseball comes natural to him. But, if you ask him, Musial will admit that there is in your inefficient body the desires of greatness. It is, as you have so often said, an intangible quality. But it is as definite as Musial's batting average. You have in you the hope of the genuine champion.

This column, written for the *Post* of 14 December 1951, is, like the others of its kind, deceptively insightful. By turning the roles of the questioner and respondent inside out – in essence, Cannon is interviewing someone who is not present and then has the subject's most personal thoughts and desires come out of the interviewer's mouth – a character profile of unsurpassed dimensionality gradually emerges. Cannon's use of the often-neglected second-person voice, resonant of the classic cadences telling of the bravery of the ancient warrior on the battlefield, ennobled his heroes in the epic manner. The repetitive, staccato refrain, almost accusatory in its frank, soul-excavating insistence – You were, You must, You have – is like a stream of steady, declaratory left jabs. Each blow rubs raw another layer of emotional defenses until the gut-level persona – the "real" person of bumps, bruises, and beauty – is revealed. As always, Cannon is offering advice – and dignity.

The essence of the style's remarkableness lies in the fact that this column has no quotes spliced in to link the descriptions. Transcending its genre, it is like a biographical documentary that makes sense on the screen despite the lack of a narrator to connect the themes and enliven the pace. Or a sculpted work of art with no glue, where only dowels and joints and other artifices are skillfully arranged to hold the showpiece together. The effort is all Cannon, but the portrait is all Stanky.

In many respects, it became hard to tell where Cannon left off and the column began. "He was married to his columns," Dave Anderson asserted in an unpublished interview. "He spent all his time devoted to them." Cannon told Holtzman that "Writing is work for me":

> I hate it. I always hated it. It's a struggle. But I have stretches. Sometimes I'm going good and it flows easily, and sometimes there's a big block there. I can get a column done in two, two and a half hours, but I've labored eight hours on a column, too. Time doesn't seem to improve it, or disfigure it. I can write a story under the gun that is just as good as when I take a long time.

Cannon's habits and sleeping difficulties (insomnia, he said in the 17 November *Time* article, is "culture's greatest ally") made it imperative that he do much of his writing at night. His struggles on his columns showed in the work he turned in, causing at least one colleague to ask how the Western Union operator was able to decipher the heavily edited copy.

He had his own strict rules for interviewing and was critical of those who pursued stories in a manner that he found insulting and ultimately fruitless. Speaking to Holtzman of a younger generation of sportswriters he called "Chipmunks" because of their incessant chatter in the pressbox, Cannon railed at their insensitivity. He was perplexed and personally affronted by what he saw as the self-defeating nature of their approach, simultaneously and characteristically asserting his own experience-emboldened, if not time-honored, method of handling the genre at which he was the acknowledged master – the interview:

> They [the Chipmunks] go up and challenge guys with rude questions. They think they're very big if they walk up to an athlete and insult him with a question. They regard this as a sort of bravery. They risk the chance of ending interviews in the first few minutes.
>
> The trick of interviewing is to start with the easy questions. I usually start with a little small talk. You warm up slowly. Then you drop the bomb, and if the guy doesn't answer, the interview will be almost completed. I save the tough question for last because I don't want an empty notebook.

In the 1960s Cannon, like many others during that decade, felt pushed aside. While critics such as Dave Anderson found his writing to be "tough, but fastidious, because he had found the words himself and he valued them, like a gangster's jewelry," others poked fun at what they saw as the redundancy and thick pastiche of his prose. The reader can "begin Cannon's columns in the middle," one anonymous critic acidly remarked in the 12 January

1959 *Newsweek.* Liebling pronounced Cannon's "daily . . . production . . . uneven," while *Newsweek* called his features "overripe." There was no question that Cannon's sentences could be dense and circular, occasionally repetitive. Sometimes they almost seemed parodies of themselves as they wound their way around and through the paragraph. One example comes from "Rocky Reverse," a boxing column written for the 17 July 1947 paper, during his first full year with the *Post:*

> The song of Indiana came full and big into the body-smelly heat of the arena where the lights hang from the ceiling on long silver-painted stems like metallic and mysterious flowers in an inverted garden. The metropolitan song dying unfinished in a grumbling discord reminded you of the last Zale-Graziano fight when Rocky fell with an unexpected suddenness after punching Tony into the aching blankness where the mind ceases to function and the body operated under the instructions of the subconscious.

Though these sentences clearly breathe life, they nonetheless exhibit a sluggish fattiness that would eventually be squeezed out of the system.

In a 13 December 1954 *Post* piece, "The Beautiful Racket," selected by the Dutton editors as the Best News-Feature Story of the year, the turn of phrase and cadence of rhythm can readily be identified as Cannon's, but it is leaner and more relaxed. And it carries along with it, now more lucid in its fervor, the flavor of the uncynical crusader. Not coincidentally, the subject matter is Cannon's favorite, boxing, what he variously and famously called "the red light district of sports" and "the swill barrel of sports":

> Listen to the boosters' lies. The fight racket is a glorious adventure. It's a force for decency. It frees the impoverished from tenement and hovel. It transports them to great public halls where strangers gladly call their names. Wealth is their reward. Fame is their prize. They're the brave and the graceful and the honored. The hell they are.
>
> They're frightened kids working at a filthy trade. It's not a game at all. It is only a game for a man sitting before his television set or leaning against a bar. It's merely a game for spectators who come whole into an arena and leave unbroken.

The anger and poignancy embodied in such passages reflected a maturity of expression. There are also echoes here of his younger years when he sat on the steps and curbs of his neighborhood and sucked in the grace and deportment of a tougher, but more personally compassionate time.

*Dust jacket for the 1978 collection of Cannon's sports articles*

Cannon's writing encompassed everything that he had to offer. "His column is his whole life," averred writer W. C. Heinz in the 12 January *Newsweek.* "He has no family, no games he plays, no other activities. When he writes, it's the concentration of his whole being. He goes through an emotional wringer. I have no idea what Jimmy would do if he weren't writing that column, he'd be so lonesome."

While the simplicity of his writing was tendered from the bowels of his being, the complexity of the man was reflected in the unbridled confidence that caused some to be wary of his companionship, including Harold Rosenthal, a *New York Herald Tribune* baseball and football reporter who worked alongside Cannon for almost forty years. "I considered Cannon . . . to be a fine word-handler, great on image, and probably the most self-centered writer I met in the business," recalled Rosenthal in a 1994 letter: "He asked once, 'Present company excepted, who do you think is writing the best column in town?' He had few press box friends, but maybe he was respecting others' privacy."

Cannon was not hesitant to criticize other journalists. In a particularly acrimonious display of the gruff patriotism he brought back with him from the war, Cannon, in his typically passionate fashion, lit into the master of iconoclastic, socially conscious invective, H. L. Mencken: "I can locate no fun in Mr. Mencken's cemetery wheezes and there can be little pity in anyone if this crusader for tolerance has been transformed by the years into a macabre clown," Cannon was quoted in the 7 July 1947 *Newsweek,* blasting the Baltimore critic after Mencken had told another *New York Post* columnist, Earl Wilson, "I'm in favor of war and hope it starts soon. . . . The country enjoys war." Ignoring Mencken's reputation for satire and sarcasm, Cannon also overlooked the apparent tongue-in-cheek and continued in the cadences that by then were familiar to his readers. "From his year-rotted platform, Mr. Mencken sounds like the voice of doom playing a kazoo. . . . There is no place in our world for a philosophy which is a subpoena to the grave. . . . Mr. Mencken has lost his identity and enlisted with all the other cranky men of the world. It is always the old men who make the war where the young men die."

Despite the genuine anger toward Mencken, Cannon would not countenance criticism of the only profession he had ever known. "Newspaper people are special to me," he said after praising Mencken to Holtzman and comparing him to Damon Runyon, "probably because I'm a newspaperman. Maybe if I was a plumber I'd talk about plumbers." Good writing was Cannon's passion, particularly when done by a newspaperman, and mere political disagreement could not obscure that fact. Twenty-five years after his outburst Cannon told Holtzman, "Of all writers maybe I revere Mencken the most. He influenced my generation more than anybody influenced any generation since I've been alive. Mencken taught a whole generation how to think, and one of the reasons I adored him was because he was a newspaperman. . . . Mencken wrote awfully good stuff." Cannon also had high regard for Chicago's Nelson Algren, calling him "the most neglected writer in this country," and Ernest Hemingway, of whom Cannon said, "I don't think he ever wrote a bad sentence."

Cannon's years with the *Post* solidified his reputation. But the lure of a higher salary and an increased readership returned him in 1959 in the midst of a newspaper strike to the afternoon rival *Journal-American,* whose own star columnist, Bill Corum, had recently died and where Cannon had begun to make a name for himself more than two decades earlier. In 1966 the *Journal-American* merged with the *Herald Tribune* and the *World-Telegram and Sun.* Cannon wrote for the resultant *World Journal Tribune* until 1967, when the paper ceased publishing. Although his column was picked up by the Hearst Headline Service and nationally distributed, it did not appear in New York again until he was rehired by the *Post* in 1972.

The last decade or so of Cannon's life found him struggling to make sense of the new wave of sportswriters who were increasingly gaining attention. While Cannon felt genuinely comradely toward his younger (and older) colleagues in the pressbox, he was bewildered by their attitude, style of writing, and apparent irreverence toward those who toiled below on the fields of play that Cannon had come to see as his own front yard. Writers such as Pete Hamill and Jack Newfield, respectful of Cannon's influence on journalists like Gay Talese, Jimmy Breslin, Tom Wolfe, and Peter Maas, as well as themselves, attempted to credit Cannon as the forerunner of the New Journalism of the 1960s and 1970s. Newspapers and magazines were filled with the colorful style, an eclectic approach that was extremely subjective and assertive both in its manner of information gathering as well as its idiosyncratic presentation on the printed page. Cannon angrily disavowed their intentions. He objected in particular to what he alleged was the New Journalists' practice of making up the truth. It violated, he said, his basic sense of reportorial fairness. "They invent action," he told Holtzman. "When I was a kid we used to call it faking and piping, smoking the pipe, opium smoking. . . . Then it stops being journalism with me. It's a betrayal of your readers." In contrast, Ring Lardner, whom Cannon placed among "the ten greatest serious writers produced by this country," was "ferociously accurate."

Cannon always felt a responsibility to his readers and became increasingly isolated and unhappy when they no longer were available to him in the crucible of his influence and fame – New York City. The times and the demographics of his readership were definitely changing. The new style of the day – questions harshly hurled at a contentious breed of television-oriented athletes from the middle of a pack of hungry, obnoxious journalists crammed together inside a sweaty locker room – dismayed him. Still, he wanted to be generous. "I think sportswriting today, as a general rule, is better than it was in my generation," he told Holtzman. "At the top, maybe, we were better. But the general

*Head for Cannon's column in* The New York Post

sportswriter is up. There are better sportswriters today. But we had a couple of flashier guys."

In May 1971, while getting ready to attend the Kentucky Derby, Cannon was paralyzed by a stroke. Unable to call for help, he was found on the floor by a cleaning woman. On the walls were autographed photos of Louis, DiMaggio, Sugar Ray Robinson, and Hemingway, among many others.

Confined to a wheelchair for the next year and a half, Cannon was visited on a regular basis by Newfield and Hamill, two brash practitioners of the fresh, in-your-face style of journalism against which Cannon had so bitterly protested. Newfield, who during these calls tried unsuccessfully to change Cannon's opinion of the writers of the 1960s, remembered Cannon's reply when asked what he was thinking as he lay helpless for more than two days before someone came upon him. "I felt like Billy Conn knocked to the canvas by Joe Louis," he told Newfield.

In 1972 Cannon began writing again for the *Post.* His pieces were back in the homes of his New York readers. But the columns never gained the strength of his earlier efforts. On 5 December 1973, living alone in his apartment on East Fifty-sixth Street in Manhattan, Cannon suffered a fatal relapse. He was sixty-three and had stopped writing his column two months before.

In a wistful salute to his fallen colleague, Red Smith gently noted in his own column two days after Cannon's death that he and "another champion," the legendary thoroughbred Count Fleet, had passed on within hours of one another. "[N]o disrespect to Jim," Smith waved in farewell.

Despite his spirited disavowal of the journalistic accolades bestowed upon him by a younger and more sophisticated generation of sportswriters, Cannon's accumulated writings, especially the columns, mark him as a distinctive, compassionate, and involved presenter of the world he witnessed. He was unrestrained in his sympathetic and ultimately sentimental portrayals of sports and other human endeavors. Much more than an observer, he wrote, and said, what he felt. He created and fleshed out a style and approach that is as assuredly a model for those to follow as Runyon's and Hellinger's had been for him.

Cannon's own cynicism, the Irish-Catholic shield standing sentinel over a tender and mournful soul, denied him the pleasures of understanding, much less accepting, his lasting place in the world of twentieth-century newspaper journalism in America. For a journalist whose primary beat was sports, he is rare in his influence beyond what he and others disparagingly have called the "toy department." Jack Kroll, the *Newsweek* book critic, praised Cannon's columns as "verbal Irish coffee of toughness laced with sentimentality, that Hemingway rhythm of stoic lyricism. It was Cannon who brought the Hemingway tone into sports writing, crossing it with Damon Runyon. And behind the mask of the New Journalism, you can still detect the old Cannon blend of hard nose and soft heart."

According to veteran *New York Times* columnist Anderson, who worked and later traveled with Cannon from 1958 to 1966, "Cannon and Red Smith were 1 and 1-A, in whichever order you want to put them." Columnist Maury Allen, one of the original Chipmunks, asserts on the book jacket for *Nobody Asked Me, But . . .* that Cannon "revolutionized sportswriting and made it noble. His style in print and his style in life will never come again. . . . He was the Joe DiMaggio of sportswriters." Tom Cannon, who, along with brother Jack, edited the posthumous collection of Cannon's work, finds it important to remember the urban soil in which their reporter brother's sensibility took flower. "He was the first street writer," Tom said in an unpublished interview. "He wrote about the men and women on the stoops."

Cannon never truly left the neighborhood behind, taking with him what he learned from the sidewalks and working people of his youth. He was perhaps the first sportswriter to display publicly his social consciousness; certainly he was among the first to be so outspoken about certain ills he encountered along his beat. Writing of Jackie Robinson's induction into the Baseball Hall of Fame in 1962, "a journey Robinson made all by himself," Cannon said he respected no one in sports more than the barrier-busting second baseman of the Brooklyn Dodgers. "He was a great player and belongs with the elite of the sport."

Cannon's comment about Joe Louis – "He's a credit to his race – the human race" – may seem trite now, even condescending. Still, it is embedded as a journalistic monument on the road to equality in athletic competition. But Cannon always denied he was anything other than a newspaperman, just doing his job: "The people who write the pieces up front in the paper consider this [sportswriting] a wasted life," he was quoted in the 6 December 1973 *New York Times*. "It's true that I have solved little of my country's dilemmas, but the statesmen also have failed. I like my life because it's a pleasant life. I function in glad places." In an uncharacteristically self-effacing moment, Cannon elaborated his passionate, unadorned approach and the purpose with which he accomplished and communicated it. "My beat is those spots to which people have come to be entertained, where the violence and dissent is all part of the joy. I've witnessed executions, sat around police courts and union halls, covered strikes and wars. But people have gathered there for somber reasons." He liked his life as a columnist because "a columnist is always permitted to get off his chest whatever is bothering him." He also told Holtzman:

> I have never felt any envy toward the ball player. . . . I never felt bad about not being a good athlete. I would much rather have been Ernest Hemingway than Babe Ruth or Red Grange. Anyway, I've never been a good team man. I don't like partners. I've gone my way by myself, and that's how it figures.

Cannon was unapologetically a devoted, passionate newspaperman. While sports may have provided the context, human endeavor, with all its vicissitudes and overarching pleasures and pains, ultimately defined the boundaries of his columns. A true child of the Big Apple – from the tenements of the family-centered laboring class to the gaudy Great White Way; from the Polo Grounds to Madison Square Garden – Cannon was endowed with the individualist's kind of misty-eyed clearheadness, underpinned by a firm determination to keep everything in perspective. In his own way, and by his own lights, he remained faithful to that vision.

**Interview:**

Jerome Holtzman, "Jimmy Cannon," in *No Cheering in the Press Box,* edited by Holtzman (New York: Holt, Rinehart & Winston, 1974), pp. 273–287.

**References:**

"Assault on Mencken," *Newsweek,* 30 (7 July 1947): 94, 97;

"Broadway Minstrel," *Time,* 40 (17 November 1952): 86–88;

Peter Golenbock, "Brooklyn Against the South," in *Bums – An Oral History of the Brooklyn Dodgers* (New York: Putnam, 1984), pp. 222–224;

A. J. Liebling, "Free-Swinging Stylist," *New Yorker,* 27 (5 May 1951): 108–110;

Robert Lipsyte, *Sportsworld – An American Dreamland* (New York: Quadrangle/New York Times Books, 1975), pp. 6–8, 25, 75–76, 105, 175–181;

"Sportswriter's Rainbow," *Newsweek,* 53 (12 January 1959): 62, 64.

# Gordon Cobbledick

*(31 December 1898 – 2 October 1969)*

James E. Odenkirk
*Arizona State University*

MAJOR POSITIONS HELD: Police reporter (1924–1927), sportswriter (1927–1944), war correspondent (1944–1945), editorial columnist (1945–1946), sports editor (1946–1964), *Cleveland Plain Dealer;* editorial writer, *Cleveland Times* (1927).

BOOK: *Don't Knock the Rock: The Rocky Colavito Story* (Cleveland: World, 1966).

OTHER: "V-E Day – Jap Steel on Okinawa Mowed Down Plenty of Americans," in *A Treasury of Great Reporting,* second edition, edited by Louis Leo Snyder and Richard B. Morris (New York: Simon & Schuster, 1962).

SELECTED PERIODICAL PUBLICATIONS – UNCOLLECTED: "Break Up the Yankees," *Collier's,* 105 (25 February 1939): 19, 62–63;

"Game Guys," *Collier's,* 107 (19 April 1941): 14, 45–48;

"So You Want to Run a Ball Club?," *Saturday Evening Post,* 21 (23 April 1949): 32–33, 141–144, 146;

"Twelfth Man in the Huddle," *Collier's,* 124 (12 November 1949): 20, 66–67;

"Happiest Kid on Earth," *American Magazine,* 158 (September 1952): 42–44, 80–84;

"Column of the Week," *Sports Illustrated,* 1 (16 August 1954): 122–123.

Gordon Cobbledick possessed two qualities necessary for a good columnist, a great sense of justice and a feeling for good writing. Cobbledick covered a Clarence Darrow lecture shortly after the famous lawyer made headlines at the unforgettable Scopes "monkey" trial in 1925. After Darrow gave a poorly organized speech, Cobbledick (affectionately called "Cobby") virtually rewrote the lecture. As he recalled in his final column in the 16 August 1964 *Cleveland Plain Dealer,* the paper for which he first wrote forty years before, he later received a letter from the grateful Darrow, thanking Cobbledick "for putting my rambling remarks into intelligible English." Early in Cobbledick's career as copy editor for the sports department of the *Plain Dealer,* he possessed a knack for substituting a superior word or phrase in the articles of his staff. Later the executive editor of the *Plain Dealer,* Philip Porter, told Cobbledick he was one of three journalists on the paper who wrote perfect English. Nathaniel R. Howard, editor of the defunct *Cleveland News,* wrote in the 24 September 1945 edition of the paper, "Good writers on sport topics apparently make good writers on anything. Ring Lardner, Charlie Van Loan, Octavus Roy Cohen, and Damon Runyon ran their bent into literature, John Kieran and Bill Cunningham wrote absolutely about anything which interested their fancy, Westbrook Pegler turned reformer and has done excellently. All these, and Cobbledick, had one thing in common; they itched, originally, for a typewriter to set down who won and who lost."

Cobbledick was born in Cleveland Heights on 31 December 1898 to William and Nina Cobbledick, formerly of Coffeyville, Kansas. William Cobbledick owned a successful butcher shop in Cleveland Heights. Gordon had two brothers, Stanley E. and Ralph. From grammar school he attended East Technical High School in Cleveland (where Jesse Owens graduated) and was a talented football player. Immediately after graduation in 1918, Cobbledick enlisted in the U.S. Marines. He demonstrated his ability as a sharpshooter, and the marines assigned him to Parris Island, South Carolina, to teach recruits the finer elements of marksmanship.

Upon his discharge in 1919, Cobbledick enrolled at Case School of Applied Science (now Case Western Reserve University), where he studied mining engineering and played football. In 1922 Cobbledick discontinued his studies at Case Tech to pursue employment. His fortunes turned upward in 1923 when he stopped at the *Plain Dealer* offices and was offered a job as police reporter for twenty-

*Gordon Cobbledick*

five dollars a week, enabling him to marry his high-school sweetheart, Doris Matthews, later that year. The Cobbledicks had two sons, William and Dorn.

In 1927, enticed by a five-dollar-a-week increase (from fifty to fifty-five dollars), Cobbledick joined a competing newspaper, the *Times,* where he wrote editorials and book reviews, among other items. Because of the precarious financial status of the *Times,* after five months Cobbledick returned to the *Plain Dealer.* A "slot man" (a person who filled in on sports events not covered by the other reporters) was needed in the sports department, and the fledgling reporter was assigned to this position.

In his 21 August 1964 article for the *Plain Dealer* Cobbledick recalled that he never intended to become a journalist, but he remarked that "every stumble on my walk through life has been a fortunate one." He was certainly fortunate in late 1927 when the Cleveland Indians baseball club was sold to local businessmen, and general manager Ernest S. Barnard was appointed president of the American League, replacing the ailing Ban Johnson. Barnard moved league headquarters to Chicago and selected Henry Edwards, then the baseball writer for the *Plain Dealer,* to supervise a newly established league service bureau. Cobbledick replaced Edwards and began a sixteen-year stint of covering the Cleveland Indians, riding the rails and gaining the respect and friendship of players and journalists alike. Recalling the sleeper jumps from Boston to Saint Louis, the recurring rush to catch five o'clock trains after three o'clock games, the spring training barnstorming trips through the South that lasted as many as twenty consecutive nights, Cobbledick admitted a preference for the convenience of air travel while recognizing that flying deterred friendships between players and writers. "There was no wandering into a club car for a modest pre-dinner libation. No leisurely dining car meal and no cards, chess or checkers, no swapping of atrocious lies. There was much to be said for rail travel," he wrote on 3 September 1963.

Cobbledick was a gifted raconteur. One of the stories he loved to tell, recounted four days after Cobbledick's death in the 6 October 1969 *Plain Dealer,* concerned Stuart Bell, a colorful, unpredictable journalist who covered baseball for the *Cleveland Press* in the 1930s. On a road trip in late May 1935 a controversy erupted between Indian manager Walter Johnson and highly regarded third baseman Willie Kamm. Aware that Bell had not been seen in several days, Cobbledick began to write the story. "Great stuff," the sports desk wired. "I kept pumping it out," Cobbledick reportedly recounted, "until past midnight. After I wrote all I knew, I started all over again writing for the Cleveland *Press* until four in the morning." To protect his friend, Cobbledick sent a second story, with Bell's byline, on the Johnson-Kamm imbroglio to the *Press*. The next day, Bell finally showed up and strutted up to Cobbledick. "Thanks pal," he said. "You saved my job and look what I got." Bell's telegram, signed by the editor of his newspaper, read: "Dear Stu, I have always considered you one of America's great sportswriters. Your work last night proves it."

Despite being a perpetual New York Yankee hater, Cobbledick admired Lou Gehrig as a player and as a human being. When he wrote a column on 17 July 1937 suggesting that Gehrig was more important than Babe Ruth to the New York Yankees, Gehrig responded with a short note thanking Cobbledick for his kind words. Of all the great athletes Cobbledick wrote about, Gehrig was the only one who wrote a letter of thanks to him. His column on 3 May 1939 detailed Gehrig's great career one day after the first baseman pulled himself from the Yankee lineup after playing a record-setting 2,130 consecutive games. "Gehrig leaves behind baseball's greatest record of grit, fortitude and stubborn courage," Cobbledick wrote. "Concerning his consecutive game performance, Gehrig says the conventional thing, that he has been lucky. . . . He has escaped broken legs and broken arms, but he has suffered nearly every other injury in the baseball catalog and kept going. And that isn't luck." Cobbledick argued that Gehrig's record properly indicated his greatness: "Records aren't always an accurate index to a player's value, but in Gehrig's case they are. No one can justly accuse him of having spared himself in order to keep his streak intact. He has played as hard, every day . . . and he has played not for Henry Louis Gehrig, but for the New York Yankees." In 1952 Cobbledick suggested to a longtime journalism colleague that the Phi Delta Theta fraternity (Gehrig's fraternity at Columbia University) establish an award to be given each year to the major league player who best exemplified the attributes Gehrig possessed. The prestigious Lou Gehrig Memorial Award was created in 1953 and has been awarded yearly since its inception.

In a 14 May 1939 column Cobbledick insightfully forecast how quickly and thoroughly night baseball would be accepted in the American League: "At 9 P.M. next Tuesday [at Shibe Park, Philadelphia] stumpy little Roy Weatherly will be the first American Leaguer in history to go to bat in a night baseball game. He'll be the first – but thousands will follow him. For night baseball has come to stay. Within ten years you'll be going to the ballpark on a summer evening as casually as you go to the movies." Noting the advent of night baseball in a minor league game in Des Moines, Iowa, and the first National League night game in Cincinnati in 1935, Cobbledick asserted that "night ball saved most minor leagues from extinction and since the majors could no more operate without the minors than a skyscraper can stand without a foundation, it saved the big league, too." He understood that "Night ball means larger crowds, larger crowds mean more money, and more money is what baseball magnates can't get enough of."

Cobbledick's affection for baseball was evident in a column published on 3 August 1939. He pointed out that the chief difference between baseball and other games is that baseball is the one team game in which every phase of individual and team performance can be reduced to figures and measured statistically. While he did not contend that baseball statistics supply an infallible measure of the worth of a player or the strength of a team, he argued that, more than any other single factor, the statistical measurements of players provided the fuel that keeps alive "hot stove" league (that is, baseball talk, mostly during the off season). Because baseball could be talked about as well as played, Cobbledick believed it to be the preeminent American pastime.

Cobbledick's fairness in reporting was well known. Athletes confided in him, knowing he would not betray their trust. The respect between Cobbledick and the Indian players was perhaps most evident when the team rebelled against their tyrannical manager, Oscar Vitt, in 1940. While Cobbledick was covering Cleveland on a road trip in Boston, an anonymous Indian player told him of the team's plans, allowing him to scoop his rival writers. On 15 June, the day after Cobbledick's story was printed in the *Plain Dealer,* he walked into the press box, where the other writers greeted their colleague with envious catcalls and epithets. "I'd rather have heard [the catcalls] than won the Pulitzer Prize," Cobbledick said. The story earned Cobbledick one of seven awards he received from the Cleveland Newspaper Guild.

Before and after World War II Cobbledick contributed articles to *Collier's, The Saturday Evening Post, Sport, Saga, Sports Illustrated,* and *American Magazine.* His first magazine article, "Break Up the Yankees," appeared in the 25 February 1939

*Collier's*. Although Cobbledick bemoaned the superior New York baseball organization that had won three consecutive pennants and World Series championships, his fondest dream went unfulfilled, and the Yankees won the championship again in 1939. Cobbledick's magazine pieces were often on subjects auxiliary to baseball. A representative article is "Happiest Kid on Earth," which Cobbledick wrote for the September 1952 *American*. The subject of the article was sixteen-year-old Bill Domhoff, who became batboy for the Indians, "the most exciting, soul-satisfying job in all the world," after winning the annual essay writing contest used to determine which boy would earn that position. Reflecting American ideology and culture of the era, Cobbledick's article touts character building and individual initiative. After failing three times to win the contest, Domhoff took an advanced journalism course to help him with "imperfect expression." He then "advanced to the attack once more. After much chewing of pencil ends and frantic rumpling of hair he produced a masterpiece." Cobbledick also wrote a column on baseball for the first issue of *Sports Illustrated* on 16 August 1954.

Cobbledick served as baseball correspondent to *The Sporting News* in the 1930s and 1940s, and he was elected president of the Baseball Writers Association of America in 1942. In late 1944 as World War II moved toward a conclusion, Cobbledick's longtime eagerness to serve as a war correspondent intensified. Sent to the South Pacific by the *Plain Dealer,* he covered the war until a few weeks short of V-J Day. Cobbledick's account of the reaction to V-E Day by troops involved in the raging Pacific war was picked up by the Associated Press. Howard referred to Cobbledick's V-E Day story and his writing skills in his 24 September 1945 column in the *Cleveland News:*

> I was in San Francisco when Cobbledick told of the bloodshed and heartache with which V-E Day had been celebrated on Okinawa. The San Francisco *Examiner,* which never heard of Cobbledick, had flung this epic, under his name, right across the top of the front page and I thought of Cobbledick's dissatisfaction with anything not nicely and pointedly written.

The column was published on front pages all over the United States and later selected for publication in the book *A Treasury of Great Reporting* (1962). Virtually every anthology of war reporting includes this biting essay.

Cobbledick was one of several esteemed sportswriters who covered World War II, including Shirley Povich of the *Washington Post;* Ring Lardner's son John; Bob "Blackie" Sherrod of the *Saturday Evening Post;* and Bob Ruark of Scripps-Howard. Among the war correspondents with whom Cobbledick associated, the most noted was syndicated columnist Ernie Pyle. The two newspapermen sat up all night just prior to Pyle's departure for Le Shima, sharing stories and a jug of wine given to them by an admiral. A few days later Pyle was killed by a Japanese sniper.

Cobbledick returned from the Pacific theater to write a provocative column, titled "Plain Dealing." Placed in a corner adjacent to the editorial page, the column's title indicated Cobbledick's blunt tone. After spending so many years as a baseball writer, Cobbledick was eager to do a different kind of writing. He enjoyed the world around him and wrote about politics, military affairs, labor issues, international problems, personalities of note, and personal experiences. Cobbledick was pleased when his sometimes controversial columns brought rejoinders from his readers. His logical mind made him a capable pundit, and he seemed happily prepared to spend the remainder of his professional career in his new assignment.

In 1946, however, *Plain Dealer* sports editor Sam Otis died suddenly. As Cobbledick recalled in the 16 August 1964 issue of the newspaper, editor Paul Bellamy called Cobbledick into his office and said, "You're the new sports editor." "No, I'm not," Cobbledick replied. "Yes, you are," declared the editor. And that was the end of the discussion. Cobbledick continued to write his column "Plain Dealing," one of the most popular features of the newspaper, which appeared six days a week. One column was reserved for letters from his readers and his tongue-in-cheek rebuttals. Cobbledick followed the fortunes of Cleveland's professional sports franchises, especially the Indians and a young professional football team called the Browns. With the success of these teams in the late 1940s, Cleveland unabashedly touted itself as the "City of Champions."

Several times in columns Cobbledick posed the question of why American literature lacked a critical novel about a baseball player, the works of Ring Lardner, Heywood Broun, and Charles Van Loan notwithstanding. Cobbledick wanted a serious study of a ballplayer as a person. Although Lardner portrayed baseball in a humorous vein, Cobbledick believed there was nothing funny about the game to the millions of fans who jammed their way into ballparks between April and October. "Why has no writer ever captured the thrills in fiction?" inquired Cobbledick in a 26 April 1946 column. "Baseball, as a purely American phenomenon

# Plain Dealing—

## Wynn Needs Eight to Join 300-Club anc Here's Hope He Gets 'Em

GORDON COBBLEDICK

By GORDON COBBLEDICK

Plain Dealer Sports Editor

I see here in the paper where it says an old chum, Early Wynn, has signed to pitch another season of baseball for the Chicago White Sox. It is a club toward which I feel no sentimental allegiance, and yet I shall be a White Sox rooter on those days when old Gus is pitching against the Yankees, the Red Sox, the Senators, the Tigers, the A's, the Orioles, the Twins and the Angels.

It may even come to pass that I shall experience a twinge of regret on the days, if such there be, when the Indians beat him.

Wynn is 42, an age at which most professional baseball players, excepting those named Leroy Paige, have long since retired to their bowling alleys or taverns.

### Last Eight Will Be Toughest

He is credited with 292 victories, notwithstanding that seven seasons of a career that has touched four decades — the 1930s, the '40s, the '50s and the '60s—were spent in the employ of one of the league's weaker teams, Washington, and one in the military service.

He needs to win only eight more games to become the 13th pitcher in major league history to rack up 300 victories. Like the final eight yards in a touchdown drive, those last eight winning games probably will be the toughest of his life to achieve.

And yet who's to say that the fiery competitor who won 22 for the White Sox at the age of 39 won't do it with something to spare?

### Mean Man on Mound

Yes, Early Wynn is a fiery competitor. A genial and even a gentle man off the field, he is known as the meanest of pitchers. He glowers wickedly at opposing batsmen. While knock-'em-down philosophy of pitching as the only avenue success.

But I have always believed that he talked, deliberatel a far meaner game than he pitched.

I recall a breakfast with him in the Tampa Terrace Hot one morning of his second spring training season with t White Sox, to whom he had been sent, along with Al Smit in one of Frank Lane's earliest trades as general manag of the Indians.

### Teaches Enemy Manners

Wynn was talking proudly about his strapping son, 1 year-old Joe Early, in whom he saw unusual promise as slugger.

"I pitched to him mornings out at the Sox park la summer," he said, "and you know something? He hit a cou of shots up against the wall in left center that Mantle wou have been proud of."

"Wonderful!" I exclaimed. "What did you do then?"

Wynn looked his disgust at what he obviously consider a foolish question.

"What could I do?" he asked. "I knocked him on duff."

In 1949, Early's first season with the Indians, there w still some residual resentment of Larry Doby, the Americ League's first Negro player. A Detroit pitcher knocked h down twice with fast balls that barely missed his head. Wh the pitcher came to bat Wynn, an Alabama boy, sent h sprawling four times.

"Doby's on my team," he explained later. "If they hu him they hurt me. I gotta teach 'em their manners."

Now I gotta pull for him to register those eight v tories he needs—mostly over the Yankees.

*A Cobbledick column (1962) on Early Wynn, who pitched for the Cleveland Indians from 1949 to 1957 and returned to the team in 1963 to record his three hundredth career victory*

and a social force of no small magnitude rates a serious novel." Cobbledick pleaded for a novel that contained all the elements that make up a major league ballplayer's life.

Cobbledick's articles were published in the E. P. Dutton *Best Sports Stories* annual on several occasions. His first appearance was a 21 September 1948 column he wrote about the Indians' Hall of Fame pitcher Bob Feller. He defended the veteran pitcher against the fans and reporters who, sometimes viciously, "turned on him when he began to lose ball games" and claimed that Feller was "as good a pitcher as ever he has been in his life, which is nearly as good as pitchers ever get." But Cobbledick did not dwell on the remarkable pitching ability of Feller, who had begun to "Hit his stride when the Indians needed him most"; instead, he focused on Feller's will and character, his perseverance through adversity and disloyalty without "any sign of impatience" or "petulant antics." He concluded that Feller was "an admirable young man who demonstrated the stuff that's in him when the going was about as rough as it gets for an athlete of his attainments."

Cobbledick's writing sometimes brought his audience uncomfortably close. On one occasion he created a satiric interview with a fictional baseball fanatic he named Joe Schmaltz. The day after the column appeared, a surly drunk drove to the Cobbledick home and demanded to see the editor. He said his name was Joe Schmaltz and his barroom buddies were making fun of him because they thought he was the man to whom Cobbledick had referred in his interview. Cobbledick's wife was frightened – the man looked mean – and she called her husband outside. The tall, erect Cobbledick took Schmaltz by the arm and talked to him calmly, assuring him that he was not the subject of his satire. Just before the Indians' appearance in the 1954 World Series, Cobbledick wrote in a 16 September column about "friends" who tried to "butter up" to him for tickets. These friends included those "who never met you but they had a brother-in-law who heard you make a speech to the Kiwanis Club of Lower Whichburg back in '37 and he'd sure get a kick out of it if they could tell him you'd fix 'em up with seats that Eisenhower himself couldn't of bought."

Cobbledick brought a high standard of professionalism to the sports department of the *Plain Dealer*. He understood that reporters were not cheerleaders, and during his tenure the *Plain Dealer* sports pages were free of sophomoric yea-saying. A reporter was not a shill for general managers and team owners; he gave the readers the facts. This policy reflected Cobbledick's personal integrity and belief in objective reporting. Cobbledick implored his staffers to keep their articles short, and he always set an example himself. Rarely would his columns be longer than two double-spaced typewritten pages.

In his own column, of course, Cobbledick was free to criticize or cheer as he pleased, but readers often mistook his tongue-in-cheek satire for pure venom. This was especially true of his golfing readers, whom he needled periodically for calling their diversion a sport. Cobbledick's talent for satire took his writing just a bit over the heads of some gullible sports-page readers. Even though he vowed to abandon this style, the devil in him kept overcoming this resolve. Often one would not suspect that those he needled were really objects of his affection. Cobbledick delighted in looking – and writing – like a grumpy old man, but his wit always gave him away.

Each year at Christmas and on New Year's Day, Cobbledick wrote columns in which he vowed to extend an olive branch to all his critics, "lubricating his typewriter with honey and becoming a sweet kid." "This is the year I'm going to do things differently," he wrote on 1 January 1953. "No more talking back to people. No more quarreling in print. No more pretending I'm the only guy who knows all the answers. A new man – that's what I'm going to be." Cobbledick detailed his projected change in behavior. "This year I'm going to be tolerant of the insufferable – I mean the high-spirited citizens – who go to football games for the express purpose of getting drunk. Boys will be boys, and so will some grownups, I guess." Cobbledick suggested, "I may even try to recapture the fondness I used to feel for basketball when a score like 26–24 meant that the defenses broke down" and added that "this year I'm going to make an honest effort to get excited about amateur sports – if I can find any that are really amateur." His column concluded with yet more good intentions:

> And that reminds me of something. In response to an overwhelming public demand I'm going to stop trying to make out that [Honus] Wagner was a good shortstop when everybody knows that the only good shortstop who ever lived was Lou Boudreau. One thing I certainly will not do . . . is refer to members of the television audience as free loaders – and not just because every such reference brings a flood of letters beginning: "Dear Sir: You bum, when did you ever pay your way into a ball game?" Which brings to mind another thing I certainly must do without delay. I must take an ax to my own television set and see if I can't rediscover the delights of reading books.

Cobbledick's insight into the sports world was especially evident in his understanding of the changes occurring in big-time intercollegiate athletics. He recognized the truth of the charge made by Frederick W. Cozens and Florence Stumpf in *Sports in American Life* (1953): "Colleges and universities in the process of fostering intercollegiate athletics have wittingly or unwittingly become engaged in the field of professional entertainment." Cobbledick often commented on college sports scandals. As early as 25 July 1946, he made his position clear: "To put 80,000 or more fans into a stadium several times each fall at $4.40 each is big business. It is commercialized sport on a scale calculated to arouse the envy of a professional baseball magnate or a boxing promoter." Like some critics of intercollegiate athletics, Cobbledick did not advocate the abolition of the programs. He simply argued that college presidents should be responsible for their athletic programs and that universities should be obligated to provide an education to their athletes.

In his 13 November 1962 column, as Cobbledick approached the end of his journalistic career, he continued to plead for a sane approach to intercollegiate athletics. His unhappy observation of trends evident in big-time collegiate football had been consistent and was best summarized by his comment concerning a college president who was fighting a losing battle against the commercialism that engulfed his school's athletic program. The president observed that he saw no more reason why he should present a football team for the entertainment of the public than train a chorus line from among the young women students to dance in night clubs. "He was a voice crying in the wilderness," Cobbledick wrote approvingly. But he observed that the president's "hucksters are still selling tickets like mad, his vast stadium is filled four or five times every autumn, and the public is still making its influence felt in fields which are properly the sole concern of the college administration." Cobbledick argued that football was best represented in smaller colleges. "It's heartening to know that it's still fun in Oberlin, Amherst, and other places where the meaning of football hasn't been lost sight of. It was fun, not show business."

One of Cobbledick's gifts was the ability to recreate vignettes and interesting stories about sporting personalities of years gone by, particularly in baseball. His research was thorough, and he retained a vast reservoir of information. He wrote frequently of players who had passed from the scene but who had made significant contributions to the lore of baseball. One of many such stories concerned Raymond "Tris" Speaker, Hall of Fame centerfielder and manager of the World Series–winning Cleveland Indians in 1920, on whom Cobbledick had written a highly regarded article in the July 1952 issue of *Sport.* Cobbledick also attended Speaker's funeral in December 1958 in the small plains town of Hubbard, Texas, and wrote a column on him in the 12 February 1960 issue of the *Plain Dealer:*

> The mortal remains of baseball immortal Tris Speaker were interred . . . on a little cedar-studded hilltop overlooking the limitless sweep of the Texas plains he loved. This is one of the two worlds in which the Gray Eagle lived his seventy years. Though [Speaker] was a big city man, an Easterner for more than half a century, he never lost contact with the town where he was born on 4 April 1888, or with its people.

Cobbledick noted that condolences were received from Ty Cobb, Speaker's principal rival as a player, and Ray Schalk, the great Hall of Fame catcher of the Chicago White Sox. Cobbledick understandably had a special attachment to many members of the Cleveland baseball franchise, but whether he wrote about the Cleveland Indians or players from other organizations, his fairness, compassion, and understanding were always discernible.

Many of Cobbledick's columns focused on the never-ending debate over who was the greatest ballplayer of all time or who did or did not deserve admission to baseball's Hall of Fame. Cobbledick did not agree with polls of sportswriters and broadcasters who named George Herman "Babe" Ruth as the greatest player of the twentieth century. In Cobbledick's opinion Ty Cobb was the greatest player. Acknowledging that Ruth was a superlative home run hitter, Cobbledick maintained that in every other category, with the possible exception of throwing, Cobb was superior to Ruth. In his 12 February 1960 column he wrote that Ruth was "as nearly a completely natural, completely instinctive ballplayer as ever lived. He needed no more teaching than a duckling requires in learning to swim." But he argued that Cobb was "the greatest to ever cut the legs from under a defensive player." Citing Ruth's career batting average of .342 and Cobb's average of .367, Cobbledick reminded readers that nearly half of Cobb's career was played during the dead ball era, whereas Ruth compiled his batting average "in the era of the lively ball when .300 hitters grew wild on every bush."

One of Cobbledick's most admirable attributes was his fairness toward minorities. This attitude was tested severely, as Cobbledick, who occasionally received letters addressed "Nigger Lover Cobbledick," noted in a 26 November 1963 column, written four days after the assassination of President John F. Kennedy. The media, he asserted, filled the airways and printed pages with consternation over the "hate element" that snuffed out human lives. " 'Love conquers all,' said Virgil, and maybe he was right," wrote Cobbledick. "But I'll tell you something. Love's getting a hell of a battle from another force, hate. At this writing it's no better than a toss-up which way the decision will go. Don't be surprised if, after the final bell is rung, the referee raises hate's hand and announces a new champion." Cobbledick was referring to vitriolic letters and phone calls he had received in response to his posture on certain sporting issues, such as his support for Larry Doby, the first black player in the American League, and for other black athletes who played for Cleveland's professional teams.

Cobbledick's stature in the community was evident when he was named chairman of the 1962 Cuyahoga County (Ohio) Cancer Crusade. "Cobby's great popularity will be one of the most significant weapons in the intensifying drive against cancer in this community," commented the president of the county unit of the American Cancer Society. Cobbledick led an army of twenty thousand volunteers who solicited throughout Cuyahoga County and brought the crusade to a successful conclusion.

When Cobbledick retired from the *Plain Dealer* at the age of sixty-four he looked for new challenges. He wrote his last column on 16 August 1964, but Cobbledick did not stay retired long. He was quickly hired to provide sports commentary for the WEWS television station in Cleveland but remained in this position for only ten months.

*Cobbledick (third from left) with former Cleveland Indian players Al Rosen, Luke Sewell, and Lou Boudreau*

Cobbledick was less than effective in this role because he had to read his notes, and the new medium was incompatible with his personality and style.

Cobbledick dedicated much of his energy in retirement to writing a biography about Rocco "Rocky" Colavito, the popular outfielder for the Cleveland Indians. *Don't Knock the Rock* (1966) chronicled Colavito's professional career from its beginning in 1951 at the Indians' minor league camp in Daytona Beach, Florida, to the conclusion of his playing days in 1965. This biography exemplified Cobbledick's strong and enduring relationship with professional athletes. Colavito wrote fondly of Cobbledick in a letter dated 4 December 1989: "Gordon Cobbledick was a true gentleman, an honest decent human being, who wouldn't write an untruth about anyone." *Don't Knock the Rock* was published by Cleveland's World Publishing and achieved strong local sales because of the popularity of both Colavito and Cobbledick.

Cobbledick continued to write a sports column twice a week for the Rowley Publishing Company, which printed newspapers for several suburban communities in metropolitan Cleveland. In 1966 the Wahoo Club, a booster organization for the Cleveland Indians, initiated the Gordon Cobbledick Tomahawk Award. This annual award was presented to an Indians' player who best exemplified the attributes of competitive spirit, skill, positive attitude, and good sportsmanship. Prominent recipients have included Jerry Kindall, Graig Nettles, Buddy Bell, Sam McDowell, and Andre Thornton.

In 1968 Cobbledick suffered two strokes that left him paralyzed. He died on 2 October 1969 at the age of seventy. On 7 August 1978 Cobbledick was honored posthumously with the J. G. Taylor Spink Award in the writers' wing of the Baseball Hall of Fame in Cooperstown, New York. The following words were inscribed upon Cobbledick's scroll:

> Personally and professionally the most stately gentleman in the press box. Forty years as a baseball writer, correspondent in World War II and sports editor of the Cleveland *Plain Dealer*.
>
> Training as mining engineer and service in the U. S. Marines in World War I were temporary. His interest was in words. His feel for them harnessed their insight and basic truth.
>
> A light strain and wit were hidden. The tall man smiled more inside than out. Hardly anyone was more personally companionable and popular. But when he

> stood up, the one-time president of the Baseball Writers Association lent a touch of royalty.

Nearly twenty years after his death, Cobbledick received another honor he would have cherished dearly. On 29 April 1988 he was enshrined in the Case Western Reserve Athletic Club's Hall of Fame for his career as a halfback for Case Institute of Technology. Cobbledick had earlier been honored as the first media person to be selected for the Sports Media of Cleveland and Ohio Hall of Fame.

Cobbledick gave authenticity to sports writing. In his columns he set the scene and placed the reader in it, with the advantage of Cobbledick as guide. Games were important to him, but he was most interested in people. A man of wit, enthusiasm, and wide knowledge outside of sports, Cobbledick had the human touch that endeared him both to wealthy club owners and to cafeteria employees who knew him as "Cobby." His aversion for showboaters in any sport certified a personal integrity that made him the confidant of a host of friends and admirers.

"No sports writer or columnist . . . could say as much in a few words as did Cobby," wrote James E. Doyle in the 4 October 1969 *Plain Dealer,* and he added, "Cobby always used the right words; he was a master of the English language. And how he respected it." He wrote in an area – sports journalism – that for a long time had been set on a low rung of the literary ladder. The aphorism that the best and the worst of newspaper journalism appear on the sports page was no doubt true during his era. Cobbledick's journalistic contributions represented the best in what he considered the most significant form of mass culture in America – sports.

**Biography:**

James E. Odenkirk, *Plain Dealing: A Biography of Gordon Cobbledick* (Tempe, Ariz.: Spider-Naps, 1990).

**References:**

Archer H. Shaw, *The Plain Dealer: One Hundred Years in Cleveland* (New York: Knopf, 1942);

David D. Van Tassel and John J. Grabowski, eds., *The Encyclopedia of Cleveland History* (Bloomington: Indiana University Press, 1987);

William Veeck, with Ed Linn, *Veeck – as in Wreck* (New York: Putnam, 1962).

# Robert W. Creamer

*(14 July 1922 – )*

Tom Gallagher

MAJOR POSITIONS HELD: Advertising manager, *Bronxville Review-Press* (1948–1949); advertising manager, *Westchester Herald* (1949); assistant editor, *Collier's Encyclopedia* (1950–1954); staff writer (1954–1955), associate editor (1955–?), senior editor (?–1985), special contributor, *Sports Illustrated* (1985– ).

*Robert W. Creamer*

BOOKS: Mickey Mantle, *The Quality of Courage,* ghostwritten by Creamer (Garden City, N.Y.: Doubleday, 1964);

*Jocko,* by Creamer and Jocko Conlan (New York & Philadelphia: Lippincott, 1967);

*Rhubarb in the Catbird Seat,* by Creamer and Red Barber (Garden City, N.Y.: Doubleday, 1968);

*Babe: The Legend Comes to Life* (New York: Simon & Schuster, 1974; Tokyo: Tuttle, 1975; Harmondsworth, U.K.: Penguin, 1983);

*The Yankees,* by Creamer, Dave Anderson, Harold Rosenthal, and Murray Chass (New York: Random House, 1979);

*Stengel: His Life and Times* (New York: Simon & Schuster, 1984);

*Season of Glory: The Amazing Saga of the 1961 New York Yankees,* by Creamer and Ralph Houk (New York: Putnam, 1988);

*Baseball in '41: A Celebration of the "Best Baseball Season Ever" – In the Year America Went to War* (New York: Viking, 1991).

OTHER: "The Old Orioles," in *The Ultimate Baseball Book,* edited by Dan Okrent (Boston: Houghton Mifflin, 1979), pp. 31–48;

"Cap Anson," in *Baseball Hall of Fame 50th Anniversary Book* (Englewood Cliffs, N.J.: Prentice Hall, 1988);

"Pop Watts, a Newspaper and a Day at the Polo Grounds," in *The Birth of a Fan,* edited by Ron Fimrite (New York: Macmillan, 1993);

"Casey Puts It On Ice" and "The Mantle of the Babe," in *Sports Illustrated Baseball* (Birmingham Ala.: Oxmoor House, 1993), pp. 34–39, 98–104;

"Thirties Baseball," in *Baseball, an Illustrated History,* by Geoffrey C. Ward and Ken Burns (New York: Knopf, 1994), pp. 260–263;

*Sports Illustrated Presents Mantle Remembered: Stories Excerpted from the Pages of Sports Illustrated,* original material by Creamer (New York: Warner, 1995).

SELECTED PERIODICAL PUBLICATIONS – UNCOLLECTED: "The Riotous Wanamaker Mile," *Sports Illustrated,* 2 (14 February 1955): 8–13;

"The Good Days and the Bad Days of Al Lopez," *Sports Illustrated,* 7 (1 July 1957): 42–47;

"Heroes of Yore Survive in Legend," *Smithsonian,* 10 (October 1979): 155;

"The Best Outfield Ever? Why, Del, Big Sam and Sliding Billy, for Sure," *Sports Illustrated,* 57 (18 October 1982): 98;

"The Daring Jackie Robinson Had the Stamp of a Very Special Ballplayer," *Sports Illustrated,* 57 (1 November 1982): 99;

"Jumping on Pop's Curveball," *Sports Illustrated,* 60 (4 June 1984): 16;

"Waite and Babe," *Sports Illustrated,* 61 (3 September 1984): 10;

"The Firebrand That Was Cobb," *Sports Illustrated,* 63 (19 August 1985): 54;

"Roger Maris, 1934–1985," *Sports Illustrated,* 63 (23 December 1985): 26;

"Bill Veeck 1914–1986," *Sports Illustrated,* 64 (13 January 1986): 10;

"Casey Stengel: Baseball's Amazin' Man," *Reader's Digest* (Canadian), 128 (May 1986): 205;

"When New York and Boston Played 74 Years Ago," *Sports Illustrated,* 65 (27 October 1986): 27;

"The Roaring Redhead: Larry MacPhail – Baseball's Great Innovator," *Sports Illustrated,* 67 (13 July 1987): 6;

"Mays, Mantle, Snider: A Celebration," *New York Times Book Review,* 24 January 1988, p. 17;

"Lip Service: Leo Durocher Belonged in Cooperstown before He Died," *Sports Illustrated,* 75 (21 October 1991): 31;

"June 29, 1954. (The Greatest Game I Ever Saw)," *Sports Illustrated,* 79 (19 July 1993): 41;

"If Your Best Friend Acts Like Angus, You Don't Need an Enemy," *Smithsonian,* 24 (August 1993): 112.

*Sports Illustrated* editor and writer Robert W. Creamer published approximately three hundred articles during the magazine's first thirty years but established a broader reputation with a biography of Babe Ruth that *New Yorker* editor and baseball writer Roger Angell called in *The New York Times* (13 October 1994) "perhaps the best portrait yet struck of an American sports hero." Five years in the making, the book benefited from the coincidence of its appearance with Hank Aaron's pursuit of Ruth's career home-run record; yet twenty years later the *New York Observer* (25 March 1994) still ranked *Babe: The Legend Comes to Life* (1974) one of the top ten baseball books of all time. Also included on the list was Creamer's *Stengel: His Life and Times* (1984), making him the only author doubly honored.

Creamer was born on 14 July 1922 to Joseph and Marie Watts Creamer in the village of Tuckahoe, which is part of the town of Eastchester, New York. Joseph Creamer, a real estate agent, once served as Tuckahoe's mayor; Marie, known as "Babe," was a schoolteacher who died when Creamer was ten. Creamer studied at Syracuse University in 1940–1941, at Fordham University from 1941 to 1943, at City College of New York in 1943–1944, and at Fordham again in 1945–1946, but he never completed a degree.

In 1946, after European combat duty in the U.S. Army, Creamer joined Grey Advertising in New York City, where he became, by his own estimation, "the worst advertising copywriter in the history of the country." On 23 August 1947 he married Margaret Schelz, a medical secretary, of Tuckahoe. They had five children: James, now an attorney in Boston; Thomas, a dramaturge for Chicago's Goodman Theater; John, a publicist in Peekskill, New York; Ellen, a nurse practitioner in Philadelphia; and Robert Jr., who is in retail sales in Eastchester. Creamer became advertising manager at the *Bronxville (N.Y.) Review-Press* in 1948, moving on in the same capacity to the *Westchester Herald* in White Plains, New York, in 1949. Later that year he became a junior account executive with the William Weintraub advertising agency in New York City. In 1950 he became the "de facto sports editor" at *Collier's Encyclopedia and Yearbook.* In 1954 he learned of Time's plans for *Sports Illustrated,* the first glossy sports weekly. He was hired as staff writer six months before the first issue appeared, becoming associate editor a year later.

A "Memo from the Publisher" in the 17 October 1955 issue describes Creamer's appearance at baseball spring training as "Tall, flaxen-haired and unduly given to sunburn. . . . Amid the tanned faces and rainbow-hued sportswear of the other sportswriters, Bob cut a solemn figure in a dark suit and darker hat, which he wore like a protective umbrella against the sun's rays. To Leo Durocher and the rest of the Giants he promptly became known as The Undertaker."

Creamer's literary and colorful style was evident from the start, as was his confidence in making bold editorial assertions. A 23 July 1956 article titled "Casey Puts It On Ice" describes the phenomenon of the New York Yankees, who in midseason appear to have locked up their seventh pennant in eight seasons. Not even a last-minute "Chicago uprising" by the White Sox, the Yankees' closest competitors, or a temporary slump on the part of the Yankees will "change the American League entry in the 1956 World Series." Creamer, who finds the July pennant race "vastly exciting," explores the reasons why the Yankees never let their lead slip away or allow other teams to narrow the gap: the Yankees are a team that goes "looking for every victory they can find." Their success is due not just to their hitting, pitching, and "museum collection of superb fielding infielders" but most of all to their

*Creamer in front of a portrait of Babe Ruth, whose biography Creamer published in 1974 (courtesy of* Sports Illustrated/*Jerry Cooke)*

manager, "the victory-greedy old fellow" Casey Stengel, who "has revolutionized baseball since he took over as manager of the Yankees in 1949 by using the players on his bench not as replacements in an emergency but as extra troops to be held in reserve in each game until the proper time came to commit them." An 18 June article, "The Mantle of the Babe," profiles a young Mickey Mantle, who was well on his way to winning the Triple Crown (an award for leading the league in home runs, runs batted in, and batting average) in 1956. Creamer daringly compares Mantle to Ruth in terms of sheer strength and raw ability; the "erratic" pace at which Ruth hit home runs; and, finally, if Mantle continues to "emulate" the Babe, "Ruth's poise and presence and competitive spark." Though Creamer puts the essay in the context of Mantle's possibly breaking Ruth's single-season record of sixty home runs, he opens the piece with a masterful description of one of the longest home runs Mantle – or any one else – ever hit in the Yankee Stadium:

> A thick-bodied, pleasant-faced young man, carrying a bat, stood at home plate in Yankee Stadium, turned the blond bullet head on his bull's neck toward Pedro Ramos, a pitcher in the employ of the Washington Senators, watched intently the flight of the baseball thrown toward him, bent his knees, dropped his right shoulder slightly toward the ball, clenched his bat and raised it to a near-perfect perpendicular. Twisting his massive torso under the guidance of a magnificently tuned set of reflexes, Mickey Mantle so controlled the exorbitant strength generated in his legs, back, shoulders and arms that he brought his bat through the plane of the flight of the pitch with a precision which propelled the ball immensely high and far toward the right-field roof, so high and far that oldtimers in the crowd – thinking perhaps of Babe Ruth – watched in awe and held their breath.

The new sports weekly format allowed Creamer to convey more comprehensive impressions than are generally found in day-to-day sportswriting, as in his profile of the manager of the Chicago White Sox, "The Good Days and the Bad Days of Al Lopez" (1 July 1957): "Spaniards have the gift of patient melancholy. The long Castilian face, the black hair, the dark eyes provide a perfect backdrop for brooding, for pondering of the inevitable cruelties of fate. Que sera, sera, the song goes. What will be, will be. The Spaniard's face grows sad, but he always shrugs his shoulders and endures." Creamer covered indoor sports as well, sometimes noting their similarities to "games" that were not sports. This article was Creamer's first to be selected for the annual Dutton *Best Sports Stories* volumes. Reporting on a track meet in "The Riotous Wanamaker Mile" (14 February 1955), Creamer observed, "There are those who say it was the best running battle New Yorkers have seen since the Democratic Convention of 1924 took 103 ballots to nominate John W. Davis. There are others who say this is not so, that there has never been anything like it before."

Creamer's first book, *The Quality of Courage* (1964), appeared without his name, since it was ghostwritten for Mickey Mantle. Published the year after President John F. Kennedy's assassination, the book had been conceived by a Doubleday editor as a sports version of Kennedy's *Profiles in Courage* (1956) – as Mantle's attempt "to get across to the boys who read it some idea of what I think courage really is." The profiles include tales having nothing to do with baseball, such as a moving account of the life of Mantle's father, who died of Hodgkin's disease at age forty-one; a story about a little boy who

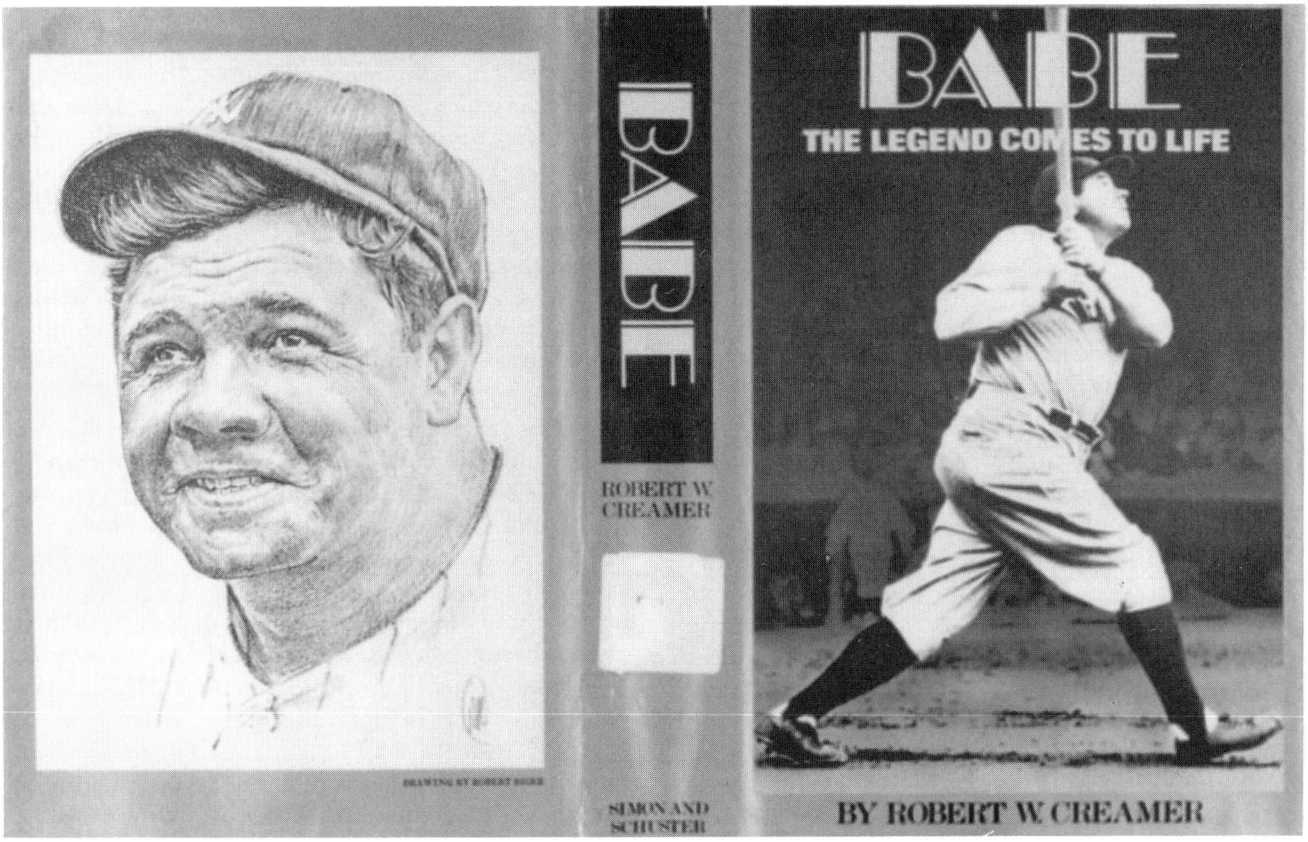


*Dust jacket for Creamer's biography of Ruth*

was brave enough to get his hair cut; and one about a man who would not drop an oil lamp and risk burning his house and family, even though the lamp was burning his hand. The book also contains the expected descriptions of Mantle's teammates and other players, such as Casey Stengel, Ralph Houk, Roger Maris, Ted Williams, and Joe DiMaggio – figures to whom Creamer would return in later books in more sophisticated fashion. Creamer found Mantle "fun to be with," although Mantle himself "didn't have many stories."

Creamer's name would not appear on a book cover until *Jocko* (1967), the story of Hall of Fame umpire Jocko Conlan. This book's subject had plenty of stories: stealing Kid Gleason's glove in 1911 or 1912, playing briefly with the White Sox in 1934–1935, ejecting Jackie Robinson ("the most difficult ballplayer I ever had to deal with as an umpire"), and retiring in 1964. Creamer later wondered if the book's reception had not suffered for its being too anecdotal. In a "co-author's note" Creamer displays the trademarks of his subsequent work – a literate appreciation of what is so special about baseball:

> But consider him . . . the chin, the mouth, the eyes, the umpire's regalia, the little blue hat, the bow tie, the cocky walk, the imperious gesture, the insistent voice. He is unique. He is Jocko Conlan. He is *the* umpire. Bernard Berenson used to talk about Man as a Work of Art, the idea that a man himself can create in himself something enduring, that he can become, in effect, a living work of art. What else is Jocko?

He combines such appreciation with an ability to distinguish the ordinary:

> It must be admitted that along with his many virtues – and they are many – Jocko has a few faults. He is vain about some things. He can be testy at times. He sometimes gives the impression that he feels he has never been wrong in his life.

Years later Creamer noted that "some of [Conlan's] political and social comments put me off a bit."

In co-authoring Red Barber's autobiography, *Rhubarb in the Catbird Seat* (1968), Creamer was finally paired with "a literate man" whose "sentences came out perfectly formed." Barber's baseball broadcasting career began in Cincinnati, but he made his mark by introducing baseball to the New York airwaves during fifteen years with the Brooklyn Dodgers; his twelve years with the New York Yankees ended during the writing of the book, when he was fired for calling attention to the small size of a crowd. The Mississippi-born Barber admitted that he "did not broadcast with a Brooklyn accent" but insisted that he "did broadcast with a Brooklyn heart." In the book's preface Creamer recalls:

> In the summer of 1941 you did not need to own a radio to hear Red broadcast. You could walk up a street and hear the game through one open window after another and never miss a pitch. You could thread your way through the crowd on a beach and get the game from a dozen different portables. In traffic you'd hear it from a hundred different cars.

Although his first three books had been collaborations, Creamer did not much care for that form of writing. In 1969 he got a contract for a biography of Babe Ruth. Creamer began writing *Babe: The Legend Comes to Life* at a time when few people foresaw Aaron's run at Ruth's career home-run record, and he finally published it in 1974, the year Aaron broke the record. Creamer's writing of the book was slowed by his work at *Sports Illustrated,* which took him to Switzerland, England, Ireland, and Australia. A self-effacing "Author's Note" says that "Some of the baseball detail may prove tedious to the casual reader, who is given permission to skip such passages. Baseball fans are expected to wade through the whole thing." But casual readers who take the author at his word do so at their own risk, for this book bears little resemblance to an earlier school of sports biography that was often only a cut above the cartoon profiles on the backs of 1950s baseball cards. At the same time, it does not indulge in the then increasingly common practice of using baseball as a metaphor. Noting that Ruth's stature was such that a *New York Times* photo caption once described Enrico Caruso as "the Babe Ruth of operatic tenors," Creamer seeks to capture "the obvious and subtle things that combined to make the man born George Ruth a unique figure in the social history of the United States." The most obvious factor was that "The home run belonged to him." Creamer further points out that "After his 1919 season, before he had played one game for the New York Yankees, Ruth was acclaimed the greatest home run hitter baseball had ever seen, even though he had only 49 for his entire career to that point." Then, "From 1926 through 1931 he averaged 50.3 a year." Ruth's performance as a pitcher has not remained as clear in the public's mind; but Creamer shows that Ruth's record of twenty-nine scoreless innings in World Series play, which lasted for forty-two years, was "of all his accomplishments in baseball . . . the one of which he was most proud." Creamer analyzes many of the legends surrounding Ruth; in regard to one of the best known, Creamer concludes that the argument as to whether Ruth actually pointed to center field before the "called shot" home run in the 1932 World Series is "an argument over nothing. . . . He did challenge the Cubs before 50,000 people, did indicate he was going to hit a home run and did hit a home run. What more could you ask?"

No prior Ruth biography had been as thorough, and Creamer's work completely overshadowed three other Ruth biographies that were published that year. The reviewer for *Time* (26 August 1974) thought that it was "really the first adult biography of the Babe, as well as one of the best, and least sentimental, books about a great sports figure ever written." In *Diamond Classics: Essays on 100 of the Best Baseball Books Ever Published* (1990), Mike Shannon notes that it is "particularly illuminating on the two phases of Ruth's life which had been the most ignored, clouded, beset by rumor, and misunderstood: that is, the periods before and after he was a great baseball star."

Creamer's "Babe Ruth and Lou Gehrig" contribution to *The Yankees* (1979), a book he co-authored with Dave Anderson, Harold Rosenthal, and Murray Chass, describes the first of "The four fabulous eras of baseball's most fabulous team." Before reprising Ruth's career and telling of Gehrig's rise through New York's High School of Commerce and Columbia University, Creamer recounts how in 1903 the original American League Baltimore Orioles moved to American League Park in the Washington Heights section of Manhattan and became the Highlanders of Jack Chesbro, Clark Griffith, and Hal Chase. Twenty years later the erstwhile Highlanders had taken the all-American name they made famous, along with their first World Series – the latter largely on the strength of having made the Boston Red Sox "a Yankee farm club." Creamer points out that "of the twenty-four Yankee players eligible for that [1923] Series, eleven had come from the Red Sox" – including Ruth, in a deal that, for a while, literally left "the Yankees holding the mortgage on Fenway Park." Creamer would remember "The Old Orioles" in a piece of that title written for *The Ultimate Baseball Book* (1979). The legendary team that invented "Inside Baseball" – "[t]ight baseball, scratching out runs" – was, writes Creamer, "a nova, a sudden manifestation in the baseball skies

DISKETTE 1941.c
Document INTRO-41.1

BASEBALL IN 1941

I

EXPLAINING THE OBVIOUS

My god, what a year 1941 was. I was eighteen [when it began] ~~at the beginning of it~~ and turned nineteen [that summer,] on the day Joe DiMaggio hit safely in his 54th straight game. He [had] kept the streak going to my birthday; it ended three days later.

1941 was the year Ted Williams batted .400 -- .406, to be precise. And the year the Dodgers, the rowdy Brooklyn Dodgers of Larry MacPhail and Leo Durocher, [survived a raucous pennant race and] won ~~the pennant~~ for the first time in twenty-one [years --] ~~seasons,~~ and then lost the World Series ~~to DiMaggio and the Yankees~~ after a game-ending third strike got past their catcher, Mickey Owen.

Is this the heritage of the hapless baseball fan, that he remembers 1941 for DiMaggio and [Williams and Mickey Owen] ~~the Dodgers~~ instead of Pearl Harbor?

Before you pass judgment, listen. I concede that if you were to ask [most] ~~any group of~~ Americans -- especially those a decade or more younger than I am -- what they associate with 1941, most would [say] ~~answer~~ "Pearl Harbor" or "The year the war began."

But when you were eighteen and nineteen and living in the year, 1941 wasn't Pearl Harbor -- not most of it, anyway. The Japanese attack on America's big naval base in Hawaii didn't take place until December. For eleven of the twelve months ~~of 1941~~, for all but ~~twenty-five~~ days ~~that year~~, most Americans my age had never heard of Pearl Harbor. When the news came over the radio that ~~gloomy~~ December Sunday

Page, with Creamer's corrections, from the typescript for his recollection of baseball's "best season ever" (courtesy of Robert W. Creamer)

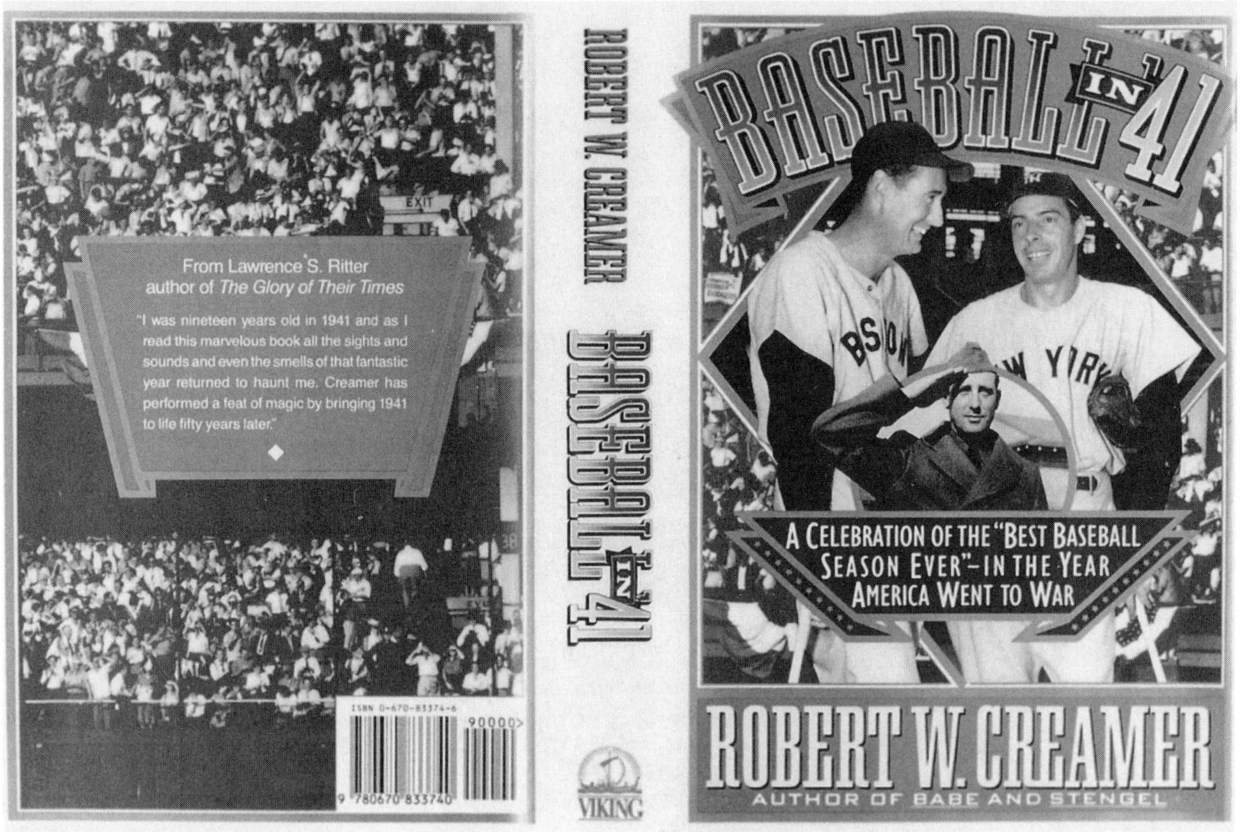


*Dust jacket for Creamer's most personal work*

that flared brilliantly for a very short time and then disappeared."

Creamer's *Stengel: His Life and Times* was written in accordance with his belief "that what readers want is 90 percent content and 10 percent style." Stengel, who describes himself in the book as "a man that's been up and down," played in the Brooklyn Dodgers' first game in Ebbets Field; hit two homers for the New York Giants in the 1923 World Series; and managed the Dodgers to three second-division finishes, the Yankees to ten first-place finishes, and the New York Mets for their first four seasons. Creamer places Stengel in the broad scope of baseball history: "It doesn't seem to be stretching the point too far to say that Ned Hanlon begat John McGraw who begat Casey Stengel who begat Billy Martin. . . ." Creamer's respect for the language he employs is evident throughout the work. For example, he notes that when Stengel's Toledo Mud Hens came home in 1927, "The city went figuratively crazy over its first pennant" (many other sportswriters would, inaccurately, have described the "*literal* madness" that swept the town). And although he claims to be "not particularly well read," Creamer roams well beyond the foul lines for his vantage point on the game:

> Leonard Woolf, Virginia's husband, wrote in his autobiography, "there develops in nearly all arts, and indeed in games like cricket, at various periods after an archaic, vague or inchoate beginning, a classical style that combines great power and freedom and beauty with a kind of self imposed austerity and restraint." Nothing could describe DiMaggio more perfectly.

The penultimate chapter, "Do not go gentle: talking, talking, talking," is a thirteen-page interview entirely in "Stengelese" – the more difficult passages are annotated – on the 1960 World Series. Loudon Wainwright, in the May 1984 issue of *Life,* praised Creamer's depiction of Stengel as "one of the most extraordinary and enchanting gabbers of recorded time." Shannon calls *Stengel,* which is also included in *Diamond Classics,* "a thorough, proportionate study of the entire lengthy life and career."

Creamer retired as senior editor at *Sports Illustrated* on 1 January 1985, but he maintains a role with the magazine as a special contributor. Most of his later essays have dealt with historical occurrences in baseball or such seminal figures as Ty

Cobb, Bill Veeck, Jackie Robinson, and Larry MacPhail. When the Yankees replaced Stengel exactly a week after the 1960 World Series loss to the Pittsburgh Pirates on Bill Mazeroski's seventh-game, ninth-inning home run, the new manager was Ralph Houk, Yankee coach and former backup catcher with only thirty-seven at bats in his final six seasons. *Season of Glory: The Amazing Saga of the 1961 New York Yankees* (1988) was a collaboration between Creamer and Houk. Creamer recalls that baseball "stories didn't come pouring out" of Houk; but he points out that ball-field contretemps might not have made a deep impression on a man who, like "the Major" (as Houk was known), had had a German sniper's bullet pass through his helmet during World War II. Creamer ultimately did not like this biography – his first of a non–Hall of Famer – "because I had to do it too quickly." Nonetheless, his account of one of the results of the Maris-Mantle "home run derby" in pursuit of Ruth's single-season home-run record is memorable: baseball commissioner Ford Frick

> said a player who hit 60 after the 154th game would be given recognition but added – much to his later chagrin – that such a record would go into the record book with "a distinguishing mark" to show that it had been made in a 162-game season. He didn't mention [anything about] an asterisk, a word that was to haunt him the rest of his life . . . [but] because Frick's ruling aroused such strong feelings on both sides of the matter, "with an asterisk" became a catch phrase that eventually entered the language as a term for a qualified success.

The book includes Creamer's disquisition on the "fractional ass" who decided that there could be such a thing as a 54 2⁄3-consecutive-scoreless-inning streak when, after all, an inning cannot be scoreless until it has been completed. As a human-interest backdrop to the season, Creamer tells how the Yankees brought fan Sal Durante, his wife, and their children to the stadium to celebrate the twenty-fifth anniversary of the day Maris broke Ruth's record: Durant had caught the ball. The book's concluding chapter reprises the careers of all thirty-six players, down to pitcher Duke Maas, whose "one batter retired" was his last of a seven-year career.

*Baseball in '41* (1991) is Creamer's most personal work; as he explains in the introduction, it is "an effort to explain to my children and their contemporaries what 1941 felt like to me." Three players dominate what the subtitle calls *A Celebration of the "Best Baseball Season Ever" – In the Year America Went to War*. The first is DiMaggio, whose fifty-six-game hitting streak broke records stretching back into the nineteenth century and was his second fifty-plus streak of the five ever recorded in the history of "organized baseball." The second player is Ted Williams, whose batting average of .406 would have been even higher had sacrifice flies not been charged as times at bat in 1941, as they are not today; Creamer estimates Williams's 1941 average under today's rules at .419. Finally, there is Hank Greenberg, the second major leaguer (after Hugh "Losing Pitcher" Mulcahy) to enter the military and "the best ballplayer in the major leagues at that time," whose four and a half years in the service account for the fact "that few baseball fans of my children's generations know how good Greenberg was." But "it was the Dodgers more than anyone else . . . who made 1941 so continually exciting." Although it had recently been written that "overconfidence may yet cost the Dodgers sixth place," Brooklyn wound up in the World Series against the Yankees (and was defeated four games to one). To get to the Series, the Dodgers battled a Saint Louis team developed in Branch Rickey's farm system. As Creamer observes, "when the standard procedure was for major league clubs to replenish their rosters by buying players from the minors, teams like the Cardinals were handicapped because richer clubs like the Chicago Cubs and John McGraw's New York Giants could outbid them."

In 1941 one might "drive to the top of a hill where there was good radio reception" to hear Red Barber or go to Lewisohn Stadium to hear the *Symphony in D for the Dodgers,* written in a lighter vein – with a part for Barber – by an otherwise eminent composer. Or one might have gone to the New York Giants' ballpark on 27 May, with Pearl Harbor still more than five months away:

> In the Polo Grounds, with the score tied at the end of the seventh inning, the umpires waved their arms and stopped the game, the way they would have halted play if it had started to rain. It was almost ten-thirty. Both teams left the field. A few moments later a voice on the public-address system said, "Ladies and gentlemen, the President of the United States," and for the next forty five minutes the 17,000 people in the Polo Grounds listened as Franklin Roosevelt spoke. It was the first and, as far as I know, the only presidential address ever piped into a ballpark in the middle of a game.

On the book's publication, Barber said in *The New York Times* (6 June 1991): "for the golden anniversary, the man and the material are perfectly matched."

Creamer has been a frequent contributor to anthologies, both of previously published pieces and of new ones. He wrote "Thirties Baseball" for *Baseball, an Illustrated History* (1994), the companion volume to Ken Burns's nine-part Public Broadcasting System production. There he recalls "the scuffed baseballs, the cracked bats, and the rough fields of my own baseball experience" and the day the exigencies of travel forced the Saint Louis Cardinals to play in uniforms unlaundered since the prior day's rain-and-mud-filled doubleheader, prompting an anonymous onlooker to immortalize them as looking "like the gang down at the gashouse." A *New York Times* reviewer (12 September 1994) considered this piece "the most evocative" of all of the book's essays. Creamer has made frequent television appearances, most notably in the Burns series, and was dubbed "The Babe Ruth Conference Distinguished Scholar" for the 1995 Hofstra University conference "Baseball and the 'Sultan of Swat.' "

Creamer has also published a history of his hometown, but he claims that "the thing I am proudest of in my life is getting a letter I wrote about James Joyce published in the *London Review of Books*. If I could finish my historical novel (about the American Revolution) and get it published, I'd die happy." As far as sportswriting is concerned, Creamer has, in the opinion of most critics, lived up to the goal he cited in the first chapter of *Babe:* "Max Eastman wrote 'The mind should approach a body of knowledge as the eyes approach an object, seeing it as a gross outline first, and then by gradual steps, without losing the outline, discovering the details.' "

**Interview:**

Mike Shannon, *Baseball – The Writers' Game* (South Bend, Ind.: Diamond Communications, 1992), pp. 33–51.

**Reference:**

Mike Shannon, *Diamond Classics: Essays on 100 of the Best Baseball Books Ever Published* (Jefferson, N.C. & London: McFarland, 1990), pp. 23–29, 342–346.

# Arthur Daley

*(31 July 1904 – 3 January 1974)*

Scott A. G. M. Crawford
*Eastern Illinois University*

*Arthur Daley (photograph by* New York Times *Pictures)*

MAJOR POSITIONS HELD: Sportswriter (1926–1942) and sports columnist (1942–1974), *The New York Times.*

BOOKS: *The Story of the Olympic Games,* by Daley and John Kieran (Philadelphia: Lippincott, 1948); revised in 1953, 1957, 1961, 1965, 1969, 1973, and 1977;

*Times at Bat: A Half Century of the National Pastime* (New York: Random House, 1950); republished as *Inside Baseball: A Half Century of the National Pastime* (New York: Grosset & Dunlap, 1950); republished as *Inside Baseball: Times at Bat, A Half Century of the National Pastime* (New York: Grosset & Dunlap, 1951);

*Sports of the Times* (New York: Dutton, 1959); revised as *Sports of the Times: The Arthur Daley Years,* edited by James Tuite (New York: Quadrangle / New York Times, 1975);

*Knute Rockne: Football Wizard of Notre Dame* (New York: Kenedy, 1960);

*Kings of the Home Run* (New York: Putnam, 1962);

*Pro Football's Hall of Fame: The Official Book* (New York: Quadrangle, 1963);

*Pageantry of Sport: From the Age of Chivalry to the Age of Victoria,* by Daley and John Arlott (New York: Hawthorn, 1968; London: Elek, 1968);

*All the Home Run Kings* (New York: Putnam, 1972).

SELECTED PERIODICAL PUBLICATIONS – UNCOLLECTED: "Another Sport Moves out of the Rain," *Literary Digest,* 119 (16 February 1935): 32;

"Water Boy," *Collier's,* 104 (26 August 1939): 17;

"He Throws Cannonballs," *Collier's,* 107 (31 May 1941): 44;

"Leaping Lester," *Collier's,* 109 (25 April 1942): 18;

"Grandstand View of the Japanese," *Reader's Digest,* 42 (June 1943): 65;

"They're Off," *Collier's,* 114 (12 August 1944): 23;

"Wizard of Green Bay," *Collier's,* 114 (25 November 1944): 15;

"Luck Guides the Puck," *Collier's,* 117 (26 January 1946): 32;

"Stumbling Down the Stretch," *American Magazine,* 142 (October 1946): 38–39;

"Football's Twelfth Man," *American Magazine,* 142 (December 1946): 44–45;

"He Makes 'em Hit It," *American Magazine,* 143 (June 1947): 48–49;

"King is Dead," *Reader's Digest,* 52 (February 1948): 107–108;

"Pop Was Right," *American Magazine,* 145 (May 1948): 44–45;

"Diamond Thieves," *American Magazine,* 147 (March 1949): 50–51;

"What They Say to the Umpire," *American Magazine,* 149 (June 1950): 46–47;

"Small Fry at Bat,"*Reader's Digest,* 57 (August 1950): 37–40;

"Superfan," *American Magazine,* 151 (June 1951): 110–113;

"Golf's Historic Moment," *Reader's Digest,* 71 (December 1957): 200–201;

"Bell, Bat and Stengel," *Saturday Review,* 42 (21 November 1959): 48–49.

Born on 31 July 1904 in New York City, Arthur John Daley thought of himself throughout his life as a New Yorker. New York was where he was raised and schooled and where he spent his entire professional career as a newspaper reporter. From an early age he was interested in a wide variety of sports. At Fordham Preparatory School he participated in baseball, basketball, football, track and field, and swimming. A skilled and competitive athlete, he seemed to enjoy the variety of his pursuits as much as his athletic success. His favorite sport, though, was baseball, which he pursued passionately as a player and later as a writer.

Daley graduated from Fordham Prep in 1922 and entered Fordham University, where he continued to be an energetic student who took part in athletics. However, it was on the baseball diamond that he left his mark as an alert and focused center fielder. Left-handed batters who pulled their drives into the open space of right field would find the long-limbed Daley making a difficult catch look easy. The notion of writing as a career had never occurred to him until his junior year. He became assistant sports editor of the *Fordham Ram,* and a year later he became the paper's sports columnist. Thus began a career of sportswriting that was to be hectic and intense until the day he died.

After completing his B.A. degree in 1926, he was hired by *The New York Times* as a sportswriter to cover track and field and basketball. He also wrote about football at a time when the professional game was beginning to capitalize on athletic heroes produced on America's college campuses. The home team was the New York Giants, and he followed them with an engaging mixture of boyish enthusiasm and dogged loyalty. In the late 1920s he doubled as an announcer, and when the Giants played at the Polo Grounds his voice was heard on the public address system. Although when Daley was hired baseball was not listed as one of his "writing briefs," he quickly displayed a penchant for baseball writing and eventually wrote as much about baseball as about all other sports combined.

Daley was a hard worker and had a good eye for detail, but his greatest asset was that he was first and foremost a "news hound." For example, early in his career when he learned that the major league owners were holding their winter meetings in the New Yorker Hotel, he followed Red Smith's advice: "Old players looking for a job, scouts and other liars sit around [at the hotel] all day telling stories. Just go over and listen, and you'll get a dozen columns."

Although Daley would always go after his story, he never fully conquered his shyness. For his entire career he remained diffident and introspective and was constantly troubled by the possibility that he did not have all the facts at his fingertips. Despite his having covered both Gene Tunney and Jack Dempsey as they prepared for their much ballyhooed 1927 world championship bout, the notorious "long count" by referee Dave Barry, which allowed Tunney to avoid a knockout and go on to win the fight, disturbed him, making him believe that merely reporting the facts was insufficient. Just as Daley hesitated to cover boxing, he balked at reporting on horse racing, which made him uneasy because he was uncertain that he possessed sufficient expertise to be an authority on horse racing.

During his first years with *The New York Times* he continued his academic studies in journalism at New York University and Columbia University. He was sent to cover the Los Angeles Olympics in 1932, and in 1936 he was the first *New York Times* sportswriter to go abroad on assignment when he reported on the Berlin Olympics. He wrote memorably of Jesse Owens's third gold medal on 5 August: "No human being ever moved faster. The Buckeye Bullet caromed around the turn from the staggered start and when he hit the home stretch he was at the head of the parade. Calmly Owens glided along – no strain, no sign of exertion, but an automaton moving on to fulfill his destiny."

By the early 1930s Daley had become a familiar figure with his clipboard and yellow notepad. His favorite haunts were spring-training camps, New York baseball parks, and the World Series. His talent was to be in the right – and best – place at the right time. For his first big league baseball match report, a Fordham Prep classmate got him a crow's-nest seat high up on a signal tower near the Polo Grounds. Smith, in *The New York Times* (4 January 1974), wrote that Casey Stengel, the former

*Ty Cobb and Daley in 1955 (courtesy of the Estate of Arthur Daley)*

New York Yankees manager, had remarked: "Some guys go to school and get an education. But other guys I know and I wouldn't want to mention any names, play hookey and watch ball games from signal towers. Know what happens to guys like that? They become sports columnists."

Daley liked to give the impression that sportswriters lived life in the fast lane. One of his observations, according to Smith in his 1974 article, was, "Doctors drink for the same reason newspaper guys do. It's the pressure we're under." In fact, Daley did not drink for the latter portion of his career, and eventually he gave up smoking.

Daley wrote on a myriad of topics for a wide variety of publications, including *Reader's Digest, Literary Digest,* and *Collier's.* His major breakthrough, however, came on 24 December 1942. Since 1927 John Kieran had written the "Sports of the Times" column, which appeared seven days a week and had made Kieran the premier sportswriter for *The New York Times.* In late 1942 Kieran quit the position to join the *New York Sun.* In Smith's 1974 *New York Times* article, Robert Lipsyte recounted the story of Daley's swift promotion at the newspaper. Daley was at the newspaper's editorial office when he received a call from Raymond J. Kelly, the sports editor:

> "What are you doing?" growled Kelly.
> "Sunday stuff," said Daley, his heart beating wildly.
> "Forget it. Write the Kieran column. It's yours until further notice."

Daley continued to write the column for the rest of his life. In the period from 1947 to 1963 he also wrote sixty-one pieces for *The New York Times Magazine.* Lipsyte jokingly recalled Daley as a self-admitted bigamist. Married to the former Betty Blake in 1928, Daley "married" *The New York Times* on Christmas Eve 1942. By the end of his life even conservative estimates put Daley's lifetime output at 10,300 columns, at a rate of 6,000 words a week. In his afterword to a posthumous collection of his father's *New York Times* articles titled *Sports of the Times: The Arthur Daley Years* (1975) Robert Daley wrote: "After many years he was told he could have Saturdays off. He would only have to write six columns a week. He took this to mean *The Times* executives were displeased with him. All the other *Times* columnists were writing only three columns a week. . . . He had no other life outside of sports and *The New York Times.*"

Over the years Daley impressed many people, including star athletes and major sports leaders, with his integrity. He wanted to generate good copy, but ethics, fair play, and loyalty were the cornerstones of his craft. In Lipsyte's 4 January 1974 article, Joe DiMaggio, a man known for his need for personal privacy, described Daley's qualities as a gentleman-reporter:

*Daley with Muhammad Ali (photograph by* The New York Times)

> I've watched the way he works. He'd be there, at a mass interview, hanging on the edges, taking it all down. Then he'd get his man in a corner, alone. He'd never come on strong. Shy and quiet and serious and friendly, he'd look you straight in the eye, with a little smile. I knew I could trust him from the first day I met him.

Daley, however, did not write only complimentary things about DiMaggio. Jack B. Moore, in his 1986 biography, *Joe DiMaggio,* mentions Daley's scrutiny of the "Yankee Clipper" during 1949 spring training. Daley criticized DiMaggio for being "strangely abrupt and almost surly."

Daley did not confine himself to the sports pages of *The New York Times.* He wrote a piece for *The Story of Our Time – Encyclopedia Yearbook* and co-wrote with Kieran *The Story of the Olympic Games* (1948), revised over the years in a series of updates. The latter work is an intelligent amalgamation of sport journalism and reputable historical scholarship. Beyond including facts and figures about each gold medalist, Kieran and Daley identify and describe the most memorable happenings of each Olympiad. In his coverage of the Olympics, Daley never concealed his love of the United States.

Daley was noticed at press conferences because he always stood erect, as if he had been groomed in the military, and he appeared taller and heavier than his six-foot-one-inch, 195-pound frame. According to Lipsyte a favorite Daley quip was a comment made to him by the three-time Olympic swimming champion Dawn Fraser of Australia: "Gee, you're a man I can look up to." Another comment that delighted Daley was made when the then relatively unknown Cassius Clay called out at a training session, "Well, there's Arthur Daley, I must be important." Stengel is reported to have described Daley fondly as "the big guy with the clipboard."

Lipsyte was one of several writers who spoke of Daley as a "steak and spuds" kind of person. Lipsyte painted a portrait of a hardworking journalist who enjoyed simple pleasures, such as the television programs *I Spy* and *The Man from U.N.C.L.E.* or an 8 A.M. breakfast in his pajamas, which, Daley said, "sets me up for the day." Smith called Daley a good man who liked a simple schedule. He recalled Daley's complaining bitterly, but with a twinkle in his eye, of a trip to Spain. In America, Daley noted, life was perfect with three good meals a day and a walk after dinner. "You know the Spaniards are barbaric," he once said after having to eat at an 11 P.M. banquet.

Daley's first solo effort as a book writer appeared in 1950. *Times at Bat: A Half Century of the National Pastime* was Daley at his best, weaving a colorful narrative around a blend of baseball history, anecdote, and personal reminiscence. Baseball legend Branch Rickey, in a review of *Times at Bat,* commented, "From cover to cover Arthur Daley has hit the ball cleanly. There is more than a good chance that what he has batted out will land safely, will be a hit."

Daley's writing was engagingly uneven. It is as if a boyish energy, a youth's overpowering love of athletics, is harnessed by the practical realities of a working journalist who had to produce his column daily. For example, on 6 May 1954 Roger Bannister broke the four-minute barrier for the one-mile run, and *The New York Times* splashed Daley's story of the feat on page one, along with articles about Sen. Joseph R. McCarthy, a politician from Texas named Lyndon B. Johnson, and an aborted flight by a Boeing Stratocruiser. Daley's account of Bannister's achievement, with its reliance on stereotype, is hardly great writing:

> A feeling of smug satisfaction follows the news that mile supremacy has returned to Britain where it originated. They don't have the coaches over there that we have and the sports attitude in general has been "jolly well done, old chap." English casualness is almost alarming to intense Americans . . . [Bannister] did it all himself with typical British doggedness . . . He was the first to reach into the realm of fantasy.

A Daley interview with Bannister at the Yale Club a week later, however, is a minor classic. "Listening to a Four-Minute Miler" grabs the reader's attention: "And Roger captivated everyone in a lot faster time than 3:59.4. He did it in nothing flat." Daley wanted to get Bannister to explain, in simple and understandable terms, the art of record breaking. He did just that, by saying that record breaking is "the ability to take more out of yourself than you've got." Indeed, the piece is good because the voice of Bannister, not Daley, dominates the narrative. And yet Daley's ability to set the scene, arrange his stage, and then exit from it is remarkable. Daley had opened the piece thus:

> Roger Bannister lounged carelessly in a chair at the Yale Club yesterday and talked with a rare graciousness, penetration and charm. He is sandy-haired, tall, slim and ever-smiling. Pictures of him in action while running a mile in 3:59.4 are a mite flattering. When he strides in a track suit, he looks barrel-chested and overpowering. Street clothes make him shrink to more boyish proportions.

Daley's conclusion was also deft. Bannister explains that at the end of his record-breaking run he had nothing left as he breasted the tape. Daley lets Bannister be both central figure and principal commentator. Daley wraps up his column with the whimsical observation, "He made it sound easy."

Daley's gifts as a writer were considerable, and his canvas was broad. His pieces could range from a humorous and charming description of golf to a sensitive yet impassioned analysis justifying why a leading American track runner, Wes Santee, should be suspended by the Amateur Athletic Union for infringing on his amateur status. Daley also possessed a vast amount of what could be called "sports sense." In June 1954 John Landy of Australia broke Bannister's one-mile record. Sporting attention was then directed to a Bannister-Landy clash at the British Empire Games in Vancouver, Canada, in early August 1954. In *The New York Times* on 22 June 1954, Daley called the race the "Clash of Titans" and noted that "The dream race will thus materialize." When the race took place, his prophecy was fulfilled. Sports historians still describe it as the greatest mile run of all time.

On 8 May 1956 the Pulitzer Award Committee awarded Daley the prize in commentary. The committee spoke of Daley's "outstanding coverage and commentary in the world of sport." The Pulitzer was hotly debated among Daley's peers, many of whom believed the award should have gone to Red Smith or Jimmy Cannon. As Richard Kluger noted in *The Paper* (1986), "Many in the New York sportswriting fraternity knew that Daley often filled his columns with blatantly fabricated quotations attributed to athletes who could not string two coherent clauses together." A review of the Daley columns listed and recognized by the committee nonetheless underscores Daley's penchant for varied topics treated thoughtfully without either humbug or pretension. The columns cited were "The Silent Mr. Stengel," "What Price Amateurism," "A Letter from Ty Cobb," "A Visit with the Challenger (Archie Moore)," "Match Race," and "Dressen's Tower of Babel."

The most valuable source for evaluating Daley's contribution to sports reporting is *Sports of the Times: The Arthur Daley Years,* edited by James Tuite, a former *New York Times* sports editor and Daley colleague. In his introduction to that volume Tuite explains the difficulty of his task:

> in embarking on the job of selecting the best of Arthur's columns, I decided first to weed out the bad from the good and then select the best from the rest. On the first go-around, I found myself stacking all of the columns in the "good" pile. I realized then that Arthur really had not written any bad columns – it's just that some were better than others. He made a believer out of me.

*Sports of the Times: The Arthur Daley Years* serves as a reminder that Daley wrote about the pantheon of American sports heroes. His columns read as a "Who's Who of Sports": Babe Ruth, Dempsey, Willie Mays, Ali, Cobb, Jesse Owens, Byron "Whizzer" White, Stengel, Jim Thorpe, Honus Wagner, Bob Jones, Joe Louis, Byron Nelson, Mickey Mantle, Pancho Gonzalez, Babe Didrikson, Rocky Marciano, Jean Believeau, Secretariat, Knute Rockne, Henry Aaron – the list could go on and on.

Daley's work is sometimes difficult to assess. There is a 1943 piece (written at the height of World War II, with American troops already girding up for the Normandy landings) in which Daley recalls his visit to Berlin for the 1936 Olympics. The writing is tactless, and critics thought that Daley was too flattering of Adolf Hitler's Germany. Daley gleefully describes gate-crashing an event, gives a respectful account of a march by the Hitler Youth, then tells a story about Hermann Goering's sense of fun. Conversely, a 1945 piece on Cobb is delightful because it shows Daley writing with a wonderful sense of banter and humor. Daley opens the piece, "Ty Cobb had the reputation of being one of the snarlingest, fightingest, ball players that the game ever has produced." The rest of the narrative is a friendly tête-à-tête with the venerable Cobb, made mellow and well mannered by age. The contrast of a meek Cobb in 1945, placed against his stormy career as a feared competitor, is a vignette that exemplifies Daley's craft.

*Sports of the Times: The Arthur Daley Years* contains the six essays (22 February 1955, 6 March 1955, 28 March 1955, 29 March 1955, 8 September 1955, and 11 September 1955) for which Daley was awarded the Pulitzer Prize. Four are about baseball, and two are about boxing. In retrospect, it is puzzling to look at these six pieces. They are far from Daley's greatest work, and it is doubtful that they would be held up in present day as either examples of good writing or intelligent sports commentary. For example, Daley writes of Rocky Marciano as a "warmly human, gentle, modest and humble man," although biographies of Marciano tend to see him as a more flawed individual. Indeed, some might suspect, particularly after seeing the many photographs of Daley with celebrity athletes, that Daley simply relished being on familiar terms with the household names of American sports. Writing in *The Era* (1993), Roger Kahn views Daley's prize-winning writing as "ordinary pieces" and considers Daley's overall work "genial" but "inept." But Daley's affection for athletes was something that he felt profoundly – his 29 January 1958 tribute to baseball great Roy Campanella is full of genuine compassion – and the simplicity and sincerity of Daley's writing still have power.

Daley was not a probing essayist who was concerned with the complex tapestry of sport interwoven with an outside world. He did not launch into in-depth examinations of racial prejudice in sport, and his interest in the female athlete as performer and personality seems to have been lukewarm at best. His tribute on 13 October 1968 to a Mexican athlete, Norma Enriqua Basilioi, who lit the Olympic flame at the opening of the 1968 Olympics, seems well intentioned but patronizing: "She raced gracefully once around the track, pattered up the 90 steps to the ramparts – no mean feat at this altitude."). Nonetheless, at his best Daley wrote extremely well. His 12 September 1972 piece on the Munich Olympics covered the deaths of the Israeli athletes at the hands of Arab terrorists. Many commentators lapsed into highly emotional diatribes about the "ills" of modern sport. Daley wanted to convey the tragic tenor of events, but his forte – clean, concise writing with an absence of cant and hyperbole – resulted in exemplary scene setting.

> It was raining when Closing Day dawned and the day was to remain damp, dark and dolorous. It is difficult to envision a more appropriate setting for the unhappiest of Olympic Games than the gray, funereal backdrop that the weatherman provided. The usual parting-is-such-sweet-sorrow motif was missing and these Olympics dragged their way into history with their honest achievement shunted aside by the mishaps and controversies that kept demanding attention.

In addition to the Pulitzer Prize Daley received the Grantland Rice Award in 1961 and the Sportswriter of the Year Award in 1963.

Over the years Daley became the master of the attention-grabbing opener. An example is his introduction to a world heavyweight boxing match in 1958: "The flaming courage of Roy Harris, the educated backwoodsman from Cut'n'Shoot, deep in the heart of Texas, carried him through twelve rounds with Floyd Paterson tonight. But blazing spirit wasn't enough." Indeed, a case could be made that Daley was a significant sportswriter simply because of his penchant for the classic lead and the telling sign-off. On 8 December 1940 the Washington Redskins were routed by the Chicago Bears 73–0. Daley's opening – "The weather was perfect. So were the Bears" – is as memorable as is his closing:

> At the end the Redskins band played "Should Auld Acquaintance Be Forgot." If said acquaintance is the Chicago Bears, it should be forgot immediately. At the moment the Bears are the greatest football team of all time.

In the 1960s Daley cut back to four columns a week and then in 1971 went to three columns a week. He told fellow journalists that to write only three pieces a week meant that his job was "like a vacation." Although his column took up the majority of his time and energy, Daley's other writing deserves attention. For example, a 1960 volume, *Knute Rockne: Football Wizard of Notre Dame,* written specifically for readers aged ten to fifteen years, stands up well to the test of time. V. K. Fitch, writing in the *Library Journal,* called it "a splendid story." In *The New York Times Book Review* (December 1960) Robert Berkvist observed: "Rockne's story is in many ways the story of Notre Dame, and Mr. Daley tells it with sensitivity and an awareness of the legacy of excitement that Rockne bequeathed to football." In 1959 Daley had published *Sports of the Times,* a collection of ninety-three of his columns, of which Arnold Hano wrote in the 6 December *New York Times Book,* "Arthur Daley obviously does not describe sports exactly as they are, but rather as he wistfully would have them." Hano thought as little about Daley's *Kings of the Home Run* (1962), an anecdotal book about major leaguers who had hit three hun-

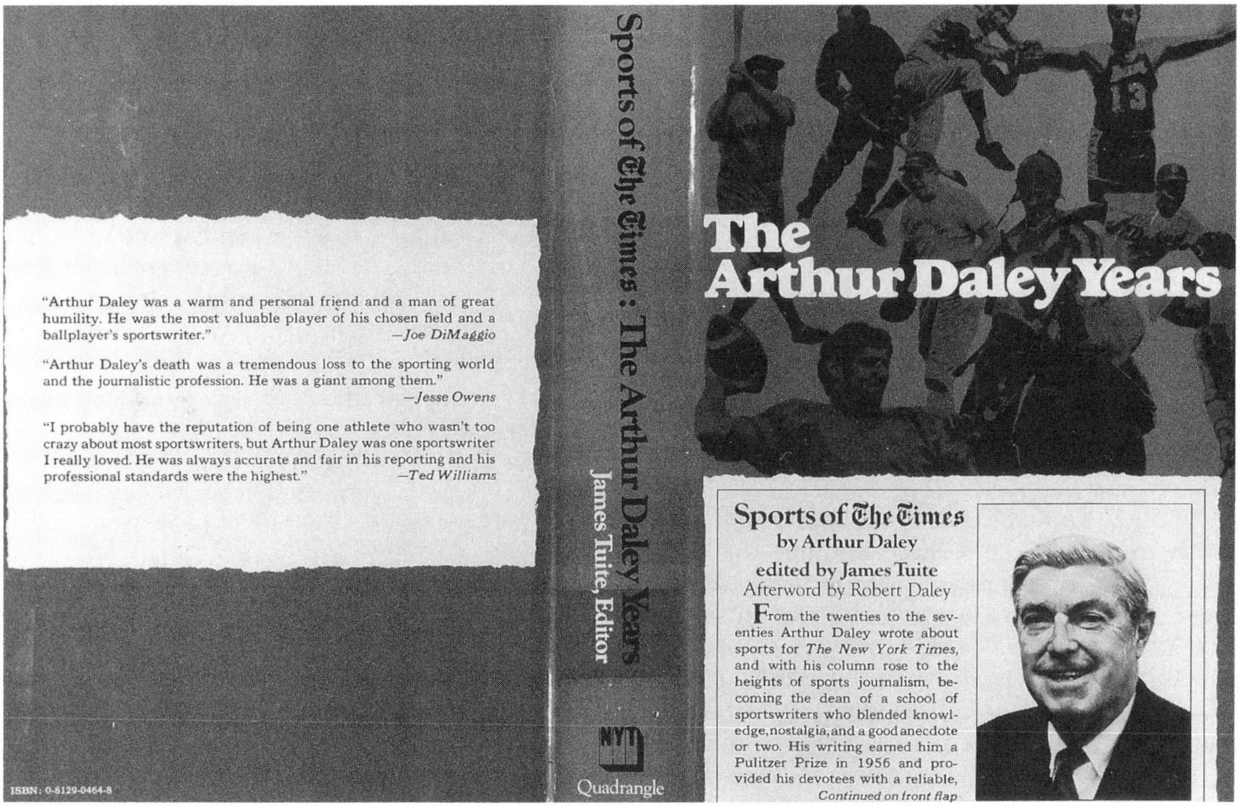


*Dust jacket for the 1975 collection of Daley's sports articles from* The New York Times

dred or more home runs, and Roger Maris, who the previous season had broken Babe Ruth's single-season home-run record.

Robert Daley was for a while a sports journalist for *The New York Times.* He recalled in a 1987 interview two incidents about growing up around his father the journalist. As a boy he remembered seeing his father deplane late at night at La Guardia. Clutched in his father's arms was a typewriter. Later, Robert Daley was allowed to use his father's typewriter with the caveat that he must use all ten fingers. "I got constant encouragement from him as I grew up, writing for the high school paper and then the college paper. I never wanted to be anything else other than a writer."

On 3 January 1974 Arthur Daley was on his way to work from his home in Old Greenwich, Connecticut. He exited the Grand Central Terminal and set off toward the offices of *The New York Times* when suddenly he collapsed and died on the sidewalk outside 11 West Forty-second Street. It seemed fitting that a man who loved the city, his job, and the Giants should die on the steps of the offices of the New York Giants. He was a close friend of former Giants president Jack Mara, and the Mara and Daley families had made much of annual pilgrimages to watch the Giants play the Redskins in Washington, D.C. Daley had planned to retire on his seventieth birthday. Robert Daley wrote a moving tribute to his father, concluding:

> His best friend was Jack Mara, who owned the Giants. When Jack died, my father wrote a column calling Jack the finest man he ever knew. But at Jack's funeral I couldn't take my eyes off my father, who was suffering so much, and whose own funeral I probably would attend next year, or the year after; so that I began to cry, weeping not so much for Jack as for my father, thinking: Jack Mara isn't the finest man I ever knew, you are.

Daley was mourned by many. Arthur Sulzberger, publisher of *The New York Times,* added that "His untimely death deprives us not only of one of the greatest sports writers of America, but also of a warm friend. His contribution to the *Times* will never be forgotten." Bowie Kuhn, then commissioner of baseball, spoke of "a great loss to us in baseball and to me personally as well as to the journalistic profession." New York mayor Abraham Beame mourned Daley's passing and said Daley would be long remembered for his "thoughtful commentaries on athletes of this city who he portrayed so well." The novelist James T. Farrell, who had been an admirer of Daley all of his life and started his day by seeing what Daley had written, observed, "He had enthusiasm and sincerity. He liked what he did. Liked most of the men he wrote about. And he had a sense of fairness." There were telegrams from DiMaggio ("my deepest sympathy") and Muhammad Ali ("I will miss him; he was always nice to me"). Robert, who delivered a homily at the service, concluded, "My father always thought the important people in the group photographs he had were the athletes, not himself. But over the years I came to think differently. If a man's worth is measured by the pleasure he gave others, he ranked very high."

While it would be true to say that the famous read his "Sports of the Times" column – he marveled that John F. Kennedy, after meeting him, sent a signed photograph with the inscription "from an avid reader" – Daley's legacy lies in his common touch. He knew sports inside and out and communicated this knowledge with enthusiasm, warmth, and a subtle sense of comedy. At Daley's funeral, Paulo Sabo of New Brunswick, New Jersey, paid his respects. For readers like Sabo, Daley's column was like a dear friend that would be sorely missed. "I've been reading him since I was 13," said Sabo. "I got a lot from him all those years. I just wanted to thank him for what he gave me."

Jesse Owens said Daley "was a giant among sportswriters," and perhaps Ted Williams came nearest to explaining why: "He was always accurate and fair in his reporting and his professional standards were the highest." The notions of good character and a sense of honor and fair play were of great importance to Daley. He wrote all those columns for all those years because he held passionately to the credo that sports was a way of life in which fairness and respect for others prevailed. He knew about the "other" side of sports but steadfastly refused to write about athletes that he did not like. Tuite recalls a Daley crack that seems to reflect Daley's real feelings: He would not publicize a "louse" because "there [are] plenty of nice, amusing guys around."

It would be difficult to argue that Daley was as good a writer as Rice, Paul Gallico, Damon Runyon, or Ring Lardner. What he did demonstrate, however, was an astonishing consistency in his writing during a career that lasted from 1926 to 1974. The hallmark of his writing, and indeed his life, was consistency. He was married for forty-five years; he wrote for *The New York Times* for forty-eight years; and in all those years he personally replied to every letter that he received at his newspaper office. Daley's ultimate accomplishment was one of longevity. At his death he was the grand old man of sportswriting.

**References:**

Roger Kahn, *The Era 1947–1957: When the Yankees, the Giants, and the Dodgers Ruled the World* (New York: Ticknor & Fields, 1993);

Jack B. Moore, *Joe DiMaggio: A Bio-Bibliography* (Westport, Conn.: Greenwood Press, 1986), p. 75.

# Allison Danzig

*(27 February 1898 – 27 January 1987)*

Adam R. Hornbuckle
*History Associates Incorporated*

MAJOR POSITIONS HELD: Sportswriter, *Brooklyn Daily Eagle* (1921–1923), *New York Times* (1923–1968).

BOOKS: *The Racquet Game* (New York: Macmillan, 1930);

*Elements of Lawn Tennis,* by Danzig and John Hope Doeg (New York: Coward-McCann / London: Eyre & Spottiswoode, 1931);

*The History of American Football: Its Great Teams, Players, and Coaches* (Englewood Cliffs, N.J.: Prentice-Hall, 1956);

*The History of Baseball: Its Great Players, Teams and Managers,* by Danzig and Joe Reichler (Englewood Cliffs, N.J.: Prentice-Hall, 1959);

*Oh, How They Played the Game: The Early Days of Football and the Heroes Who Made It Great* (New York: Macmillan, 1971);

*The Winning Gallery: Court Tennis Matches and Memories* (Philadelphia: United States Court Tennis Association, 1985).

OTHER: *Sport's Golden Age: A Close-up of the Fabulous Twenties,* edited by Danzig and Peter Brandwein (New York: Harper, 1948);

*The Greatest Sports Stories from the New York Times: Sport Classics of a Century,* edited by Danzig and Brandwein (New York: Barnes, 1951);

*The Fireside Book of Tennis,* edited by Danzig and Peter Schwed (New York: Simon & Schuster, 1972).

SELECTED PERIODICAL PUBLICATIONS – UNCOLLECTED: "Chief Jester of Tennis," by Danzig and George M. Lott, *Literary Digest,* 117 (28 April 1934): 24;

"How the Davis Cup Can Be Won in 1935," by Danzig and Lott, *Saturday Evening Post,* 207 (12 January 1935): 33, 104–105, 109, 111;

"There's Many a Slip," *Collier's,* 99 (10 April 1937): 12;

*Allison Danzig*

"Are Amateurs Hypocrites?," *North American Review,* 246 (September 1938): 198–205;

"Why I'm Turning Professional," by Danzig and Donald Budge, *Saturday Evening Post,* 211 (19 November 1938): 8–9, 82–83, 85;

"The Football Brainbuster," *New York Times Magazine,* 2 November 1941, pp. 10, 28;

"Sports Carry On, but the Picture Changes," *New York Times Magazine,* 23 May 1943, pp. 16, 38;

"Nine Men and a Boat," *New York Times Magazine,* 12 June 1949, pp. 48, 50;

" 'Little Mo' – Precision-Made Champion," *New York Times Magazine,* 16 September 1951, p. 26;

"The Little Girl with the Big Racquet," *New York Times Magazine,* 23 August 1953, p. 18;

"Anyone for Tennis? Yes, Gonzales," *New York Times Magazine,* 19 May 1957, p. 14;

"The Greatest Tennis Player of All Time," *New York Times Magazine,* 22 December 1963, p. 19.

Allison Danzig began covering sports for the *Brooklyn Daily Eagle* in 1921. For the next forty-seven years, forty-five of them at *The New York Times,* he specialized in writing about lawn tennis and indoor racket sports such as court tennis, squash racquets, and squash tennis, making himself, in the words of tennis authority Bud Collins, "the widest regarded literary voice of the game." Danzig wrote about tennis, as well as about football, crew, and the Olympic Games, in a straightforward and authoritative style that was backed up by a keen knowledge of the sports and their players. "If I didn't know what I'd seen," he remarked to Joy Duckett in a 1981 *World Tennis* magazine interview, "I shouldn't have been writing about it." Danzig's opinion of a tennis match carried special authority: like actors reading theatrical reviews, players would gather at newsstands the day after a match to see what he had written about their performance. Bill Talbert, a top-ranked American player of the 1940s and 1950s, observed to Barry Lorge in *Tennis Magazine* (September 1987) that Danzig's "understanding of the game was so complete" that he could advise players on how to improve their game as soon as they walked off the court. Some, such as Tony Trabert, the 1955 Wimbledon champion, even sought his advice *during* their matches: when Trabert suffered a shoulder separation during the finals of the 1955 Southampton Invitational on Long Island, he consulted Danzig rather than the officials before deciding to default. Danzig also distinguished himself from other sportswriters by showing an interest in sports history well before the topic became an acceptable area for academic analysis in the 1970s.

Of the twentieth century's tennis correspondents, none has been more revered and respected by fellow reporters and players than Danzig. Dave Anderson, one of his colleagues at *The New York Times,* recalled for Lorge that there was "nothing stiff or superior at all about his manner" and that Danzig was always "very approachable, very friendly, and very helpful." Reporters, especially the younger ones, consulted Danzig on every aspect of the game – especially its history, since he had witnessed so much of it. Although of slight build, Danzig was tough and determined – traits that carried him through two years of varsity football at Cornell University. "If there were any problems in the press box at Forest Hills," recalled Anderson, "he would be the first one to stand up for the writers in a firm, reasonable way." Collins, who remembered Danzig as a fastidious dresser and a crack chili chef, said that he "did not chase players for interviews. They respectfully sought him."

Danzig was born in Waco, Texas, on 27 February 1898, the fourth of six children of German immigrants Morris Danzig and Ethel Harvig Danzig; his father was a timber merchant. A fan of the Waco minor league baseball team, Danzig was arrested at age nine for sneaking into a game. In 1909 the family moved to New York City, where Danzig's father went into the knitting-goods manufacturing business. They later moved to Albany, New York. Danzig, who was five feet six inches tall and weighed 117 pounds "soaking wet," played quarterback on the Albany High School football team. A baseball player and track and field performer as well, he "played a bit of tennis, but was never good enough for competition." He graduated in 1917 and won a Regents Scholarship to Cornell University, where he majored in history and English with the goal of becoming a journalist. During his freshman year he competed against E. B. White in a contest for a position on the editorial board of the student newspaper, the *Cornell Sun.* The competition, which lasted from the fall to the spring, was to be decided on the basis of the amount of space filled in the paper by each competitor. Although Danzig produced the most copy, White was declared the winner and given the seat on the board. Disgusted, Danzig left Cornell and enlisted in the army. White, who later remarked that he "did not feel any guilt" about his success, nevertheless "sympathized with Danzig" and thought that he got a "raw deal." White acknowledged that, as Danzig had suspected, his victory rested on fraternity politics: one of his fraternity brothers had been a member of the editorial board.

Danzig was stationed at Camp Grant near Rockford, Illinois, and earned the rank of second lieutenant. While awaiting orders to embark for France, his entire unit fell ill during the worldwide influenza epidemic. World War I ended in November 1918, and Danzig returned to Cornell the following spring. In 1919 and 1920 he played substitute halfback on the Cornell football team to Edgar "Eddie" Kaw, a two-time All-American who, according to Danzig, possessed a "peculiar style of running" with "quick, short steps" to elude tacklers.

According to George McGann, who wrote a profile of him for *World Tennis* in 1956, Danzig, at 125 pounds one of the smallest players on the team, was also one of the most determined: "he never missed a day's practice and got the hell beat out of him."

Graduating from Cornell in 1921, Danzig went to New York City in search of a job as a foreign correspondent, editorial writer, or book critic. "I had pretty fuzzy notions of how newspapers were run," he confessed to McGann in 1956, not realizing "that I was asking for the very top jobs." He inquired about a position at the *Brooklyn Daily Eagle,* a small but leading newspaper in the New York metropolitan area. The editor indicated that there was an opening in the sports department and asked Danzig if he would be interested in it. Danzig later told McGann that he "nearly flipped": "I would never have had the nerve to ask for a sports-writing job. I regarded [sportswriting] as the top job on the paper."

His career began immediately, as the *Eagle* needed someone to cover a Long Island junior championship golf tournament that afternoon. Although he knew nothing about golf, Danzig told the sports editor, Thomas Rice, that he did, and he got the assignment. A "goodhearted telegraph operator" wrote most of the story for him, and Danzig remembered that he "got a bigger thrill out of seeing that story in the paper" the following day than from earning five dollars for it. He remained on space rates for the next two months, earning between eight dollars and twelve dollars a week reporting on baseball, billiards, boxing, golf, football, and tennis. Danzig's break as a sportswriter came with an article on the Metropolitan Lawn Tennis Singles Championship at the Crescent Athletic Club in the Bay Bridge section of Brooklyn. Rice liked the story so well that he gave Danzig his own byline and promoted him to full-time correspondent at twenty-five dollars a week.

In 1923 Danzig joined *The New York Times* as chief tennis correspondent. On 9 July he married Dorothy Charlotte Chapman of Chicago, whom he had met while he was stationed at Camp Grant, and they settled in Roslyn, New York. In 1924 their daughter Dorothy was born; a second daughter, Marion, followed in 1929, and a son, Allison, in 1932.

In his first year with the *Times* Danzig began an annual pattern of sports reporting that he would maintain throughout his tenure at the newspaper. After the outdoor tennis season concluded with the men's United States Lawn Tennis Association (USLTA) Championship in September, Danzig covered college football, traveling to Ithaca to report on his alma mater. In the winter and early spring he wrote about indoor tennis, court tennis, racquets, and squash – obscure sports that were played and followed by the social elite. In the spring he reported on crew, concentrating on the major Ivy League regattas and the intercollegiate championships.

When Danzig entered journalism, American sport was in the midst of its "golden age." Fans filled stadiums to witness previously unimagined feats of skill, speed, and endurance in baseball, boxing, football, golf, tennis, swimming, and track and field. Each sport had its special hero – Babe Ruth in baseball, Jack Dempsey in boxing, Red Grange in football, Bobby Jones in golf, Bill Tilden in tennis, Gertrude Ederle in swimming, Charley Paddock in track – and sportswriters described their exploits in language that made the athletes seem invincible. The unassailable hero and heroine, whether in sports or other endeavors, suited the mood of a nation that had emerged victorious and vibrant from World War I and offered a public weary of the collective effort and sacrifice of the war years the opportunity to cheer individual triumphs.

Tennis had been confined to country-club courts since the late nineteenth century, but in the 1920s the brilliant play of Tilden, Bill Johnston, and Helen Wills, vividly described by sportswriters such as Danzig, caught the public eye. In a 17 September 1922 *Daily Eagle* story Danzig extolled Tilden, who won his third consecutive USLTA men's singles championship by coming from behind to defeat Johnston three sets to two, as "a player who rises to heights unassailable when in the shadow of defeat and breaks the spirit of his opponents, confounding them with his illimitable powers and variety." By contrast, Johnston, who had won the championship in 1915 and 1919, was on a journey down the "lane of defeat, a lane that will have no turning for him so long as Tilden continues to play tennis." Danzig reported on five more USLTA championship contests between Tilden and Johnston during the 1920s, in all of which Tilden triumphed.

Although Danzig wrote in the exaggerated hyperbole characteristic of the "Gee Whiz" school of sports reporting, he set himself apart from other correspondents by his authoritative analysis of strategy and technique. Danzig, who confessed to Duckett that he "didn't have much of an appreciation for tennis at first," learned about the game by studying the work of other tennis writers, particularly the British, who developed the drama of the match by examining the strengths and weaknesses

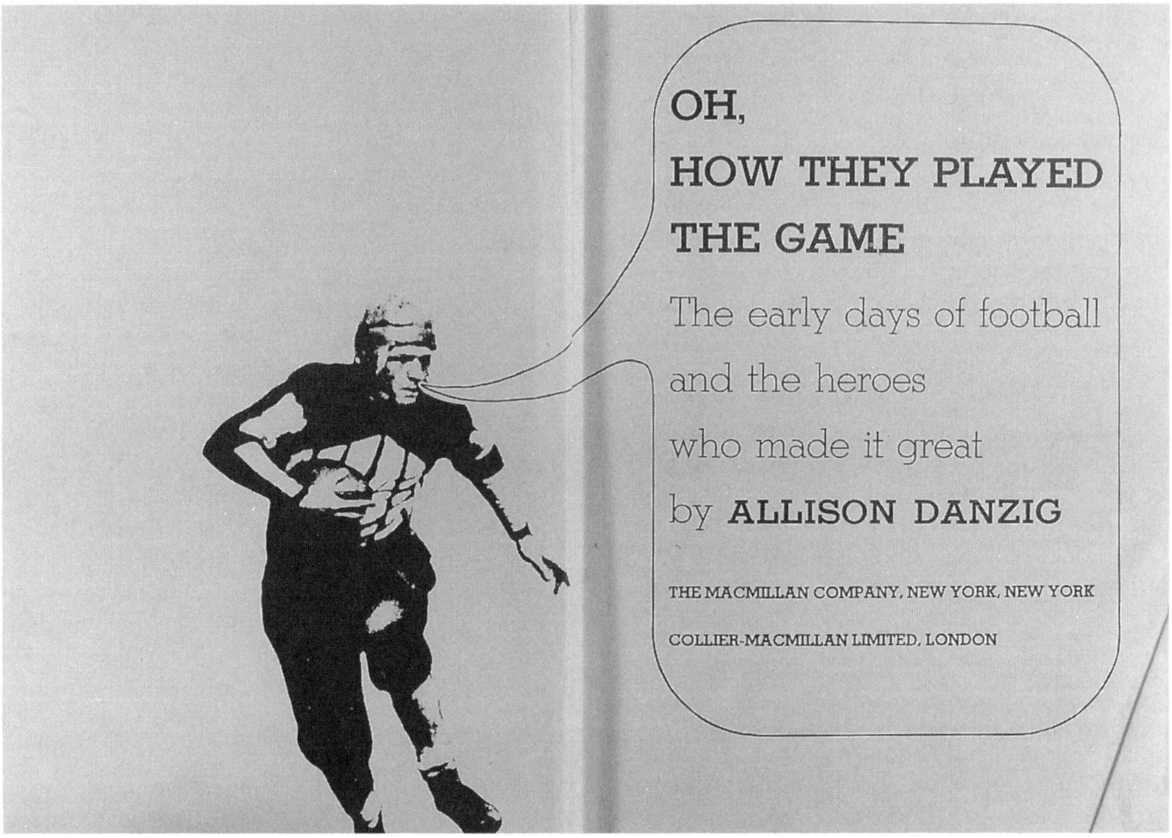


*Frontispiece and title page for Danzig's sentimental memoir of the great players and coaches of college football*

of the competitors. In a 22 August *Daily Eagle* article on the 1922 Davis Cup tournament at Forest Hills, Danzig described Johnston's triumph over James Anderson of Australia:

> The Californian's work was a rare exhibition of the science of tennis. So cleverly did he mix up his shots that Anderson was continually being caught yards and yards away from the ball. If the Australian showed a tendency to hug the baseline, "Little Bill" trapped him with a drop volley just over the net or angled sharply to the side; if he sought a middle court position, his opponent put the ball at his feet; if he advanced to closer quarters, Johnston passed him overhead or sent a streaking drive down the sides.

The 1920s, a period of significant social liberation for American women, witnessed the rise of female athletes in tennis, golf, and swimming. Wills, whom Danzig would recall in *Sport's Golden Age: A Close-up of the Fabulous Twenties* (1948) as a master at "scientific placement and control of the ball," was the top American woman tennis player of the decade. Danzig remembered her "unbelievable accuracy as she put the ball precisely where she willed and maneuvered her opponent until the opening was offered for the finishing thrust." Wills exercised such precise play to capture seven American, eight British, and four French singles titles in the 1920s and 1930s. "The great player personality" of the decade in women's tennis, however, was Suzanne Lenglen of France, who never secured an American singles title but won many Wimbledon, French, and world championships and several gold medals in the 1920 Olympics. Danzig described her in *Sport's Golden Age* as "a prima donna with a figure that would have adorned the stage, as she moved about the court with a glowing grace and ease of movement that was beyond compare."

Danzig also distinguished himself from other tennis correspondents of the 1920s by reporting on lawn tennis's concealed cousins: court tennis, racquets, squash tennis, and squash racquets. These indoor games continued in the obscurity that had shrouded them since the mid nineteenth century because they were played by the nation's wealthy elite, many of whom spent fortunes to build and maintain their own private courts. For example, Jay Gould, who captured the U.S. court tennis singles championship from 1906 to 1925, learned and practiced the game on a court built by his father, George Gould, a wealthy industrialist, in Lakewood, New Jersey.

In 1930 Danzig established himself as the world's leading authority on court tennis, racquets, and squash with the publication of *The Racquet Game,* the first book-length history of these sports. "Mr. Danzig's knowledge of his subjects," according to the review in the *Times Literary Supplement* (1 May 1930), "is only equalled by his modesty in writing of what should be a classic." Danzig worked on the book for five years, using the libraries of racquet and tennis clubs throughout North America as well as private magazine and newspaper collections of past and current players and aficionados. "It is doubtful," he points out early in the work, "if there is any other game quite as fast as racquets and squash tennis, in which the play is sustained at so terrific a clip as to make the fastest game of lawn tennis seem almost mild by comparison."

In 1931 Danzig co-authored with John Hope Doeg, the 1930 American singles and doubles champion, *Elements of Lawn Tennis,* a primer on the game's accoutrements, rules, and techniques. Danzig and Doeg explain, using the tennis elite of the 1920s as examples, how to select the proper clothes and racket; develop a game-winning forehand, backhand, lob, volley, serve, smash, and chop; formulate a game-winning strategy; and play a decent doubles game. Richly illustrated, *Elements of Lawn Tennis* shows Johnston making a forehand drive, Wills demonstrating a correct service, and Tilden executing a smash. Although the authors emphasize that many of the world's outstanding players "were winning laurels on the courts by the time they were in their teens," they assure their readership that not "every player is doomed to mediocrity who does not cut his teeth on a racquet instead of a rattle." Written to encourage ordinary people to take up tennis for health and enjoyment, *Elements of Lawn Tennis* reflects the popularity that tennis had acquired during the 1920s and Danzig's fundamental understanding of the game and its players. Danzig and Arthur Daley covered the Summer Olympic Games in Los Angeles in 1932, the first occasion on which the *Times* sent a team of correspondents to the West Coast to cover a sporting event.

Throughout the 1930s Danzig's reporting style became increasingly precise, assuming the authoritative quality that would become his trademark for the next thirty years and losing the exaggerated hyperbole characteristic of the previous decade. Although he continued to analyze the personalities, strategies, and techniques of the players, Danzig also began to provide a historical perspective on current events that was unique among sportswriters. In reporting on Jack Crawford's thwarted attempt at winning all four major national titles in 1933 (as he lost the American championship to Fred Perry), Danzig was the first to apply to tennis the term *grand slam,* which had been used previously to describe Bobby Jones's winning of four major international amateur golf titles in 1930. Donald Budge's victory in the 1938 USLTA men's singles championship, which gave him tennis's first grand slam, provided Danzig an opportunity to demonstrate his appreciation of the past: by winning the American, Australian, British, and French titles in a single year, Budge took "precedence over William Tilden, the Frenchmen, Henri Cochet and René Lacoste; Ellsworth Vines, Wilmer Allison, Fred Perry, and all the other great modern champions," Danzig wrote in the *Times* of 25 September.

Along with other journalists, Danzig followed the conflict between amateurism and professionalism that threatened to undermine the integrity of American sports. In *Farewell to Sport* (1938) Paul Gallico sharply criticized amateur sports in the United States for hypocrisy, pointing to examples of athletes who traveled, trained, competed, and lived comfortably with no visible means of financial support. Danzig, reviewing Gallico's book for the *North American Review* (September 1938), argued that amateur sports, especially lawn tennis, had become a "big business enterprise" and that the directors of the USLTA and the International Lawn Tennis Federation (ILTF) would face great monetary losses if the players, instead of the associations, collected the gate receipts. He pointed out that the "Australian Association profited to the extent of eight thousand pounds from the tour of Budge and Baron Gottfried von Cramm" in 1938.

The rise of professional tennis in the 1930s robbed the amateur game of its most acclaimed performers, including Tilden. Although Danzig covered professional matches, he harbored great fondness for the previous decade of amateur play. Danzig's sentimentalism for the 1920s first emerged in an article for *The New York Times Magazine* (23 May 1943) on the impact of World War II on American sport. The war had hardly made a mark on the nation's games, as "the homefires of competition were kept burning in baseball, boxing, tennis, track and, to a less degree, in football and golf"; baseball fans still crowded into the nation's ballparks to watch "the four hundred 3A's and 4F's who make up the roster of the major leagues battle for the 1943 world's championship." He predicted, however, that after 1943 there would be "a considerable reduction in sports activity," especially in professional sports, as more athletes joined the armed forces. His outlook for sports after the war was optimistic because "World War I was followed by the golden age of American sports. Never before did so many all-time greats appear on the sporting scene at one time as in the early 1920s."

*Sport's Golden Age,* a collection of essays by the decade's premier sportswriters edited by Danzig in collaboration with Peter Brandwein, marked the crescendo of Danzig's affection for the 1920s. Grantland Rice, the dean of American sportswriters, contributed the lead essay, "The Golden Panorama," in which he attributed the success of sports during the decade not to the athletes' skill but to their personalities and crowd appeal. These factors, according to Danzig and Brandwein's introduction, "challenged the most gifted writers of the nation's sporting press to rise to flights of descriptive English never before nor since attained."

Danzig wrote the chapter on tennis in *Sport's Golden Age,* recapitulating the memorable matches and analyzing the outstanding performers of the 1920s. For Danzig, Tilden was the top performer not only of the decade but of all time (an opinion he would hold well into the 1980s). Tilden's greatness was, however, not based only on his ability but also on the factors defined by Rice. Tilden, writes Danzig, "was so overwhelmingly the favorite in every match and the crowd was for the underdog, he would sometimes play upon the heartstrings of the gallery by purposely getting himself in a hole until the fans were worked up to a high pitch of excitement. Then he would come on with an onslaught that blew his rival off the court and leave the crowd limp and marveling at his performance." Tilden, who also excelled as a stage actor, executed this script repeatedly as he captured seven national lawn tennis singles titles and led the United States to seven consecutive Davis Cup championships.

In 1956 Danzig published *The History of American Football: Its Great Teams, Players, and Coaches;* he would reveal to Duckett in 1981 that the game had been his "lifelong passion." Herman Hickman, the Yale football coach, reviewed *The History of American Football* for *The New York Times Book Review* (9 December 1956), hailing it as "the most ambitious and best book ever published on the subject of college football." Other histories of football, wrote Hickman, "were dull and dry, dealing primarily with the inception and early history of the collegiate game in the East." Danzig's history, based on interviews and correspondence with former gridiron greats, as well as on his own observations and interpretations, offers a vivid and comprehensive analysis of the rise and development of the game. It remains the standard work on the topic, as a scholarly analysis of American football has yet to be published.

Danzig co-authored with Joe Reichler, the baseball editor for the Associated Press, *The History of Baseball: Its Great Players, Teams and Managers* (1959). The strength of the book, according to Arnold Hano in *The New York Times Book Review* (22 November 1959), rests on its "refutation of the Doubleday legend, a careful report of the White Sox scandal of 1919, and an exposure of the club owners as men of greed." The volume was, according to *The New Yorker* (19 December 1959), "the first general history of baseball to be published in nearly fifty years, and it will almost certainly remain the definitive work of its kind for a long time to come." Unlike *The History of American Football,* however, *The History of Baseball* has not remained "the definitive work" but has been supplanted by extensive scholarship.

Although his histories of football and baseball tended to be sentimental narratives rather than critical analyses of sport and society, Danzig's reporting of the Olympic Games examined the impact of the Cold War on international sport, especially in exacerbating the tensions between amateurism and professionalism. In his stories on the Olympics in London in 1948, Helsinki in 1952, Melbourne in 1956, and Rome in 1960 he discussed the implications of the Soviet Union's participation. Unprepared to compete in 1948, the Soviets entered the Olympics in 1952 and finished second to the United States in the unofficial team score. The Soviet performance in Helsinki, observed Danzig, "is a portent of what to expect from them in the future. The belief is that with their intensive athletic training program under government direction and financing they will constitute a real menace in the 1956 games at Melbourne, Australia, if their amateur qualifications are not challenged." The Soviets, indeed, dominated the 1956 Olympic Games, winning both more gold medals and more total medals than the United States. "How can nations," asked Danzig, "whose teams do not have unlimited government funds behind them and which draw a line of demarcation between professional and amateur sports and make an honest effort to maintain it hope to compete on even terms with such subsidized athletes?"

The dominance of Soviet state-supported Olympic teams throughout the Cold War drama-

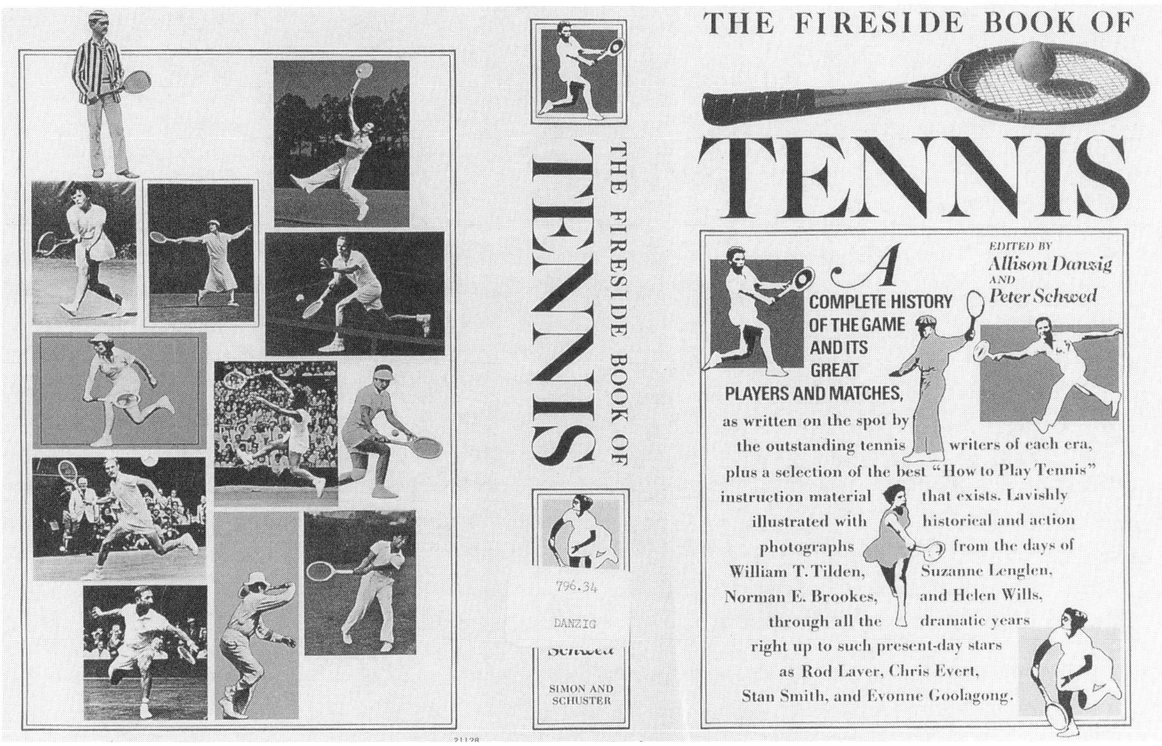


*Dust jacket for the second of the three books Danzig wrote or edited after his retirement from* The New York Times

tized the conflict between amateur and professional sports. Since the late 1920s the rift between amateurs and professionals had divided tennis into separate camps. Although Danzig covered professional tennis, he believed that the division damaged the sport. As the nation's leading tennis writer, Danzig served on a five-member committee appointed by the USLTA in 1957 to examine the possibility of permitting amateurs and professionals to play against each other as a way to curb "the payment of expense money to amateurs" and halt the exodus of the best amateur players to the professional ranks. The committee decided unanimously in favor of the proposal, but the USLTA rejected the committee's recommendation and did not open its tournaments to professionals until 1968.

Danzig had been elected to the Board of Governors of the National Lawn Tennis Hall of Fame and Tennis Museum in 1954, and in 1958 he was elected to the Board of Directors of the National Lawn Tennis Hall of Fame (now the International Tennis Hall of Fame). In 1960 Columbia University established the Allison Danzig Cup, which is awarded to the winner of the Columbia-Cornell tennis match. He received the first Allison Danzig Award from the Longwood Cricket Club in Chestnut Hill, Massachusetts, for distinguished tennis writing in 1963 and was named to the National Football Foundation Hall of Fame in 1965. He also received the prestigious Grantland Rice Award for Outstanding Sportswriting. In May 1968, following his retirement from *The New York Times,* the New York Cornell Club held a testimonial dinner for Danzig at which media colleagues and athletes paid tribute to him. In August of that year he became the first journalist to be inducted into the National Lawn Tennis Hall of Fame.

During his retirement Danzig completed three books. *Oh, How They Played the Game: The Early Days of Football and the Heroes Who Made It Great* (1971) is another sentimental look back at the great players and coaches of college football. Joseph Durso used such terms as "Romantic" and "hero worship" to describe the book in his review in *The New York Times* (9 January 1972). *The Fireside Book of Tennis* (1972), edited by Danzig and Peter Schwed, is an anthology of articles by various tennis writers; Neil Amdur, a sportswriter, hailed it in *The New York Times* (30 July 1972) as a "history book to satisfy the insatiable appetites of patrons who long to relive the sport's countless court treasures," a work that chronicles "every significant match since the turn of the century, from the first Davis Cup match to Jaroslav Drobny's Wimbledon." *The Winning Gal-*

*lery: Court Tennis Matches and Memories* (1985), financed by the U.S. Court Tennis Association, examines the history of court tennis and the game's outstanding players.

Danzig's death of a heart attack on 27 January 1987 at the Valley Hospital in Ridgewood, New Jersey, marked the end of an era in sports journalism. In Lorge's *Tennis Magazine* article Dave Anderson describes Danzig as an "old-school, formal writer." Whereas the modern sportswriters blend players' comments into their stories, Danzig, according to Lorge, seldom quoted players: "he believed that his responsibility was to give readers his observations and interpretations, not those of the participants." Nevertheless, Danzig established a standard in sportswriting that remains a source of guidance and inspiration, especially to tennis correspondents. When Lorge started reporting tennis in 1967, he remarked to Duckett in 1981, he read everything he could by Danzig because he was "the standard." For Lorge, Danzig's writing was "vivid, the voice of authority and authenticity"; he was simply "the god of tennis writers."

**Interview:**

Joy Duckett, "A Fireside Chat with Allison Danzig," *World Tennis,* 28 (May 1981): 40.

**References:**

Bud Collins, "Danzig's Strokes Were Elegant," *Boston Globe,* 9 July 1987, p. 8;

Collins and Zander Hollander, eds., *Bud Collins' Modern Encyclopedia of Tennis* (Detroit: Gale Research, 1994);

Dorothy Lobrano Goth, ed., *Letters of E. B. White* (New York: Harper & Row, 1989), p. 18;

Barry Lorge, "Chronicler of the U.S. Championships," *Tennis Magazine* (September 1987): 107–108;

George McGann, "The Gentleman from the Times," *World Tennis* (March 1956): 15.

# Charles Dryden

*(10 March 1860[?] - 11 February 1931)*

William H. Wood
*Community College of Philadelphia*

MAJOR POSITIONS HELD: Reporter, *Chicago Sunday Times* (1889–1890); sportswriter, *San Francisco Examiner* (1890), *New York Evening Journal* (1898–1899), *Philadelphia North American* (1899–1905), *Chicago Tribune* (1906–1908), *Chicago Examiner* (1908–1917[?]), *Chicago Herald and Examiner* (1918–1921).

BOOKS: *Swanson, Able Seaman; or, The Melancholy Fate of a Hapless Mariner* (Asheville, N.C.: Bandar Log Press, 1901);

*The Champion Athletics: Personal Sketches of the Men Who Have Just Landed the 1905 American League Pennant after the Greatest Fight in Base Ball History* (Philadelphia: *North American,* 1905);

*On and off the Bread Wagon: Being the Hard Luck Tales, Doings and Adventures of an Amateur Hobo* (Chicago: Star, 1905).

One of the earliest sportswriters to be seriously recognized, Charles Dryden was hailed by *The Sporting News* (19 February 1931) as the "creator of [a] new baseball literature." Celebrated for his wit, he was widely regarded, according to *The New York Times* (12 February 1931), as "first and perhaps the greatest of all baseball humorists." In the introduction to his *The Champion Athletics: Personal Sketches of the Men Who Have Just Landed the 1905 American League Pennant after the Greatest Fight in Base Ball History* (1905), the editors of the *Philadelphia North American,* one of the newspapers for which he worked, called him "concededly the best base ball writer who ever took up a pencil." Dryden took baseball reporting beyond its limits at the start of the twentieth century, when two or three short paragraphs often sufficed for a featured game, and expanded it to columns of amusing, accurate details about the players on and off the field, the umpires, fans, writers, ballparks, and even the streets of the cities.

His humor was the talent most appreciated by his contemporaries, followed by his ability to sum up a player or situation with an apt nickname or telling phrase. According to Wayne K. Otto, with Ring Lardner "he made satirical baseball writing an art." His pen (only late in his career did he use a typewriter) was dipped in corrosive lime, and he gave equal attention to errors and great plays, to scrubs and stars, to the gaffes and the pageantry of the game – all of which he helped make known to a much wider audience.

One of eight children produced by his mother's two marriages, Charles Dryden is believed to have been born on 10 March 1860 – in his autobiography he says "the late '60's," which he calls "close enough for the purpose of this confession" – in Monmouth, Illinois, to William A. and Harriet Cabeen Dryden. He may have inherited what he called his "hobo instinct" from his father, a traveling salesman; the first of Dryden's many travels occurred at age five, when his family moved in a bobsled from a farm to "a little mud-streaked village in Western Illinois." The young Dryden was, by his own account, "a natural born promoter" who worked a deal with the town's elderly librarian for whom he carried coal in exchange for library privileges. He "thirsted for choice miscellany, useful facts worth knowing and light summer reading in winter." This knowledge bore immediate fruit in schemes to bilk small sums from friends and gullible adults, and later, riper fruit in promoting himself from job to job. Much of the "choice miscellany" and many of the "useful facts" would turn up in his columns as background and colorful digressions on anyone or anything involved with baseball.

Dryden ended his formal schooling ("time sacrificed in the academic halls of Mudville") at fourteen, when he began work as an iron molder in his hometown. He soon grew restless and began fourteen years of wandering that included stints as a

*1905 cartoon by Hy Gage showing Charles Dryden with Finley Peter Dunne, editor of the* Chicago Sunday Times

hobo, stagecoach driver, sailor, boat pilot, and longshoreman; some of these experiences would provide material for his booklet-poem *Swanson, Able Seaman* (1901), and all of them would provide anecdotes and images for his sports reports. In 1887 a reporter friend had almost persuaded him to write for the *Chicago Mail,* but on a winter walk with Dryden the friend spotted "a ragged laborer" whom he said he envied because the man wore socks. "I did not yearn for a profession that could not haberdash itself," Dryden declared.

His friends persisted, and in 1889 they talked him into writing down one of his travel yarns, which was printed in the *Chicago Sunday Times;* the humorist Finley Peter Dunne, the editor of the paper, hired him immediately as a free-roving reporter. Having come to journalism somewhat late in life – in his late twenties – he seems to have worked out his methods, and certainly his sentiments, beforehand; he says in his autobiography, "My habit for years was to read the daily papers from top to bottom, and I was, therefore, loaded with language, spelling, padding, punctuation, syntax, wind and other essentials without knowing it." His earliest sports story, in 1889, covered "a baseball insurrection" of orphans who had run off for a game of sandlot ball. He named their leader after the well-known Chicago pitcher Cap Anson, "and spun out the adventures of the three young players." The story, which was "more or less upholstered" with his own terms and inventions, "passed muster" with Dunne.

A year later Dryden was writing sports for the *San Francisco Examiner.* In 1898 he moved to the *New York Evening Journal,* where he got into trouble with the Giants' owner for his irreverent reporting and was banished from the Polo Grounds. Thereupon he began watching Giants games from the top of a telegraph pole outside the park (one of his previous jobs had been stringing telephone wire in Texas). "The Giants look just as bad from out there," he is supposed to have said. In 1899 he moved to Philadelphia, where he covered the home games of the Phillies and, beginning in 1901, of the Athletics for the *North American.* By the start of 1902 he might still spend as much as half his report on routine inning-by-inning coverage of plays, but as his editors came to trust him more he began looking to the human-interest side of the game instead: "As this thing of losing with [George Edward] Rube [Waddell] on the slab is extremely dismal business, we will turn to peanuts as a more cheerful theme upon which to dilate" (1 July 1903); most of the game report then deals with the duties of Mr. Somers, Cleveland's full-time peanut taster. Other good topics for digression were fashion at the ball yard, a player's origins and quirks, or the fact that his family was in the crowd. The day Waddell's father came out was one to remember: "Every time the southpaw mowed down a man the old gentleman laughed until his whiskers swayed like a weeping willow in the summer breeze. His joy was of the infectious sort." Suddenly Rube charged into the stands to remove a heckler "by the face," and "after the battle Mr. Waddell lighted a fresh cigar and beamed down upon the scene. 'I see my boy got what he went after, and Ed isn't hurt, either' " (18 July 1903).

This "spun-out" style, from early on, relied on specific details, balanced sentences, assorted current and classical references, and the thing that most set Dryden apart from other journalists: his imagery. No other sportswriter has invented so many vivid images to create a scene or make a point. Sometimes they recalled his days at sea or in the foundry: in the outfield, trying repeatedly to pick up a ball in play, "Ralphy Orlando sailed up like a schooner bucking a head breeze" (20 July 1904). "All hands and the cook took a whack at the closing game" (30 July 1903) of a long season. In his great 1905 Series, "Matty [Giants' pitcher Christy Mathewson] put the lid on the can and soldered it fast" (15 October) for his third win. At an exhibition game on the Fourth of July, "the place was hotter than a puddler's furnace" (5 July 1903).

A booklet, *The Champion Athletics: Personal Sketches of the Men Who Have Just Landed the 1905 American League Pennant after the Greatest Fight in Base Ball History,* was printed by his paper in 1905. A series of one-page biographical accounts of the team's top players, written with an absurdist touch – "when in condition Rube eats with both hands and pitches with his left" – it was a tremendous success with the fans of Connie Mack's first dynasty. Dryden's autobiography, *On and off the Bread Wagon: Being the Hard Luck Tales, Doings and Adventures of an Amateur Hobo,* was published the same year. An episodic, sometimes inspired account of his adventures in the West up to the time he took his first newspaper job, it is a sort of latter-day version of Mark Twain's *Roughing It* (1872).

The *Chicago Tribune* hired Dryden away from Philadelphia at a record salary in 1906. There he met the young Lardner, who was working at his first large daily newspaper. Lardner learned much from Dryden: according to Lardner biographer Donald Elder, "the Cubs players called [Lardner] 'Charlie's hat' because he went everywhere with Dryden, who probably had a greater influence on his early career than any other writer." Dryden's reporting techniques, style, and attitude all made an impact on Lardner. The focus of a sports column, Lardner learned, did not have to be the game itself but could be the events surrounding it. Conciseness was important, and slang could be used. Most of all, the reporter did not have to be a cheerleader. He could note the flaws in a player's performance and satirize pretension everywhere. "The three most prominent Chicago baseball writers of the time [were] Hugh Fullerton, I. E. Sanborn and Charles Dryden," says Lardner biographer Jonathan Yardley. "It was Dryden who had the greatest influence on his writing style."

The bulk of Dryden's work consists of his daily reports of the thousands of games he covered – and of the games he did not cover: rainouts, off-day pieces, and at least one "snowout." In all of these pieces his inventiveness, powers of observation, and humor reveal themselves.

Reports of games, for Dryden, organized themselves into three unequal parts: the lead, with an overview of the contest; descriptions of and digressions on the players and other human-interest elements; and the routine details of the game. The digressions are often the longest and most interesting part of each report (they are severely cut in his few anthologized articles). Dryden digresses, muses, creates metaphors and motifs, re-creates dialogue, fills in details, and makes colorful observations and overstatements at whatever length his editor would allow – sometimes it was considerable – but always gets back to the play-by-play description of the game.

"Hoodooed by his own spit ball and the affectionate pitching of Mr. Plank, the oozy Howell lost the game he had vowed to win," from the *North American* (5 October 1905), is a typical lead for Dryden on an unimportant game – this one between the Athletics and the Saint Louis Browns – and displays several of his stylistic traits: the internal rhyme of *hoodooed* and *oozy*; the unexpected description of "affectionate" pitching (Eddie Plank allowed nine base runners, including four walks and a hit batter, for five runs); and the ironic outcome for Harry Howell. He could write an equally good lead for a significant game, as he did on the following day: "This was for sure a red-letter day in the career of our noble red man. Charles Albert Bender, an American Indian, virtually landed the American League race by winning two games this afternoon, backed up by the lusty slap sticks of his mates." Here are wordplay, with *red-letter* and *red man* (an unfortunate expression from a more sensitive late-twentieth-century perspective); puns, with *slap sticks* for bats; and an intentionally inflated style, using the player's full name (or, as in the previous day's lead, *Mr.*) at a time when professional athletes were considered roughnecks. Dryden pokes affectionate fun at the players and the proceedings – "at all of them indiscriminately," according to Edgar G. Brands in *The Sporting News* (19 February 1931) – all the while informing his readers of the essentials: who played and how.

His leads telegraph his iconoclasm. When the 1906 World Series shifted from the Cubs' park to the White Sox's he wrote in the *Tribune* for 11 October: "Over in the stockyards district, where gentle deeds and smells are rare, the Cubs dragged the Sox around their own killing beds and slaughtered them to a finish. Score: 7 to 1." When a 1919 Sox team reached the Series as an overwhelming favorite, he began his report: "Dear Old Cincy, Sept. 30. – On the eve of battle one is oppressed with the notion that having a series here is like staging Barnum and Bailey's circus on a tennis court" (*Chicago Herald and Examiner,* 1 October 1919). Dryden avoided the bucolic "green grass, blue sky" clichés found in much baseball writing then as now.

These are not the short, punchy openings that, by conventional journalistic standards, make the best leads. (Dryden could write those, too, however: "Beaten by a nose!" is the opening paragraph of his 12 October *Tribune* story on game three of the 1906 Series, in which the White Sox's Fred Hahn was hit in the nose by a pitch, setting up a game-winning triple.) They are indicative of a slower, more relaxed era when most games were played at three o'clock on a weekday afternoon, the gentlemen in the stands wore starched collars, and a crowd of fifteen thousand was extraordinarily large. There was time for the writer to develop his thoughts in long, ornate sentences, with a rich vocabulary and telling details, and for the public to read the column at a leisurely, literate pace.

In many ways, the long center sections of Dryden's columns are his most impressive work. At a time when there was no tradition of pre- or post-game interviews the writer was left to his own resources to amuse and enlighten his readers. It was here that Dryden catalogued the sights, sounds, and feel of the game and of the era – the ferryboats and trolleys; the look and smell of the ballpark; the excitement of the fans in the big industrial cities at the turn of the century over a burgeoning game with *two* major leagues; and the colorful characters who played that game.

That they were "characters" is in part because of him. With wit and a talent for summing up a habit or significant trait in a phrase, Dryden created memorable portraits and epithets that became legend. Ed Walsh of the White Sox was "the Big Moose" and "the only man who could strut standing still." Frank Chance became the "Peerless Leader" of the Cubs. The White Sox were "the hitless wonders" in 1906, and their owner, Charles Comiskey, was "the Old Roman." Joe McGinnity, the Giants' hurler who always led the league in innings pitched, was "Iron Man" McGinnity. Brands says that Dryden made the players "the idols of the multitude."

Creating idols was not Dryden's goal, however; he wanted to report the game accurately, even when doing so meant criticizing players or smashing idols. Nor were all the epithets he created intentional. In a *Tribune* story for 24 September 1908 one can see the origin of the unfortunate tag *bonehead* for rookie Fred Merkle:

> Minor league brains lost the Giants' game today after they had it cleanly and fairly won by a score of 2 to 1. In the ninth round, Fred Merkle did a bonehead base-running stunt. [Joe] Tinker's home run in the fifth was all the Cubs had in the way of visible assets. . . . Then came the bonehead finish. . . . [ Merkle, on first, saw the winning run cross the plate on a single and, instead of going all the way to second, turned off and headed for the clubhouse.] Capt. [ Mike] Donlin realized the danger about to overtake the Giants, so he set off after the fat-headed Merkle. [Solly] Hofman fielded [Al] Bridwell's knock and threw to Evers for a force play on the absent Merkle. But McGinnity, who was not in the game, cut in ahead and grabbed the ball before it reached the eager Trojan [Evers]. . . . Tinker wedged in and the ball was conveyed to Evers for the force out of Merkle. . . . Some base runners should have a groove cut for them so they couldn't go wrong.

Typically, Dryden is angry here at ineptitude, even though it led to his own team's winning: because of the crowd that poured across the Polo Grounds at the apparent New York victory, the game had to be replayed; the Cubs won, and the Giants lost the pennant by one game. The nickname seems well earned for that game, though Merkle went on to have a solid career. But *bonehead,* used twice, stuck with him for good, while other possibilities such as *fatheaded* or *absent* did not. *The Trojan* had already stuck with Johnny Evers, who is referred to later in the article as "the well-known Troy shoe dealer." Clearly, which label creates the legend is only partly under the control of the writer. Dryden invented hundreds; some stuck, but many more deserved to.

Dryden's character sketches and commentary were irrepressible. In a *Tribune* story for 9 October 1906 the National League president was "the Hon. H. Pulliam, the human haberdashery. Without saying a word Harry can make himself heard above the hum and turmoil of traffic which grows pretty loud in this city." Character sketches such as this one took Dryden's creativity and humor a step further than the nicknames and sometimes contained an epithet into the bargain. When Browns pitcher "Handsome Harry Howell" threw the spitball, he became "the slobber artist" (*North American,* 1 April 1905). Phillies catcher "Mr. C[harles] Dooin, social favorite, is in the pink of condition from his hair down" (5 October 1905). In attendance at a 1905 Series game was Yankees pitcher "Clark Griffith, worn from a long and fretful campaign" (10 October) – another weak season for the pre-Babe Ruth Yankees. Also present was one of Dryden's favorite umpires: "Mr. Hank O'Day trotted out dressed within an inch of his ears in a suit of the best blue serge money can buy. There is nothing tight about Hank when it comes to clothes, unless it be the fit across the hips." Two days later, at the next game, Dryden comes back to O'Day in a continuing series-within-the-Series (another device that had an impact on Lardner) on umpires: "He was arrayed in the new suit, too. Hank moved the left foot forward until the leg was bent at the angle greatly affected by the gentlemen of wax who pose in the front window of the clothing store." Dryden was a writer with an eye for detail of every kind, inventing similes and metaphors on a daily basis.

The character sketches are peppered with puns and effective exaggeration. There were two younger, unrelated "versions" of Cy Young – Irv Young ("Cy the Second") and Harley Young ("Cy the Third") – and Irv "displayed remarkable control. . . . Were it not for the catcher and grand stand, Young Cy's curve would eventually come floating back and bump him in the solar plexus. A stocky youth is Cy; built like the gentleman who tosses cannon balls at the circus" (29 April 1905). Chief Bender was "the lean and sinewy Charles" (11 October 1905). Another player, whose career was suddenly over, found "his grand right arm turned to pig iron," and a staggering Boston Braves base runner "arrived at the bag like the leading lady of the beef trust falling in a faint" (4 July 1904). During the 1907 Series, poor attendance was perhaps because of the slump of one player: "T. Cobb was not living up to his picture on the front page of Sporting Life" (*Tribune,* 11 October). The eyewitness immediacy of these detailed and sharp portraits often makes them iconoclastic rather than heroic: Dryden reported what he really saw.

Manager John J. McGraw, who labeled the Athletics "white elephants" when the American League first challenged the National on the field in 1903, was presented a ceramic white elephant before his Giants faced the A's in the Series two years later. "When the elephant came home to roost, John J. was game, but he did the proper thing. He wiggled and blushed some; then, snatching off his cap he placed the little pale beast on his head and faced the crowd. By this action, McGraw conceded the elephant a place" (*North American,* 10 October). This portrait is quite different from the familiar image of McGraw; rather than tough, grim, and unyielding he is conciliatory, even shy. His best pitcher was Christy Mathewson, "the dreaded Factoryville blonde" (12 October 1905), "the greatest pitcher of his day and generation" (15 October 1905).

Dryden's spontaneity and flow of words, his enjoyment of baseball and of his job, obvious in these sketches, belie his appearance in a 1905 photograph that shows a smallish, sharp-featured man whose face seems a cross between Ty Cobb's and the farmer's in Grant Wood's painting *American Gothic.* But he was sociable, and he evidently enjoyed few things more than talking with the players before games or off the field. He was said to be whimsical and quiet, and many of the best passages of his autobiography describe solitary bliss – driving a stagecoach ("sometimes . . . I had the entire Northwest to myself"), fishing, camping out. His umbrella in the photograph is tightly rolled, but his almost elegant suit is rumpled; and he is wearing a Stetson hat.

One of Dryden's favorite players was Philadelphia's Waddell, a great pitcher of the era and a radically nonconformist individual. Season-long coverage in 1905 gave readers the chance to accumulate telling details about the man as he became a legend. When Waddell made his first pitching appearance of 1905, "he waited until the rest of the cast had assembled. Then he strode with flying steps to his place, while the populace reared up and shrieked." Waddell was the league's top pitcher that year, but he injured his arm before the Series. Someone "gave Rube a bottle of prize-fighter liniment to rub on the sore arm. Rube toted the turpentine into the grand stand and showed it to all who cared to look." It did not work, and Waddell had to watch the Series from the sidelines. Before game two, "the

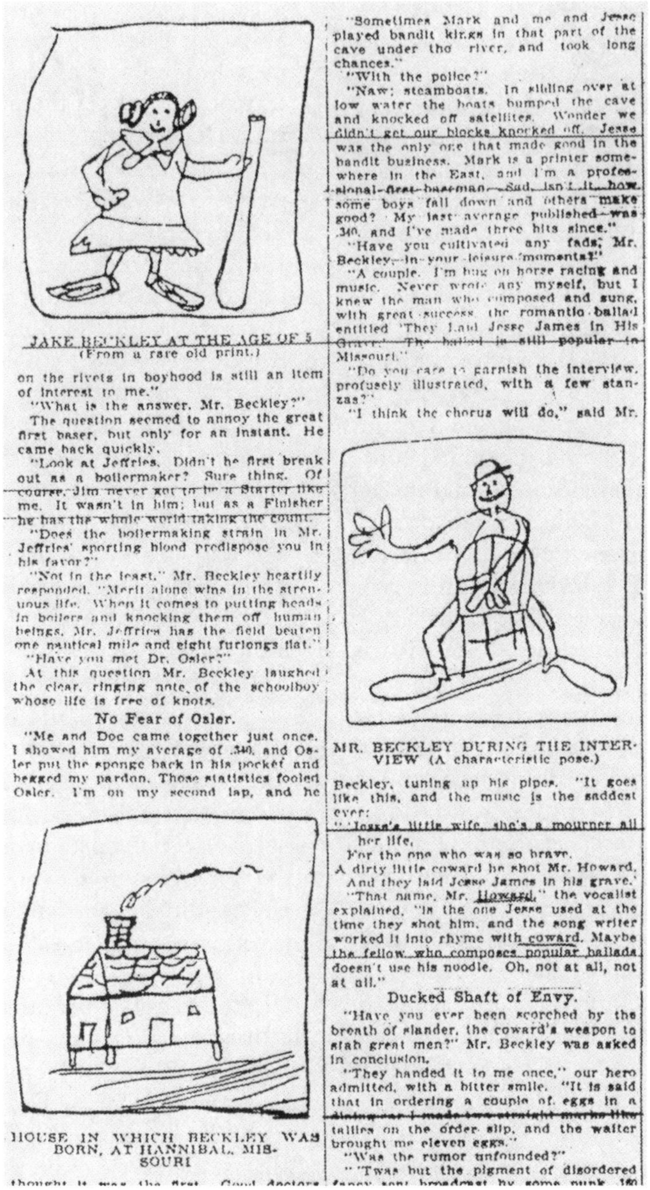

JAKE BECKLEY AT THE AGE OF 5
(From a rare old print.)

on the rivets in boyhood is still an item of interest to me."

"What is the answer, Mr. Beckley?"

The question seemed to annoy the great first baser, but only for an instant. He came back quickly.

"Look at Jeffries. Didn't he first break out as a boilermaker? Sure thing. Of course, Jim never got to be a Starter like me. It wasn't in him; but as a Finisher he has the whole world taking the count."

"Does the boilermaking strain in Mr. Jeffries' sporting blood predispose you in his favor?"

"Not in the least," Mr. Beckley heartily responded. "Merit alone wins in the strenuous life. When it comes to putting heads in boilers and knocking them off human beings, Mr. Jeffries has the field beaten one nautical mile and eight furlongs flat."

"Have you met Dr. Osler?"

At this question Mr. Beckley laughed the clear, ringing note of the schoolboy whose life is free of knots.

**No Fear of Osler.**

"Me and Doc came together just once. I showed him my average of .340, and Osler put the sponge back in his pocket and begged my pardon. Those statistics fooled Osler. I'm on my second lap, and he

HOUSE IN WHICH BECKLEY WAS BORN, AT HANNIBAL, MISSOURI

"Sometimes Mark and me and Jesse played bandit kings in that part of the cave under the river, and took long chances."

"With the police?"

"Naw; steamboats. In sliding over at low water the boats bumped the cave and knocked off satellites. Wonder we didn't get our blocks knocked off. Jesse was the only one that made good in the bandit business. Mark is a printer somewhere in the East, and I'm a professional first baseman. Sad, isn't it, how some boys fall down and others make good? My last average published was .340, and I've made three hits since."

"Have you cultivated any fads, Mr. Beckley, in your leisure moments?"

"A couple. I'm bug on horse racing and music. Never wrote any myself, but I knew the man who composed and sung, with great success, the romantic ballad entitled 'They Laid Jesse James in His Grave.' The ballad is still popular in Missouri."

"Do you care to garnish the interview, profusely illustrated, with a few stanzas?"

"I think the chorus will do," said Mr.

MR. BECKLEY DURING THE INTERVIEW (A characteristic pose.)

Beckley, tuning up his pipes. "It goes like this, and the music is the saddest ever:

"'Jesse's little wife, she's a mourner all her life,
For the one who was so brave.
A dirty little coward he shot Mr. Howard,
And they laid Jesse James in his grave.'

"That name, Mr. Howard," the vocalist explained, "is the one Jesse used at the time they shot him, and the song writer worked it into rhyme with coward. Maybe the fellow who composes popular ballads doesn't use his noodle. Oh, not at all, not at all."

**Ducked Shaft of Envy.**

"Have you ever been scorched by the breath of slander, the coward's weapon to stab great men?" Mr. Beckley was asked in conclusion.

"They handed it to me once," our hero admitted, with a bitter smile. "It is said that in ordering a couple of eggs in a dining car I made two straight marks like tallies on the order slip, and the waiter brought me eleven eggs."

"Was the rumor unfounded?"

"'Twas but the pigment of disordered

*Part of a page from the 18 May 1905* Philadelphia North American *featuring one of Dryden's Rainy Day Interviews. The crude sketches by Dryden always accompanied these pieces.*

horde of snapshotters picked out prominent victims on both sides, and took their maps for the papers. Waddell was featured in various poses. . . . Rube cannot play ball, but he played to the gallery just the same" (*North American,* 11 October 1905). These details reveal a simpler time, when professional athletes and their fans might actually get to know each other.

In an off-day article on the all-Chicago World Series of 1906, Dryden turns a nonevent into a mock epic. His lead sets the tone: "Civil war breaks loose here today in Mr. Murphey's ball yard, and this morning sporting blood flows red and warm in the veins of the jungle." The war motif continues to the end. "If weather permits, the Spuds [later Cubs] and the Sox begin a battle for the world's horse blankets – the emblem of baseball supremacy these days. . . . All the choice seats . . . have been sold, and the populace clamors fiercely for more. Final instructions to athletes, umps, and the local and visiting war correspondents filled in time yesterday." After twenty-two Series rules had been read aloud,

> the bronzed athletes had grown restless and were casting uneasy glances at the door. [Yet] . . . their loyalty

> stood firm and unmistakable. The White Sox, headed by President Comiskey, camped on the south side of the room in a bunch. Along the west wall the Spuds massed behind their beloved Murphey.... Ranged along the rear wall, the muckrakers listened.... The main idea was to insure correct deportment on the battlefield..... There are said to be athletes who coax wild shrieks from the crowd.

There were even spoils of war: "The Spuds had united on diamond studded cuff buttons ... provided they won. As yet the Sox have not submitted ground plans and specifications for an emblem. Being too near the stockyards it is thought they would esteem an oxidized silver wienerwurst to dangle from their watch chains" (*Tribune,* 9 October). Dryden turns potential filler into an amusing look at the game off the field that preserves a sense of the rivalry between the teams, the importance to these professionals of the Series but not of the niceties that sometimes upstage it, and the players' relationship with their "beloved" owners.

Highly literate effects and allusions, both classical and modern, are pointedly placed in many of his columns. Napoleon Lajoie "glared at the pitcher like Polyphemus at Aeneas" (quoted in *The New York Times,* 13 February 1931). Mounted police were brought into the Polo Grounds in 1905 for crowd control, and "the sword-bedecked cavalry steeds cavorted, with bleacher to the right of them, to the left and behind" (*North American,* 11 October 1905). During one game "Plank fanned Sandow, and much tumult smote the air" (10 October 1905). "Baseball, like war, is all that General Sherman said it was, and more, too" (4 April 1905). There are less-high-flown, but equally effective, metaphors. "The drizzle was stronger than the will of the people" (12 October 1905) – the game was called. In the same article, a league official gives a tedious speech: "the words rumbled out of his face like a train of empty coal cars emerging from a tunnel," and groundskeeper Schroeder provided "soft drinks for hard drinkers." With three errors in one game, C. D. Dexter "played his job like a blind pig hunting for acorns" (5 September 1903). Another player, after sliding in the mud, was "decorated with the brown badge of courage" (4 September 1903). The language is literate, direct, and concise, the vocabulary simple and solid. After a home loss, "it was a dumb and dismal throng that wiggled to the L and subway, when the big show dumped its contents from the yard" (11 October 1905).

Along with his impressive standard vocabulary, Dryden experimented over the years and helped to work out a new language for baseball. *The New York Times* labeled him "the father of baseball slang" (13 February 1931). Occasionally his word choice seems stilted; the ninth *round* or *chapter* sounds awkward now. But he coined some of the terms that today are taken for granted: *mound* (instead of *pitcher's box*), *ball yard, fanned;* and he helped to standardize others: *skied* to left, *sidewheeler* for "sidearm," *short* for "shortstop."

Dryden broadened the scope of sports reporting: the crowd in and around the park is often as much a part of the game for him as are the players and umpires. "Cruel disappointment jarred the 15,390 beauty and chivalry that made undue haste to be among those present at the ball yard. Those Phillies lost, 5–4, in a corking good game, and there wasn't a single fight inside the lot, nor any attempt to scalp Mr. McGraw's Mastodonic Metropolitan Marvels outside the breastworks" (*North American,* 25 April 1905). To Dryden, the game was played for the fans; during his time the park was not yet a stadium, and folks sat close to the field, on it, or on fencetops or rooftops across the street, active participants rather than passive spectators. The fans are "this gorgeous display of humanity in bulk" (25 April 1905) – a tongue-in-cheek, but affectionate, description.

"The crowd at first series prices filled the structures, roped spaces and fence tops.... Late comers with the coin and football training could find a peeping place and there was room on the soft grass for a few more. All the $1 and $2 seats were grabbed early.... Scores of local and visiting scribes were placed on the graveled pavilion roof, because it is hard to keep the working man down." Two strikeouts came at a crucial point, and "a dull ache gripped the heart of the fanatics" (10 October 1905). In the next day's game "A riot of sound broke out in the seventh following Doolin's two-bagger among the grass-pressers in right." Everyone and everything gets due notice, and Dryden and his "working man" colleagues are needled into the bargain. Finally, "When the third day's battle ended a sad and sticky throng oozed out of the ball yard," and "Mr. Bausewine [the umpire] went home in a trolley car, same as the rest of us" (11 October 1905).

In the spring the parks were full of promise. At Columbia Field, "Everything about the works looked bright and gorgeous. New signs, beer and otherwise greet the eye of the thirsty. A new green drop crtain [*sic*] in deep centre makes a break in the pleasing vista of literature on the walls, and the structures everywhere glisten with fresh paint and varnish." For the Phillies' opener against Baltimore in 1902, "across the green field, marked by the newly chalked diamonds, marched a noble array of knights of the national game" (19 April).

The contours of the park affected the game, as did the weather, and readers always got enough details to re-create the events in their minds: "A high wind baffled the fielders some and a higher screen topping the right field fence blotted out a few prospective home runs" (2 April 1905). At the Polo Grounds, "In right field the sun seats slope sharply upward to a tall slatted fence. Reckless boys, unable to find seats shinned up the slats and hung on by their teeth" (11 October 1905). Some days were beautiful: "Monte tore off a triple that pushed the clouds off the face of the sun" (2 April 1905). Others were dismal: "Score, 6–2 below zero, partly cloudy, variable winds" (15 April 1904). Sometimes April was too cold for baseball: "Attendance about one gross, most of whom kept moving constantly, on the theory that the process of freezing to death is thereby delayed" (15 April 1904). Time could be a factor: "Gentle night came on and stopped the maddened efforts" (*Tribune,* 1 October 1907). During a rain delay "enough sawdust to carpet the infield was spread" on the basepaths, and "during the brief time they played the outfielders stood in water" (*North American,* 10 September 1905). As for the psychological effect: "These damp days fill the soul of the umpire with ennui and he has a dismal time of it loafing around hotels in alien towns" (29 April 1904). The crowd, the park, and the weather give a broad sweep to these reports, and an intensity that survives today. Dryden obviously delights in the excitement of the game and the skill displayed in it, and in bringing them to the reader.

Prominent people and contemporary events, such as Teddy Roosevelt at San Juan Hill, the Russo-Japanese War, and Adm. Robert Peary, and other topics that would normally fall outside the purview of a baseball writer, such as Tammany Hall politicians or a gas company scandal ("we cannot afford to waste much gas on those Phillies owing to the imperiled state of the city's supply, which is being so ably defended in other parts of the paper. Hence this brief synopsis" [*North American,* 29 April 1905]), are frequently drawn in for metaphors and local color. The 1918 World Series, played during World War I, featured a key two-run "roaring" hit in game four and made the Cubs' right fielder think "the shot came from the supergun the Germans used to have in the forest of St. Gobain. But it was only a triple from the bat of Babe Ruth" (*Chicago Herald and Examiner,* 10 September). The references urge the reader to balance the importance of the sport (and its writers) against larger events. Dryden gives all aspects of the sport their due but keeps them in perspective.

Dryden consistently refused to adulate or promote. The American League Senators were the target of one of his best lines: "Washington: first in war, first in peace, and last in the American League." They were the "cellar champions," but they eventually lost the title to "those Phillies," "champions of the lower levels." Those Phillies "emerged from the cellar for the second time this season. . . . The deed was accomplished yesterday because Boston did not play, though they went through the motions" (*North American,* 5 September 1905). In an early look at the finances of the game he was even blunter. "In the centre of the upper grandstand overlooking the busy pencil pushers sat the National Commission, mentally figuring on the 10% rake-off. On either side ran the boxes in a sweeping crescent, filled with the bonton, gazing at the National Commission" (11 October 1905). Dryden kept a running public record of the World Series gate, the receipts, and the percentage for each player, thereby reminding readers that baseball was a business as well as a game. There is no pretense or boosterism here, and one can see why a club owner might be hostile to this kind of reporting.

The third section of Dryden's game-day reports was the play-by-play, which was kept lively and concise. During the third game of the 1905 World Series,

> Coakley lost the game, or the Giants won it, in the first round. Andy soaked the ball into the fat back of [Roger] Bresnahan with sufficient force to melt the marrow in his spinal column. The next gent skied, [Mike] Donlin dented the scoreboard with a freshly laundered single, and the man with the lame spine stumbled to third base. [ Dan] McGann's hit to another part of the fence scored Bresnahan (*North American,* 13 October).

Articles in Dryden's Rainy Day Interview series – pieces to tide fans over until play resumed – were always accompanied by three childlike sketches: a house with curlicue smoke rising from the chimney, a small child holding a large baseball bat, and a seated man in a derby talking to an interviewer. Only the captions changed with the subject: "Connie Mack at the age of five") (*North American,* 6 April 1905), "Mr. O'Loughlin at the age of five [From a photo]" (20 April 1905). In addition, the "interviews" often turn out to be partly or wholly invented. It seems that Mack was once involved in a six-day

> walking match. "But it was not for the glory," [Mack] hastily added, to correct a wrong impression. "The gate receipts looked good to me. . . . My physique, being a growing boy, was thinner then than now. I looked like a fellow who had trained at one of those get-sick-quick lunch places."

These are Dryden's words and sense of humor, and the pieces are really anti-interviews. "Any athlete, active or retired . . . ," he announces, "will be exploited in the Rainy Day series" (6 April 1905). He ridicules space fillers for a century to come, and there is a cumulative effect in seeing the simple drawings reused and recaptioned to illustrate their supposedly important subjects, who are then "quoted." The series makes hero worship impossible.

At the end of 1908 Dryden was hired away to the *Chicago Examiner,* where he would be joined by Lardner in February 1912. The two "would be the star reporters the *Examiner* would put up against the *Tribune,*" says Yardley. Sportswriters were now among the most valuable reporters a paper could offer. In 1918 Dryden continued working for the merged *Herald and Examiner.*

Dryden's only sporting love besides baseball was fishing; it had been important to him since his early vagabond days. "If the stream appeared to have the goods," he says in his autobiography, "I hung my pack on a willow tree, got out my trusty fish line, captured a grasshopper and added fried trout to my regular fare of bread and bacon. . . . Being in no hurry, I spent a profitable lot of time fishing." In his middle years he often spent the off season at it, and late in his career, as Otto notes, he sometimes sent in "a humorous gem" of a fish story for the Sunday *Herald and Examiner.*

His professional detachment remained steady over the years. At the beginning of what would turn out to be the fixed World Series of 1919 he wrote: "In the first bout of the big series here today Dutch Reuther, the boy wonder, trampled on Ed Cicotte, the waning veteran, 9 to 1. If the White Sox had not absent-mindedly scored a run in the second round it would look like a forfeited game" (2 October). He probably had a sense of what was happening, and he made some references elsewhere to gambling, as did his colleagues, but there was no proof as yet; there was little investigative reporting in sports, and Dryden stuck to detailing the games and the color. He had no great love of team owners in general, but he restrained himself and said little about them; a notable exception is a comment that Giants owner John T. Brush "had slipped into town disguised as a cockroach" (*Tribune,* 10 October 1906). Unlike many of his contemporaries and immediate successors, Dryden never baited umpires. If he had a bias, it was for the fans over the owners of a "base ball trust that would send prices up 50 per cent" (*North American,* 19 April 1902).

The same objectivity extended to his personal life. When, in his forties, he encountered an old flame whom he had rejected years before, and saw her settled family life and children, he had regrets. "There was I old and beefy and bald, plugging on alone through life," he says in his autobiography. "Twenty years of bliss gone forever, and years of misery, perhaps, yet to come. . . . Either I am not a ladies' man or else I lacked the nerve to take a chance." He is as tough and uncompromising about himself as he was about baseball; and he was to be proven right about the misery to come.

Dryden's writing helped to expand baseball coverage and move its reports to the front page; it made Dryden, in salary and reputation, one of the most celebrated journalists of his time. The travels and adventures that had begun at age fourteen had continued with his reporting, as he got paid to travel the country with various ball clubs for thirty years. His production decreased in the late 1910s – one suspects illness – and the humor was draining out of his writing. By 1920 he was covering only Cubs games at home, and some of his reports were shorter than half a column. A stroke in 1921 partially paralyzed him; another incapacitated him for the last decade of his life, during which he lived under the care of his younger sister in Mississippi, his agile mind blocked from creating the sweeping reports that influenced how fans appreciated games and how writers covered them. He was the subject of glowing accounts in obituaries after his death on 11 February 1931, and three decades later he was given the Spink Award by *The Sporting News* for distinguished baseball writing. His accounts had noted changes in the game's strategy ("scientific" baseball), equipment, ballparks, ticket prices, and personnel, and in Stanley Frank's opinion he was "the prime mover in making baseball the best documented of all sports." During the following decades, though he was largely forgotten, his influence remained. His techniques "are embodied in many articles without realization on the part of the authors that Mr. Dryden was the father of their style," as *The New York Times* obituary noted.

His work influenced contemporaries who are now better known, well beyond Lardner, and his dry humor whetted their appetites for more. Red Smith's biographer Ira Berkow quotes Smith's praise of Dryden's "scholarship and wit" and

Smith's account of how Dryden, along with Westbrook Pegler and William McGeehan, "were deflating windbags of sports from coast to coast. . . . Hell, Charlie Dryden had New York howling with laughter." Damon Runyon's biographer, Tom Clark, notes Dryden's influence on Runyon and says that Dryden produced prose with "an artist's sense of form and a barroom comedian's sense of humor." Grantland Rice's biographer, Charles Fountain, is equally laudatory: "Dryden had a novelist's eye for detail and irony, and Rice admired him for his felicity with the language. . . ." Furthermore, "For sheer talent . . . [Hugh] Keough and [Hugh] Fullerton had to bow to Charley Dryden." Dryden established a new level of excellence in sportswriting, and he prepared the way for new writers. And, counter to the dominant "Gee Whiz" movement, he was a founding member of the debunking "aw nuts" school, which was to include Lardner, Runyon, and Smith, as well as others who might never have heard of him but who owed him a debt for approach and attitude. Much of his writing deserves a modern audience: "Of all the pioneer chroniclers of sport, Dryden and his copy best survive the passage of time," writes Frank. It is not too much to say he was the Shakespeare of baseball writers, dominant in the early modern era. Prolific, perceptive, irreverent, funny, popular, topical, with an enormous range of images and references, and dramatic in depicting his themes and the characters he observed, Dryden set a high standard for sportswriting in the early modern era.

**References:**

Ira Berkow, *Red: A Biography of Red Smith* (New York: Times Books, 1986);

Edgar G. Brands, "Charles Dryden – Game's Greatest Humorist," *Sporting News* (19 February 1931): 5;

"Charles Dryden Dies; Once Sports Writer," *New York Times,* 13 February 1931, p. 17;

Tom Clark, *The World of Damon Runyon* (New York: Harper & Row, 1978);

Donald Elder, *Ring Lardner: A Biography* (Garden City, N.Y.: Doubleday, 1956);

"Father of Baseball Slang," *New York Times,* 13 February 1931, p. 16;

Charles Fountain, *Sportswriter: The Life and Times of Grantland Rice* (New York: Oxford University Press, 1993);

Stanley Frank, ed., *Sports Extra* (New York: Barnes, 1944);

Frederick G. Lieb, "Dryden, Ex-Chicago Scribe, Wins Fourth Taylor Spink Award," *Sporting News* (13 November 1965);

Wayne K. Otto, "Charles Dryden, Noted Writer of Humorous Baseball, Dies," *Chicago Herald and Examiner,* 12 February 1931, p. 19;

Irving Vaughn, "Charles Dryden, Humorist and Baseball Writer, Dies at 74," *Chicago Tribune,* 12 February 1931, section 2, p. 1;

Jonathan Yardley, *Ring: A Biography of Ring Lardner* (New York: Random House, 1977).

# Hugh Fullerton

*(10 September 1873 - 27 December 1945)*

Tom Nawrocki
Worth *Magazine*

MAJOR POSITIONS HELD: Sportswriter, *Chicago Tribune* (1897-1907, 1912-1913), *Chicago Herald* (1907-1912), *Chicago Herald and Examiner* (1913-1919); sports editor, *New York Mail* (1919-1921); associate editor, *Liberty* (1923-1928).

BOOKS: *Touching Second,* by Fullerton and Johnny Evers (Chicago: Reilly & Britton, 1910); republished as *Baseball in the Big Leagues* (Chicago: Reilly & Britton, 1910);

*Jimmy Kirkland of the Shasta Boys' Team* (Philadelphia: John C. Winston, 1915);

*Jimmy Kirkland of the Cascade College Team* (Philadelphia: John C. Winston, 1915);

*Jimmy Kirkland and the Plot for a Pennant* (Philadelphia: John C. Winston, 1915);

*Tales of the Turf* (New York: A. R. DeBeer, 1922);

*Racing Yarns* (New York: A. R. DeBeer, 1924);

*Two-Fisted Jeff* (Chicago: Consolidated, 1929).

OTHER: *"By Hek" In the Wake of the News: A Collection of the Writings of the Late Hugh Edmund Keough,* edited by Fullerton (Chicago, 1913).

SELECTED PERIODICAL PUBLICATIONS – UNCOLLECTED: "Fine Points of the Game" *American Magazine,* 68 (May 1909): 29-40;

"Deciding Moments of Great Games" *American Magazine,* 68 (June 1909): 107-116;

"The Inside Game," *American Magazine,* 70 (May 1910): 3-13;

"Between Games," *American Magazine,* 72 (July 1911): 321-331;

"Physics of Baseball," *American Magazine,* 74 (October 1912): 754-761;

"Baseball on Trial," *New Republic,* 24 (20 October 1920): 183-184;

"The Fellows Who Made the Game," *Saturday Evening Post,* 200 (21 April 1928): 18-19, 181-186;

"Baseball's Best," *North American Review,* 229 (May 1930): 499-506.

Hugh Fullerton entered the world of sportswriting when its slangy, colorful style was brand new, and he brought to the field a deep affection for and knowledge of baseball. While dispassionate analysis and statistical research was in its infancy, he developed an objective stance and accurately predicted the outcome of World Series contests. He became one of the first sportswriters to present "inside baseball" – what the players did on and off the field to help win ball games – to a wide audience. Fullerton's reverence for the game led him to crusade against its corruption when the scandal no one wanted to face erupted at the World Series of 1919.

Hugh Stuart Fullerton was born on 10 September 1873 in Hillsboro, Ohio. While still in high school, he worked for the Hillsboro newspaper; he began working for the *Cincinnati Enquirer* in 1889 at the age of sixteen under editor Harry Weldon, whom Fullerton described in "The Fellows Who Made the Game," a 21 April 1928 article in *The Saturday Evening Post,* as "a tremendously fat man who knew everybody and was an authority on everything in sport." When the paper's regular baseball reporter got himself barred from the Reds' park, Fullerton was the natural choice to fill in.

After graduating from high school in Hillsboro, Fullerton entered Ohio State University in the fall of 1891 with a planned major in electrical engineering. He also was a catcher for the Ohio State baseball team, switching to second base during the summer, when he played in the Iron and Oil League's franchise in Olean, New York. John McGraw, later a longtime New York Giants manager and bitter enemy of Fullerton, also began his professional career in Olean, playing six games in 1890 before being cut. The Fullerton and McGraw feud developed partly because of Fullerton's partiality to the Cubs during their rivalry with the Giants and partly because Fullerton resented McGraw's heavy ties to gamblers.

Fullerton did not return to school after the 1892 baseball season, a circumstance he later blamed on his role in the theft of a professor's pig. He had played football on one of Ohio State's early teams: "We didn't have a coach," he recalled in the March 1922 *Ohio State University Monthly,* "none of the players knew the game, we had no uniforms, and when we went up to Otterbein to play them, they beat us." Fullerton later tried to make some money by playing for a couple of Eastern schools without enrolling in them but broke his collarbone and headed back to Ohio. He returned to the *Cincinnati Enquirer* and tried vainly during the summers to attract some notice for his baseball-playing ability.

In 1893 Fullerton left Ohio for Chicago, where he worked on the *Chicago Record* for a year under the tutelage of George Ade, later famous for his *Fables in Slang* (1900). Chicago in the early 1890s saw the invention of modern sportswriting, as writers such as Charles Dryden, Len Washburn, Charlie Seymour, and Irving Sanborn plied their trade. (Finley Peter Dunne of Mr. Dooley fame had turned in a memorable season on the baseball beat in 1887 before graduating to bigger things.) Most of the nation was still reading laborious Henry Chadwick–style baseball stories, in which the reporter commonly began with a description of the crowd and weather and indulged in an inning-by-inning recap of the game. Humor and drama were nonexistent in the typical sports story.

Already noted for its vigorous, competitive newspaper industry, Chicago was the first city to develop a healthy competition in its sports pages. In "The Fellows Who Made the Game" Fullerton reflected on the era:

> The papers of Chicago in those days [the late 1880s] were unlike any printed anywhere else. They were written largely in the language that the wild growing young city understood. They had individuality. The *Herald,* which was owned by a cooperative crowd of newspapermen, set the pattern, and the *Inter Ocean* and the *Times* rivaled it in presenting the news in an entertaining manner. There was nothing sedate or dignified about them except the editorial pages and the stockyards reports. They were boisterous, at times rough; they lacked dignity, perhaps, but they were readable, entertaining and amusing.

This style – playful, accessible – reigned in Chicago when Hugh Fullerton came to town. Never the humorist that Dunne or Dryden or Ring Lardner was, he compensated by his immersion in the baseball world and his knowledge of the sport. Among the prominent Chicago sportswriters of the day, Fullerton alone had played professionally. His game stories were knowledgeable, workmanlike, and thorough, though his style was a little stiff for the times. "He was the game's greatest dopester," his friend and fellow sportswriter Fred Lieb wrote in the 7 November 1964 *Sporting News.* "He perhaps had a lot to do with the use of the slang word *dope,* now happily discarded by most writers, for analysis of inside baseball. Fullerton often referred to the reams of copy that flowed from his busy typewriter as 'my dope.' "

Fullerton excelled in the newly emerging field of sports feature writing. A Republican, he moved to the *Chicago Tribune,* the city's preeminent newspaper that also fit his politics, in 1897. In addition to his daily reporting he contributed colorful stories that often appeared in the Sunday paper. The 1899 Sunday feature titled "Stories of the Diamond," though unsigned, bears the unmistakable stamp of the tall tales for which Fullerton would become famous. His first signed pieces ran in 1900 under the heading "Gossip of the National League," with the tagline H.S.F. (game stories still ran without a byline at this point). The *Chicago Tribune* was successful in promoting the expertise of its contributors in other sports, such as Henry Ten Eyck on horse racing and George Siler on boxing, each of whom had their names standing prominently over their columns. Fullerton eventually became known as one of the paper's baseball experts.

Fullerton also found time during the 1900 baseball season to marry Edith Zollars of Fort Wayne, Indiana. The marriage lasted until Fullerton's death and produced two children, Dorothy and Hugh Stuart Jr. Fullerton's son later went into his father's business and worked as a sportswriter for the Associated Press for thirty-eight years.

While at the *Chicago Tribune,* Fullerton made attempts to break into the hard-news side of reporting. During the baseball off-season, he traveled around the country reporting on such matters as Indian affairs and irrigation. In 1897, under the tutelage of legendary *Chicago Tribune* editor Joseph Medill, Fullerton found himself for the first time the recipient of a hot, breaking story. Despite his treading on the toes of some powerful people, Fullerton was not afraid to pursue the scandal, as he later recounted to Philip Kinsley for *The Chicago Tribune: Its First Hundred Years* (1946):

> The Fullertons had some rather prominent relations living on the north side and once a year I had to make a duty call on them. I had to attend a dinner up there one night and during the dinner they spilled one of the most sensational scandal stories I ever heard. It seemed that all the North Shore society people knew all about it – and kept it covered. I got all the details and sprang it on the city editor and wrote a big yarn about it. A couple of days later I got orders to see Mr. Medill. I found the old fellow in his office.
>
> "Your name Fullerton?"
>
> "Yes, sir."
>
> "Any relation to the Rev. Thomas Fullerton?"
>
> "Yes, sir, he is my uncle."
>
> "That so. I went to school with Tom Fullerton. How is he?"
>
> I reported what little I knew and then he asked:
>
> "Did you write that story?"
>
> "Yes, sir."
>
> "We are sued for $50,000 for libel."
>
> "Is that so?" I asked, scared to death.
>
> "Yes, praecipe is filed. Is the story true?"
>
> "Yes, sir, it is true."
>
> "Well, then libel him for about $100,000 more tomorrow."
>
> That ended the matter. They never pushed the case.

When Fullerton began his sportswriting career interviewing the athletes involved in the games was unheard of. Sportswriters would not even bother quoting their subjects. (As Larry Merchant later pointed out, the controversy over Babe Ruth's called home run would have been neatly resolved had anyone thought to ask him about it after the game.) Although his attempts to bring the players' views to his readers would not find favor in the pontifical writing of Grantland Rice and his many imitators, Fullerton was able to get to know the players both on the field and off. His access got him stories that other writers missed, and he reciprocated by letting the players speak in their own voices.

Fullerton's relationship with athletes paid off in such stories as the one he called "Steinfeldt's Star Story," which was first published on 29 October 1905 and later reprinted in the *Greatest Sports Stories from the Chicago Tribune*. The story highlights the voice of Harry Steinfeldt, a third baseman the Cubs had acquired:

> "The gamest man that ever broke into the game," said Steiny during a fanning bee in a hotel one night, "was the second baseman we had down in Dallas in the Texas League. We were playing against Galveston. In the first inning the center fielder of that Galveston team got to first and started to steal. Our catcher threw him out a block and, instead of sliding, he just took a flying leap at the second baseman and came down on his feet with his spikes."
>
> "The game fellow limped around for a minute, then went on playing. That afternoon he made four putouts, eight assists, and four hits, including a double and a home run. After the game he and I were walking over to the clubhouse together when he said, 'I believe there's something in my shoe.' He stooped down, took off his shoe, and shook out two toes."

Fullerton, whose interests ranged from travel to gourmet cooking, loved to talk to the players about their lives off the diamond. When Hugh Jennings was playing shortstop for the old Baltimore Orioles in the 1890s, he led a second life as a law student at Cornell. During one of the endless train trips that were a feature of turn-of-the-century baseball, Fullerton began talking law with him. "I confess I was merely idling to see whether he really was studying for the bar, or whether it was 'press agent' stuff," Fullerton later wrote in the July 1911 issue of *American Magazine*. Jennings proceeded to illustrate his prowess as an attorney by holding a mock prosecution of the Sparrow for the killing of Cock Robin.

Fullerton felt that players responded to newspaper criticism well as long as they thought the writer knew the game thoroughly. Rice in the June 1912 *American Magazine* said Fullerton was "one of the few baseball writers who can 'pan' a ball player in picturesque detail and still hold his friendship; for the player knows that Fullerton is fair and will be quick enough to change when there is cause for praise." In "The Fellows Who Made the Game" Fullerton repeated a conversation he claimed to have overheard between one Dr. Phil McCarthy and Tim Donohue, a catcher for the Cubs of the 1890s. McCarthy asked Donohue, "Did you see what Fullerton said about you this morning?"

> "That blank – blank – blank," Donohue responded reflexively.
>
> "What are you talking about?" McCarthy said. "He says you're the best backstop in the league."
>
> "Well," said Donohue, "me and him always was good friends."

In the fall of 1906 Fullerton made the leap from local favorite to nationally known expert. Chicago suddenly became the most important sporting town in the country when the city's two major league teams were set to square off in the World Series. The consensus was that the Cubs (or Spuds, as they were commonly called) could expect an easy victory. Frank Chance's Spuds had won a major-league record 116 games that season (a record that still stands) and were considered one of the most powerful teams in history. The White Sox, by contrast, were thought to be an anemic team that had lucked into the American League championship despite being last in the league in both batting average and home runs. The thousands who bet on baseball games were intensely interested in the odds to be given in what most thought would be a lopsided series.

Fullerton was one of the few who doubted the National Leaguers' superiority. Shortly before the regular season ended, Fullerton carefully examined the way the two teams matched up and came away with a startling conclusion. "Perhaps the most amazing thing about the series for the World's Championship between the White Stockings and the Spuds, which begins on Tuesday, is the fact that the White Sox will win by taking four out of the first six games," Fullerton wrote. "What makes it more surprising is the way in which they will win it, taking their games honestly and by sheer merit and winning decisively, although outhit during the entire series."

A legend grew up around Fullerton's prediction and was reported as true by Lieb in the 3 January 1946 obituary he wrote for his friend in *The Sporting News. Newsweek* and various wire services also reported the myth. Lieb asserted that Fullerton's editor had refused to print the story because he felt no one would believe it. To placate the editor Fullerton supposedly had thrown in a prediction for a day of rain, thus turning the whole story into a joke. According to Lieb and others, though, not only did the White Sox win the series and prove Fullerton to be the consummate baseball expert but also it rained one day just as Fullerton predicted. The problem with this story– undoubtedly propagated by Fullerton himself – was that there was no rainout during the 1906 World Series. Furthermore, Fullerton's forecast did not actually see print until after the series was completed.

The *Chicago Tribune* published the prediction that made Fullerton's reputation in a special section full of salutes to the two local teams on 15 October 1906, the day after the Sox beat the Cubs 8–3 to take the series. Above the article, an editor included this note: "When it was known that two pennants would come to Chicago, H. S. Fullerton of 'The Tribune' wrote an analysis of the two teams, telling why the White Sox would win the world's championship. A man in authority refused to take a chance in printing Mr. Fullerton's forecast, but it has been

*Hugh Fullerton*

verified so remarkably by the games as played that it is now printed just as it was written." This *post facto* prediction was apparently never subject to rumors of chicanery. Any doubt about Fullerton's predictive abilities was quelled over the following years, when he picked the winners of three consecutive World Series and five of the next six. The 1906 pick made Fullerton famous as the game's preeminent dopester, a reputation he maintained to his death in 1945. That this standing was founded on a prediction printed after the games had been played somehow faded from the story.

But Fullerton's prediction of the upset was significant for more than just establishing his career. His story was one of the first careful, clearheaded, dispassionate analyses of baseball. "I took a large Faber lead pencil, nine sheets of white glazed copy paper," he wrote, "and figured it out." Fullerton considered such factors as the expanded crowds of the World Series, which would overflow the stands and cut into the fielders' range:

> The Sox will gain in the outfield. In fielding balls [Frank] Schulte and [Jimmy] Sheckard [of the Cubs] undoubtedly outclass [Edgar] Hahn and [Patsy] Dougherty [of the Sox], while [Sox center fielder] Fielder Jones is head and shoulders above [Cub center fielder Jimmy] Slagle. But center field will be clear, leaving Jones free to make his great catches, and at the same time both Schulte and Sheckard will be handicapped by the restricted space until neither can cover more ground than their White Sox rivals.

Fullerton also did some statistical homework, discovering that the type of pitcher that had had some measure of success against the Spuds was exactly the type that was working for the Sox: "Take your dope books for the year and study the defeats of the Chicago National league team. They have been beaten by good left handers and by spit ball pitchers. . . . [Sox ace Ed] Walsh is the superior of any spit ball pitcher in the National league."

Fullerton's research-based work prefigured the writing of later baseball analysts such as Leonard Koppett and Bill James. In the dawn of baseball's statistical age, Fullerton reached beyond the basics in order to understand what made winning teams win. The White Sox were known as the Hitless Wonders for good reason: They hit just .230 as a team for the season. Fullerton noticed that the Sox "score more runs on fewer hits and hold their opponents down to the smallest number of runs on the greatest number of hits of any team in the world." Fullerton argued that the Hitless Wonders did not need a lot of hits to win, a point he proved with statistics: "The Sox do not hit anyone hard – nor do they have to. They get runs in proportion to 1 1/3 runs to 2 hits throughout a year. The Spuds get one run to every 2 3/7 hits, roughly calculated. [The White Sox] don't expect to slaughter [Cub ace Mordecai] Brown, simply to beat him." They did.

A baseball writer in Fullerton's day was sometimes treated to less than cordial treatment when he followed his team on the road. In 1908 Giants owner John T. Brush barred Fullerton from the Polo Grounds, and his seat for the famed play-off between the Giants and the Cubs was occupied by Louis Mann, an actor friend of the owner's. Mann would not budge, so Fullerton filed his story while seated on the actor's lap. (Like so many Fullerton stories, this one seems to have been exaggerated over the years. More contemporaneous accounts have Fullerton pushed to a corner of the press box by a coterie of actors and forced to file his copy with his feet hanging out the edge of the box.) After further indignation was piled on the writers at that fall's World Series in Detroit – the out-of-town reporters had to climb a ladder to the roof of the pavilion down the right-field line and watch the games from the rain-soaked rooftop – Fullerton, along with Cy Sanborn, Tad Dorgan, and others, helped to form the Baseball Writers Association of America. In "The Fellows Who Made the Game," a significant examination of sportswriting at and after the turn of the century, his respect for his colleagues is evident as he recalls the talents and adventures of three generations of baseball writers, from Henry Chadwick to Rice.

By 1907 Fullerton had been removed from the daily coverage of Chicago's two baseball teams, leaving that duty to a pair of legends, Sanborn and Dryden. Fullerton concentrated on his Sunday features, contributing pieces such as "Some Excuses of Ball Players" and "Freak Plays That Have Won Games." In 1908 he was lured to the *Chicago Herald and Examiner,* William Randolph Hearst's paper in Chicago, but returned to the *Chicago Tribune* in 1912, when Hugh Keough, better known as HEK, passed away. Fullerton was asked to take over "In the Wake of the News," the column HEK had originated as "Sidelights on Sports" in 1904. "The last time we saw him," Fullerton noted in his first "Wake" column, "HEK urged 'the other Chooey' to 'come home to "The Tribune." ' " It is good to be 'home,' but there will always be a vacancy." However, the light, humorous style of the column was far from Fullerton's strength, and he moved on after less than a year in the post.

In May 1909 Fullerton began a series of stories for *The American Magazine* in which he took an in-depth look at various facets of baseball. Between 1909 and 1920 he wrote several dozen articles for *The American* on seminal figures such as Honus Wagner and Charles Comiskey as well as historical pieces on great games and plays. He provided inside dope in articles such as "Fine Points of the Game," "Between Games," and "Physics of Baseball."

Fullerton collaborated with Cub second baseman Johnny Evers on a book, *Touching Second* (1910), which combines history and strategy with some of Fullerton's anecdote-laden pieces for *The American Magazine*. Evers's contribution to the book, aside from the relevance of his position to the title, cannot be easily assessed. "The manuscript was originally written by the reporter and was rewritten, added to, corrected and revised, by the player," the introduction reads. "In it are included several articles originally printed in The American Magazine under the reporter's name, with corrections and additions by the player." But Evers is referred to throughout as "Evers" rather than "I."

The *American Magazine* stories were the typical Fullertonian mix of inside baseball and tall tales. In "Deciding Moments of Great Games" (June 1909) he told a once-famous story about Chicago center fielder Bill Lange, often called the greatest athlete in the game during his heyday in the 1890s. Already in trouble with manager Cap Anson for missing a couple of trains, Lange had a chance at redemption in a game against Washington. The Cubs had scored in the top of the eleventh inning to break a tie, but Washington had a man on with two out in the bottom of the eleventh and one of the team's top hitters, Kip Selbach, up. Selbach delivered a terrific drive over Lange's head. "The hit seemed a sure home run but Lange, a man weighing 225 pounds, turned and without looking, sprinted desperately straight out toward the fence, racing the flying ball," Fullerton wrote. "At the last instant, as the ball was going over his head, Lange leaped, stuck up both hands, turned a somersault and crashed against the fence. The boards splintered, one entire panel crashed outward, and out of the wreckage crawled Lange, holding the ball in his hand, and the crowd went mad." The kicker was that Anson promptly rescinded Lange's $200 worth of fines for missed trains.

While Fullerton was probably at the game in question – or if not, knew intimately many of the participants and eyewitnesses – Arthur Ahrens, writing in the 1980 *Baseball Research Journal,* refutes Fullerton's version of the story at some length, arguing that he conflated two separate incidents in the account, which did not see print until seven years after the supposed play. Yet the tale as Fullerton tells it undoubtedly makes a better story than Ahrens's. As reported by Leib in his 3 January 1946 article, Bill Hanna, a New York sportswriter, once told Fullerton, "Hughie, you would sacrifice accuracy for the sake of a good story." Fullerton told him, "Bill, you would sacrifice a good story for the sake of accuracy."

Fullerton was extremely popular with his fellow writers, who called him "Hughie." He took under his wing Cy Sanborn, C. E. Van Loan, and the young Ring Lardner, whom Fullerton got hired at the Chicago *Inter Ocean,* Lardner's first sportswriting job in Chicago. Lardner, who eventually succeeded Fullerton as the "In the Wake of the News" columnist at the *Chicago Tribune,* called Fullerton "the Old Master," and the two men drank together often. During one session at Stillson's, a tavern near the *Chicago Tribune* offices, Lardner asked Fullerton to stay and have dinner with him, but Fullerton begged off, claiming that he and his wife were expecting guests. Lardner promised to get him out of it and called Edith Fullerton at home; the excuse he came up with was that Hughie was drunk. Edith told Lardner that if that was the case, her husband should not bother coming home. Lardner returned to the bar and told Hughie that it was fine with his wife if he stayed. "What did you tell her?" Fullerton asked him. Lardner told the truth, whereupon Fullerton dashed home.

Fullerton covered the 1912 World Series between the Red Sox and the Giants for the *Chicago Tribune*. The memorable series concluded with Fred Snodgrass's muffed fly ball in the tenth inning of game seven, leading to a Boston victory. Despite the Red Sox's triumph, much of Boston did not celebrate. "There is a bitter taste remaining in the greatest series of games ever played, for today Boston boycotted its ball club and the rumor that the series was fixed and all prearranged ran through the town," Fullerton wrote in the 16 October 1912 *Chicago Tribune*. "Half the people believed it." Fullerton was evidently not one of them.

Seven years later, however, there were again rumors that the World Series was not being played on the level, and this time Fullerton, now with the *Chicago Herald and Examiner,* pursued them doggedly. Before the first game of the 1919 World Series between the Reds and the White Sox, Fullerton ran into former White Sox pitcher Bill Burns. Ten years earlier, he and Burns had passed the night of a long train ride talking of the various breeds of dogs Burns had seen in his native Texas and the Far East. Now retired, Burns was known as a professional gambler and passed along some advice to Fullerton: All his money was going on the underdog Reds. "Get wise, Hughie, and do the same," Burns urged his friend. To Fullerton the advice confirmed the rumors he had heard. He sent a telegram to all the newspapers that were using his game stories:

"Advise all not to bet on this series. Ugly rumors afloat." Fullerton made a pact with recently retired pitching great Christy Mathewson, who was covering the games for the *New York Evening World,* to keep an eye on any untoward business on the part of the White Sox.

It did not take them long to spot something suspicious. In game one, Chicago starter Eddie Cicotte hit the Reds' lead-off man, Morrie Rath, in the back with his second pitch. Then, in the fourth inning, with one out, one on, and the game tied 1–1, Cicotte fielded what seemed to be a perfect double-play ball. But his throw to second was late and high, and the batter was safe at first. Fullerton noticed Cicotte's hesitation and circled the play in his scorebook. A small, almost imperceptible mistake, it nevertheless aroused Fullerton's suspicions. His concern was confirmed when the next five Cincinnati batters all hit safely, scoring five runs. The game was lost.

The Reds eventually jumped out to a 4–1 lead in the nine-game series. But the favored White Sox came back to win a pair, renewing Fullerton's confidence in the team, and the Sox looked to even the series in game eight. As he walked to the press box before the pivotal game, a gambler friend of his stopped Fullerton and asked if he had his money down on the Reds, a sure thing. Fullerton asked the gambler what he knew; he responded, "It'll be the biggest first inning you ever saw." As the gambler suggested, Chicago starter Lefty Williams allowed four runs in the first, and the game was never close.

At first Fullerton tried to deny what he knew to be the truth. After the series, he wrote, "This series was lost in the first game, and lost through overconfidence. Forget the suspicious and evil-minded yarns that may be circulated." But Fullerton could not forget those yarns. Through the winter, almost alone in the face of a baseball establishment that wanted the nastiness to go away, Fullerton sought the truth. White Sox owner Charles Comiskey had promised a $20,000 reward to anyone with knowledge of a fix, but the well-known skinflint had no intention of parting with his money. At the league meetings in December, the possibility of the World Series having been rigged was not discussed.

Fullerton, though, did take action, preparing a piece that he hoped would urge the game to address the problem of gamblers sullying its most important event. Fullerton's home paper, the *Chicago Herald and Examiner,* shied away from his charges, but the *New York Evening World,* which ran his syndicated work, printed them: "Is Big League Baseball Being Run for Gamblers, with Ballplayers in the Deal?" the front-page headline screamed. Fullerton's attack was oblique, mindful of libel laws. "Charges that ballplayers are bribed and games are sold out," he wrote, "are made without attempts at refutations by men who have made their fortunes in baseball." The piece turned into a series, eliciting a series of denials on behalf of baseball's establishment. *Baseball Magazine* mocked Fullerton in verse:

A dopester whose first name was Hugh
Said the White Sox would win five to two.
But the Reds with a rush
Put the dope on the crush
Now the dopester still bellows "bugh-hugh!"

Fullerton waited almost a year for his vindication.

The following September, several members of the White Sox testified to a grand jury that they had indeed conspired to throw the World Series. Although Fullerton recognized the culpability of the eight players involved, he saw as much rottenness in the upper levels of the sport. Fullerton presented his case as the critic of a corruption seen to transcend the sport in the 20 October 1920 edition of *The New Republic.* He had become the semiofficial voice of baseball's antiestablishment. "The officials, to whose welfare the business and sport of 'organized' baseball was entrusted, have been exposed as incompetent," he wrote bitterly. "Through it all the officials in charge of baseball adhered to their policy of curing an evil by declaring it did not exist and by using their influence over consciously or unconsciously subsidized sporting writers to suppress the accusations and punish those who demanded an investigation."

In the early 1920s Fullerton shifted to New York full-time, serving as sports editor for the *New York Mail* before joining *Liberty* magazine as associate editor. He stopped predicting World Series winners, claiming that the lively ball threw his statistical system out of whack, although his difficult experience in the 1919 Series was as much a contributing factor. His interests branched out in these later years. His first foray into fiction had been a trio of books for boys published in 1915, on the baseball adventures of Jimmy Kirkland, an orphan raised by a wealthy uncle in Oregon; the three books covered Jimmy's baseball career as an adolescent, *Jimmy Kirkland of the Shasta Boy's Team*; college student, *Jimmy Kirkland of the Cascade College Team*; and major league rookie, *Jimmy Kirkland and the Plot for a Pennant.* Fullerton published a short collection of horse racing fiction, *Tales of the Turf,* in 1922, seemingly in the manner of his onetime protégé, Ring Lardner.

While Lardner wrote in the vernacular of farm boys and small-town rubes, Fullerton attempted a story in black dialect with less success. The main character of "Toutin' Mistah Fox" was described as having a "mood . . . two shades darker than his skin, the color scheme of which was that of the ace of spades" and said things like "Dat yaller gal too brash wif her mouf!" In 1929 Fullerton published the autobiography of boxer Jack Jeffries, *Two-Fisted Jack;* although the book is presented in Jeffries's first-person voice, Fullerton is given full credit as the author. He also wrote for various papers, including his old forum the *Chicago Tribune,* on boxing, golf, and horse racing.

After six years at *Liberty* magazine, Fullerton returned to newspaper work full-time, writing for the Bell Syndicate before going back to his sometime hometown of Columbus, Ohio, and the *Columbus Dispatch* in 1928. He also wrote Sunday features for the *Philadelphia Enquirer.* In addition, he contributed frequently to *The Sporting News* and other national magazines. In May 1930 Fullerton published a piece in the *North American Review* in which he chose baseball's best-ever team. His surprising pick was the Boston Beaneaters of 1897 and 1898, who featured Hall of Famers Jimmy Collins, Hugh Duffy, Billy Hamilton, and Kid Nichols. He also named his all-time team: Hal Chase at first, Napoleon Lajoie at second, Hans Wagner at short, Collins at third, Babe Ruth in left, Bill Lange in center, Ty Cobb in right, John Kling and Mickey Cochrane behind the plate, and, to pitch, Christy Mathewson, Walter Johnson, and Rube Waddell, whom Fullerton once caught in a semipro game.

Hugh Fullerton died in Dunedin, Florida, on 27 December 1945 at the age of seventy-two. He is remembered for his passion for baseball and workaholic's thirst for further knowledge of the game. His service to the sport at its darkest hour left a legacy for objectivity and honesty. His work was recognized two decades later, when he became the third recipient of the J. G. Taylor Spink Award for outstanding writer, thus enshrining him in the writers' wing of the Baseball Hall of Fame (following Lardner by one year). He also left behind his Ten Commandments of Sports:

1) Thou shalt not quit.
2) Thou shalt not alibi.
3) Thou shalt not gloat over winning.
4) Thou shalt not sulk over losing.
5) Thou shalt not take unfair advantage.
6) Thou shalt not ask odds thou are unwilling to give.
7) Thou shalt always be willing to give the benefit of the doubt.
8) Thou shalt not underestimate an opponent or overestimate thyself.
9) Remember that the game is the thing, and he who thinks otherwise is no true sportsman.
10) Honor the game thou playest, for he who plays the game straight and hard wins even when he loses.

**References:**


Arthur R. Ahrens, "Bill Lange's Classic Catch Re-Classified," *Baseball Research Journal* (1980): 87–90;

Eliot Asinof, *Eight Men Out* (New York: Holt, 1963);

William Lucius Graves, "The Crow's Nest," *Ohio State University Monthly* (February 1927): 203;

Philip Kinsley, *The Chicago Tribune: Its First Hundred Years,* volume 3 (Chicago: Tribune Company, 1946), pp. 329–330;

Frederick G. Lieb, "Fullerton, Famed Forecaster, Named Spink Award Winner," *Sporting News,* 7 November 1964, p. 10;

Lieb, "Hugh, Who Was Game's Original Dopester, Dies at 72," *Sporting News,* 3 January 1946, p. 7;

Tom Nawrocki, "The Chicago School of Sportswriting," *National Pastime,* no. 13 (1993): 84–86;

Grantland Rice, "Interesting People: Hugh S. Fullerton," *American Magazine,* 74 (June 1912): 160–162;

Peter D. Vroom, "Chicago's Baseball Writers," *Baseball Magazine* (October 1908): 29–34;

Jonathan Yardley, *Ring* (New York: Random House, 1977).

# Paul Gallico

*(26 July 1897 – 15 July 1976)*

William J. McGill
*Lebanon Valley College*

See also the Gallico entry in *DLB 9: American Novelists, 1910–1945.*

MAJOR POSITIONS HELD: Motion picture critic (1922), sportswriter (1923–1924), sports editor, columnist, and assistant managing editor, *New York Daily News* (1924–1936); war correspondent, *Cosmopolitan* (1943).

BOOKS: *Farewell to Sport* (New York & London: Knopf, 1938);

*Adventures of Hiram Holliday* (New York: Knopf, 1939);

*The Secret Front* (New York: Knopf, 1940);

*The Snow Goose* (New York: Knopf, 1941; London: M. Joseph, 1942);

*Selected Stories* (New York: Editions for the Armed Services, 1941);

*Lou Gehrig: Pride of the Yankees* (New York: Grosset & Dunlap, 1942);

*Golf Is a Friendly Game* (New York: Knopf, 1942);

*Confessions of a Story Writer* (New York: Knopf, 1946);

*The Lonely* (London: M. Joseph, 1947; New York: Knopf, 1949);

*The Abandoned* (New York: Knopf, 1950); republished as *Jennie* (London: M. Joseph, 1950);

*The Small Miracle* (London: M. Joseph, 1951; Garden City, N.Y.: Doubleday, 1952);

*Trial by Terror* (New York: Knopf, 1952);

*Snowflake* (London: M. Joseph, 1952; Garden City, N.Y.: Doubleday, 1953);

*The Foolish Immortals* (Garden City, N.Y.: Doubleday, 1953);

*Love of Seven Dolls* (Garden City, N.Y.: Doubleday, 1954; London: M. Joseph, 1954);

*Ludmila: A Legend of Liechtenstein* (London: M. Joseph, 1955); republished as *Ludmila* (Garden City, N.Y.: Doubleday, 1959);

*Thomasina: The Cat Who Thought She Was God* (Garden City, N.Y.: Doubleday, 1957; London: M. Joseph, 1957);

*Paul Gallico (photograph by Carl van Vechten)*

*The Steadfast Man: A Biography of St. Patrick* (Garden City, N.Y.: Doubleday, 1958; London: M. Joseph, 1958);

*Mrs. 'Arris Goes to Paris* (Garden City, N.Y.: Doubleday, 1958); republished as *Flowers for Mrs. Harris* (London: M. Joseph, 1958);

*Too Many Ghosts* (Garden City, N.Y.: Doubleday, 1959; London: M. Joseph, 1961);

*The Hurricane Story* (London: M. Joseph, 1959; Garden City, N.Y.: Doubleday, 1960);

*Mrs. 'Arris Goes to New York* (Garden City, N.Y.: Doubleday, 1960); republished as *Mrs. Harris Goes to New York* (London: M. Joseph, 1960);

*Further Confessions of a Story Writer: Stories Old and New* (Garden City, N.Y.: Doubleday, 1961); republished as *Confessions of a Story-Teller* (London: M. Joseph, 1961);

*Coronation* (Garden City, N.Y.: Doubleday, 1962; London: Heinemann, 1962);

*Scruffy: A Diversion* (Garden City, N.Y.: Doubleday, 1962; London: M. Joseph, 1962);

*Love, Let Me Not Hunger* (Garden City, N.Y.: Doubleday, 1963; London: Heinemann, 1963);

*The Day the Guinea-Pig Talked* (London: Heinemann, 1963; Garden City, N.Y.: Doubleday, 1964);

*The Silent Miaow: A Manual for Kittens, Strays, and Homeless Cats,* by Gallico and Suzanne Szasz (New York: Crown, 1964);

*The Hand of Mary Constable* (Garden City, N.Y.: Doubleday, 1964; London: Heinemann, 1964);

*The Day Jean-Pierre Was Pignapped* (London: Heinemann, 1964; Garden City, N.Y.: Doubleday, 1965);

*Mrs. 'Arris Goes to Parliament* (Garden City, N.Y.: Doubleday, 1965); republished as *Mrs. Harris, M.P.* (London: Heinemann, 1965);

*The Golden People* (Garden City, N.Y.: Doubleday, 1965);

*The Day Jean-Pierre Went round the World* (London: Heinemann, 1965); republished as *The Day Jean-Pierre Went around the World* (Garden City, N.Y.: Doubleday, 1966);

*The Man Who Was Magic: A Fable of Innocence* (Garden City, N.Y.: Doubleday, 1966; London: Heinemann, 1966);

*The Revealing Eye: Personalities of the 1920s,* text by Gallico, photographs by Nikolas Muray (New York: Atheneum, 1967);

*The Story of Silent Night* (New York: Crown, 1967; London: Heinemann, 1967);

*The Lost Christmas* (Berkeley, Cal.: Lederer, Street & Zeus, 1968);

*Manxmouse* (New York: Coward-McCann, 1968; London: Heinemann, 1968);

*The Poseidon Adventure* (New York: Coward-McCann, 1969; London: Heinemann, 1969);

*The Day Jean-Pierre Joined the Circus* (New York: Watts, 1969);

*Matilda* (New York: Coward-McCann, 1970; London: Heinemann, 1970);

*The Zoo Gang* (New York: Coward, McCann & Geoghegan, 1971; London: Heinemann, 1971);

*Honorable Cat* (New York: Crown, 1972; London: Heinemann, 1972);

*The Boy Who Invented the Bubble Gun: An Odyssey of Innocence* (New York: Delacorte, 1974; London: Heinemann, 1974);

*Mrs. 'Arris Goes to Moscow* (New York: Delacorte, 1974); republished as *Mrs. Harris Goes to Moscow* (London: Heinemann, 1974);

*Miracle in the Wilderness: A Christmas Story of Colonial America* (New York: Delacorte, 1975); republished as *Miracle in the Wilderness: A Story for Christmas* (New York: Delacorte, 1975);

*Beyond the Poseidon Adventure* (New York: Delacorte, 1978);

*The House That Wouldn't Go Away* (London: Heinemann, 1979; New York: Delacorte, 1979).

OTHER: "A Large Number of Persons," in *The Omnibus of Sport,* edited by Grantland Rice and Harford Pruel (New York: Harper, 1932), pp. 620–624;

"I Like the Circus," *Vogue's First Reader* (New York: Messner, 1942), pp. 203–210;

"What We Talked About," *While You Were Gone: A Report on Wartime Life in the United States,* edited by Jack A. Goodman (New York: Simon & Schuster, 1946), pp. 28–63;

Helen R. Hull, ed., *Writer's Book,* contributions by Gallico (New York: Harper, 1950);

"Mainly Autobiographical," *A Reader for Writers,* edited by William Targ (Hermitage, N.Y.: 1951);

"Thicker than Water," in *Greatest Collier's Sports Stories,* edited by Tom Meany (New York: Barnes, 1955), pp. 194–199;

Ian Fleming, *Gilt-Edged Bonds,* introduction by Gallico (New York: Macmillan, 1961);

Don Congdon, ed., *The Thirties: A Time to Remember,* contributions by Gallico (New York: Simon & Schuster, 1962);

"Aim for the Heart," *The Writer's Handbook,* edited by Abraham S. Burack (New York: Writer, 1962), pp. 3–5.

SCREENPLAYS: *No Time to Marry,* Columbia, 1937;

*Pride of the Yankees,* by Gallico, Jo Swerling, and Herman Mankiewitz, RKO, 1942;

*Joe Smith, American,* M-G-M, 1942;

*The Clock,* M-G-M, 1945;

*Never Take No for an Answer,* Souvaine Selective, 1951;

*Lili,* by Gallico and Helen Deutsch, M-G-M, 1953;

*Merry Andrew,* M-G-M, 1957;

*Big Operator,* M-G-M, 1959;

*The Three Lives of Thomasina,* Walt Disney Productions, 1963;

*The Snow Goose,* National Broadcasting Company, 1971.

During his thirteen years as a sportswriter, sports editor, and columnist for the *New York Daily News,* Paul Gallico wrote, by his own estimation, 5.11 million words. With those words he described and commented on the feats of a generation of giants – including Babe Ruth, Bobby Jones, Jack Dempsey, Helen Wills Moody, Johnny Weissmuller, and Gar Wood – during what has been called the Golden Age of Sports. He became one of the highest-paid sportswriters in the country. Then, in 1936, at the age of thirty-nine, he walked away from that position to pursue a freelance writing career. Little of what he wrote during the next four decades concerned sports, and few of those who enjoyed his novels, stories, and screenplays knew that he had gotten his start as a sportswriter. But the daily experience of trying to capture in words, under a deadline, the games people play gave vitality to all his prose; and the qualities that brought him success in his first career sustained his second career as well.

Paul William Gallico was born in New York City on 26 July 1897 to Paolo and Hortense Erlich Gallico, who had immigrated to the United States two years before. His father, an Italian of Spanish extraction, was a concert pianist, composer, and music teacher whose distinguished career as a performer enabled the family, though not prosperous, to take annual summer trips to Europe. Gallico showed no sign of inheriting his father's musical aptitude, but from an early age he displayed an interest in writing that his Austrian-born mother encouraged. He would later maintain that he had always dreamed of becoming a writer.

When he graduated from New York public schools he entered Columbia University as a premedical student; becoming a physician, he thought, would provide a secure income while he pursued a writing career. After a year and a half in the navy at the end of World War I he returned to complete his degree in science, but he decided to seek means other than medicine to support himself. He worked his way through college with a variety of jobs, from gym instructor to longshoreman to usher at the Metropolitan Opera House. He also rowed on the Columbia crew, of which he was captain his senior year.

Between his release from the navy and his return to Columbia he had worked for a newspaper syndicate for six months; during that time he had sold some fiction sketches. That experience showed him that it might be possible to earn a living by writing, and on graduating in 1921 he took a job with the National Board of Motion Picture Review. On 5 September of that year he married Alva Thoits Taylor, the daughter of the columnist Bert Leston Taylor of the *Chicago Tribune;* they would have two children: William Taylor, born in 1927, and Robert Leston, born in 1930.

In 1922 Gallico became motion picture critic for the *Daily News.* That job lasted, by his own count, for four months and twenty-nine days: "On the thirtieth day I was removed from the chair by Captain [Joseph Medill] Patterson because my reviews were too smart-alecky," he recalled in *Further Confessions of a Story Writer* (1961). "I was not fired from the *Daily News;* this was not the custom of the publisher. He merely, in a fit of exasperation, told his managing editor, Phil Payne, to get 'that man' out of the movie department. Payne hid me in the sports department as an anonymous and un-by-lined sports reporter, odd-jobs and rewrite man."

One of his assignments as a sportswriter was to cover Dempsey's training camp in Saratoga Springs, New York, as the Manassa Mauler prepared for his fight with Luis Angel Firpo. Gallico persuaded Dempsey to spar with him so that he could "write more convincingly what it was like to have the fist of the world champion connect with one's chin in battle." In one minute and thirty-seven seconds Gallico had acquired the knowledge he sought; he also had his first byline. The story caught the eye of Patterson, who, one year after removing Gallico as movie critic, promoted him to sports editor and assigned him a regular column.

Later, as a writer of fiction, Gallico would get an idea, play with it, and try to get inside the characters he was imagining before embarking on the actual work of writing. As a sportswriter he seldom had the luxury of time, yet all the stages of this process were present in his early career, albeit in compressed form – with the added step of learning by experience what it was like to see, hear, and feel a champion in fighting mode. The immediacy of the reward for the Dempsey article – the first of his "researches" into "getting the feel" – also confirmed him in his conviction that storytelling was the core of the art to which he aspired.

The device of putting himself in the position of a sports star's opponent served him well. Gallico "challenged" stars of three dozen sports: he played

*Gallico (in left rear, with binoculars and notebook) covering a golf match between Bobby Jones and Walter Hagen (photograph by FPG)*

golf with Bobby Jones, swam against Johnny Weissmuller, batted against Dizzy Dean, skied on an Olympic course, accompanied Gar Wood in a speedboat, tried to catch passes from Benny Friedman, and rode in an Indianapolis race car with Cliff Bergere. Each encounter provided grist for his writing – not only the immediate description of the encounter, but for all of his subsequent writing about that person and that sport. As he would write of himself in *Farewell to Sport* (1938): "I was always a child who touched things and I have always had a tremendous curiosity with regard to sensation." That characteristic motivated his "researches," as he called them. He said that he never regretted the experiences, though no amount of money could induce him to repeat them. And there were some things he would not even do a first time: a high dive and ski jumping, for example. "I know when to stop. After all, I am entitled to rely upon my imagination for something." The vitality of his writing derives from the interplay of experience and imagination, as it does for all good storytellers. George Plimpton, who would expand and refine the device in the 1960s and 1970s, readily admitted that when he first began writing he read the essay "The Feel" in *Farewell to Sport* – according to Plimpton, the essay was the essential primer for all sportswriters – and drew his inspiration from it.

Readers of Gallico's column seemed to get to know him personally, and his stories often emerged as dialogues with them. On 28 September 1927 he answered readers who angrily wrote to him about his account of the Jack Dempsey–Gene Tunney rematch, the "long count" bout of 26 September, in which, Gallico insisted, Dempsey hit Tunney "with punches to the groin" and head butts. Although Gallico had predicted that Dempsey would reclaim his title, he conceded that Tunney was the better fighter when he was on his feet – but "he was off his feet for fourteen seconds, which is four more seconds than Marquis of Queensberry rules allow." The article demonstrates the importance of trying to report such emotionally charged sporting events fairly, honestly, and critically, since, in spite of the decision in favor of Tunney, Dempsey had actually won the fight while fighting dirty, and Tunney had really lost it while fighting "a clean, courageous, winning fight." Gallico refused to "scent up both parties" and could not understand why some readers expected him to do so. Dempsey got breaks while he was champion, Gallico said, and Tunney got one now. "Dempsey," Gallico observed, "understands things like that. He's been around." Nonetheless, Gallico admired the bloodied Dempsey's "gallant stand in the final rounds" and admits to enjoying "those fourteen seconds in which Genie was on the deck."

His column for 21 June 1932 reveals Gallico's eye for detail and his intimate knowledge of the sport he was reporting. Jack Sharkey upset Max Schmeling in "one of the dullest heavyweight fights in the history of the ring" because Sharkey was "cool, accurate. He never lost his head. He made only one mistake, a missed right lead that got him a good punching. He kept his hands in Schmeling's face all the way. He didn't whip him or hurt him. He merely annoyed him."

Gradually Gallico's sports stories took on the feel of light essays and narratives. On 1 August 1934, after the Cubs and the Cardinals replayed two and a half innings of a protested game, Gallico speculated on "the cosmic design" after the Cardinals walked Babe Herman with first base open rather than pitch to him as they had in the previous version of the game, in which he singled. Gallico wondered about this rare opportunity "to live life over again" and "profit by your mistakes of the time before" – "the sort of thing people dream about." The outcome of the game did not change, but, miraculously, "twenty minutes had been snatched from the unreclaimable past."

The powers of description that would become the hallmark of Gallico's mature fiction first appeared in his sports stories. Of a World Series game in which the Cardinals' Joe Medwick was ejected for his own safety by the commissioner, Judge Kenesaw Mountain Landis, Gallico wrote on 10 October 1934:

> I watched the crowd and Medwick and the pelting missiles through my field glasses, and it was a terrifying sight. Every face in the crowd, women and men, was distorted with rage. Mouths were torn wide, open eyes glistened and shone in the sun. All fists were clenched. Medwick stood grinning with his hands on his hips, just out of range of the bottles. A green apple rolled to his feet, and he fielded that, too. Umpires and attendants rushed out to left field and began picking up the mess.

Gallico's newspaper columns brought to his readers a sense of excitement and anticipation, as if to say – as, in fact, he did say in an article on 28 September 1934 – "Oh no, my friends, we can't miss that."

Being sports editor and writing a daily column gave Gallico's life a peripatetic, even frenetic quality. He had to move from sport to sport and event to event, and, of course, his beat was not just the local but the national sports scene. On one occasion he covered five major sporting events in a single day by flying his own plane from one to the other. The flowering of sports in this golden age produced a whole social dimension that drew him on and in, and it inevitably affected his family life. As he would later say, "It was a fast, gay, wonderful, and completely thoughtless life." In 1934 Gallico's marriage ended in divorce. A second marriage, on 12 April 1935, to Elaine St. Johns, the daughter of the journalist Adela Rogers St. Johns, dissolved the following year.

During the first eight years of his sportswriting career Gallico had written no fiction; but, as he later observed in an interview with Jerome Holtzman (1973), "I always knew that eventually I would write fiction. That was always at the back of my head. I wasn't using sports as a go-between, but the thing about sports-writing which is so marvelous is that it's constant drama – and drama is the basis of fiction." He wrote some articles for *Vanity Fair,* and in 1931 he hired an agent. In 1933 his first published short story, "McKabe" – which was not about sports but about journalism – appeared in the 12 August *Saturday Evening Post.* That first breakthrough into the popular magazine market whetted his appetite, and in 1936 he resigned from the paper to become a freelance writer.

He later gave two explanations of his decision to leave the *Daily News.* The first concerned a fight that he attended, at which he took a seat at ringside and was joined by a colleague. The colleague did not sit down immediately but fussed with taking off his jacket and arranging it on the back of his chair. A paying customer behind them finally called out, "Siddown, you're nothing but a sportswriter." Though Gallico was one of the highest-paid sportswriters, a syndicated columnist, and an editor, the kind of writing he had done over the previous decade was not the kind he really wanted to do. The story the *Post* accepted was more like it, and it reminded him of his earlier hopes of creating his own stories rather than telling other people's. He also remarked in *Confessions of a Story Writer* (1946) that "After thirteen years of looking at every kind of athletic contest, I had nothing more to say. Better to get out while the getting was good and before the readers made the same discovery I had." He had an abiding fear of becoming a "veteran sportswriter," a label that, to him, suggested both age and caricature. In his new career sports would provide material for few of his stories, though after his first sale to the *Post* he did have a couple of sports stories accepted by the magazine. George Horace Lorimer, the editor, told him that the *Post* had great need of good stories with sports backgrounds and that he was looking to Gallico as a good source of such material. Gallico hesitantly replied that he wanted to write other kinds of stories. Lorimer said: "Young man, let me tell you something; I don't care what your background is, or your characters, or where it is set if you will only tell me a story." That was Gallico's emancipation proclamation. While the first group of stories he published tended to draw on his experience in journalism, he increasingly let his imagination prevail over his experience.

When he left the *Daily News* he moved to England and bought a house on the Devon coast. "The decision [to leave the *Daily News*] coincided with an emotional crisis brought on by the break-up of my second marriage and a hangover from the break-up of the first. I needed to change my life and my way of living completely," he would recall in *Confessions of a Story Writer.* He turned down an offer of $75,000 a year to write sports for a newspaper syndicate, but after four months in Devon he returned to New York to write cityside copy for the *Daily News* at minimum scale. Increasingly the paper called on him to write features, but he knew that he could get more money for the same articles as a freelancer; so after eight months he quit again. He worked briefly for the International News Service as a feature writer, then returned to England in 1938.

His first book, the collection of essays *Farewell to Sport,* was published that year. He justified the book by saying that one thing he had missed in all the years of sportswriting was the opportunity to reflect on the people and events he was covering. In the essay "Inside the Inside," a classic exposition of the beautiful intricacies of baseball, Gallico wrote,

> All games are alike in form and intent. One man tries to beat another man, or one group of men try to worst another through skill, courage, and physical condition. It is merely the materials with which they are provided for this purpose, the rules, and the playing grounds that differ. The more intricate the game and tangled the rules and complicated the materials, the more difficult it is to understand, but the more fascinating it becomes when you do understand it.

In literature games have often provided metaphors for life. For Gallico, the process of writing about games also reinforced his way of viewing life, his eye for the story. Also, games are always people-centered; the athletes are at the heart of the contest. It was, finally, the athletes themselves who most fascinated Gallico.

Of all the athletes he covered, three held a special place in his hall of fame: Dempsey, Ruth, and Jones. Their appeal to him was varied and complex, but in his mind they ultimately shared a quality he treasured: they were the only three golden people he encountered in the sports-rich era who, in his mind, remained unspoiled, natural, and fundamentally true to their own natures in the midst of victory and celebrity. His writings about them demonstrate not only Gallico's authorial skill but also his perceptiveness and thoughtfulness. Gallico was no mere hero-worshiper, though he sometimes called himself one. "The Sportswriter," he observes at the beginning of "One Hero," his elegant paean to Jones in *Farewell to Sport,* "has few if any heroes. We create many because it is our business to do so, but we do not believe in them. We know them too well." He saw the human flaws of the athletes and the unsavory elements that were often intertwined with organized sports. He did not ignore those negative aspects, but he could see through them to the central quality of the games and to the skill, bravery, and determination it took to excel at them. Thus, he always retained an enthusiasm for and an admiration of the men and women who provided the stuff of his early writing.

Characteristic is his treatment of Dempsey. His stunning first impression of the fighter always remained with him. He retold the story of their brief encounter in the ring many times over the years, and in all the recountings he maintained his respect for Dempsey without glorifying or glossing over the crude reality that the trade of a fighter is to injure his opponent. Both as a columnist and in the more leisurely retrospective of *Farewell to Sport,* Gallico viewed Dempsey as a thoroughly natural man whose inner self survived the corrupt environment of the sport that gave him wealth and fame and also survived the wealth and fame themselves. It was not that Gallico always liked Dempsey or what he did, or that he regarded him as a friend – he knew that a reporter could not be a friend to his subjects and remain true to his own craft. But he understood Dempsey's nature, had a sense of

Dempsey's place in his sport and in his time, and retained his admiration for the qualities, however primitive, that enabled Dempsey to be a champion. Those perceptive strengths are demonstrated in the essay "Shoot the Million," which focuses on the two epic Dempsey-Tunney fights and vividly re-creates a critical period in the history of boxing – indeed, of sports generally: a period that saw "the opening up of the columns of newspapers to a type of publicity and ballyhoo that differed entirely from any that had ever before featured prizefighters or ball-players or six-day bicycle-riders." It was the beginning of the era of personal publicity. "By personal publicity is meant the close watch kept by the sports pages – and the news side, as well – upon every single phase of the life of an athlete. He was no longer news only when he was fighting or preparing to fight. He was always news."

Gallico manifests a peculiar love/hate relationship with the fight game. He was clear-sighted about the rampant corruption that seemed inherent in it, and he was critical of the ways in which moneymen and mobsters manipulated and used the warriors. His essay "Pity the Poor Giant" is an unrelenting and compassionate recital of the exploitation of Primo Carnera. As sports editor of the *Daily News,* however, he had created the New York Golden Gloves amateur boxing program and was directly involved in the annual competition between the New York and Chicago Golden Gloves teams. He accompanied the New Yorkers and personally managed the trip when they went to Chicago for matches.

A substantial part of his five million words on sports were about boxing. Its appeal, as he writes in *Farewell to Sport,* was elemental:

> When two men face each other in a boxing ring with gloves on their hands and begin to fight when the bell rings, and stop fighting when it rings again after three minutes, it is reasonably obvious to anyone what is going on and what they are trying to do to each other and what means they are employing. The struggle is a simple hand-to-hand trial for complete mastery within certain time limits, and because the struggle sometimes gets atavistic, abysmal, and terrifying, with show of blood, it is arresting and arousing.

A boxing bout readily lends itself to the most basic of all story lines: the good guy versus the bad guy. For someone like Gallico who viewed the world as a storyteller, it could not help but have a continuing fascination. Years later, in his *The Golden People* (1965), Gallico would reveal the persistence of that fascination in his recounting of the careers of Dempsey and Tunney.

In contrast, Gallico saw baseball as a complex and subtle game with many elliptical plots and subplots. Compared to boxing, its appeal is more elusive and more intellectual; consequently, the pleasures it gives are more diverse and enduring. But the characters are no less dramatic, and of baseball characters none was greater than George Herman "Babe" Ruth. To Gallico, Ruth was not only a great ballplayer who had revolutionized the game – who had made it the national pastime – but he was also "a wonderful person and a great human being" whom Gallico genuinely loved and admired.

Gallico wrote about Ruth frequently. His own career with the *Daily News* roughly paralleled Ruth's career: Ruth joined the Yankees in 1920, three years before Gallico started with the *Daily News,* and left New York after 1934. He retired from the Boston Braves in the midst of the 1935 season, a year before Gallico departed the *Daily News.* In the essay "His Majesty the King" in *Farewell to Sport* Gallico wrote: "I learned to love him because he was all man. . . . He was discovering, living and enjoying this wonderful thing called life with all his senses, enjoying it more than anyone I have ever known." Yet Gallico also saw that Ruth never grew up, that he remained in some ways a child trapped in a man's body. The combination of raw talent and natural exuberance, however, created a figure of mythic proportions. Had he been only the caricature of the drinking, wenching, cigar-smoking glutton that forms part of the legend, he would have flamed out, exhausted by his own excesses. He was more, however, because for twenty-two years "he played ball on the same enormous scale on which he lived his life, intensely, fervently, and with tremendous sincerity and passion. It was impossible to watch him at bat without experiencing an emotion."

For all his faults, Ruth was a genuine American hero; Gallico understood why, and he was not reluctant to contribute to the legend, recounting visits Ruth made to the hospital rooms of sick children and affirming the story of the called home run against the Cubs in the 1932 World Series. "Ruth could do those things, take those chances and get away with them, because he was The Babe and because his imagination told him it was a fine, heroic, and Ruthian thing to do. . . . I suppose in fifty or sixty years the legend will be that Ruth could call his shots any time. But once is sufficient for me, and I saw him do that."

Like his treatment of Dempsey, Gallico's perception of Ruth says almost as much about Gallico

*Gallico in 1947*

and his vision of his own craft as it does about the subject at hand. The easy assumption that only two approaches to sportswriting (or, for that matter, to any form of journalism) are possible – uncritical "gee whiz" idolatry or self-righteous iconoclasm – fails to comprehend either the real nature of the craft or the essential humanity of athletes. Gallico frequently dismissed himself as a genuine "gee whizzer," a romantic who always fell for the "good guys versus the bad guys" cliché in sports. But that was an affectation. Gallico's ability to tell a good story depended on his ability to understand his characters, whether they were his own inventions or the real men and women who played games in the 1920s and 1930s. He knew their faults, knew that their excesses and imperfections could affect their performance. He also knew, however, that he was writing sports, not exposés; that he was telling stories, not engaging in social or psychological analyses.

Gallico admired Dempsey; he loved the Babe; but he had only one real hero: "In all the years of contact with the famous ones of sport I have found only one that would stand up in every way as a gentleman as well as a celebrity, a fine, decent, human being as well as a newsprint personage, and who never once, since I have known him, has let me down in my estimate of him. That one is Robert Tyre Jones, Jr., the golf-player from Atlanta, Georgia." While Gallico focused sharply on the characters of sports and their stories, he had an instinct for recognizing the social context and for making deft observations of that context without pomposity. This instinct comes through clearly in his treatment of Jones, who was hardly as gargantuan or dramatic a figure as Dempsey or Ruth but was one whose accomplishments in his sport are unmatchable. The combination of Tex Rickard and Dempsey had made boxing an enormously popular sport, but it was a spectacle, a contest of gladiators that people merely watched. Ruth had increased the popularity of baseball, but it was a boy's game raised to a level of perfection by boys with the physical skills of men. Golf was a participant's game, and the 1920s saw the first great leap in its popularity, a rapid expansion in the number of courses in the United States and in the number of players. Jones was just one of them – an amateur, a small man, not noticeably muscular – except that he played the game better than anyone else and beat all comers, professionals and amateurs alike, and accomplished what no one had ever accomplished before and no one is likely ever to accomplish again: winning the British and American Amateurs and the British and American Opens – the Grand Slam.

Reading Gallico on Jones, Ruth, Dempsey, or many other athletes of the 1920s and 1930s leaves one with a sense of an era that is past, but also with a consciousness that in both positive and negative ways there is much about the world of sports that has not changed. Some of his columns could appear in the sports section today and appear newly minted – although the quality of the writing might give them away. His observations on the hypocrisy of big-time college football are apt; the problems that he identified in the 1920s and 1930s have still not been solved: the lure of money and ego corrupting the relationship between athletics and education; the system's encouraging, even requiring, moral blindness on the part of the institutions, the coaches, and the athletes themselves; and the point that the only difference between professional and amateur football is that in "amateur" football the management keeps the income. This last observation is a particular application of his insistence that the distinction between amateur and professional sports in general makes no sense: people who spend the better part of their time practicing and competing in their sports are not amateurs; they may not be paid directly, but someone is supporting them. The erosion of the distinction today – the virtual disappearance of amateur tennis and golf on the national and international levels, the professionalization of Olympic basketball, and the periodic proposals to "reform" Division I college athletics by paying the players directly – confirms the accuracy of Gallico's observation. His dismissal of wrestling as a sport that has been attenuated into theater underscores the dangers all sports face in a media-driven world. Of hockey, he writes in *Farewell to Sport:* "I love a good fight, but why call it hockey?" College basketball, he says in the same essay, is headed toward a major scandal because "it is making too much money, and the gamblers have taken it up as a betting proposition."

Gallico was clearly ahead of his time in his view of black athletes:

> I do not bring any particular knowledge or incisive philosophy to this problem, beyond a frequently outraged sense of justice. To the Negro-hater's eternal question: "But do you want to sit down at table with a nigger and invite him into your home?" I have no particular answer. I have sat down to dinner with some very strange people and races without coming to grief. If it were a matter of compulsion, I might be upset, not by the company, but by the compulsion. As it is a matter of choice, I keep to the company of my own kind. But I have always felt that where we make demands upon the Negro and exploit him as we do in sport, there is no longer any choice possible.

More sharply, he equated American attitudes toward blacks with the views of Germans in the 1930s toward all non-Aryans: "We differ merely in our methods." Gallico detested exploitation, whether of black athletes, of college players, of boxers by boxing operators, or of Ruth, who was discarded by organized baseball when his playing days were over.

He was not without his blind spots. Gallico himself fell into the stereotype of using animal analogies to describe Joe Louis's movements and qualities. In the 1950s some libraries removed *Farewell to Sport* from the shelves because of its supposed anti-Semitism. The charge resulted from one paragraph that clumsily invoked a stereotype: in describing the appeal of basketball to the New York Jewish population he wrote that "the game places a premium on an alert, scheming mind and flashy trickiness, artful dodging, and general smart-aleckness." One can also criticize his treatment of women athletes. While he praised such stars as Moody and Gertrude Ederle, said all the right things about their courage and dedication, and amply publicized their feats, his comments on women athletes in general tend toward sexist stereotyping. He devoted a whole chapter of *Farewell to Sport* to women, titling it "Farewell to Muscle Molls, Too." In *Sportswriter* (1993) Charles Fountain, after repeating a story told by Grantland Rice in which Babe Didrikson beat Gallico in a footrace during a golf match, describes Gallico as "the quintessence of what in later decades would come to be known as male chauvinism." Fountain says that the essay was Gallico's revenge for that galling defeat.

In Mexico City in 1939 Gallico married Baroness Pauline Gariboldi of Budapest, but that marriage, too, ended in divorce. After leaving sportswriting Gallico wrote a few stories with sports settings – including "The Roman Kid" (1938), about a sportswriter who bears a striking resemblance to the author. That story reveals that he had never forgotten the lessons learned in one minute and thirty-seven seconds in the ring with Dempsey. And he wrote a series of humorous short stories about golf for *The Saturday Evening Post,* which were later published in book form as *Golf Is a Friendly Game* (1942), and *Lou Gehrig: Pride of the Yankees* (1942), a biography that provided the basis for the Gary Cooper movie of the same year, for which he co-authored the screenplay. "Thicker than Water," published in *Collier's* magazine in July 1944, is a short story about a cowardly boxer, Joey White, who learns that his now dead brother, Tommy, a war hero and a former boxing champion, had donated blood for a transfusion when Joey was sick. Joey finally gains the confidence to win a big match and help support his struggling family when he realizes that Tommy's blood is "coursing through his veins. A part of Tommy's life was alive inside of him." The story was reprinted in *Greatest Collier's Sports Stories* (1955).

Gallico's plunge into freelance writing did not take full effect until after World War II, when his stint as a war correspondent for *Cosmopolitan* ended. He turned increasingly to writing novels – though he would call them "long stories" rather than "novels," which he thought sounded too "literary." The popularity of his works provided him with a sufficient income to spend most of the rest of his life in England, Italy, Switzerland, Liechtenstein, and France. He did not regard himself as an expatriate, but as "an American who likes to live in Europe."

Six years after Ruth's death in 1948, Gallico wrote the short story "Saint Bambino," in which he makes Ruth "the patron saint of baseball and of the small boys who try so hard to play it well." The story, collected in *Further Confessions of a Story Writer,* was, he admitted, unmitigated nostalgia for a profession and an era he had left behind. But for the most part, the material for his fiction came from other sources. His novel *The Snow Goose* (1941) became a Christmas classic; Gallico reveled in telling the story of a chance meeting at the Stork Club when Ernest Hemingway said, "You know, Gallico, *Snow Goose* – I wish I'd written that." The Leslie Caron movie *Lili* (1953), for which Gallico cowrote the screenplay, was based on his novel *Love of Seven Dolls* (1954). His children's books *Ludmila: A Legend of Liechtenstein* (1955) and *Thomasina: The Cat Who Thought She Was God* (1957), which was made into the Walt Disney film *The Three Lives of Thomasina* in 1963, are enduring favorites. His novel *The Poseidon Adventure* (1969) was the basis for the 1972 film of the same title that started the fad of catastrophe movies. His four *Mrs. 'Arris* novels (1958, 1960, 1965, 1974) created an enormous following in both the United States and England.

Gallico returned to sports in 1965 with *The Golden People,* which begins with an essay on Ruth and ends with one on Jones, "the King himself of the Golden People." Gallico's esteem for Jones is based on three factors. The first is Jones's personal qualities: while he was one of the giants of the age, like Dempsey and Ruth, he remained unspoiled by his success. But Dempsey and Ruth were natural men who had the faults of the unsocialized – even the uncivilized. Jones, by contrast, was quintessentially civil, a gentleman who had an unfailing "instinct for doing the right thing." But he was no prig, and he had a genuine sense of humor, including, most important, the ability to laugh at himself. The second factor is the nature of the game at which Jones excelled: "The fact remains that of all the sports, golf in one sense is the most taxing in that it calls for absolute self-control. Those adrenalins which in the boxer, ballplayer, or footballer can whip him to a fighting frenzy and produce a sometimes seemingly miraculous, winning performance are of no use whatsoever to the championship golfer. If anything they are a handicap." Golf elicits the best of human instincts – moderation, control, and reasonableness – and can be played well by amateurs. Jones was the quintessential amateur, whom every man or woman can become, at least for one shot, one hole, or even perhaps one round. Third, Jones's winning of the Grand Slam in 1930 was an incomparable achievement that had grown even more remarkable with the passing of the years.

When Gallico left the *Daily News* to pursue broader fields, he left behind a significant legacy. The quality of his sportswriting set a standard that only occasionally has been matched since, though he had equals in his own time: Rice, Bill Corum, Damon Runyon, and W. O. McGeehan. He writes about them in the concluding chapter of *The Golden People,* "Men with the Golden Pen": "And if that era gave rise to the greatest concentration of champions and sports personalities in our history, there was also a similar outpouring of men fit to sing of their deeds." He dismisses criticism of the flowery rhetoric in which they (and he) indulged: "This was a florid era and called more than occasionally for florid reporting to do it justice."

In 1967 Gallico collaborated with the Hungarian photographer Nikolas Muray on *The Revealing Eye: Personalities of the 1920s.* The book includes only three portraits of sports figures; of course, Muray's work constrained Gallico, but the proportion also reflected Gallico's understanding that the culture of the 1920s and 1930s was far more varied and complex than the fields of play. The book provided the occasion for him to observe:

> Sportswriters seemed to be immediately accepted in all circles with first-name camaraderie. If they had read us, they knew us – taxi-drivers, socialites, bootleggers, gang executioners, merchants, actors and directors. There were very few celebrities in almost any field who did not have some interest in some game.
>
> But over and beyond these personal involvements in whatever it might be . . . the spread of interest in the major sports was colossal and unprecedented. Boxing, baseball, football and tennis widened their audiences not so much because of the contests but rather because of the characters who emerged from them, such as Babe Ruth, Dempsey, Tunney, Sharkey, [Bill] Tilden, Helen Wills, Bobby Jones, Walter Hagen, who not only were dramatic figures as competitors but whose private lives were a continuing serial more thrilling than any to be found in the magazines of the day. . . . We were their Boswells.

To accompany a striking portrait of Ruth, Gallico wrote:

> We not only reported his activities in the ball yard, but wallowed emotionally in his personal life, his problems with gluttony and other vices which brought him into conflict with his team manager and fellow players. Once even the Mayor of New York, James J. Walker, made a tearful plea to the Babe to reform for the sake of the dirty-faced kids in the street. When he died of cancer on 17 August 1948, an entire nation mourned as though a king had passed away. Well, one had.

Gallico's text for Muray's portrait of Jones captures the quality of the subject, but with a personal touch that underscores Gallico's respect and admiration for him: "For us who belonged to this fraternity [of amateur golfers], there was always Jones, who, too, once had thrown clubs and wept at his own failures and now had it made – the exquisitely perfect golfing machine, sportsman and one of the best-loved sports figures of the era." The contrast between Jones, on the one hand, and Dempsey and Ruth, on the other, also points up Gallico's range. Seeing in all games the same fundamental element meant that he found appeal in all, or almost all of them. It also aroused his curiosity as to the differences, both obvious and subtle, among them, sharpening his eye for detail.

On 19 July 1963 Gallico married Baroness Virginia von Falz-Fein. He died in Monaco, the last of his foreign retreats, on 15 July 1976, leaving his wife, a stepdaughter, two sons, and three grandchildren – and the eloquent legacy of his essays.

Gallico was successful as a sportswriter at a time when there were giants in that profession. He had an eye for detail, and he instinctively understood that the central story of games is the people who play them. The games had plots, and his analyses of those plots – in boxing in "Shoot the Million" and in baseball in "Inside the Inside" – remain fresh and alive. But as a storyteller he knew that the athletes carried the action and that they were more than automatons. In his writing he captured both their physical skills and their humanity: "You learn eventually that, while there are no villains, there are no heroes either. And until you make the final discovery that there are only human beings, who are thereafter all the more fascinating, you are liable to miss something." Of his own early prose Gallico said, "I find my pen was dipped into the sheerest heliotrope hysteria. My bill for purple ink must have been appalling." Nevertheless, the vigor and clarity of the writing and of the ideas remain fresh. If one wants to understand the personality of the Golden Age of Sport, one could do much worse than reading Gallico's columns from that era.

**Interview:**

Jerome Holtzman, *No Cheering in the Press Box* (New York, Chicago & San Francisco: Holt, Rinehart & Winston, 1973), pp. 61–80.

**References:**

Charles Fountain, *Sportswriter: The Life and Times of Grantland Rice* (New York: Oxford University Press, 1993), pp. 245–246;

Richard Goldman, *Sportswriters Choice* (New York: Barnes, 1958);

George Plimpton, *The Best of Plimpton* (New York: Atlantic Monthly Press, 1990), p. 4;

Grantland Rice, *The Tumult and the Shouting* (New York: Barnes, 1954).

**Papers:**

There are sixty boxes of uncatalogued Gallico manuscripts in the Columbia University Manuscript Library.

# W. C. Heinz

*(11 January 1915 - )*

Edward J. Tassinari
*S.U.N.Y. Maritime College*

MAJOR POSITIONS HELD: City reporter (1939–1943), war correspondent (1943–1945), sports reporter (1945–1949), sports columnist, *New York Sun* (1949–1950); guest sports columnist, *New York Daily News* (1950).

BOOKS: *The Professional* (New York: Harper, 1958);
*The Surgeon* (Garden City, N.Y.: Doubleday, 1963);
*Run to Daylight!*, by Vince Lombardi and Heinz (Englewood Cliffs, N.J.: Prentice-Hall, 1963);
*M.A.S.H.*, by H. Richard Hornberger and Heinz as Richard Hooker (New York: Morrow, 1968);
*Emergency* (Garden City, N.Y.: Doubleday, 1974);
*Once They Heard the Cheers* (Garden City, N.Y.: Doubleday, 1979);
*American Mirror* (Garden City, N.Y.: Doubleday, 1982).

TELEVISION: *Biography of the Fight*, preview of the second Patterson-Johansson heavyweight championship fight by Heinz and Alvin Boretz, ABC, 19 June 1960;
*Run to Daylight*, ABC, 9 September 1964.

OTHER: *The Fireside Book of Boxing*, edited by Heinz (New York: Simon & Schuster, 1961);
"One Throw" in *The Third Fireside Book of Baseball*, edited by Charles Einstein (New York: Simon & Schuster, 1968), pp. 201-203.

*W. C. Heinz (photograph by G. B. H.)*

SELECTED PERIODICAL PUBLICATIONS – UNCOLLECTED:

FICTION

"The Head and the Heart," *Cosmopolitan*, 125 (19 July 1948): 52–53, 107–112;
"Win to Go Home," *Collier's*, 122 (4 September 1948): 20–21, 72–73;
"Man's Game," *Collier's*, 124 (15 October 1949): 18–19, 34, 36–38, 40;
"Half a Buck," *Esquire*, 33 (February 1950): 32, 123–127;
"Fixed Fight," *Saturday Evening Post*, 224 (7 July 1951): 24, 78, 80–82.

NONFICTION

"Parallel: A Revolution in Skiing," *Esquire*, 15 (February 1941): 37, 114–115;

"Home Maid Ski Champ," *Collier's,* 107 (1 March 1941): 22, 44;

"If Branca Weren't So Nice," *Baseball Digest,* 8 (April 1949): 15–17;

"Last Day," *Life,* 28 (16 January 1950): 40, 42;

"Most Fighters Are Scared," *Saturday Evening Post,* 222 (24 June 1950): 36–37, 84–88;

"Why Don't They Like Robinson?," *Saturday Evening Post,* 223 (9 December 1950): 30–31, 178, 180, 182;

"Private Life of a Fight Referee," *Saturday Evening Post,* 223 (31 March 1951): 36–37, 127–129;

"So He's Going to Clean Up Boxing," *Saturday Evening Post,* 224 (16 February 1952): 29, 97–100;

"Yankees Crafty Ace," *Saturday Evening Post,* 224 (10 May 1952): 29, 139–141;

"Strange Case of Ezzard Charles," *Saturday Evening Post,* 224 (7 June 1952): 34, 127–129;

Chuck Davey, "I Like to Fight," as told to Heinz, *Saturday Evening Post,* 225 (7 February 1953): 25, 92, 94–95;

"I Scout for the Yankees," *Collier's,* 132 (11 July 1953): 18–22;

"When Women Invade the Racetrack," *Cosmopolitan,* 136 (January 1954): 46–50;

"Baseball Players Dream Town," *Cosmopolitan,* 136 (May 1954): 88–93;

"Knockout," *Collier's,* 134 (17 September 1954): 94–97;

"Yankee Batboy," *Cosmopolitan,* 137 (October 1954): 122–131;

"They Called My Husband Hard Luck," *Cosmopolitan,* 138 (May 1955): 26–29;

Rocky Marciano, "How It Feels to Be Champion of the World," as told to Heinz, *Collier's,* 135 (13 May 1955): 23–27;

"Mystery of Archie Moore," *Saturday Evening Post,* 228 (17 September 1955): 26–27, 88, 90;

"The Two Faces of Carmen Basilio," *Saturday Evening Post,* 228 (19 November 1955): 46–47, 163–164, 166;

"Boss of the Behemoths," *Saturday Evening Post,* 228 (3 December 1955): 47, 72, 74;

Eddie Arcaro, "You Have to Ride Rough," as told to Heinz, *Look,* 20 (17 April 1956): 116–123;

Arcaro, "How to Win the Derby," as told to Heinz, *Look,* 20 (1 May 1956): 90–96;

"Boxing's Human Windmill," *Saturday Evening Post,* 229 (26 January 1957): 24–25, 96;

"Is Boxing Crooked – Or Isn't It?," *Sport,* 12 (July 1957): 18, 76–78;

"Battler from the Backwoods," *Saturday Evening Post,* 230 (5 October 1957): 31, 69–71;

Nathan Hare, "What Makes a Man a Fighter?," as told to Heinz, *Saturday Evening Post,* 230 (8 March 1958): 27, 75–76, 78;

"Visit with Jack Dempsey," *Saturday Evening Post,* 230 (17 May 1958): 32–33, 101–102;

"And Now There Are Eight," *Life,* 44 (26 May 1958): 109–110, 113–114;

"New York's Revolution in Horse Racing," *Reader's Digest,* 73 (December 1958): 118–121;

"Most Beat-Up Man in Football," *Life,* 45 (1 December 1958): 57–58, 60, 77;

"Visit with the Champion," *Saturday Evening Post,* 231 (28 February 1959): 19–21, 59–60;

"A Visit with Ingemar Johansson," *Saturday Evening Post,* 231 (20 June 1959); 26, 79–80;

"Dodger Who Never Left Brooklyn," *Coronet,* 46 (June 1960): 149–153;

"King of the Horse Traders," with Clyde Hirt, *Saturday Evening Post,* 233 (30 July 1960): 22–23, 51–52;

Vince Lombardi, "Day of the Game," as told to Heinz, *Sports Illustrated,* 19 (9 September 1963): 94–104;

"Stan Musial's Last Day," *Life,* 55 (11 October 1963): 96–98;

"Golden Boy Is Back," *Life,* 57 (28 August 1964): 60–74, 77–78;

"Countdown for Joe Namath," *Life,* 59 (20 August 1965): 37–40;

"I've Been Living with It a Long Time," *Saturday Evening Post,* 239 (8 October 1966): 112–117;

"Last Campaign of Boxing's Last Angry Man," *Saturday Evening Post,* 240 (11 February 1967): 38–42, 47–48;

"Game for Madmen," *Look,* 31 (5 September 1967): 85–90;

"Green Bay's Coach Reveals His Secrets for Winning Football," *Look,* 31 (19 September 1967): 70–75;

"I Miss the Fire on Sunday," *Life,* 65 (27 September 1968): 121–123;

"Quick, Thomas, a Brick," *TV Guide,* 24 (21–27 August 1976): 28–30, 32;

"When Archie Tried to Break Up the Rock," *Sport,* 71 (September 1980): 74–76;

"When the Bell Rings, Know What to Look For," *TV Guide,* 30 (25–31 December 1982): 28–30;

"I Remember Joe DiMaggio, the Pro," *Fifty Plus,* 25 (July 1985): 59;

"Sketching John Groth," *Fifty Plus,* 26 (November 1986): 42–47;

"In Covering Sports: Why Can't TV Encourage Class?," *TV Guide,* 35 (7–13 November 1987): 16–19;

"It Was Like a Light Bulb Busted in Your Brain," *TV Guide,* 36 (19–25 November 1988): 10–11, 14.

Sportswriter W. C. Heinz is considered by many of his peers to be one of the best pure writers of his time. From 1937 to 1950 Heinz enjoyed several successful writing careers with the *New York Sun.* Progressing from copyboy to local reporter, feature writer, and war correspondent, he turned to sportswriting in September 1945. When in 1950 the *Sun* was acquired by Scripps-Howard, publisher of the *New York World-Telegram,* Heinz rejected several other newspaper job offers to pursue a career as a freelance writer. While writing on various other subjects, he continued to emphasize sports. With the publication of his boxing novel, *The Professional* (1958), Heinz began a fiction writing career. He produced three medical novels, including *M.A.S.H.,* a best-selling 1968 novel about the lives of U.S. Army surgeons during the Korean War that he wrote with H. Richard Hornberger under the pseudonym Richard Hooker. In *Once They Heard the Cheers* (1979) Heinz paid a final nostalgic visit with sports performers whom he had most admired during their prime.

Heinz grew up in the 1920s and 1930s and has written about the powerful impact that the athletic heroes of his youth held upon him. He longed to duplicate the feats of Babe Ruth, Red Grange, and Jack Dempsey, but, burdened with a slender physique and modest athletic talents, he resolved upon a career in sportswriting as the best alternative. As a contemporary and close friend of Red Smith and Frank Graham, Heinz was a perfectionist constantly striving to enhance his craft and was influenced, as were so many World War II correspondents, by Ernest Hemingway. Heinz viewed sport as a means of self-expression. At its height, he saw it as an art form in which courage, craftsmanship, wisdom, and professionalism – an amalgam of study, effort, imitation, and adaptation – could be found. Heinz thought it was the writer's task to present the scene, locate the character, and convey an understanding through the eyes of the subject.

Although producing consistently fine writing about all sports and in all sportswriting genres from newspaper features and columns to magazine pieces and books, Heinz wrote best about boxing, a sport in which he passionately extolled the honest workmen such as welterweight contender Billy Graham and strategist manager Jack Hurley, who were the models for Heinz's fictional characters in *The Professional.* He also exposed the hopes, fears, and inner turmoil of his characters, from champions and challengers to referees and club fighters (and their wives), while he preserved the language and flavor of such bygone boxing haunts as Stillman's Gym in Manhattan.

Wilfred Charles "Bill" Heinz was born on 11 January 1915 in Mount Vernon, New York, the only child of Frederick Louis Sylvester Heinz, a salesman, and Elizabeth Thielke Heinz. He graduated in 1933 from Mount Vernon High School and in 1937 from Middlebury College, where he completed an bachelor of arts degree in political science while having been sports editor of the college newspaper and yearbook. In the fall of 1937 Heinz landed a job as a messenger boy at the *New York Sun,* where he made fifteen dollars a week. Having advanced to copyboy two years later, Heinz impressed Keats Speed, the city editor, by writing a piece on cleaning women who trudged to work in Manhattan in the early mornings from Mount Vernon and the Bronx en route to the elevated subway lines, and he was placed on staff as a city reporter. On 18 January 1941 Heinz married Elizabeth B. Bailey in Middlebury, Vermont. They had two daughters, Barbara in 1947 and Gayl in 1951.

As a city reporter, rewrite man, and feature writer, Heinz reported on various New York stories until fall 1943 when he became a *Sun* war correspondent. He spent seven weeks aboard a U.S. Navy aircraft carrier in the Atlantic Ocean on patrol for submarines and then one month in Brazilian waters with the Fourth Fleet. On 1 April 1944 he arrived in London to report on Allied preparations for the invasion of Europe, and on 31 May 1944 he departed on his D-day assignment aboard the battleship USS *Nevada,* as he covered the bombardment of Normandy in support of Allied troop landings. Then in August 1944, when the Nazis captured the chief European war correspondent of the *Sun,* Heinz replaced him and remained with the First Army through the final days of the war in Europe.

Heinz has said that he learned to write through his World War II experiences. In a 1961 *Newsweek* interview he described the war as, "a patsy for learning writers . . . the perfect foil, the perfect sparring partner. It was so dramatic, you couldn't write it badly." In a revealing passage from *Once They Heard the Cheers* Heinz writes of suffering in late September 1944 while covering the fighting of the First Army in the Huertgen Forest. He agonized about the guilt and fear that he felt impeded him from continuing to write of the sacrifice and loss of young men dying and about his struggle to get their stories right, a struggle that he felt he was losing. He survived that crisis, and his evolution and development as a writer were better for it.

Although his earliest pieces of wartime reporting attend consistently to detail, present the facts, and set the scenes, these works are infused with a measure of patriotic rhetoric. Yet Heinz's initial wartime assignment makes this understandable. Distant from the main theaters of combat, he wanted to convey the belief that fighting and winning the war was to be a collective effort in which the efforts of cooks, clerks, and grease jockeys were as significant as the exploits of bomber pilots hunting German submarines. As Heinz shifted to covering ground-combat troops in Europe, his writing increasingly used contrast, understatement, and irony effectively. He was particularly adept in describing the utter confusion and emotional turmoil that envelops combat, the impact of war's devastation on civilian populations, and the toll exacted on men who lead others into combat and see them die. He was skillful at painting miniature word pictures: his vignettes of troops on the march convey the simplest scenes with a razor-sharp clarity, combining sight and sound as a fused unit. His most powerful piece of wartime reporting – a stark, gripping account of the execution of three captured German spies in U.S. Army uniforms and with an American jeep behind American lines on 23 December 1944 – could not appear for censorship reasons until December 1949, when *True* magazine published it as "The Morning They Shot the Spies."

Shortly after V-E day Heinz returned to the United States, where the *Sun* gave him a three-month vacation, a $1,000 bonus, and a promotion. He was assigned to the newspaper's Washington, D.C., bureau as a feature writer. Before going on vacation, however, Heinz informed the managing editor that he wanted to write sports. After spending the rest of the summer with his wife in a cabin on Lake Champlain, he returned to the *Sun* in September 1945 and was assigned to the sports department, where he remained until the newspaper published its final edition on 4 January 1950.

Heinz has written about his first days and early sports assignments in the opening chapter of *Once They Heard the Cheers*. The *Sun* was an evening newspaper with a large sports staff and no Sunday edition. Heinz's writing for the *Sun* took on the nature of features, sketches of athletes in training and teams in preparation, wrap-ups of earlier contests, and general assignments. One of his first sportswriting bylines was an atmospheric piece about things heard and seen as he left Yankee Stadium. Heinz also became a habitué of Stillman's Gym in Manhattan, then the most famous boxing gymnasium, where he soaked up the argot and mystique of the fight game.

On 10 January 1949 he began his own sports column, "The Sport Scene," which appeared five times a week. His columns used a mixture of styles and were often conversation pieces (such as those popularized by Frank Graham), sketches of sports figures, technical analyses of an athlete's skills, or atmospheric pieces linked to boxing, horse racing (another of his favorite subjects), baseball, or other sports. He wrote few opinion pieces, but when he did, he avoided noisy verbal crusades and was content to pursue salient issues in a calm, coherent fashion.

Heinz began to contribute articles to magazines such as *Collier's, Saturday Evening Post, Cosmopolitan,* and *True*. His first magazine article, an interview with Fritz Loosli, the Swiss-born skiing instructor who popularized the parallel skiing instructional method, appeared in *Esquire* in February 1941. Other works by Heinz appeared often in the E. P. Dutton series Best Sports Stories series. One of Heinz's stories appeared in the second volume of that series in 1945; Heinz was to win five E. P. Dutton Awards between 1948 and 1959 for the best magazine sports story. Four of these pieces pertain to boxing, and all five appear in *American Mirror* (1982), an anthology of his best work.

"The Day of the Fight" was a detailed account of the activities of middleweight challenger Rocky Graziano from the moment he awoke on 27 September 1946 to the second he stepped into the ring at Yankee Stadium that evening to fight champion Tony Zale. In "Fighter's Wife" Heinz walked neighborhood streets with Graziano's wife, Norma; her mother; and a family friend while Graziano was fighting Charley Fusari on 14 September 1949. Too nervous to listen to the fight, Norma revealed her thoughts to Heinz as the sounds of the radio broadcast drifted through windows and doorways. Expertly interweaving accounts of how the Grazianos had met, what their married life was like, and what strains his profession placed on their marriage, this innovative article, described by Quentin Reynolds in *Best Short Stories* (1950) as "a beautiful story of people under tension," was one of many in which Heinz emphasized the human qualities of the athlete and the concerns of the athlete's wife.

Rocky Graziano appeared again in Heinz's work during the latter's years with the *Sun,* notably when Manhattan district attorney Frank Hogan and the New York State Athletic Commission (NYSAC) investigated Graziano early in 1947 for an alleged bribe offer that he had failed to report. Heinz covered the NYSAC hearings that led to the suspension of Graziano's license to fight in New York, and he later wrote of how he had failed in his obligation to objectivity when, the night before the second Zale-Graziano fight in 1947, he had admonished Graziano to win in order to strike back at politicians, boxing commissioners, and reporters who, Heinz believed, had contrived to vilify the inarticulate fighter and poison public opinion against him. In a piece for *Sport* magazine ("Good-bye, Graziano," March 1952) Heinz again criticized himself for having lost his reportorial balance, especially since Zale was a decent man and a courageous fighter, one whom Heinz admired greatly.

Such personal writing would lead Jack Newfield, writing in *Dutton Review* in 1970, to list Heinz together with Jimmy Cannon as one of the forerunners of the New Journalism genre that had become influential in the 1960s, a style that challenged objectivity, adopted a participatory stance, and often broke traditional molds through innovative approaches in language, structure, and subject to get to the essence of reality. While Cannon would ultimately disavow the excesses of certain New Journalists (while accepting credit for his influence) Heinz never sought to propagate a school of adherents but kept his focus on the humanity of his subjects, saving his skepticism for the self-styled crusaders who sought to foist their grandiose plans to expunge the evil influences within boxing.

Shortly after the demise of the *Sun,* Heinz wrote his first piece for *Life,* a straightforward account of the newspaper's final day ("Last Day," 16 January 1950) that underscored the staff's professionalism even in the newspaper's final hours of existence. For four weeks during July and August 1950, Heinz filled in for vacationing sports editor Jimmy Powers as sports columnist at the *New York Daily News.* For a time in 1951–1952, he did promotional work for "Meet the Champs," an armed forces boxing telecast on the ABC network. Throughout the 1950s he worked as a freelance writer, contributing articles to *Life, Collier's, Look, Saturday Evening Post, Sport, True, Argosy, Esquire, Cosmopolitan,* and other magazines. After receiving a substantial fee from *Look* for a two-part series on jockey Eddie Arcaro in 1956, Heinz spent eleven months working on *The Professional.*

*The Professional* is the story of middleweight boxing contender Eddie Brown, preparing for his first title fight, as seen through the eyes of sportswriter Frank Hughes, working on a magazine article on the fighter. Much of the novel takes place during the three weeks Brown spends in an upstate New York training camp prior to the fight. Widely praised for its authenticity, honest dialogue, and deep understanding of the fight game and its denizens, *The Professional* served as a vehicle for several of the writer's essential beliefs. Hughes, explaining his love of boxing to Eddie Brown, describes it as "The basic law of man. The truth of life. . . . I find man revealing himself more completely in fighting than in any other form of expressive endeavor." Heinz ultimately saw boxing as the most scientific of all sports and the most self-expressive as an art form in that man could "vent and divest himself of his hatred and his anger." Yet he also considered boxing, when elevated to a higher creative plane (as in a vintage Sugar Ray Robinson performance in 1950 against a game but outclassed Charley Fusari) comparable to an art – becoming "one with the other arts and thus a subject to flatter one's talent."

Through the character of Doc Carroll, Brown's canny manager, Heinz further expressed his distaste for the "amateurs" he felt were ruining the sport, notably dilettante fight managers, inept trainers, and venal promoters. To Hughes, struggling to maintain his objectivity in light of his emotional involvement with the success or failure of the fighter while striving to comprehend fully the essence of what takes place in the sporting arena, Eddie Brown represents the consummate professional in a world of mediocrities masquerading as first-rate. The book's denouement, a stunning first-round knockout at the hands of the champion, is a shocking anticlimax.

Contemporary reviewers found much to praise in *The Professional.* Robert Cromie in *Booklist* described the novel as "a strong, expressive story free of stereotypes and melodrama." An anonymous *New Yorker* reviewer referred to it as, "precise, poignant and absolutely honest . . . its climax, if devastating, is not in the least pitiless." Herbert Kupferberg, writing in the *New York Herald Tribune* book review section, called Heinz "a stylist among sports writers," alluding to Hemingway's work by praising the "affection, insight and lyricism one usually finds in novels about bullfighting."

Not all reviews were uniformly positive. Although praising the novel's sincerity and citing several particular scenes, Robert Daley in *The New York Times* was critical of the use of the noncombatant narrator and found the emotion of the leading character obscure and the driving narrative force lacking. Charles Fenton, writing in the *Saturday Review,* lauded the author's skill and perception but ultimately felt the novel's characterization was too one-sided and that the author was unsuccessful in his "insistence that a skilled professional athlete is in the last analysis something remarkable among human beings." One acerbic *Time* critic found fault with Heinz for using the historical present and for addressing the reader as "you." The critic, although noting that Heinz had crafted a "reasonably effective story" with some "wonderful examples of tough prose," concluded, "*The Professional,* in short, is a classic example of the Hemingwayward conviction that small words must be used to denote big things," finding the book lacking in "the quality of thought."

Heinz's good friend Red Smith devoted a column to a review of *The Professional.* Smith was lavish in his assessment: "*The Professional* is first-rate – clean, honest, spare, exciting and in this view an extraordinary performance by a man doing his first novel." Smith made the point that the boxing milieu here ought not be interpreted as circumscribing the significance of the novel. As he indicated: "Eddie Brown and Doc Carroll are symbols of all who strive and encounter hardship, frustration, ignorance. What happens in the microcosm of boxing, this author says, happens in life, but he does not say it loudly and he doesn't set up red flags to mark his point." To Smith, the novel's weakness was that the reader would accept it solely as an absorbing tale of a fighter and nothing more, something the author never intended.

*The Professional* sold about fourteen thousand copies in the original hardcover edition and was reprinted in paperback by Arbor House in 1984. In his introduction to the paperback reprint, George Plimpton cited a 1958 visit to Havana with Hemingway, who praised the book as the only good boxing novel he had ever read. Contributing the foreword to *American Mirror,* Red Smith related how Heinz, after writing the first sentence leading up to the decisive fight, had to take an extended walk in the woods to summon up the courage to have Brown knocked out, despite knowing full well how the book would end at its inception. A. J. Liebling, a penetrating boxing observer in his own right who had been critical of Heinz's writing about Graziano in a 1946 assessment of New York City sportswriting in *The New Yorker,* wrote Heinz personally to praise him for *The Professional,* telling Heinz that he had finished reading the book at 3:00 A.M. and adding, "all praise in varying degrees . . . high and extra high" with further encomiums for "avoiding chances for melodrama and avoiding irrelevant hokum."

When he compiled *The Fireside Book of Boxing* (1961), an anthology of boxing pieces, Heinz had lost his earlier passion for boxing. In a 7 January 1961 article for the *Saturday Evening Post,* "The Twilight of Boxing," a bittersweet final return to little-used Stillman's Gym in its final throes before demolition, Heinz ruminated with Nat Fleischer, Irving Cohen, and other fight game veterans on the reasons for the decline of the sport. The article reflected a strain in Heinz's later writing that would become more pronounced with time – a tendency to recall the past and individuals who had strongly impressed Heinz but who, together with the values and virtues they represented, would eventually disappear in the America of the 1960s and beyond.

Heinz's preface to *The Fireside Book of Boxing* reflected sentiments similar to those conveyed in *The Professional:* criticism of the second-rate, which gets undue publicity while better quality material (appealing to a select audience) fails to survive. In his brief introductions to each selection, Heinz was generous in his praise. A *Newsweek* review of *The Fireside Book of Boxing* recalled Heinz's ability as a *Sun* columnist to capture the essence of the fight crowd and found the book a reflection of his "integrity, compassion and affection for the sport." Heinz's own contributions were "The Fighter's Wife" and an as-told-to piece by Rocky Marciano on how it feels to be heavyweight champion. Heinz told the *Newsweek* reviewer that he found it more difficult to sustain interest in boxing because of the declining quality of fighters and was now seeking new writing challenges, no longer eager to rewrite. He was described as presently at work on a book about a cancer operation.

Before that book appeared, Heinz contributed a notable piece of straight reporting for *Life,* "The Man With a Life in His Hands" (20 January 1961), that elaborated in meticulous detail a groundbreaking lung-removal operation that would form the basis of the climactic chapter in his second novel, *The Surgeon* (1963), a chronological account of eight hours in the working life of thoracic surgeon Matthew Carter. In Carter (a composite of several surgeons), Heinz created an idealized figure who could be compared with Eddie Brown, in that becoming a great surgeon, as with becoming a top-flight boxer, was the result of years of hard work and sacrifice while gradually perfecting one's craft through trial and error and instruction from capable, even inspiring, teachers.

In 1962 Heinz spent several weeks working with one of the best and most successful "teachers," Green Bay Packers football coach Vince Lombardi. The result of their collaboration was *Run to Daylight!* (1963), a seminal work and the first of many such

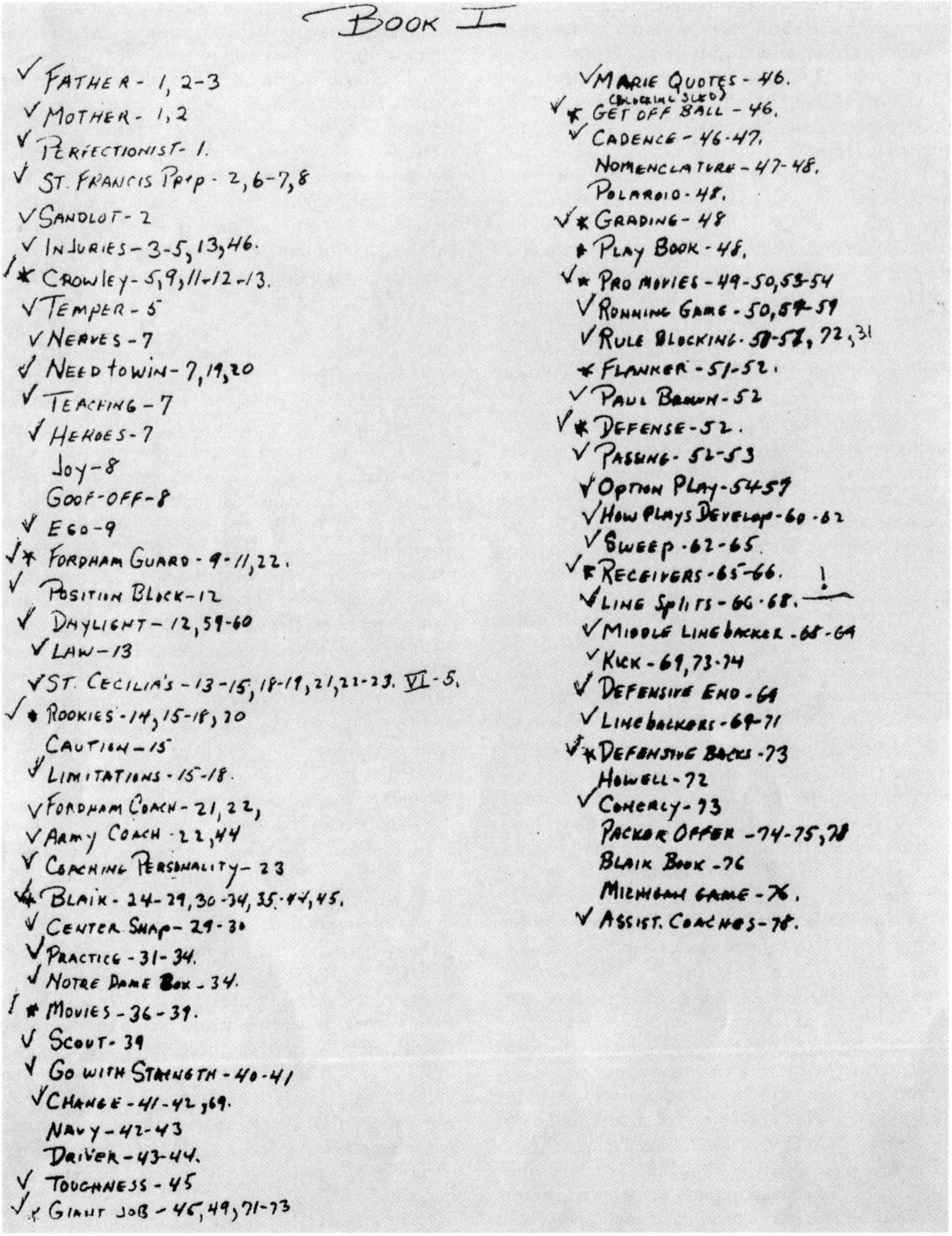

BOOK I

✓ FATHER - 1, 2-3
✓ MOTHER - 1, 2
✓ PERFECTIONIST - 1.
✓ ST. FRANCIS PREP - 2, 6-7, 8
✓ SANDLOT - 2
✓ INJURIES - 3-5, 13, 46.
✓* CROWLEY - 5, 9, 11-12-13.
✓ TEMPER - 5
✓ NERVES - 7
✓ NEED TO WIN - 7, 19, 20
✓ TEACHING - 7
✓ HEROES - 7
JOY - 8
GOOF-OFF - 8
✓ EGO - 9
✓* FORDHAM GUARD - 9-11, 22.
✓ POSITION BLOCK - 12
✓ DAYLIGHT - 12, 59-60
✓ LAW - 13
✓ ST. CECILIA'S - 13-15, 18-19, 21, 22-23. VI-5.
✓* ROOKIES - 14, 15-18, 20
CAUTION - 15
✓ LIMITATIONS - 15-18.
✓ FORDHAM COACH - 21, 22,
✓ ARMY COACH - 22, 44
✓ COACHING PERSONALITY - 23
✓* BLAIK - 24-29, 30-34, 35-44, 45.
✓ CENTER SNAP - 29-30
✓ PRACTICE - 31-34.
✓ NOTRE DAME BOX - 34.
✓* MOVIES - 36-39.
✓ SCOUT - 39
✓ GO WITH STRENGTH - 40-41
✓ CHANGE - 41-42, 69.
NAVY - 42-43
DRIVER - 43-44.
✓ TOUGHNESS - 45
✓* GIANT JOB - 45, 49, 71-73

✓ MARIE QUOTES - 46.
✓* GET OFF BALL - 46. (BLOCKING SLED)
✓ CADENCE - 46-47.
NOMENCLATURE - 47-48.
POLAROID - 48.
✓* GRADING - 48
* PLAY BOOK - 48.
✓* PRO MOVIES - 49-50, 53-54
✓ RUNNING GAME - 50, 57-59
✓ RULE BLOCKING - 50-58, 72, 31
* FLANKER - 51-52.
✓ PAUL BROWN - 52
✓* DEFENSE - 52.
✓ PASSING - 52-53
✓ OPTION PLAY - 54-57
✓ HOW PLAYS DEVELOP - 60-62
✓ SWEEP - 62-65
✓* RECEIVERS - 65-66. !
✓ LINE SPLITS - 66-68.
✓ MIDDLE LINEBACKER - 68-69
✓ KICK - 69, 73-74
✓ DEFENSIVE END - 69
✓ LINEBACKERS - 69-71
✓* DEFENSIVE BACKS - 73
HOWELL - 72
✓ CONERLY - 73
PACKER OFFER - 74-75, 78
BLAIK BOOK - 76
MICHIGAN GAME - 76.
✓ ASSIST. COACHES - 78.

*Pages from the notebooks Heinz kept while working with Vince Lombardi on* Run to Daylight! *(collection of W. C. Heinz)*

MASTER I

√ADDERLEY-II-42, 63-64. III-3. IV-7. V-30

√AUSTIN-II-55-56. III-42.

√BLANK-I-24-29, 30-34, 35-44, 45. II-79. IV-45.

√BLOCKING SLED-I-46. III-37, 38, 73. IV-8, 25.

√CARPENTER-III-58. IV-14, 49.

√COCHRAN-III-42. IV 1-3.

√CROWLEY-I-5, 9, 11-12-13. III-36, VI-41, 6.

√CRUICE-III-71-73. IV-41.

√CURFEW-III-60, 63 IV-45.

√CURRIE-II-42, 59. III-7. IV-52.

√DAVIS-II-59, IV-54-56. 1960-1

√DEFENSE-I-52. II-8. 1962-5

√DEFENSIVE BACKS-I-73. II-11. III-55-57, 74. IV-4, 8, 21, 41, 46-47, 78. 1961-31

DISCIPLINE-II-79 III-54. IV 18-19.

√DOWLER-II-13-14, 32-33, 45, 48, 75-76. III-5, 51-52. IV 7, 11, 23-24, 53. 1960 32. V-1 [illegible]

√DRAFT-II-17-22, 29-30. III-24-37.

√FEARS-II-2. IV-15, 16.

√FINES-II-53. III 57.

√FLANKERS-I-51-52. II-73-75.

√FORDHAM GUARD-I-4-11, 22. III-23. VI-5

√GIANTS-I-45, 49, 71-73. II-56.

√GRADING-I-48. II-1-2.

√GREGG-II-42-43, 43-44, 69. III-5. IV-26-27.

√GREMMINGER-II-61, 61-62. III-8.

√HANNER-II-58. III-29. 43, 55, 60, 71. IV-5-6, 32, 42-43, 57. 1961-11 (12-13, TEARS)

√HORNUNG-II-43, 46-48. 49. 78-79 III-1-3, 20, 57, 64. IV-5, 54, V-2

√JERRY-II-44, 67. III-24, 53-54, 58-59. 71. IV 26, 53, 530.

√JORDAN-II-12, 43, 58-59. III-7. IV-4, 1960-30.

√KNAFELC-II-31. III-7. IV-27. 1960-24

"inside" books detailing the workings of a coach's mind. Universally praised for its fine, even poetic writing ("When I turn onto Oneida Avenue the deserted practice fields to the left are like a lush, damp pasture under the low, gray skies, and the gray, canvas-padded seven-man and two-man blocking sleds, standing there alone, look like lonely forgotten cattle"), deep understanding, and seamless collaboration between narrator and ghost (and benefiting from Robert Riger's excellent photos and sketches), the book, written in the first person from Lombardi's vantage point, concentrates on the week of preparation culminating in an early-season 1962 game between the Packers and their archrivals, the Detroit Lions. Heinz, while deftly explaining the technical jargon, was most effective in revealing the roots of Lombardi's coaching philosophy and the coach's often piercing observations of the personalities, strengths, and weaknesses of his players and how, as a coach, teacher, and psychologist, he strove to exact the utmost from each player every Sunday.

*Run to Daylight!* has gone through fifteen printings and was, according to Robert Lipsyte of *The New York Times,* perhaps the first among contemporary sports books to find a larger audience. Critical plaudits were effusive. Writing in *Book Week,* Al Silverman (then editor of *Sport* magazine) described *Run to Daylight!* as "The book that defines it [the football boom] best and with the most literary distinction." In *The New York Times* Rex Lardner said the book was "tightly written and sharply focused," with "splendid insights into the complexity of the pro game." A *Newsweek* reviewer referred to Heinz as "a matchless reporter and brilliant author," while stating that the narrator and ghost "make an eerily perfect pair." Well-known sportswriters including Dan Parker, Jimmy Cannon, William Wallace, and Shirley Povich were enthusiastic in their praise. *Esquire* columnist George Frazier wrote of the "poetry in W. C. Heinz's prose."

Heinz's early difficulties with the often impatient Lombardi nearly led him to scrap the entire project. Eventually Heinz hit upon the idea of a chronological narrative and filled six notebooks with technical details, Lombardi's penetrating analysis, and his astute assessments that, when distilled, produced a book that can be classified as both a basic primer for better understanding professional football and a humanizing look into a great coach and a team able to perform at the championship level he demanded of them.

*Run to Daylight!* led to further collaborative efforts between Heinz and Lombardi, including the celebrated ABC television network documentary *Run to Daylight,* adapted for television by Heinz and produced by Howard Cosell; a two-part series for *Look* in September 1967 that disclosed Lombardi's disenchantment with the emotionally draining aspects of coaching professional football; and, one year later after Lombardi had left coaching, a piece in *Life* that revealed his second thoughts about his decision.

Heinz continued to freelance during the 1960s. Two of his finest pieces were for the *Saturday Evening Post:* an introspective return to the European battleground of World War II ("After Twenty Years; the G.I.'s War Fades Away," 12 December 1964), and a final attempt by veteran boxing manager Jack Hurley to mold a heavyweight champion from a promising novice ("Last Campaign of Boxing's Last Angry Man," 11 February 1967). The Hurley piece, about a man Heinz would later describe as the individual in sport who most fascinated him, is a perceptive yet poignant portrait, devoid of false sentimentality. Heinz depicts Hurley as a boxing original, one of the last of a dying breed.

Heinz collaborated with Dr. H. Richard Hornberger who, as Richard Hooker, was credited with sole authorship of the 1968 novel *M.A.S.H.* This fast-paced, episodic account of three surgeons of the 4077th Mobile Army Surgical Hospital during the Korean War was the basis for the popular 1970 film and the even more popular television series (1972–1983). Heinz's touch is evident in the pacing and characterization and in the depiction of the surgeons as effective teachers, having not only displayed their talents at "meatball" surgery but in passing on their techniques to their replacements before they depart for the United States and a return to civilian life.

Heinz spent three years working on his final novel, *Emergency* (1974), a plotless, episodic account of several days in the life of a county medical center emergency room, related through the thoughts and words of medical personnel, patients, and others affected by what transpires there. *Emergency,* like *The Surgeon,* was praised for its realistic medical detail and its sincere efforts to deal with moral and ethical issues within the medical field; but it lacks a central focus, serving more effectively as a loose collection of character sketches.

Entering his sixties, Heinz spent several years in the late 1970s revisiting the athletic heroes he most admired. His interviews would form the basis of *Once They Heard the Cheers,* a then-and-now selection of portraits of individuals he first wrote about in their heydays. The concept for the book was suggested to Doubleday by Hornberger, and Heinz received a $45,000 advance on royalties. In crisscrossing the country to reconnect with those men who basically represented the beliefs, values, work ethic, courage, and class he esteemed decades before, Heinz sought to know if they still retained those attributes long after their halcyon days had passed. His personal odyssey led him to reaffirm what they represented to him and what he now believed to have been lost in a nation gone soft, lacking character, pride, and respect for hard work.

Heinz prefaced his portraits with autobiographical recollections of his boyhood, wartime reporting experiences, and tentative start as a sports reporter for the *Sun*. The book contains some of his most perceptive writing and serves as a testament to certain fundamental aspects of Heinz the writer and man: a respect and admiration for the dedicated professionals of sport – generally decent, hardworking men of basic values – a spare, incisive prose typical of his best work; and (linked closely to his war reporting) the influence of Hemingway's profound respect for sheer physical courage.

*Once They Heard the Cheers* succeeds on many levels and garnered considerable critical acclaim. Roger Kahn wrote, "*Once They Heard the Cheers* should be read not only by people who like sport, but just as surely by everyone who likes writing. If I taught writing, this would be high on my required reading list." In *The New York Times* Red Smith described it as "a beauty of a book, a warmly nostalgic wedding of yesterday and now." Robert Lipsyte, while put off somewhat by the author's emphasis on toughness "in barroom brawls, hospital rooms and World War II combat," did contend that "Mr. Heinz's then-and-now portraits are vivid, informative, subtle and often moving."

Heinz is adroit in his ability to combine bits of earlier conversations he drew from his previous articles on his subjects with his more recent interviews. Thirteen of the book's nineteen chapters concern boxers or others in some way connected to boxers. Heinz is at his best in recounting the stories of those athletes who had suffered for a variety of reasons in their later lives, whether financially (former lightweight champion Beau Jack, whose fights grossed millions of dollars but who, when interviewed by Heinz, was a bootblack at the Fontainbleau Hotel in Miami Beach) or physically (baseball players Joe Page and Pete Reiser, suffering from illnesses that would, within a few years, claim their lives) or emotionally (the family of Rocky Marciano, still struggling to accept his 1969 death in a plane crash and the uncertainty connected to potentially enormous sums of money supposedly secreted by Marciano and as yet undiscovered). It is here also that he displays the breadth of his appreciation and depth of his understanding for the many styles and personalities he found in the fight game. Heinz, while fully understanding the elemental nature of the sport, could appreciate within its confines the supreme sensitivity and priestlike compassion in Floyd Patterson ("the most miscast of fighters") and the creative artistry of Willie Pep ("he turned boxing contests into ballets, performances by a virtuoso in which the opponent, trying to punch him out, became an unwitting partner in a dance, the details of which were so exquisite that they evoked joy, and sometimes even laughter"). Heinz also caught the stolid, uncomplicated primitivism of Beau Jack; the overarching greatness of Sugar Ray Robinson; and the brutal fury of Rocky Graziano. Heinz even devoted one chapter to a long-forgotten fighter named Norman Rubio ("The Opponent"), who never achieved greatness but displayed his class while being interviewed by Heinz and other boxing beat writers, patiently answering their questions after absorbing a one-sided defeat.

However, it is in writing about Billy Graham that Heinz finds the closest kinship and underscores the essence of his life's work. Graham, a good welterweight (whom Liebling once described as being "as good as a fighter can be without being a hell of a fighter"), was, to Heinz, the epitome of the honest workman, the ultimate professional, having achieved considerable success in his chosen profession, but never greatness. Heinz saw in Graham the best of himself, another true professional trying to create his masterpiece, yet never quite having "the art transcend the science." While he often wrote of boxing as an art form, Heinz, citing the German poet Rainer Maria Rilke, made the connection between works of art and danger, between the need to risk and to extend one's capabilities. In a self-revealing passage Heinz conveys his own inability in that respect:

> So the Himalayan Peaks are scaled by those who dare, while the rest of us frequent the foothills. We, too, have the desire to stand atop the world, but there is something in the psyche that saps the will, a timidity that tethers us to what we know, and so our work is never the product of having gone to that very end in our experience.

Heinz then relates how Graham, in a memorable title fight against Kid Gavilan, went to the end of his experience, only to lose a disputed split decision that many at ringside insisted he had won.

During the 1970s Heinz contributed articles on medical matters for *Today's Health* and in the 1980s to *Sport, Fifty Plus,* and *TV Guide,* essentially looking back on his memories of World War II and a career in sports journalism. In one of his last pieces for *TV Guide* ("In Covering Sports: Why Can't TV Encourage Class?," 7–13 November 1987), Heinz upbraided the medium for pandering to the "loutish behavior" and "egregious egomania" of certain contemporary athletes as well as their general insensitivity to the defeated opponent. In citing specific actions by football players Mark Gastineau and Lawrence Taylor as well as boxer Muhammad Ali (among others), Heinz offered in contrast the modest, self-effacing heroes of an earlier time, who, he contended, reflected class, sportsmanship, and what Hemingway would have considered grace under pressure.

Heinz's last book was *American Mirror,* a collection of thirteen of his best magazine and newspaper pieces. The original articles were briefly updated by the author. Written between 1947 and 1965, nine of the thirteen selections in *American Mirror* are from the world of sport. Seven of the nine concern boxing. Every prizewinner from the Best Sports Stories series is included. One of the prizewinners was "The Rocky Road of Pistol Pete" (originally published in *True* magazine, March 1958), a probing look at Pete Reiser, Heinz's favorite baseball player, whose great talent and fearless style of play coruscated brilliantly for a time in the 1940s until serious injuries curtailed his playing career and nearly cost him his life. By 1957, when Heinz located him managing a class D minor league team in Kokomo, Indiana, Reiser was an old man at age thirty-nine, moving slowly and haltingly, suffering from a heart ailment that required hospitalization during the season. In Reiser, Heinz saw "perhaps the purest ballplayer of all time," a man who put his life on the line every day in pursuit of his craft and ultimately destroyed his skills in the process – a man totally true to the essence of his being.

"Brownsville Bum" (originally published in *True* magazine, June 1951) was an atmospheric portrait of the boxer Al "Bummy" Davis, an inarticulate, misunderstood Brooklyn battler from a poor immigrant background who was often depicted as the bad guy during his lifetime but was seen by Heinz as shy, sensitive, and fiercely loyal to his friends. The influence of Damon Runyon is evident here in the style, substance, and turns of phrase. Jimmy Breslin, who wrote sports during the 1950s, once referred to this piece as the best magazine sports article of all time. Heinz personally thought more highly of his piece about a boxing contemporary of Davis, Lew Jenkins, whom he wrote about in "G.I. Lew" (originally published in *Argosy,* November 1952). The only newspaper piece included in *American Mirror* was "Death of a Racehorse," a 28 July 1949 account of the events surrounding the broken leg suffered by a promising two-year-old thoroughbred at the Jamaica racetrack in New York, a piece Heinz spoke of as his only newspaper work worthy of preservation in such a collection. Factual, concise, yet moving, it emanated from his desire to report, for the record, the human side of an occurrence in sport otherwise relegated to one sentence within a story.

*American Mirror* was well received critically and benefited from a typically perceptive Red Smith foreword, describing Heinz as "a dedicated craftsman and a penetrating observer who never gives half measure." The one complaint voiced by Smith (and other reviewers) was that there were too few pieces in the collection.

Heinz's nonfiction writing beyond the realm of sport benefited from this reportorial training in getting the facts right and making the story understandable to the reader. As a *Sun* reporter before World War II, he covered a variety of New York stories from pushcart fires and roller-coaster track walkers to political campaigns and the New York World's Fair. While a freelance writer, he wrote about such subjects as the struggles of a stand-up comedian trying to crack the big time; actor Peter Falk's sweating out the initial season of a dramatic series; and the operations of the American Bank Note Company, a little-known Bronx commercial plant responsible for printing $250 billion in currency, traveler's checks, and stock certificates yearly. What is striking in assessing Heinz's cumulative work in this sphere is his consistency; through a combination of factual information and explication, even the mundane (for example, an article on the year-round preparations of Macy's department store in New York City for the Christmas shopping season) engages the reader's interest.

In several short stories written early in his career, Heinz treated well-known sporting themes (the football coach under fire from the college president and win-hungry alumni; a mediocre young fighter struggling in the shadow of his former fighter father; an honest fight referee who is confronted by a clever boxing fix) with his deep knowledge of the subjects and the fortitude of the code hero, who may suffer for his honesty but adheres to an abiding respect for the sport that prohibits him from profaning its essence. Heinz's fight referee, Eddie Brown (a name representing the generic professional working man in Heinz's fictional writing), in "Fixed Fight" foils the fix at the expense of his relationship with his girlfriend and his career while being vilified by an unknowing press. Years later, he reveals the story to the narrator (an alter ego for Heinz) as they are watching a televised bout. In "Man's Game" football coach Eddie staves off his detractors by outthinking his coaching counterpart, upsetting a previously unbeaten team. Along the way, one comes to understand Eddie as teacher, scholar, and professional; he is as knowledgeable in his own sphere (and thereby no less worthy of respect) as any faculty member at the college and, in trying to instruct a young man in the techniques of line play, facing as demanding a task as a mathematics professor teaching that same young man in a different type of classroom. What Heinz makes evident here is the cerebral side of coaching; even the coach's wife understands that her husband must express himself on the football field as he and his coaching rival are engaged in a mental chess game the strategy and tactics of which only they can fully appreciate. Unwilling to continue in a "win-or-else" environment and lacking the respect accorded college educators, coach Eddie quits after the big win, his integrity intact, having gained the respect of his coaching assistants and players. Fully aware of his creative abilities, he congratulates himself with the remark "With a piece of chalk and a blackboard I'm an artist." In "The Win to Go Home" veteran boxing trainer Eddie devises the unexpected strategy that enables a mediocre young fighter to upset a superior opponent and gain a measure of self-respect before he returns home to start a new life. Later the fighter's manager has second thoughts about his fighter's retirement when he realizes that a possible rematch (and inevitable defeat for his fighter) might mean a lucrative payday for all. In "One Throw" the inner fiber of a promising but troubled young minor league shortstop, Nick Maneri, is tested by Yankee scout Eddie Brown who, concealing his true identity, befriends him. Brown suggests that Maneri, who thinks he is being ignored by the parent club, make a couple of errant throws during a game to "get the top brass curious." In a crucial situation, however, Nick makes the game-saving play. Eddie then discloses his real identity to Nick and assures Nick that he will soon be a big leaguer because he "couldn't throw that ball away." This simple story was considered by Heinz to be completely true as to dialogue and characterization.

The influence of Heinz's wartime experiences is evident throughout his fiction. His athletic heroes, like his doctors, surgeons, and police detectives, are often combat veterans, tempered in life-and-death situations, able to understand the unique bond shared by men in war, where teamwork and mutual trust assume transcendent meaning. His medical protagonists are basically heroic – idealized figures with athletic backgrounds (often college football stars) and working-class origins; they are "professionals" in the same fundamental fashion as the athletes he most admired in the real sports world. These fictional characters achieved a measure of nobility and dignity through their conduct in war but are now plagued by doubts and cynicism and are growing into middle age, slowly losing their ongoing struggle to sustain decency, civility, and competence in a world that has lost respect for the values and code of conduct they had prized so highly.

Heinz was an introspective writer who plumbed the depths of his own experience to better understand himself and capture the essence of his subjects. He wrote, while covering World War II, of the war correspondent's desire to fit in, to play a more significant role in the fighting, and, in moments of self-doubt, of being "a pretty pathetic thing" in a place where "the guy with the gun . . . is of course the only one who really belongs." More than thirty years later, after a telephone conversation with an ailing Joe Page in 1976, he recalled a previous occasion in 1950 when he last had seen Page, then a fading star, in his final, losing relief appearance with the New York Yankees. Speaking of that past memory with a friend, Heinz was unsparing about the negative side of his profession:

> That's what's wrong with this business. We're a lot of hustlers. We latch on to someone because he's in the public eye and we need to make a living. We plumb his background and pick his brain. We search out his motivations and his aspirations, and if he's a good guy, an association, even a friendship forms. Then we say goodbye and good luck, and if his luck runs out, where are we? We're long gone and on to somebody else.

Such inner self-assessment, even to the point of questioning his own profession, is perhaps not unusual for a man once described by one of his editors as "a sufferer," striving to get every word right. Red Smith related how Heinz, when advised by a physician concerning a stomach complaint brought on in his quest for literary perfection, explained to the physician, "I'm only trying to be the best writer I can be."

Heinz the purist described the beauty, grace, science, artistry, and courage in sport while Heinz the realist was aware of the dark corners, ugly manipulators, and nefarious influences (especially in boxing). He found more to fault, though, in the critics and naysayers who, he believed, overemphasized the sport's seamy side in film, television, and fiction. But Heinz essentially was not a detractor; his criticism of individuals is brief and pointed when warranted. Early on he understood the difficulty facing the athlete in doing his job well, a realization he retained throughout his writing career. Heinz's failing may have been that he cared too much, that his passion for a sport like boxing may have blinded him to the extent of the mob influence in the fight game that came to light with more-aggressive congressional investigations into boxing in the 1960s when Heinz was writing less about the sport.

Heinz has said in an unpublished interview that good writing is heard and that bad writing simply does not sound right. He has compared the writing process to the building of a dry stone wall, noting how words (and strings of words) can impinge on each other and how, in his novels, he sought initially to attract the reader by a distinctive piece of descriptive writing. Using a boxing analogy, he has compared the writer's task to that of facing a boxer (the opponent-reader) in the ring and needing to sustain the reader's attention and move him, whether the writer is crafting a newspaper column (which Heinz likens to a four-round bout) or a novel (a fifteen-round contest), each an art form in itself. According to Heinz, writing a novel is analogous to painting an enormous mural; the writer-artist finds it impossible to take a figurative step back and assess the work in progress and must instead carry it all in his head. To Heinz, ultimately, "writing becomes more real than what you are doing."

Roger Kahn has written that to be fully enjoyed W. C. Heinz must be read often and carefully. In his best work Heinz brought a perception, sensitivity, and keen ear for realistic dialogue to sportswriting. He reveled in the efforts and achievements of true professionals, men who went about mastering their craft with deep understanding and respect for its complexities. His stylistic contributions to the genre and the influence his writing had on the New Journalists who came to flower during the 1960s have been cited by Newfield and David Halberstam, among others. In emphasizing the human side of the athlete (and by breaking literary ground to convey the emotional turmoil within, say, the athlete's wife), Heinz at times, self-admittedly, lost his reportorial objectivity. But by infusing his writing with passion and feeling, he gave the reader a real sense of the subject's inner core and an understanding of the complexity of the athlete both in and outside the sporting arena.

If Grantland Rice can be said to have popularized sports for the masses and lauded the virtues of sport and its heroes in a style that emphasized positive entertainment, then Heinz saw in sport the art, science, self-expression, and instinctive elements that when shaped and molded by a skilled teacher, coach, manager, or trainer produced the calibre of individuals (and teams) Heinz was compelled to explore and eager to describe. Heinz lives in retirement with his wife in a southwestern Vermont village. He no longer writes professionally but retains an active interest in sport and sportswriting. His reputation remains strongly entrenched among the literary fraternity; he is a writer best appreciated by his peers and the many he influenced as a reporter, columnist, and freelance writer.

**Interview:**

"Out of the Ring," *Newsweek,* 58 (9 October 1961): 96–97.

**References:**

Howard Cosell, *Cosell on Cosell* (Chicago: Playboy Press, 1973), pp. 99–101, 152;

David Halberstam, ed., *The Best American Sports Writing, 1991* (Boston: Houghton Mifflin, 1991);

A. J. Liebling, "The Scribes of Destiny" (The Wayward Press column), *New Yorker,* 22 (28 September 1946): 50, 52, 54–61;

Jack Newfield, "Journalism: Old, New and Corporate," *Dutton Review,* 1 (1970): 151–166;

Harry F. Walters and George Hackett, "Real Hawkeye Pierce," *Newsweek,* 101 (28 February 1983): 101.

# Roger Kahn

*(31 October 1927 – )*

Richard Orodenker
*Pennsylvania State University, Ogontz*

and

Andrew Milner

MAJOR POSITIONS HELD: School sports reporter, general assignment reporter, sportswriter, *New York Herald Tribune* (1947–1955); contributing writer, *Sports Illustrated* (1955); sports editor, *Newsweek* (1956–1960); editor at large, *Saturday Evening Post* (1963–1969); columnist, *Esquire* (1969–1975), *Time* (1976).

BOOKS: *The 1955 Mutual Baseball Almanac,* by Kahn and Harry Wismer, edited by Paul Lapolla (Garden City, N.Y.: Doubleday, 1955);

*The 1956 Illustrated Mutual Baseball Yearbook,* by Kahn and Wismer, edited by Lapolla (Garden City, N.Y.: Doubleday, 1956);

*Inside Big League Baseball* (New York: Macmillan, 1962);

*The Passionate People: What It Means to Be a Jew in America* (New York: Morrow, 1968);

*The Battle for Morningside Heights: Why Students Rebel* (New York: Morrow, 1970);

*The Boys of Summer* (New York: Harper & Row, 1972; revised edition, New York: New American Library, 1973);

*How the Weather Was* (New York: Harper & Row, 1973);

*A Season in the Sun* (New York: Harper & Row, 1977);

*But Not to Keep: A Novel* (New York: Harper & Row, 1979);

*The Seventh Game* (New York: New American Library, 1982);

*Good Enough to Dream* (Garden City, N.Y.: Doubleday, 1985);

*Joe and Marilyn: A Memory of Love* (New York: Morrow, 1986);

*Pete Rose: My Story,* by Kahn and Pete Rose (New York: Macmillan / London: Collier-Macmillan, 1989);

*Games We Used to Play: A Lover's Quarrel with Sports* (New York: Ticknor & Fields, 1992);

Roger Kahn (photograph by Gordon J. Kahn)

*The Era: 1947–1957, When the Yankees, the Giants, and the Dodgers Ruled the World* (New York: Ticknor & Fields, 1993).

OTHER: *The Mutual Baseball Almanac,* edited by Kahn and Al Helfer in cooperation with the Mutual sports staff (Garden City, N.Y.: Doubleday, 1954);

"Soldiers Are His Game," in *Best Sports Stories 1959 Edition: A Panorama of the 1958 Sports Year,* ed-

*Kahn about age six*

ited by Irving T. Marsh and Edward Ehre (New York: Dutton, 1959), 127–130;

*The World of John Lardner,* edited by Kahn (New York: Simon & Schuster, 1961);

"Football's Taking Over," in *Best Sports Stories: A Panorama of the 1960 Sports Year,* edited by Marsh and Ehre (New York: Dutton, 1961), pp. 115–122.

"The Time of the Hustler," in *Best Sports Stories 1963 Edition: A Panorama of the 1962 Sports Year,* edited by Marsh and Ehre (New York: Dutton, 1963), 255–263;

"The Ten Years of Jackie Robinson," in *The Twentieth Century Treasury of Sports,* edited by Al Silverman and Brian Silverman (New York: Viking, 1992), 334–351;

"The Life and Hard Times of Jim Bouton," in *Best Sports Stories 1971: A Panorama of the 1970 Sports World,* edited by Marsh and Ehre (New York: Dutton, 1971), 247–256;

"Where Have All Our Heroes Gone?" in *Best Sports Stories 1975: A Panorama of the 1974 Sports World,* edited by Marsh and Ehre (New York: Dutton, 1975), 254–265;

"He's a Yankee Doodle Dandy," in *Best Sports Stories 1982: A Panorama of the 1981 Sports World,* edited by *The Sporting News* and Ehre (New York: Dutton, 1982), 15–23.

SELECTED PERIODICAL PUBLICATIONS – UNCOLLECTED: "Oklahoma's Mickey Mantle: Can the Young Yankee Beat the Babe?," *Newsweek,* 44 (25 June 1956): 63–67;

"Halfbacks Carry the Mortgage," *Nation,* 183 (22 December 1956): 539–541;

"Rookie of the Year," *Cosmopolitan,* 144 (June 1958): 74–77;

"Control is Seventeen Inches Wide," *Coronet,* 44 (September 1958): 135–139;

"Man with the Brainy Arm," *New York Times Magazine,* 31 May 1959, pp. 22–25;

"Advantage, Mr. Kramer," *Nation,* 192 (18 February 1961): 146–147;

"Our Davis Cup Runneth Over," *Saturday Evening Post,* 237 (26 September 1964): 66–69;

"Case of the Dead Bookie," *Saturday Evening Post,* 238 (13 March 1965): 34–38, 40, 42, 43;

"The Game without the Ball," *Saturday Evening Post,* 240 (16 December 1967): 79–83;

"Hurry, Hurry Hurry, the Dodgers Are in Town," *Sport,* 70 (June 1980): 30–37;

"Whitey's Redbirds," *Sport,* 72 (July 1981): 26;

"Pray for a Great World Series," *Sport,* 72 (November 1981): 77–81.

Few authors have achieved as many levels of success as has Roger Kahn. When his *New York Herald Tribune* salary reached $10,000 a year in 1955, he was the highest-paid and most-sought-after baseball writer in the city. Turning to freelance magazine writing, he won five Best Sports Stories prizes between 1960 and 1982. A "serious" book, *The Passionate People: What It Means to Be a Jew in America* (1968), was published as Kahn turned forty. Kahn has published eleven books – two of them novels – since then, including the landmark *The Boys of Summer* (1972), which contrasted the young, diverse, talented Brooklyn Dodgers of the 1950s with their middle-aged selves facing personal agonies twenty years later; *A Season in the Sun* (1977); *Good Enough to Dream* (1985); and *The Era: 1947–1957, When the Yankees, the Giants, and the Dodgers Ruled the World* (1993). Though he has also written about the 1968 Columbia University student protests, the poet Robert Frost, the violinist Jascha Heifetz, and sports other than baseball, Kahn is best known for his works on the national pastime and its role in American life.

Kahn was born in Brooklyn on 31 October 1927 to Gordon Jacques and Olga Rockow Kahn, both of whom were teachers at Thomas Jefferson High School. Gordon Kahn – who in 1938 would be one of the creators of the *Information Please* radio quiz show – passionately loved baseball, while the learned Olga found the game unintelligible and never understood the hold it had on her son. Kahn's father, who claimed to have played third base for City College of New York, taught Kahn the fundamentals of baseball. As Kahn told James S. Kunen in 1983, "Baseball was my first gateway into the world of men. My father couldn't talk to me about sex, but we could talk about a Dodger centerfielder and how he had run down a ball. Baseball was a metaphor for talking about life." Kahn's mother would have preferred that her son appreciate James Joyce's *Ulysses* (1922), but he was, in his own term, "enslaved" by the partly ghostwritten prose of the great New York Giants pitcher Christy Mathewson's *Pitching in a Pinch* (1912). Over time Kahn would absorb the works of Virgil, John Keats, Thomas Hardy, and Fyodor Dostoyevsky. Olga Kahn, though delighted that her son, at least, was not "selling shoes," would never quite reconcile herself to his becoming a sportswriter. She had hoped that he would write novels, or perhaps something about her grandfather – anything, besides sports, "that I could understand."

After three years of intermittent study at University College of New York University, Kahn joined the *New York Herald Tribune* in 1947 and was assigned to cover school sports; his first story was about a walking race. When New York City public-high-school coaches went on strike in 1950, Kahn's article on what the athletes were doing in lieu of playing garnered him his first professional praise, and he was promoted to general assignment reporter. On 14 July 1950 he married Joan Rappaport; they had a son, Gordon Jacques II. Kahn was assigned to the Brooklyn Dodgers beat in 1952, the season after the club blew a thirteen-game lead in mid August and lost the National League pennant to the Giants in the last inning of the last game of a playoff. Writing about that game in his October 1960 *Sports Illustrated* retrospective "The Day Bobby Thomson Hit His Home Run," Kahn would call it one of those occurrences whose "impact thrusts them into the private lives of millions of people, who forever remember these events in personal terms."

In the early 1950s Kahn became, as he tells Peter Golenbock in *Bums – An Oral History of the Brooklyn Dodgers* (1984), Jackie Robinson's "pale ghost" for a series of articles in a magazine for African Americans called *Our Sports:* "After [Leo] Durocher went to the Giants [in 1948], Jack became the editor of the magazine. . . . And the first piece we wrote was called 'My Feud With Leo Is Over.' And so I talked with Jack a while, and we talked back and forth, and I wrote the piece under Jackie's byline. Durocher had said publicly maybe fifteen times that Robinson was his kind of ballplayer, so I had the piece ending, 'He says I'm his kind of ballplayer, well he's my kind of man.' And Robinson crossed out the word 'man' and wrote 'manager.' "

Kahn was taken off the Dodgers beat in 1953, ostensibly for calling the Brooklyn clubhouse a "sarcophagus" after the team suffered its fifth World Series loss to the Yankees since 1941. The *Herald Tribune* editors decided, Kahn later said, that he was getting too personally involved with the team. He was assigned to the Giants, managed by the combative Durocher, who promised Kahn "a variety of exclusive stories, if I wrote them from his point of view," and featuring a young center fielder named Willie Mays. In a 27 February 1954 article Kahn expressed skepticism about the effusive praise the Giants front office was bestowing on Mays, who was about to return to the team after twenty-one months in the army: "It's only human to wonder if this is man or superman coming to join the Giants. In my case the wonder takes the form of questions." A few days later Kahn got to see Mays play at spring training in Phoenix, Arizona. After watching Mays hit a four-hundred-foot, fifth-inning pinch-hit home run, Kahn reported on 2 March: "He went to

*Kahn (far right) and other New York sportswriters with Giants manager Leo Durocher in Phoenix in 1954. From left: Milton Gross of the* Post, *Louis Effrat of the* Times, *Jim McCulley of the* Daily News, *Joe King of the* World-Telegram, *Barney Kremenko of the* Journal-American, *Mel Woody of the* Newark Star-Ledger, *Charles Feeney of the* Long Island Press, *Arch Murray of the* Post, *and Ken Smith of the* Daily Mirror.

center field and had no chances until the seventh inning. Then, with Billy Gardner on first, Harvey Gentry slammed a long drive to right center. Willie raced to the right center-field fence, speared the ball one-handed, whirled, threw a strike to first base, and doubled Gardner. And then, when Bill Taylor blasted a tremendous wallop to dead center, Willie galloped fifty feet back and caught the ball over his shoulder." Kahn, who calls Mays "the most exciting player I've ever seen," would follow his career to the end.

Kahn's Giants beat ended abruptly during the 1955 season. He had quoted Durocher as having said early in the season that the Giants' "spirit is shot," and Durocher had demanded a retraction. "Durocher denied that he said any such thing and stopped speaking to me," Kahn said in an unpublished 1996 interview, "not even telling me who would pitch tomorrow. The *Herald Tribune* sports editor, Bob Cooke, said he had talked to Giants owner Horace Stoneham, and everybody knew Durocher really had said it, but the politics of the situation required me to apologize. I resigned." Although he left the paper on a matter of principle, Kahn acknowledges in *The Era* that he was ready for a career change: "For four years I covered ballgames pretty much every day from spring training through the World Series, about 750 major league games in all. Since the *Tribune* published a variety of editions, each requiring fresh material, I wrote 2,000 baseball stories for the paper. I moved on after [the] squabble with Durocher. The daily beat was grinding down my fingers and my nerves. I was ready to proceed to something else."

While spending a year at Time-Life to help launch *Sports Illustrated,* Kahn wrote "The Ten Years of Jackie Robinson," a penetrating look at the growth over a decade – nearly the whole of Robinson's major league career, as it would turn out – of a controversial and enigmatic figure who could "converse with Eleanor Roosevelt and curse at Sal Maglie with equal intensity and skill." Kahn depicts a man who "endured outrage and vituperation with an almost magic mixture of humility and pride"; the attacks from players and the press – some of them physical – were solely on account of his race, since his ability was never in doubt. The experience made Robinson an obstreperous, combative individual who really "doesn't need any excuse. If the man rugged enough to break baseball's color line turns out to be a thoroughly rugged man, no one has any license to be surprised." His newspaper editors had criticized Kahn for writing about race relations; in his magazine pieces in the 1950s, however, he was determined to remain on the cutting edge of the issue.

Kahn joined *Newsweek* in 1956 as sports editor. The magazine introduced him to readers by noting that "for relaxation he reads modern poetry (his favorite is Wallace Stevens) or attends symphony concerts (he likes Sibelius, Brahms, and Beethoven). But whether it's baseball, track, tennis, or golf, or symphonic music, Kahn applies himself diligently and soon becomes an expert." *Newsweek* sports columnist John Lardner became his good friend and, along with the sportswriters Red Smith and Stanley Woodward, his mentor. While working at *Newsweek* Kahn began writing freelance articles for other magazines, including *Sports Illustrated, Sport, The Nation, Esquire, Life,* and *The Saturday Evening Post.*

In a 1956 article for *Sport* Kahn paints a vivid portrait of hockey great Howie Morenz, even though Morenz had died when Kahn was ten:

> He is wearing the *bleu-blanc-rouge* of Les Canadiens, number 7 on his back, the puck at the end of his stick as he glides swiftly across the ice of Montreal's famous Forum. First he is circling behind his own goal. A little hop signals the start of his charge. Skating from the hips with long, easy strides, Morenz gains speed. The crowd at the Forum rises in waves as he skates past, and the cheering comes in waves, too, rolling down the tiers of seats louder and louder. As if to match the crescendo's rise, Morenz drives himself still faster toward the goal. In his path, two defensemen brace their heavy frames. Suddenly Morenz is upon them, hurling his body into the air as though he would jump over the two men. The crowd is on its feet and screaming. Morenz breaks through. A final feint, a goalie's lunge, a hard quick shot into the open corner of the cage. Morenz has scored. While the Forum rocks with sound, he skates easily back to center ice for the next face-off.

In a 1957 essay for *The American Scholar,* "Intellectuals and Ball Players," Kahn notes that "The romance between intellectuals and the game of baseball is, for the most part, one-sided to the point of absurdity." For "the imaginative intellectual" major leaguers are "demigods," while "the ordinary ballplayer" ranks intellectuals "several notches below umpires." The intellectual uses baseball's "leisurely afternoons of summer" to call up "an anecdote from his treasury of baseball stories" or dream of striking "out Williams, Mantle, and the boss on nine pitches." For the average major leaguer, who was "once a star in his neighborhood or in his town years ago," however, realizing his dream of playing professional ball has been "a tragedy of fulfillment," for he has "made the major leagues and discovered that in spite of it, life remains distressingly short of ideal." The essay was reprinted in Irving T. Marsh and Edward Ehre's *Best Sports Stories 1958* and in Charles Einstein's *Second Fireside Book of Baseball* (1958). For *Newsweek* Kahn wrote a memorable 1958 piece on the legendary Army football coach Red Blaik, who cut a figure "more like that of a dedicated chess player than that of a traditional back-slapping coach," as Blaik was preparing to meet Navy, a team he had not defeated since 1955. That story marked Kahn's second appearance in the *Best Sports Stories* series. Also in 1958 his first short story, "The Perfect Ball Player," appeared in *Cosmopolitan.*

Kahn won his first Best Magazine Story prize in 1960 for "The Real Babe Ruth." In the article, originally published in *Esquire* in 1959, Kahn became one of the first writers to demythologize the Yankee great while still appreciating Ruth's "unique, unrivaled, and unchallenged" legend. As one of the judges, John Chamberlain, noted in the preface to *Best Sports Stories 1960:* Kahn "makes Babe Ruth seem very much like one of those ancient Greek deities who could combine immoral behavior with a godlike capacity for seeming to be above mere human standards."

Sometimes Kahn would use a sporting event to assay the larger meaning of the sport itself, as he did in a 1959 piece for *The Nation* on the first Ingemar Johansson–Floyd Patterson fight:

> Boxing's harshest critics ignore . . . the fact that a great fight crams courage and strength, fear and weakness, into a tiny cockpit. That fight at Yankee Stadium might have given even Aristotle catharsis. The business of boxing is changing, but the heavyweight division still generates the most electric instant in sport, when a great right fist crashes home.

By 1960 Kahn was becoming somewhat disenchanted with major league baseball. In "Football's Taking Over," a 1960 article for *Sport* that was reprinted in *Best Sports Stories 1961,* he recognized football's growing popularity. Without counting baseball entirely out, he concluded, "In the sweep of time a sport comes and flourishes and almost al-

*Kahn with Jackie Robinson at Robinson's home in Stamford, Connecticut, in 1970*

ways goes. Perhaps baseball's greatest flourishing lies behind."

For *The Saturday Evening Post* Kahn wrote "A Visit with Robert Frost" (19 November 1960), which appeared in somewhat different form in *The Nation* under the title "A Pride Ferocious." Kahn would say in *The Boys of Summer* that his encounter with Frost was one of the factors that led him to write that book:

> It was then, in Frost's last decade, that I began to see the Dodgers not as baseball players but as baseball-playing men, as some are poetry-writing men, or painting men or men who make decisions of state. The experience of traveling with them was not something to dismiss, nor to let anyone dismiss in arrogant ignorance. . . . Not ten years had passed, but already the boys of summer were aged. Frost could compose with wit at eighty-six; they slumped in hitless decline at thirty-five. As the first longing to seek them stirred, one thought settled on me like a headstone. If they were old, why then I was old myself.

Kahn left *Newsweek* in 1960. In 1961 he edited and contributed an epilogue to *The World of John Lardner,* a collection of writings by his friend, who had died the year before. In a *Sports Illustrated* profile that was reprinted in *Best Sports Stories 1962* Kahn depicted Roger Maris's "fierce, combative integrity that is unusual in baseball as it would be unusual anywhere." He also described the constant media assault on Maris, who, in his pursuit of Ruth's single-season home-run record in 1961, "realized the price one pays for being a hero." The next year, in "The Time of the Hustler," published in *Show: The Magazine of the Arts* and included in *Best Sports Stories 1963,* Kahn attacked the growing commercialism of the World Series as agents began representing covetous ballplayers and television promised lucrative contracts:

> It would be nice to believe that the Series is what it pretends to be, a beery kind of Hellenic ideal, an Olympic for the populace. It would be nice even to think of it as a monster, but innocent. The World Series is neither Hellenic nor innocent. It is a quick, incredibly successful grab for money and attention. It is a monument to the fast buck. It has become the time of the hustler.

"The Time of the Hustler" and other philippics, such as "Where Have All Our Heroes Gone?," published in *Esquire* and reprinted in *Best Sports Stories 1975,* would set the tone for Kahn's more recent essays on baseball in the age of what he calls "the moaning millionaires."

Kahn's book *Inside Big League Baseball* (1962) was aimed at the young adult market but was described by Marsh in *Books* (13 May 1962) as filled with "enough inside information to make it fascinating reading for any follower of the game no matter what his (or her) age." In 1963 Kahn became editor at large at *The Saturday Evening Post.* His marriage ended in divorce the same year. In 1964 he married Alice Russell; they had two children, Alissa Avril and Roger Laurence.

For *The Saturday Evening Post* he covered the unsuccessful presidential campaign of Barry Goldwater in 1964. Also that year he published in the magazine an important nonsports essay, "White Man, Walk Easy." Though his focus for the *Post* was not exclusively on sports, he wrote an exceptional 1965 profile of the scholar-athlete Frank Ryan, the Cleveland Browns quarterback who had a Ph.D. in mathematics. Another football piece, "The Game without the Ball" (16 December 1967), focuses on the tackle position: "Tuesday is a rotten day for tackles. A hip or thigh or shoulder still aches from Sunday's game, and another Sunday distantly approaches. Tuesdays mix old pain with impending demands, a trying circumstance." "A Hour or So of Hell" (1968) asks why the hockey goalie Glen Hall, "whose distaste for play is overwhelming" and causes him physical and mental distress, continued to subject himself to the continual assaults of the slap shot, "a full, prodigious, ferocious swipe" that "propels the puck, a six-ounce disc of hard, frozen rubber, at incredible speed – as fast as 120 miles an hour." The piece won Kahn his second Best Magazine Story prize; as one of the judges, John Hutchens, remarked in *Best Sports Stories 1969,* Kahn had captured Hall as "an artist who has come to dislike his art but stays with it – for reasons of money, to be sure, but also out of pride so deep that occasional failure sickens him." Kahn covered another losing presidential campaign, that of Eugene McCarthy, for the *Post* in 1968. Also in 1968 he published *The Passionate People: What It Means to Be a Jew in America.* The book includes an intriguing profile of former Cleveland Indians great Al Rosen, who had learned early in life that "If he wanted to play, he would have to play among men who hadn't known Jews, or didn't want to know Jews, or who figured that if this big young Jewboy was going to play with them, he better have it, and have it all. He did." Kahn left *The Saturday Evening Post* in 1969. His next book, *The Battle for Morningside Heights: Why Students Rebel* (1970), analyzed the 1968 student protests at Columbia University.

Two years later *The Boys of Summer* appeared. The title of the book comes from a 1939 Dylan Thomas poem, "I See the Boys of Summer"; Kahn used the phrase to refer to the aging players of the old Brooklyn Dodgers, but it has entered the American lexicon to refer to all baseball players, past and current. The book had its inception when Kahn met former Dodger third baseman Billy Cox, who was then tending bar at an American Legion hall in Newport, Pennsylvania, and decided that he would like to catch up with some of Cox's old teammates. Another idea was tugging at him, as well: the Brooklyn Dodgers, after all, were not just another ball club. "The integrating of this country," Kahn has said in an unpublished 1994 interview, "begins with Jackie Robinson, not Rosa Parks. We were all made larger, closer, better because we integrated the United States of America. Even the most mean-spirited of us." *The Boys of Summer* would become, in James A. Michener's words, "a work of high purpose and poetic accomplishment," perhaps "the finest American book on sports."

Kahn acknowledges the fragility of what he had experienced in Brooklyn in the essay that introduces *The Boys of Summer,* "Lines on the Transpontine Madness":

> The team grew old. The Dodgers deserted Brooklyn. Wreckers swarmed into Ebbets Field and leveled the stands. Soil that had felt the spikes of Robinson and Reese was washed from the faces of mewling children. The New York *Herald Tribune* writhed, changed its face and collapsed. I covered a team that no longer exists in a demolished ball park for a newspaper that is dead.

The first half of *The Boys of Summer* is Kahn's memoir of his Brooklyn childhood and his career as a young reporter. The intellectual atmosphere of his family life is set off against the rambunctious Dodger teams of the 1930s. Baseball serves as a point of reference between Kahn and his father: in one memorable scene Gordon Kahn teaches the thirteen-year-old Roger the facts of life during a ritual game of catch. In "Ceremonies of Innocence," the chapter on the two years he spent covering the Dodgers for the *Herald Tribune,* Kahn captures the contrasting personalities of the Brooklyn team members: during a rough flight in a DC-9, pitcher Carl Erskine calms an anxious Kahn by reciting from memory Robert Service's poem "The Cremation of Sam McGee"; on another occasion manager Charlie Dressen tells Kahn, "Ah, [expletive] . . . I got this far without readin' a book. I ain't gonna start now."

Kahn spent three years tracking down the thirteen former Dodgers whose profiles comprise the second half of *The Boys of Summer.* Three prominent themes emerge in this part of the work. The first concerns athletes coming to grips with middle age and their return to a working-class life. By the early 1970s Cox was a bartender; George Shuba was a postal clerk-typist; Preacher Roe owned a grocery store; Carl Furillo was working on the construction of the World Trade Center; and Roy Campanella was still recovering from the 1958 car crash that had left him paralyzed. The second theme is the promise and tensions of integration. Robinson, "a bent, gray man" for whom Kahn must slow his own pace as they walk together, wants to put his past

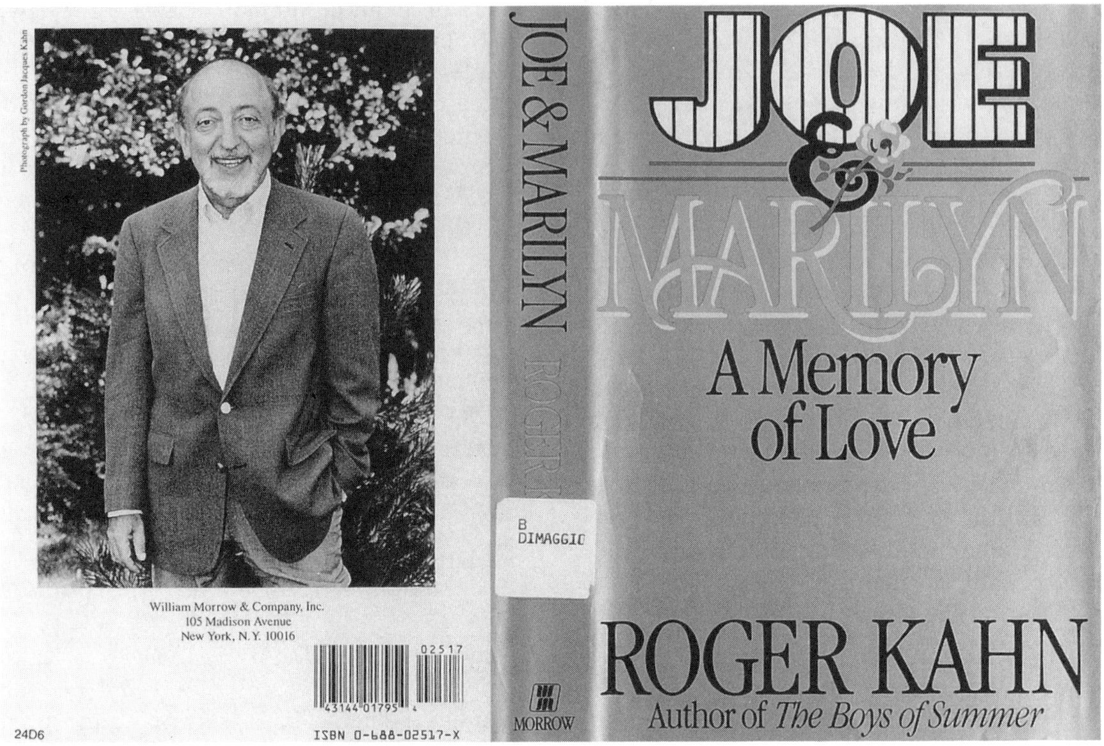


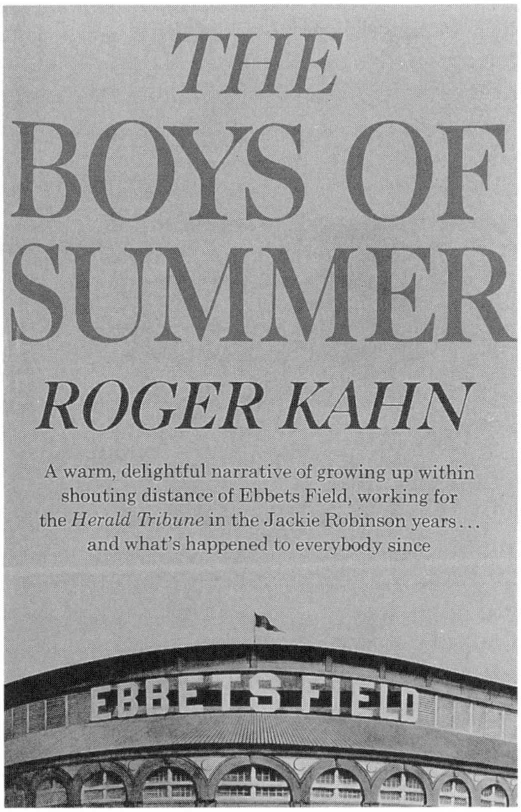


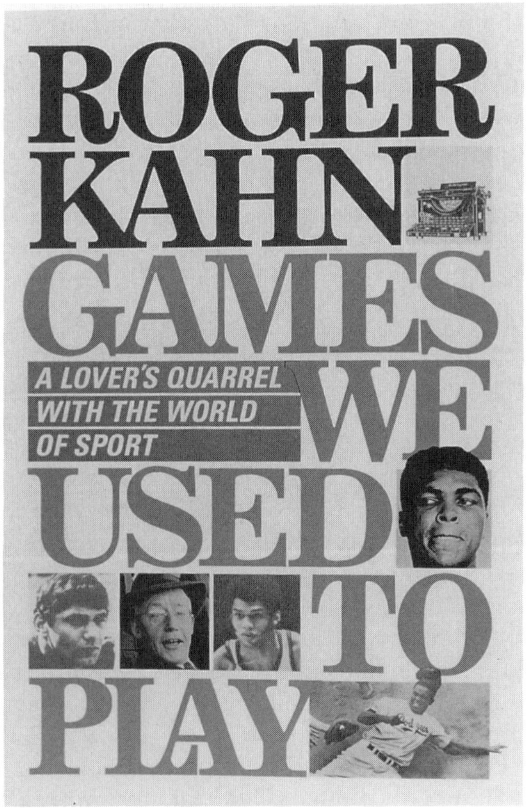


*Dust jackets for three of Kahn's books*

achievements in perspective and help integrate the Republican Party. Kentucky-born Pee Wee Reese recalls that he had conflicting emotions on learning that the Dodgers had signed Robinson in 1945 but refused to sign a players' petition demanding that Robinson be barred from the team. Joe Black tells of his work with young African Americans, insisting that it is important for them to speak proper English while taking pride in their heritage. The third theme is the relationships between the players and their sons. Clem Labine's Catholic faith was shattered after his son Jay lost a leg in Vietnam. Erskine discovered his own inner strength in raising his son Jimmy, born with Down's syndrome. Robinson stood by Jackie Jr. during the latter's recovery from heroin addiction, only to lose his son in an automobile accident in June 1971.

Kahn concludes the book by interviewing blustery team owner Walter O'Malley, who, Kahn believed at the time, moved the Dodgers from Brooklyn to Los Angeles out of sheer greed. (He would revise his estimation of O'Malley in *The Era,* noting that Robert Moses, New York's commissioner of parks, shared the blame.) In *The Boys of Summer* Kahn contrasts the middle-class existence of the former Dodgers with the lifestyle of the wealthy O'Malley, who owned not only the Dodgers but also several hundred acres of downtown Los Angeles real estate.

The death of Robinson in October 1972, after the book's publication, further cast the Dodgers as a tragic team. Concluding the afterword to the paperback edition (1973), Kahn wrote of the formidable task Robinson had had before him in 1947: "The courage with which he met that challenge and the years beyond is a saga of our times. And if his death has robbed my wife, my children, and me of an admired friend, we are still fortunate for having had the friendship of Jackie Robinson. So are the boys of summer. So is all the family of man." Twenty years later Kahn would tell Mike Shannon, in an interview published in *Baseball: The Writer's Game* (1992), "I was pleased that my book had made him current again in the last summer that he would know."

*The Boys of Summer* became a best-seller. It received generally favorable reviews; Peter S. Prescott said in *Newsweek* (13 March 1972): "Kahn's book is knowledgeable, leisurely, and anecdotal, as good informal baseball writing must be. But it is more: Kahn never forgets that he is writing about men in relation to a certain level of achievement, a certain process of decline, and as such his book acquires a cumulative power." In reviewing Roger Angell's *The Summer Game* (1972) in *The New York Times Book Review* (11 June 1972), however, Ted Solotaroff would attribute the success of Kahn's book to a "contemporary penchant for candor or nostalgia or both." The reputation of *The Boys of Summer* has grown since its publication, no doubt because of its universal message: as Kahn puts it in the book, the skills of the players "lifted everyman's spirit and their defeat joined them with everyman's existence." *The Boys of Summer* helped to spur a renaissance in sports-book publishing during the 1970s and 1980s and helped baseball reestablish its place in American popular culture. A television documentary based on the book and narrated by comedian Sid Caesar appeared in 1983, and a motion picture script written by Ring Lardner Jr. has been long in the works.

The collection *How the Weather Was* (1973) includes some of Kahn's magazine articles and excerpts from three of his previous books, including *The Boys of Summer.* In addition to several nonsports pieces – most notably "A Masque of Genius," a profile of Heifetz that had originally appeared in *Life* – *How the Weather Was* includes an acerbic piece on Kahn's old nemesis Durocher, who in 1967, at sixty-two, had become manager of the Chicago Cubs. Another article profiles the almost-forty Mays in what may be his final seasons of baseball with the San Francisco Giants: alternating the narrative between the present and Mays's glorious early years with the Giants in the mid 1950s, the piece also details the great friendship between the author and the player. Kahn concludes: "I don't say that Willie today is as exciting as Willie in '54, but what he does now is immeasurably harder. . . . Willie stands up to time defiantly and with dignity, and one is fortunate to write baseball in his generation." Originally published as "Willie Mays, Yesterday and Today" in *Sport* in 1969, the essay won Kahn his third Best Magazine Story prize.

Kahn contributed articles to *Esquire* from 1969 to 1976; many of them are collected in *Games We Used to Play: A Lover's Quarrel with Sports* (1992). He ruminated on Mickey Mantle in 1971, the twilight of Mantle's career: "It all passed so swiftly for Mickey Mantle, the vaulting home runs, cheers like thunder, and the dark devil's wine of fame, that he cannot believe it is done." In a piece remembering Robinson near the first anniversary of his death, he says: "I loved Jackie Robinson well enough not to have to deify him." In 1974 he wrote of Notre Dame football, "you can cheer, cheer for [coach] Ara Parseghian's warriors, while your operative, who prefers pro football anyway, will reserve his highest

Notre Dame enthusiasm for English professors who risked distinguished careers to say things that had to be said about Richard Nixon." In a 1975 piece he says of boxing promoter Don King, "He learned street lore, the look of whores, the flash of pimps, as suburban children learn the rooflines of a Mercedes." In 1976 he wrote of former heavyweight champion George Foreman, who two years earlier had lost his title to Muhammad Ali, "God, Napoleon reminds sophomore history students, is on the side of the army with the most cannon. Ali slipped, moved, covered up against the ropes until finally George Foreman, the indestructible, had punched himself weary. Foreman's artillery was silenced, and Ali knocked him out in round eight."

Kahn and his second wife were divorced in 1974; later that year he married Wendy Meeker. He began writing a column for *Time* magazine in 1976 but soon abandoned it because he found the 750-word limit too restrictive. He spent the summer that year following baseball in its various forms for "Baseball in America," a three-part series for *Sports Illustrated* that was expanded to become his next book, *A Season in the Sun.* The book's prologue includes perhaps Kahn's most evocative writing on what the game of baseball means to him:

> I do believe this: that baseball's inherent rhythm, minutes and minutes of passivity erupting into seconds of frenzied action, matches an attribute of the American character. But no existential proclamation, nor any tortured neo-Freudianism, nor any outburst of popular sociology, not even, or least of all, my own, explains baseball's lovelock on the American heart.
>
> You learn to let some mysteries alone, and when you do, you find they sing themselves.

The chapter "The Children of Roberto" describes Puerto Rican youths emulating the late Hall of Famer Roberto Clemente. The John Brown University team in Siloam Springs, Arkansas, coached by former 1950s slugger Wally Moon, is profiled in "The Town Where Someone Drives a Kaiser Car." The travails and triumphs of Cincinnati Reds catcher Johnny Bench, facing injuries, a divorce, and a second straight world championship, are detailed in "The Great J. B." "The Franchise Business" compares two National League West teams, the unsuccessful Houston Astros and O'Malley's highly profitable Los Angeles Dodgers. Kahn's piece on the Eastern League, "The Country of the Poor," juxtaposes minor-league poverty with big-league wealth. "William the Unconquerable" is a portrait of innovative Chicago White Sox owner Bill Veeck. The most telling chapter, "Golden Triumphs, Tarnished Dreams," focuses on three retired stars: Stan Musial, Artie Wilson, and Early Wynn. Musial has parlayed his enormously successful playing career into equally fulfilling work with the Saint Louis Cardinals front office; Wilson, though grateful for his opportunity in the Negro Leagues, has had a markedly less lucrative post-playing career. Kahn's portrait of the three-hundred-game-winner and Hall of Famer Wynn as an iconoclast and contentious former player echoes his *The Boys of Summer* interview with Carl Furillo, who sued the Dodgers when they released him in 1960. Though not as commercially successful as *The Boys of Summer, A Season in the Sun* garnered superb reviews. Donald Hall in *The New York Times Book Review* (3 July 1977) provides a typical example of the praise: "for baseball's characteristic stories – anecdotes and lore the game is rich with – Roger Kahn is best of all, with his sweet ear for the cadence of baseball talk." Howard Means in the *Washington Post* (31 July 1977) was in the minority, finding the book "too loose and rambling"; also, according to Means, it failed to show how baseball's inherent rhythm actually "worked its way into the lives of men who played the game."

Kahn's first novel, *But Not to Keep,* appeared in 1979 and received wide acclaim. According to the review by Michael J. Bandler in *The Christian Science Monitor* (18 July 1979), the work depicts "the interrelationship between two evolving institutions, one threatened, the other on the advance. The institution seemingly on the verge of collapse – or at least somewhat battered – is marriage. The one that appears to be holding its own is fatherhood, especially of the single parent variety." Articles for *The New York Times* around this time included a description of Kahn's last lunch with O'Malley, who is now seen in a more sympathetic light, and a debunking of Joe Namath as a "promoter, hedonist, and unemployed quarterback." Kahn's article "Past Their Prime" in *Playboy* won him his fourth Best Magazine Story prize; it is an examination of the way many athletes age "into other men's prime time," some with grace, others not. These players "soar almost from puberty" but burn out quickly. "It also hurts more. The human body was not designed to play catcher from April to October" or "designed to fight for the heavyweight championship at thirty-six." True, Kahn observes, some individuals last beyond what should have been their primes and achieve new heights or even build new careers upon the old ones, but "the complete athlete measures pain against glory, risk against profit. He considers what is left of his body and then . . . subconsciously decides whether or not he wants to go on."

Kahn and his third wife were divorced in 1982, the year his second novel, *The Seventh Game,* appeared. The protagonist is forty-one-year-old John Lee Longboat, who is ending his big-league career by pitching in the seventh game of the World Series for the New York Mohawks against the Los Angeles Mastodons. Kahn provides a nonlinear narrative that moves from crucial minutes in the present before, during, and after the Sunday game – 11:58 A.M., 2:17 P.M., 3:12 P.M. – to events of the past, such as Longboat's impoverished boyhood in Trail o'Tears, Oklahoma; his minor-league experiences in Pittsfield, Massachusetts; and his strained relationship with his son at home in Ridgefield, Connecticut. The novel displays Kahn's gift for crisp, well-observed dialogue and witty characterization, and it risks a complex ending: Longboat loses a no-hitter in the eighth inning and loses the game in the ninth, giving up a grand slam homer with a "perfect" curveball that causes him writhing, blinding shoulder pain; but he salvages his shaky marriage when he abandons his longtime mistress, who is the wife of the baseball commissioner's shady brother and also the mistress, it turns out, of the team owner. Critics who panned the book for its episodic quality missed the fact that in many ways the novel was a throwback to earlier models of baseball fiction, such as Mark Harris's verisimilitudinous "Henry Wigenn" novels and John R. Tunis's young adult books, that focus on characters who must make difficult choices. Kahn's "racy-fine" narrative, according to Budd Schulberg's dust-jacket blurb, gives the novel a contemporary feel. The characters – coach Lefty Levin, reporter Rob Brownell, and players "Jap" O'Hara, Rocco Lombardo, "Raunchy" Kauff, and Saviour Santo Domingo (Santo Domingo, in the Dominican Republic, was where Robinson and the Dodgers trained in the spring of 1947 rather than risk racial confrontations in Florida) – are not clichés; they are archetypes. Even the team name, the Mohawks, calls to mind Harris's New York Monarchs, as do Kahn's ribald and funny dugout and locker-room scenes. Longboat is a flawed but compelling character who "tried to elevate himself from ballplayer attitudes with inconsistent success." Though he is destined for the Hall of Fame, he must come to terms with his knowledge of the frailty of existence – "ballplayers at fifty are old men. They crammed the whole damn thing into maybe ten years. . . . After that, there's nothing but a mess of memories" – and with a sense that in life, as in the major leagues, "even perfect pitchers can be beaten." Finding renewed strength in the "infinite tenderness" of his relationship with his devoted wife, Longboat will strive to achieve something that was written about him years ago: "We know, too, what Johnny Longboat's place in the world will be, when his pitching days are done. At the top. Or pretty close."

Kahn won his fifth Best Magazine Story prize (one he shared with Thomas McIntyre of *Sports Afield*) for a 1981 *Sport* profile of Yankees owner George Steinbrenner, who "has become a thundering presence, a water buffalo in [Baseball Commissioner] Bowie Kuhn's path of summer flowers." In *Best Sports Stories 1982* editor Ehre wrote about this piece, "He pictures Steinbrenner as a complicated, multifaceted individual who will offend the public in one breath, perform charitable acts in another. . . . Kahn's assessments often are regaling and imaginative."

During the 1983 season Kahn owned the Utica Blue Sox, the only team in the minors with no major league affiliation; his players were castoffs from other minor league organizations. Kahn was financially responsible for everything from bats and balls to toilet paper in the Blue Sox restrooms; the season ended with Kahn's team winning the New York–Pennsylvania League championship but Kahn losing $20,000. He wrote about the experience two years later in his next book, *Good Enough to Dream,* which won a Casey Award for best baseball book of the year. *Good Enough to Dream* is the kind of participatory journalism that Paul Gallico suggested in his essay "The Feel" (1936), where he recommended that a writer get into the ring with a boxer or stand at the plate to try to hit a major league pitcher's curveball. Kahn becomes wrapped up in the lives and fortunes of his players – the whole team, he says, became "an extension of your essence, your values, your competence, your very manhood." The book, however, is not just a straightforward account of an extraordinary season; throughout the narrative Kahn weaves the larger themes of baseball dreams, boyhood memories, and human relationships. Kahn brought his sixteen-year-old daughter Alissa to Utica for the summer to serve as his office assistant and to hawk programs during the games. When she is lewdly targeted by some Blue Sox players, the Sox's star pitcher counsels her:

> "If you give your body to a ball player," Roy said, "you'll be in love and the ball player will be gone."
>
> Such a comment from me . . . might have triggered hot words. Alissa felt surfeited with fatherly counsel. But the remark coming from cool, handsome Roy Moretti, the best pitcher in the New York–Penn League,

> so impressed Alissa that she repeated it to me many times. As well she might. It was, I thought, a wise and caring observation that did not wither in its repetition.

The novelist George V. Higgins, reviewing the book in *Newsday,* remarked: "This is the way stories are supposed to be. Especially stories about baseball. They should have people struggling against adversity. They should have people conquering weaknesses. They should have dedication and courage . . . and they should have splendid writers on the scene. . . . All of which the Utica Blue Sox had in 1983. And readers have today."

After publishing *Joe and Marilyn: A Memory of Love,* a biography of Joe DiMaggio and Marilyn Monroe, in 1986, Kahn was approached by associates of the Cincinnati Reds' Pete Rose, who the previous year had broken Ty Cobb's lifetime record for most hits, to collaborate on Rose's autobiography. Rose, who also served as the Cincinnati manager, was on the verge of retiring as an active player. Though Kahn disliked ghostwriting, the large advance was difficult for him to refuse; and he was, as he later said, looking "forward to studying what made the great Pete Rose tick, or anyway hit." The collaboration between Rose and Kahn did not prove to be successful, although it produced *Pete Rose: My Story* (1989). As Kahn wrote in his compelling account of the experience, "Story without a Hero," which concludes *Games We Used to Play,* from the start Rose would not allow Kahn to "reach for depths beyond television chatter," and Kahn could not turn Rose into the brooding, introspective character he was not. Rose, he says, was "hyperactive, self-absorbed, brilliantly disciplined on the field, wildly undisciplined beyond," even though his "harsh edges" were "beginning to soften." Kahn was three chapters from finishing the book, which he had wanted to title "Head First," when he learned of the investigations into whether Rose, whose extracurricular gambling activities had never been a secret, had bet on major league baseball games, including Cincinnati games. By then Rose, surrounded by his agent, his lawyers, and various enterprises, was all but unapproachable, even by his ghostwriter.

Still, *Pete Rose: My Story* breaks the mold – as it was intended to do – of the typical ghostwritten account: the dominant voice is Kahn's, with Rose providing "a recollection here, a vignette there, an insight into a ball game or a ballplayer somewhere else. But with no more literacy than he truly possessed." Kahn ends the narrative poignantly and personally as he recounts his own son Roger's drug addiction and Roger's suicide, at twenty-two, in 1987. Watching Pete Rose Jr. take batting practice in early 1989, Kahn tells Pete Sr., "After I lost Roger, whenever I'd see a good young athlete, I'd feel a knife of sadness. . . . With Petey, this is the first time I've been able to see a young athlete and just feel good, without the sadness." The book also yielded a significant piece of journalism: Kahn worked diligently through what he believed were layers of deception in the widely acclaimed Dowd report, which charged that Rose had probably bet on baseball. Critiquing that report, which Kahn calls "persuasive in spots" but "overall . . . an unconvincing mix of allegation and distortion," was, Kahn later said, "the hardest reporting job of my life."

Kahn was disappointed by Rose's refusal to promote the book, but even more by the neglect of the media to read carefully his critique of the Dowd report. "I made decent money working with Rose and my dispute is not really with him," Kahn acknowledged in the unpublished 1996 interview. "The man was going to jail. My complaint is with the reporters and reviewers who attacked the book in the mindless feeding frenzy that followed Rose's suspension from major league baseball. I like to think I was a better newspaperman than that."

On 18 June 1989 Kahn married Katherine C. Johnson. In 1993 he turned to social history with *The Era: 1947–1957, When the Yankees, the Giants, and the Dodgers Ruled the World,* which earned widespread critical raves and a Pulitzer Prize nomination. In his prologue Kahn explains eloquently why that period in New York baseball remains so beloved:

> I know two reasons why the Era remains magnificent in our memories. The first is simple. The Era was magnificent. The second is more complex. The Era came to a dramatically proper finale.
>
> It ended when the time was right; its end was sad but suitable. The drama was played out. By the time the ballclubs moved West, the great Dodger team was waning. . . . The Giants had fired Leo Durocher. The Yankees' decline took longer, but eventually even the Yankees collapsed. Along that route, someone finally fired Casey Stengel.
>
> The Era possesses an inevitability and a rightness. It begins with Jackie Robinson arriving and ends with Walter O'Malley departing. Hero and brigand, who despised each other.

The book attempts to place New York baseball of the post–World War II and Eisenhower eras within the cultural and historical context of the age. It also provides a chance for Kahn to reconsider, reflect on, and report definitively many significant events of that time, including Durocher's suspension in 1947 and his firing the next year, the Saint Louis Cardinal players' threat to go on strike when Robinson joined the Dodgers in 1947, the Eddie Waitkus shooting in 1949, and O'Malley's decision to move the Dodgers west in 1957. It also allows him to consider such issues race relations and the notion that while "many things today *are* better than ever . . . many things are not better than ever: epic poems, violins, presidents, concert halls, blondes, plays about royalty, and – to put a point on this – managers and ballplayers."

Despite his own success, Kahn is skeptical about the current state of publishing. "There aren't enough book people" in the industry, he lamented in an unpublished interview. "Why don't they sell galoshes instead?" He points to the infusion of MBAs into publishing houses and mass-market bookstores, people who lay down specifications for a publishable book. "Once we have these specs," Kahn contends, "we'll have a writer write to specs. That scenario will be an American disaster. I don't want a [bookstore chain] executive telling me what to write."

Kahn has never let anyone tell him what to write. He has always written for self-expression. Writing is a personal exercise for him; in an age of word processors and floppy disks he still often uses a manual Olympia typewriter "crafted during the late Jurassic age." Though his career of more than forty years has spanned three significant outlets for sportswriting – newspapers, magazines, and books – and he has been at the top of the game in each of them, *The Boys of Summer* will likely stand as his best-known work. He has said that he does not mind being constantly associated with the book, which has sold some three million copies: "It's like hearing someone say, 'nice backhand'; you can never hear enough of that."

Kahn served as an assistant professor of journalism at Long Island University in 1967, as an adjunct professor of creative writing at Colorado College in 1972, and as director of the nonfiction workshop at the University of Rochester in 1974, 1975, and 1977. He was the recipient of the Distinguished Alumni Sesquicentennial award of New York University in 1981. In 1995 he was appointed to the board of trustees of the Authors Guild Foundation. He has narrated a television series, *The Original America's Team: The Brooklyn Dodgers,* written, produced, and directed by Mark Reese, the son of Dodgers great Pee Wee Reese.

Kahn's next book is to be published by Hyperion in 1997. Though it is as yet untitled, Kahn says that the subtitle will be "The Days When Baseball Was a Game and Writing It an Art." He is trying to find the right publisher for his as-yet-unfinished manuscript "Rescuing Roger," about his late son.

Kahn, like his friend Lardner and another great sportswriter, W. C. Heinz, reaches beyond the baseball diamond or the boxing ring to write prose that makes people think, feel, laugh, and sometimes cry. If the national pastime has informed Kahn's works more than any other sport it is because baseball, as Kahn told Kunen, "is a marvelous frame of reference for my writing. At times in my life when I've wanted something stable, I seem to go back to baseball as something that hasn't changed very much from the days when I was small."

**Interviews:**

Willard Bailey, "Writer's Digest Interview," *Writer's Digest,* 56 (March 1976): 18–23;

Peter Golenbock, *Bums – An Oral History of the Brooklyn Dodgers* (New York: Putnam, 1984);

Mike Shannon, "Roger Kahn," in *Baseball: The Writer's Game* (South Bend, Ind.: Diamond Communications, 1992), pp. 153–173;

Craig Little, "PW Interviews: Roger Kahn," *Publishers Weekly,* 240 (4 October 1993): 49–50;

Steve Gietschner, "Covering the Bases in This Era," *Sporting News,* 216 (1 April 1994): 10–11.

**References:**

Allen Barra, "Pre-Armageddon, All-Star Panel Chooses Favorite Baseball Books," *New York Observer,* 25 April 1994;

Peter C. Bjarkman, "Bums Lit: The Literary Boys of Summer," *Cooperstown Review,* 1 (1993): 55;

James S. Kunen, "Sportswriter Roger Kahn Employs the Boys of Summer and Doubles as the Homer of the Minors," *People,* 20 (29 August 1983): 93–95;

Charles Leerhsen, "Waiting for the Final Chapter," *Newsweek,* 113 (10 April 1989): 71;

James A. Michener, *Sports in America* (New York: Random House, 1976);

Mike Shannon, *Diamond Classics: Essays on 100 of the Best Baseball Books* (Jefferson, N.C.: McFarland, 1989);

Michael Sokolove, *Hustle: The Myth, Lies and Life of Pete Rose* (New York: Simon & Schuster, 1990).

# Hugh Edmund Keough

*(24 January 1864 – 9 June 1912)*

Joel Sternberg
*Saint Xavier University*

MAJOR POSITIONS HELD: Reporter(?), *Hamilton* [Ontario] *Spectator* (1881); editor, *Sporting Journal* (1888–1890); reporter/sports editor, *Chicago Times* (1891–1894); sports editor, *San Francisco Chronicle* (1895), *New Orleans Item* (circa 1896); managing editor, *Lake County* [Hammond, Indiana] *Times* (1900–1905); sports reporter and columnist, *Chicago Tribune* (1905–1912).

BOOK: *"By Hek" in the Wake of the News: A Collection of the Writings of the Late Hugh Edmund Keough,* edited by Hugh S. Fullerton (Chicago: Regan Printing House, 1912).

OTHER: Arch Ward, ed., *The Greatest Sport Stories from the Chicago Tribune,* includes contributions by Keough (New York: Barnes, 1953): 68–69, 72–73, 106–109, 110–112.

SELECTED PERIODICAL PUBLICATION – UNCOLLECTED: "Johnson Winner By Knockout Blow," *Chicago Tribune,* 5 July 1910, fight supplement, pp. 23, 26.

Hugh Edmund Keough was one of the brightest wits in American sportswriting during the early part of the twentieth century. The first editor of the "In the Wake of the News" column in the *Chicago Tribune* – thought to be the oldest continuous sports column in the United States – Keough was cited by his newspaper as a "powerful and unique influence for high standards, sanity, and sportsmanship in athletics." He made "In the Wake of the News" the most brilliant feature of the *Tribune* sports section, creating a tone and style that would be followed by sports journalists for generations to come.

Keough, who would come to be known as "Hek" from his byline, "By Hek," was born on 24 January 1864 in Hamilton, Ontario, Canada. While information about his early life is sketchy, it is known that he began his newspaper career with the *Hamilton Spectator* at the age of seventeen. He continued his journalistic pursuits in Indianapolis and Logansport, Indiana, and in the early 1880s he ventured to Chicago – a fiercely competitive newspaper center with more than a dozen major dailies. In 1891 Keough's sharp and incisive writing style earned him a position as reporter and sports editor for the *Chicago Times.* He joined the Whitechapel Club, whose ninety members consisted of leading Chicago writers, cartoonists, and reporters. There Keough brushed shoulders with such newspaper luminaries as George Ade, John T. McCutcheon, Bert Leston Taylor, Hugh S. Fullerton, and Finley Peter Dunne.

In 1893 Keough married Bertha Atherton of Dubuque, Iowa. In 1895 they began moving about the country as he pursued his career in sports journalism with the *San Francisco Chronicle* and the *New Orleans Item.* Around 1898 they returned to the Midwest, where, for the next couple of years, Keough served as an official at every important racetrack in that part of the country. When horse racing suffered a decline in popularity, Keough focused his attention once again on journalism, and in 1900 he became managing editor of the *Lake County Times* in Hammond, Indiana. Shortly thereafter he began contributing a column to the *Chicago Tribune* titled "Sidelights on Sports." Impressed with his work, sports editor Sherman Duffy of the *Chicago Journal* attempted to hire Keough to write a column of sports humor for fifteen dollars a week; but *Journal* publisher John C. Eastman found nothing amusing in Keough's samples and turned down Duffy's recommendation.

Keough's association with the *Tribune* evolved into a Sunday column titled "Some Offside Plays." His good-natured, witty criticism of popular figures in the news was so successful that in 1905 he was offered, and accepted, a permanent position with the *Tribune.*

Columns and special services had proven successful in various sections of the *Tribune.* The legendary editor and notorious curmudgeon James

Keeley had created the popular "How to Keep Well" column by Dr. William A. Evans, a charm and beauty column by Lillian Russell, Laura Jean Libbey's "Advice to the Lovelorn," and the immensely successful editorial-page feature, "A Line o' Type or Two," written and edited by Taylor under the byline "B. L. T." In sports, Keeley launched several columns; one of them was "In the Wake of the News," to which he assigned Keough as the first editor around 1907.

From the outset, as Charles Bartlett notes, Keough did not limit himself to "mere typewriter punching." According to his colleague Harvey T. Woodruff, who would later edit the column, Keough was "the greatest sports columnist of all time": he "could be vitriolic or tender, he could be subtle or obvious, he could write verse or prose. His greatest forte was his gift for concise wit, often so subtle that it was over the heads of a majority of his readers." In his introduction to his 1912 compilation of Keough's writing, *"By Hek" in the Wake of the News,* Fullerton remarked that his friend was unique in that "he wrote classics in the language that men understand." When "classroom" English proved inadequate to his ends, he achieved "clarity and force by adopting the language made by the people among whom he lived. This philosophy he learned in the school of sport, where human nature is vivisected by men trained in studying each other for profit." Liberal and "forgiving toward human frailties," Keough was impatient with hypocrites, on whom he turned his gift for satire. "His vocabulary," wrote Fullerton, "ranged from thieves' slang to the race track tout's patois and, with his keen, whimsical humor, he brought the lightning flash of meaning from the words."

Much of what Keough wrote was ephemeral, but his miniportraits of both real and fictitious people still hold their bite and charm. For example, in an early "Some Offside Plays" column (2 July 1905) Keough reported:

> The Hon. Joseph Cannon, speaker of the house of representatives, attended the races at Seattle last week, and in a speech to the populace admitted that he had surrendered to the human impulse to bet. Those old gentlemen get quite giddy when they are allowed to stray away from home on these wholly unnecessary summer excursions. We should not be surprised if we were told that the old boy had been seen riding in the smoker.

On 16 July 1905, commenting on the state of heavyweight boxing after Jim Jeffries had given up the heavyweight championship and the less-than-desirable Marvin Hart had won the crown in a fight with Jack Root, Keough pointed out that fighters such as Paddy Ryan and John L. Sullivan were honored and worshipped by boxing fans: "Jim Corbett was generally respected. Bob Fitzsimmons was an artistic triumph. The world had to give it to Jeffries for what nature had done for him. But Marvin Hart? Well, really, has it come to this?" Later in the same column Keough again tendered his displeasure in no uncertain terms: "The pugilistic grave has given up its dead since Marvin Hart woke up and found the championship in his stocking." Still grumbling about heavyweight boxing, on 30 July 1905 Keough noted that "Marvin Hart's folks are at a loss to account for his progress in the prize ring. Here, too."

Five years later Keough covered Jeffries's return to the ring as a "Great White Hope" in a championship bout with the black fighter Jack Johnson. In an era when writers and editorial cartoonists often expressed racist views, Keough has nothing but praise for Johnson's skill and effort. Keough filed his story on 4 July 1910 from Reno, Nevada, the only state that permitted prizefights:

> The king is dead. Long live the king, even though he be a Negro. No dissenting voice rises tonight in pentup Reno to question John Arthur Johnson's right to the throne.
>
> The white man's hope, which never should have been encouraged, flickered out pitifully in the fifteenth round of a battle that never should have been fought.
>
> There was no demonstration of resentment by the crowd of 20,000 that gathered around the ring under the blinding sun of the desert in the "last free state of the American union."
>
> An awed hush fell over that most remarkable assemblage when it saw Jeffries going to his inevitable finish, and it was not broken when the referee laid his hand on the black man's shoulder and proclaimed him the heavyweight champion pugilist of the world.
>
> **Cheers for the Victor.**
>
> Later, when the champion's head appeared above the throng that surged about him in his corner, he was given the reward of fair play and the recognition of merit in the form of a lusty cheer in which all who were not dazed by the overwhelming defeat of their pride joined heartily.
>
> Johnson was hailed as a great man, the greatest man probably who ever stripped in a prize ring, and Jeff was shown only pity for his folly and egotism in ever thinking he could return after five years of inactivity and indulgence and beat a man like this. He was even denied the credit of being able even at his best to perform the task laid out for him today.
>
> All this world loves a winner and that part of it which goes prize fighting does not care whether the champion is white or black.

*Drawing by cartoonist Clare Briggs that accompanied the obituary for Hugh Edmund Keough in the* Chicago Tribune *(11 June 1912)*

Keough's "In the Wake of the News" frequently consisted of short citations, quips, light verse, responses to readers' letters, and witty sayings – usually twenty to thirty items of varying length and seriousness. His epigrams are classic bits of humor. On any given day one might find interspersed throughout his column such one-line items as: "The losses we take philosophically are those the other guy pays"; "A game that everybody is good at cannot be a hard game to play"; and "Ill fares the game that is governed by the box office." In a column titled "Made Over Maxims" he gave a new and cynical twist to some old sayings:

> Hell hath no fury like a discharged caddy.
>
> Never let your left hand know what your right hand doeth, unless you have declared it in.
>
> Cast your bread upon the waters and it will return to you. This applies to the waters of Salt Lake, where there are no fish to beat you to it.
>
> Your flowing tide of easy money is ever offset by the undertow.
>
> The blatant fan usually mixes his business cards with his rooting.
>
> It is easier to convince a loser than it is to square his friends.

Incorporating accounts of both actual and fictional events, a typical "Some Offside Plays" or "In the Wake of the News" column would often find Keough going to illogical and funny extremes. In response to a reader's suggestion that "if one's horse is left at the post one should get one's money back," Keough replied on 10 July 1904: "After long experience turf authorities have agreed that such a provision to the rules of racing would be subject to abuse. For example, if two's horse were left at the post two might insist on getting his money back twice." In a more cynical and satiric mood, under the subhead "Recent Decisions" Keough reported on 9 July 1905 that "Judge McStravick of Iowa has established the legal status of English walnuts. They are strictly within the law until the kernels are removed. Then they become gambling implements."

Poetry was a regular staple of Keough's "In the Wake of the News" columns, and, like his epigrams, his poems highlighted his intellectual insight and curiosity about a wide range of subjects – for example, the perils of aviation, Teddy Roosevelt's sentiments on peace and war, the pleasures of flirting, and, of course, sporting events. On 7 October 1906 Keough mused that no matter how the crosstown Chicago World Series turned out, "Chicago is satisfied to see its entry run one-two." To celebrate this situation he wrote a poem:

> Now glory to the gallant band
> That brought us home the flags!
> And glory to the stout limbed lads
> That wear the champion tags.
> Praise be unto our Capt. Chance
> And, eke, our Capt. Jones,
> Who led their gamesome athletes
> To the bunting and the bones.
> Praise be to them who stood behind
> And them who filled the box,
> And glory to the white haired chief,
> Comiskey of the Sox!
>
> Now glory to the hoarse voiced fan
> Who constant was always
> (It will cost him now two dollars,
> But he'll gladly stand the raise).
> And praise be to the two bit bleach
> Who helped the thing along,
> By hurling empty bottles
> At the umpire who was wrong.
> If they haven't got the money
> They'll hypothecate their duds
> To demonstrate how much they think
> Of Murphy of the Spuds.

In the same column, noting that umpires had been given authorization by baseball's national commission to eject all players indulging in vulgar language during the World Series, Keough commented: "Coming just at a time when we are deifying the national game, this is rather disenchanting, don't you think?" And, besides, this "provision for punishment for the use of vile and indecent language seems rather irrelevant outside of Cincinnati."

Keough's concerns about the future of baseball often centered on his perception of greed among the game's participants. On 29 December 1907 he noted that during "these times of uproarious reform Chicago has nothing to do with its spare change but support baseball teams, but isn't there a slight danger of pressing it a bit?" He concluded by wondering "what designs commercialized baseball has upon us." His prophetic verse "The Baseball Situation," included in Fullerton's 1912 compilation of his writing, speaks more directly to the issue:

Sing a song of dollar marks reaching to the skies;
When they talk of millions now it causes no surprise.
Magnates on the inside, counting up their tin;
Hustlers on the outside, honing to get in;
Stories of the big wads backing other leagues;
Darkly hinting treachery, treason and intrigues;
Guesses that are crazy, statements that mislead;
Just an ounce of sportsmanship to twenty tons of greed.
Money! Money! Money! till you cannot rest;
Reaching for the kale seed; tell with all the rest.

Continually demonstrating his dexterity with language and his ability to create unusual words and phrases, Keough often concocted playful columns that gave his readers a humorous perspective on a news story or sports event. His "Autumn Reflection" is an example:

Gowf should no be played in lang breeks.
It's doonricht sacreleegious.
The cutty yins air bad enow, but th' lang yins are an abomination.
Kilts, mon, are the only habit for a Gowfer. The skirl o' the skairts at the tap o' the drive maks the bonny pechter. Indeed, aye, Macphearson!

Items such as these led Taylor to offer his colleague a poetic tribute, under the title "HEK!," in his "A Line o' Type or Two" column for 24 July 1909:

O, singer of the track and ring,
Pan's pipe is yours – likewise a pen;
How wild and weird the slang you sling!
The gay Heksameters you fling
Are often sweet as note of wren,
O, singer of the track and ring! –
As often pungent, with a sting
Of sauce tabasco or cayenne.
How wild and weird the slang you sling!
But ever fresh the note you sing –
Sometimes a bit beyond our ken –
O, singer of the track and ring!
You sling the slang that's slung – but spring
A new one every now and then.
How wild and weird the slang you sling!
Hekstatic Hek! The laurel bring
And bind it on his Number Ten.
O, singer of the track and ring
How wild and weird the slang you sling!

Keough's "In the Wake of the News" began appearing on a daily basis in 1907, along with a feature column on Sundays that typically focused on a range of sports and civic topics. A keen eye, gentle humor, and satiric point of view were still his trademarks on this expanded schedule, and his colleagues were not spared his acerbic comments. On 3 January 1911 he offered his opinion on America's racing correspondents:

Curly Brown, the czar of the Jacksonville racetrack, invited the "turf correspondents" to a New Year's banquet and then got up on his hind legs and told them the dire things that would happen to them if they dared write and cause to have published anything that impinged on the notion of proper criticism. Did they resent it? What a foolish question! They meekly stood for it, or sat through it and partook of the free feed. Having constituted themselves a part of the degrading machinery they must needs bow to the tyrant who furnishes the vehicle of their employment. And the free and independent American press keeps those fellows there and pays telegraph tolls on the product!

Keough aimed his next barb at writers who covered baseball:

It is indeed sad – or does it affect you that way? – that the opportunity to preface the baseball story with those immortal words, "This being Friday, the thirteenth," etc., will be furnished only once during the baseball season this year, and then only away back in October, when the world's series, possibly, will be on. Do you remember how you used to watch for that beautiful, compelling thought and speculated on how many baseball scribes in the range of your reading would evince enough originality to pass it up for once?

Despite his penchant for cynical commentary, Keough was quick to help a friend in need. Charles H. Hermann recalls in his *Recollections of Life & Doings in Chicago* (1945) that Ade's hometown baseball team in Brook, Indiana, had been totally humiliated by the Rensselaer, Indiana, team in a local championship game in 1905. Ade learned that the Rensselaer team had used semiprofessional players during the game. The following year Keough and some other friends imported six members of the National League champion Chicago Cubs to play for the Brook team. None of the professionals was recognized by the Rensselaer team or fans, and Brook avenged its loss with a lopsided victory.

Keough was noted for his compassion, especially for children. At Christmas 1910 he created one of his better-known poems, "Be a Goodfellow," which includes the lines:

Be a good fellow, if revel you must;
But set a small portion apart
To buy trinkets and goodies for poverty's kids,
Who've been given the worst of the start.
It sets yourself right when you know you have done
Something to share just a part of your fun
With those who have nothing to do nothing with –
Show them that Santa Claus isn't a myth!

Using candor and a sentimentality that, according to *Tribune* reporter Charles Bartlett, "rang true, without any semblance of straining," Keough published words of sympathy and appreciation in his columns on the deaths of his friends and comrades that were often reprinted by relatives of the deceased for memorial cards. An example is the verse he wrote on the death of Cleveland pitcher Addie Joss in April 1911:

He pitched good ball – and what he was beside
He did not say; but showed in gentle acts;
No braggart he, nor puffed with empty pride –
A model for his kind in simple facts.
Just what he was, he was, not ever tried
With vain acclaim to be what he was not;
No strength he bragged, nor weakness he denied;
The best he had to give was what you got.
An honest tribute this, from one and all;
He pitched good ball.

Perhaps Keough's best-known poem of sympathy was written on the death of his good friend Charles F. Spalding. He subsequently repeated this poem, as did future "In the Wake of the News" editors, on the occasion of other deaths:

TO CHARLES F. SPALDING.
HIS MEMORY.

Let's lay aside the lighter thought
And pause a little while;
Let's give a voice to feelings pent –
Let tears usurp the smile.
Let grief reign where the laughter was,
And let us cease to doubt
That friendship lives beyond the grave –
A good old pal's gone out.

To Keough's friends, coworkers, and readers, "a good old pal went out" on 9 June 1912. Woodruff reported that Keough had had two operations to relieve an "affection [*sic*] of the throat which had prevented him from swallowing." After the second operation, on 3 June, Keough was said to be improving; but an embolism developed, and he died on 9 June. A wreath was laid across Keough's rolltop desk at the *Tribune,* and Clare Briggs, a cartoonist for the paper, paid tribute to his late friend with a drawing. Twenty-eight honorary pallbearers were named, including Keeley, Ade, Taylor, *Tribune* sportswriters Charles Dryden and I. E. Sanborn, Duffy and Eastman of the *Journal,* left fielder and White Sox manager James Callahan, White Sox owner Charles Comiskey, and Ban Johnson, president of the American League. Comiskey and Johnson were longtime enemies, and Keough had used his column in a continuing but unsuccessful attempt to bring about a reconciliation between them.

Hundreds attended the funeral. Father C. J. Quill, who had visited the White Sox training camp in the spring with Keough on the writer's last out-of-town assignment, said in his eulogy that the keynote of Keough's life and work was a hatred of falsity:

> This quality often lead "Hek" in his column to attack fearlessly and with courage many shallow pretenses. His barbed shafts of wit sometimes left a sting and even resentment in their wake, but they were evidence of his stand for the genuine, for the best things in life. Hugh lived as he wrote, modestly, genuinely, and without sham. . . . In the old Irish tongue, the word "Keough" meant "Who are you?" That was what Hugh wanted to know. If he numbered you among his friends, it was a rare privilege. The world is better for his writings and his life. His words and his thoughts will live in the memories of his friends.

For several days after the funeral the *Tribune* printed comments on Keough's death, using "A Good Old Pal's Gone Out" as the column title. Condolences and poetry came in from across the country. A committee that included such notables as Comiskey, Woodruff, and the detective agency executive William A. Pinkerton scheduled a theatrical benefit at Chicago's Colonial theater for Sunday, 21 July, to create a memorial fund for Keough's widow; the idea proved so popular that a second theater, the Garrick, was reserved in anticipation of an overflow crowd. The theaters and the printing of the programs were donated.

On 14 July "In the Wake of the News" was taken over by the *Tribune* sportswriter Fullerton. In his first column Fullerton wrote that he felt comparable to "the bush leaguer who was sent to bat for Cobb with the bases filled and two out." He noted that in tribute to his departed friend it had been "decided to keep the name over the column as a sort of memorial. HEK christened the Wake, and none can improve upon it." Promoting the Keough memorial benefit, Fullerton recalled on 21 July that Keough was the best of good fellows, one who "left the world richer by thousands of laughs. He never dum-dummed his shots of sarcasm even when he aimed them at the false or crooked." Keough would help "wear out the streets" to the cemetery "carrying other good fellows to their home. If it were for some other good fellow's family and he were still here, he'd expect to meet you there this afternoon." The performances were held that night at the Colonial and Garrick theaters with nearly thirty performers, including Will Rogers, working without compensation. Receipts from the sale of tickets, programs, and program space netted approximately $8,500.

Fullerton quickly realized he was not the appropriate successor to Keough at "In the Wake of the News." A fine sports reporter, he had little talent for humorous writing and no liking for it. He recommended Ring Lardner for the job, and almost a year to the day after Keough's death Lardner took proprietorship of the column. At first he was not confident that he could fill his friend's shoes. As Donald Elder relates in his biography *Ring Lardner* (1956), Lardner believed that Keough's work was unique and that "a merely fair imitation of it would not do." Some unenthusiastic letters convinced Lardner that he was failing to maintain the quality of the column. Adding to his feelings of inferiority, one day, when delivering copy to the copydesk, he overheard someone say, "Does he think he's as good as Keough?" For weeks, reports Elder, the sensitive Lardner stayed away from the copydesk and sent his columns up by messenger.

In the introduction to his compilation of Keough's writings, Fullerton wrote: "many times I chided him for not writing something for posterity; something 'worth while.' He could have done something that perhaps would have brought more lasting fame, but he preferred to write for his own people – the 'good fellows' he loved so well. And, perhaps, judging from the depth and sincerity of the grief they showed at his death, he was right."

**References:**

Charles Bartlett, "A City and a Sports Department (The Tribune Grows up with Chicago)," *Chicago Tribune,* 24 January 1942, II: 20;

Donald Elder, *Ring Lardner* (Garden City, N.Y.: Doubleday, 1956), pp. 46, 60, 81, 85, 94–95, 138, 316;

John E. Erickson, "Newspapers and Social Values: Chicago Journalism, 1890–1910," dissertation, University of Illinois at Urbana-Champaign, 1973;

"H. E. Keough Funeral Today," *Chicago Tribune,* 11 June 1912, p. 11;

Charles H. Hermann, *Recollections of Life & Doings in Chicago: From the Haymarket Riot to the End of World War I* (Chicago: Normandie House, 1945), pp. 130, 136;

"Keough Fund Started by General Committee of 350," *Chicago Tribune,* 30 June 1912, p. 10;

"Many Mourn for Hugh E. Keough," *Chicago Tribune,* 12 June 1912, p. 14;

John J. McPhaul, *Deadlines & Monkeyshines: The Fabled World of Chicago Journalism* (Englewood Cliffs, N.J.: Prentice-Hall, 1962), p. 163;

Karl E. Meyer, *Pundits, Poets, and Wits: An Omnibus of American Newspaper Columns* (New York: Oxford University Press, 1990), p. xxxiii;

Alson J. Smith, *Chicago's Left Bank* (Chicago: Regnery, 1953), p. 52;

Bert Leston Taylor, "A Line o' Type or Two," *Chicago Tribune,* 24 July 1909, p. 8;

"Tribute to 'HEK' Fills 2 Theaters," *Chicago Tribune,* 22 July 1912, p. 10;

Arch Ward, "In the Wake of the News," *Chicago Tribune,* 14 June 1937, II: 21;

Lloyd Wendt, *Chicago Tribune: The Rise of a Great American Newspaper* (Chicago: Rand, McNally, 1979), p. 359, 363;

*The WGN* (Chicago: Tribune Company, 1922), p. 164;

Harvey T. Woodruff, "Hugh E. Keough Taken by Death," *Chicago Tribune,* 10 June 1912, p. 10;

Jonathan Yardley, *Ring: A Biography of Ring Lardner* (New York: Random House, 1977), pp. 90, 101, 142.

# John Kieran

*(2 August 1892 – 10 December 1981)*

William Curran
*Silverton, Oregon*

MAJOR POSITIONS HELD: Sports reporter, *The New York Times* (1914–1917, 1919–1922); baseball writer, *New York Tribune* (1922–1925); columnist, *New York American* (1925–1926); sports columnist, *The New York Times* (1927–1941); nature columnist, *New York Sun* (1941–1945).

BOOKS: *The Story of the Olympic Games: 776 B.C.–1936 A.D.* (New York: Frederick A. Stokes, 1936);

*John Kieran's Nature Notes* (Garden City, N.Y.: Doubleday, Doran, 1941);

*The American Sporting Scene* (New York: Macmillan, 1941);

*An Introduction to Birds* (Garden City, N.Y.: Doubleday, 1946);

*Footnotes on Nature* (Garden City, N.Y.: Doubleday, 1947);

*An Introduction to Wildflowers* (Garden City, N.Y.: Doubleday, 1948);

*An Introduction to Trees* (Garden City, N.Y.: Hanover House, 1954);

*John James Audubon,* by Kieran and Margaret Kieran (New York: Random House, 1954);

*A Natural History of New York City* (Boston: Houghton Mifflin, 1959);

*Not Under Oath: Recollections and Reflections* (Boston: Houghton Mifflin, 1964).

OTHER: *Poems I Remember,* edited by Kieran (Garden City, N.Y.: Doubleday, Doran, 1942);

*Information Please Almanac, Atlas and Yearbook,* edited by Kieran (New York: Simon & Schuster, 1947);

*Treasury of Great Nature Writing,* edited by Kieran (Garden City, N.Y.: Hanover House, 1957);

*Books I Love,* edited by Kieran (Garden City, N.Y.: Doubleday, 1969).

SELECTED PERIODICAL PUBLICATIONS – UNCOLLECTED: "Red Grange," *St. Nicholas* (October 1925): 1266–1267;

*John Kieran (photograph by Erich Hartman)*

"Rockne Made Football What It Is Today," *American Magazine* (November 1929): 26–31;

"Big-league Business," *Saturday Evening Post* (31 May 1930): 16–17;

"Tyrants of the Turf," *American Magazine* (September 1937): 52;

"There Goes the Ball Game!," *American Magazine* (October 1937): 61–62;

"Not Merely a Schport," *Scribner's* (November 1937): 86;

"Good Skate," *American Magazine* (March 1938): 46–47;

"Horse and Buggy Day," *American Magazine* (August 1938): 32–33;

"Screwballs," *American Magazine* (October 1938): 32–33;
"How I Got That Way," *American Magazine* (March 1940): 26–27;
"Flights of Fancy," *Audubon Magazine* (November 1942): 360–361;
"Lost and Found," *Collier's* (25 December 1943): 18–19;
"Inspiration of Dr. Chapman," *Audubon Magazine* (May 1946): 287–291;
"About Edwin Way Teale," *Audubon Magazine* (May 1952): 150–155.

In the late 1930s anyone bold enough to challenge the proposition that John Kieran of *The New York Times* wrote the most learned and agreeable prose ever seen on a sports page would have faced the wrath of some heavyweight Kieran devotees, among them Yale University's William Lyon Phelps, America's doyen of humane letters. Moreover, at the popular level Kieran enjoyed a national reputation not only as an authority on sports but also as the man who knew something about almost everything, America's "walking encyclopedia." In large measure Kieran's apotheosis from respected sports columnist to the country's most celebrated polymath was the product of his tenure on one of the favorite radio programs of the period, *Information Please*. Although not the first quiz show to be aired, *Information Please* was the first to adopt a format of inviting listeners to stump a panel of experts rather than having a master of ceremonies query random volunteers from a studio audience.

In 1927 Kieran, who had been reporting sports in the New York press for more than a dozen years, was invited by *The New York Times* to create a daily column called "Sports of the Times," the first signed column of any kind to appear in that old-line newspaper. With "Sports of the Times" Kieran helped raise sports reporting to a new level of journalistic excellence. In his first column, which appeared on 1 January 1927, he invoked the opinions of historic jurists Sir William Blackstone and James Kent and used such expressions as lèse-majesté and "writs of replevin." It was not done pretentiously – Kieran was never pretentious in his writing. Rather he introduced occasional learned references into his columns in a spirit of fun, often achieving a gentle irony reminiscent of Joseph Addison and other writers of the familiar essay. Of Kieran's contemporaries only the sophisticated Heywood Broun of the *New York Tribune* and later the *New York World*, a man of wide reading, had earlier ventured to challenge the locker-room conventions of sports prose.

Kieran accepted in a broad sense the charge from *The New York Times* that in the new column he might write about anything he wished. Although in fact he devoted most columns to the principal sport of the season – baseball in summer, football in autumn, hockey and basketball in winter – he exercised his privilege to range among all sports and occasionally came close to departing altogether from the subject of sport. Kieran felt free to discuss almost any matter of the moment and on occasion to indulge his fondness for light verse, a taste he may have acquired from his longtime friend Grantland Rice. As early as 1925, when Kieran was still with William Randolph Hearst's *New York American*, he reported the victory of Scots golfer Will Macfarlane in the U.S. Open in a four-stanza parody of Robert Burns. Kieran's opening stanza must have ranked among the most novel leads in the history of sportswriting:

> Come a'ye Scots who skirl the pipes, bedeck't in tartan plaid,
> The gowfing crown adorns the brow of our ain canty lad.
> Sing, hey, the land of Bobbie Burns, age strang upon the green,
> Sing, ho, for Will Macfarlane, from the town of Aberdeen.

Describing his years doing "Sports of the Times," Kieran wrote:

> I rambled scandalously, and touched on topics that rarely found their way into any reputable newspaper. This included ancient history and modern art, organic and inorganic chemistry, popular astronomy, free translations from *L'Echo des Sports* and *Tutti Gli Sporti*, quotations from John Keats, Robert Browning and Virgil, occasional book reviews and a discussion of Heisenberg's Theory of Probabilities as applied to the game of 3-cushion billiards.

It was precisely the discursive quality, of course, that helped make Kieran's column a favorite among readers of *The New York Times*, which was not, after all, a newspaper directed at a mass readership.

For all the playful interjection of classical references and literary allusions, Kieran's prose offered a refreshing change from the baroque excesses and blizzard of clichés that marked much of American sportswriting even as late as the 1930s. Long before Kieran had taken up the considerable challenge of writing a daily column in the country's most prestigious daily newspaper, he had arrived at a plain, clear, and easy manner of reporting, a style that won him ardent readers, more in faculty clubs, he

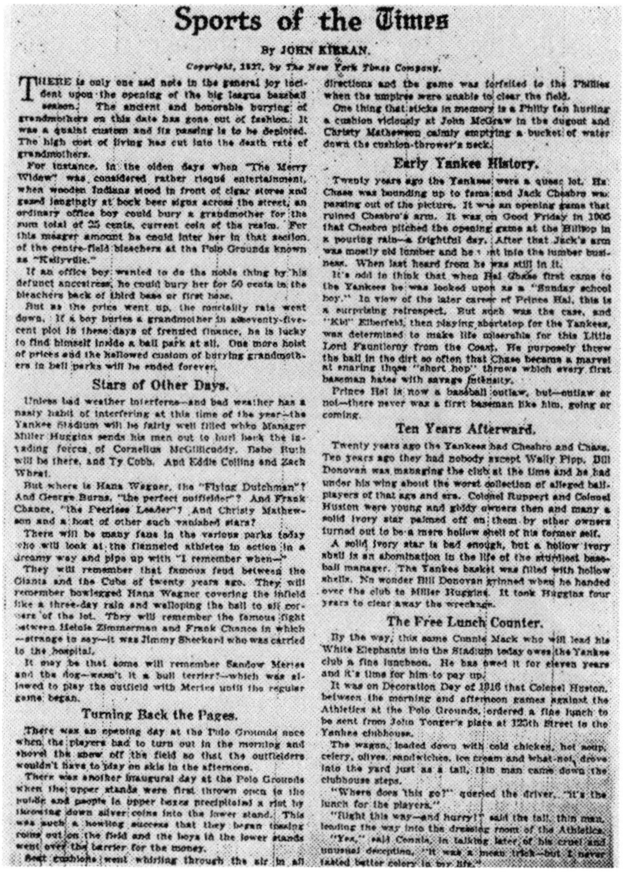

## Sports of the Times

By JOHN KIERAN.

Copyright, 1927, by The New York Times Company.

THERE is only one sad note in the general joy incident upon the opening of the big league baseball season. The ancient and honorable burying of grandmothers on this date has gone out of fashion. It was a quaint custom and its passing is to be deplored. The high cost of living has cut into the death rate of grandmothers.

For instance, in the olden days when "The Merry Widow" was considered rather risqué entertainment, when wooden Indians stood in front of cigar stores and gazed longingly at bock beer signs across the street, an ordinary office boy could bury a grandmother for the sum total of 25 cents, current coin of the realm. For this meager amount he could inter her in that section of the centre-field bleachers at the Polo Grounds known as "Kellyville."

If an office boy wanted to do the noble thing by his defunct ancestress, he could bury her for 50 cents in the bleachers back of third base or first base.

But as the price went up, the mortality rate went down. If a boy buries a grandmother in a seventy-five-cent plot in these days of frenzied finance, he is lucky to find himself inside a ball park at all. One more hoist of prices and the hallowed custom of burying grandmothers in ball parks will be ended forever.

### Stars of Other Days.

Unless bad weather interferes—and bad weather has a nasty habit of interfering at this time of the year—the Yankee Stadium will be fairly well filled when Manager Miller Huggins sends his men out to hurl back the invading forces of Cornelius McGillicuddy. Babe Ruth will be there, and Ty Cobb. And Eddie Collins and Zach Wheat.

But where is Hans Wagner, the "Flying Dutchman"? And George Burns, "the perfect outfielder"? And Frank Chance, "the Peerless Leader"? And Christy Mathewson and a host of other such vanished stars?

There will be many fans in the various parks today who will look at the flanneled athletes in action in a dreamy way and pipe up with "I remember when—"

They will remember that famous feud between the Giants and the Cubs of twenty years ago. They will remember bowlegged Hans Wagner covering the infield like a three-day rain and walloping the ball to all corners of the lot. They will remember the famous fight between Heinie Zimmerman and Frank Chance in which—strange to say—it was Jimmy Sheckard who was carried to the hospital.

It may be that some will remember Sandow Mertes and the dog—wasn't it a bull terrier?—which was allowed to play the outfield with Mertes until the regular game began.

### Turning Back the Pages.

There was an opening day at the Polo Grounds once when the players had to turn out in the morning and shovel the snow off the field so that the outfielders wouldn't have to play on skis in the afternoon.

There was another inaugural day at the Polo Grounds when the upper stands were first thrown open to the public and people in upper boxes precipitated a riot by throwing down silver coins into the lower stand. This was such a howling success that they began tossing coins out on the field and the boys in the lower stands went over the barrier for the money.

Seat cushions went whirling through the air in all directions and the game was forfeited to the Phillies when the umpires were unable to clear the field.

One thing that sticks in memory is a Philly fan hurling a cushion viciously at John McGraw in the dugout and Christy Mathewson calmly emptying a bucket of water down the cushion-thrower's neck.

### Early Yankee History.

Twenty years ago the Yankees were a queer lot. Hal Chase was bounding up to fame and Jack Chesbro was passing out of the picture. It was an opening game that ruined Chesbro's arm. It was on Good Friday in 1906 that Chesbro pitched the opening game at the Hilltop in a pouring rain—a frightful day. After that Jack's arm was mostly old lumber and he went into the lumber business. When last heard from he was still in it.

It's odd to think that when Hal Chase first came to the Yankees he was looked upon as a "Sunday school boy." In view of the later career of Prince Hal, this is a surprising retrospect. But such was the case, and "Kid" Elberfeld, then playing shortstop for the Yankees, was determined to make life miserable for this Little Lord Fauntleroy from the Coast. He purposely threw the ball in the dirt so often that Chase became a marvel at snaring those "short hop" throws which every first baseman hates with savage intensity.

Prince Hal is now a baseball outlaw, but—outlaw or not—there never was a first baseman like him, going or coming.

### Ten Years Afterward.

Twenty years ago the Yankees had Chesbro and Chase. Ten years ago they had nobody except Wally Pipp. Bill Donovan was managing the club at the time and he had under his wing about the worst collection of alleged ballplayers of that age and era. Colonel Ruppert and Colonel Huston were young and giddy owners then and many a solid ivory star palmed off on them by other owners turned out to be a mere hollow shell of his former self.

A solid ivory star is bad enough, but a hollow ivory shell is an abomination in the life of the sturdiest baseball manager. The Yankee basket was filled with hollow shells. No wonder Bill Donovan grinned when he handed over the club to Miller Huggins. It took Huggins four years to clear away the wreckage.

### The Free Lunch Counter.

By the way, this same Connie Mack who will lead his White Elephants into the Stadium today owes the Yankee club a fine luncheon. He has owed it for eleven years and it's time for him to pay up.

It was on Decoration Day of 1916 that Colonel Huston, between the morning and afternoon games against the Athletics at the Polo Grounds, ordered a fine lunch to be sent from John Tonger's place at 125th Street to the Yankee clubhouse.

The wagon, loaded down with cold chicken, hot soup, celery, olives, sandwiches, ice cream and what-not, drove into the yard just as a tall, thin man came down the clubhouse steps.

"Where does this go?" queried the driver. "It's the lunch for the players."

"Right this way—and hurry!" said the tall, thin man, leading the way into the dressing room of the Athletics.

"Yes," said Connie, in talking later of his cruel and unusual deception, "it was a mean trick—but I never tasted better celery in my life."

*Part of Kieran's 12 April 1927 column for* The New York Times

conceded in his memoirs, "than in the right-field bleachers." A paragraph from Kieran's story of the U.S. Open Golf Championship of 1923 for the *New York Tribune* serves to illustrate the tasteful, precise style of reporting he had achieved early in his career as a newsman:

> The Inwood course was a fine test for a hard tournament. Four qualifying rounds were held in as many days, and many valiant knights of the niblick fell by hill and hollow on the way to the final green. When the seventy-two holes of championship play had been finished it was discovered that Gene Sarazen, 1922 Open champion, had been robbed of his crown; that Walter C. Hagen had somehow wandered off into the rough; that Jock Hutchinson had gone up in a blaze of brilliancy and come down in a cloud of smoke, and that Bobby Jones and Bobby Cruickshank had tied for the Open title at 296 strokes each.

John Francis Kieran (he never used the Francis) was born on 2 August 1892 in the Kingsbridge section of the then semirural Bronx, New York. With the impish wit that became a mark of his public personality, Kieran once described his family to a reporter as "poor but Irish." His parents, James Michael and Kate (Donohue) Kieran, both teachers in the New York City system of public education, were in fact American-born, though of Irish ancestry. Moreover, the Kierans lived in what would be viewed today as modest affluence. When Kieran was young, the family occupied a large, rambling Victorian house and kept a couple of carriages and a staff of servants. In his autobiography, *Not Under Oath,* published in 1964, Kieran observes that "as the number of Kierans increased the number of horses and helpers decreased." The Kierans also owned a "gentleman's farm" in Dutchess County, New York, to which the family regularly retired for summers and holidays. Kieran acknowledged that it was the summers spent at the farm that laid the foundations for his late-life career as a self-taught naturalist. The young Kierans were not only numerous – seven in all – but also high achievers. John's brother James, a *New York Times* reporter, helped found the American Newspaper Guild and became secretary to New York mayor Fiorello LaGuardia. A sister, Helen Kieran Reilly, was a popular mystery novelist. Another brother, Leo, was also a *Times* reporter, and a third, Larry, was a legal aide in the Manhattan Surrogate Court. In fact, the children of James Michael and Kate Kieran were sometimes referred to in bantering fashion in the New York press as "the seven smart Kierans."

Kieran's father rose through the New York City system of education to become president of Hunter College, then the largest women's college in the country. Kieran's mother, who retired from teaching as her young family began to grow, was an excellent musician and had a particular fondness for poetry, interests which she shared enthusiastically with all her children. Kieran once observed of his mother that "she wrote poetry but otherwise showed no mark of derangement." The remark was another bit of Kieran waggery from a man who was himself passionately fond of poetry and had committed to memory extraordinary amounts of verse.

Kieran was educated in the public schools of New York City and graduated from prestigious Townsend Harris High School, a school for gifted students drawn from throughout the city. He moved on to City College of New York, which he attended for three years, but eventually he took his degree of bachelor of science, cum laude, from Fordham University with the class of 1912. Small of stature, Kieran was nevertheless a good athlete. In college he participated in several sports but always liked baseball best. After college Kieran limited his athletic activities to golf, tennis, and hiking. He was also a good horseman. Although Kieran received his college degree cum laude, his educator father noted wryly that had young John been less of a shortstop he might have been more of a scholar.

Kieran was both wise and fortunate in recognizing early in life that he was unsuited by temperament to pursue the conventional office-bound occupations – law, medicine, teaching, or business. He was lucky, too, in having parents and siblings who understood and supported his bent. And so upon his leaving Fordham – to the puzzlement of friends and teachers if not his family – Kieran moved to the Dutchess County farm with the idea of supporting himself by raising chickens and growing apples. Years later he wrote in his autobiography, "Though I was born in a great city and grew up there, I was really a country boy all the time." Although young Kieran's venture into chicken breeding and apple growing was personally gratifying, it proved not very profitable. Yet here again the recent graduate was fortunate. He landed a part-time job teaching at a country school near the farm at forty dollars a month and was able to support himself. Despite days busy with farm chores and teaching, Kieran discovered that life at the farm afforded him time to read as well as to study nature. Between feeding chickens and gathering apples, he was able, for example, to read through Blackstone's *Commentaries on the Laws of England* (1765–1769) – twice. Although Kieran had no special interest in the law, eclectic reading of this kind underscored a strong autodidactic leaning.

After a year on the farm Kieran reluctantly returned to New York City. Still, he continued to resist the idea of confining himself to a downtown office. Through a Kingsbridge neighbor, James Pilkington, who ran a large engineering and construction business, Kieran acquired a job as timekeeper and handyman on a construction project that was within walking distance of the Kieran family home. In Kieran's view the job was ideal, and it diminished his amour propre not a bit if occasion demanded that he put on boots and overalls and get down in the muck with the Italian laborers, most of them recent immigrants to America, many illiterate. In characteristic fashion Kieran soon acquired a conversational knowledge of Italian. In addition to the absence of confinement there were unexpected advantages to his job. To begin with, he learned something about civil engineering. And when his assignment was shifted to a major construction project in midtown Manhattan, Kieran took advantage of the long subway rides to study French.

It may have been when he was riding the subway to work that Kieran first began the practice of stuffing his pockets with books so that he would never be short of reading. This custom led to joking among his relatives and friends about his customary inelegance of dress, since his suits and coats were often made shapeless by the diverse volumes stuffed into the pockets – collections of French poetry, studies of plant life, or treatises on astronomy. As long as Kieran had a book handy there could not be such thing in his life as an idle moment. His guiding motto was "improve each shining hour."

By 1914 it was plain to Kieran that he must settle into an occupation appropriate to his education and family background and one at which he could make a living. Since newspaper reporting appeared to offer flexibility in working hours, a certain freedom of movement, and interesting work, he applied for and obtained a job at *The New York Times* as a cub reporter for general news. Once again luck played a role in his life. A providential rainstorm enabled him to circumvent the news department and get assigned to cover a golf tournament. Golf was a game that he knew thoroughly and played well. He also wrote well and was eventually assigned permanently to cover the sport. Kieran viewed the job as an idyllic task, since it involved summer days spent at secluded, landscaped country clubs working with pleasant and interesting people. It was while covering golf for *The New York Times* that Kieran first became acquainted with sports reporter Grantland Rice, who had just joined the *New York Tribune.* The two became lifelong friends. Although Kieran had accepted a salaried job and was doing well at it, the spirit of entrepreneurship – perhaps it was at bottom an urge for independence – moved him to bid successfully in 1916 for the rowboat rental concession for Central Park's three lakes. He also found a proficient on-site manager for the concession and so was able to continue writing for *The New York Times.* The Central Park adventure involved risk and a substantial investment, but when Kieran surrendered the concession after five years, he found that he had made a profit.

When the United States entered World War I in April 1917, Kieran drew upon his brief experience in the construction business to enlist in the army's combat engineers, an outfit that was promising in its recruiting posters to be "First to France." By early July, Kieran was seaborne for Europe with Company F of the Eleventh Engineering Regiment and was among the first twenty thousand American troops to land in Europe. He did not seek a commission but in time rose to the rank of staff sergeant. Kieran spent twenty-five months in the army, twenty-two of them overseas. Although Kieran was viewed by the men of his company in France as a noncom who would not be pushed around, who got things done, and who was always supportive of those under him, all who knew him then and throughout his life attest to his remarkably even temperament. Of his experience in World War I Kieran commented in later years, "I did no harm to anybody on either side in the War, collected no wounds but was scared to death many times."

In an unpublished interview Kieran's son John Kieran Jr. could recall only a single instance in which his father engaged in angry contention with anyone. It was a brief dustup with gossip columnist Walter Winchell, a notoriously prickly and bellicose man who spent much of his professional life involved in controversy. The literary critic Clifton Fadiman, who was moderator for *Information Please* and Kieran's longtime friend, said in an unpublished interview that "John's erudition, while it was marked, was not the most significant quality that he had. What struck me most was his kindness. There was no malice in him. John could be amusing without ever being malicious."

An incident from Kieran's years on *Information Please* illustrates the way he could poke fun without being mean-spirited. One night on the program the panel was challenged by a listener to identify the counties of New York State that produced the greatest volume of each of a series of agricultural products – apples, potatoes, milk, cheese, grapes, and so forth. Kieran soberly and correctly identified the counties with top production, but when moderator Fadiman came to raspberries, Kieran mischievously responded, "Kings County" (coterminous with Brooklyn, home of the Dodgers and Ebbets Field). In his career as a baseball reporter he had doubtless heard many times the authentic Ebbets Field "raspberry" directed at umpires and visiting players. Listeners everywhere, including Brooklyn, enjoyed a good chuckle, and the author of the question was rewarded for having stumped the expert.

Within days of his return from France, Kieran married Alma Boldtmann, who had been chief of telephone operators at *The New York Times*. Kieran acknowledged that he had been interested in Boldtmann from the day of his arrival at the newspaper in 1914, but shyness kept him from pressing his suit. They remained married until her death in 1944 and reared three children: James, John Francis Jr., and Beatrice. Immediately after his honeymoon Kieran returned to his position as golf writer at *The New York Times*.

In 1922, once again through the good offices of a friend, this time "Granny" Rice, opportunity beckoned Kieran. He moved to the *New York Tribune* as a beat writer for baseball with a byline, a perquisite as yet unknown at the conservative *New York Times*. In addition to getting to cover his favorite sport daily, Kieran had his name known to his readers for the first time. Something that he especially enjoyed about his new baseball assignment was the opportunity to travel with the team. By custom, when on the road a baseball writer had his mornings free. Instead of sleeping late as many writers and ballplayers did, Kieran made use of the early hours to learn as much as he could about the cities where his team was playing. He visited the art galleries, libraries, parks, zoos, botanical gardens, and historic sites. Kieran used some of these mornings on the road to lay the foundation for his later career as a nature writer. "Just over the outfield fence at Forbes Field lay Schenley Park through which I roamed regularly with my field glasses at the alert," he reports in his autobiography. "It was a good place to look for migrant warblers in May and September."

In his first season on the New York baseball beat, Kieran, through no fault of his, became embroiled in one of the major scoring controversies of the era. That season he served for the first time as official scorer for games at the Polo Grounds, home park for both the New York Giants and the New York Yankees. At the close of the season American League president Bancroft "Ban" Johnson, to insure that Detroit star Ty Cobb finished with a batting average above .400, set aside a scoring decision that Kieran had made against Cobb back in August. Over Kieran's protests and the somewhat milder objections of the Baseball Writers Association, Johnson's decision stood and Cobb was credited with an additional hit for the season, just enough to put his average above .400 for the third time in his career. Although Kieran pointedly omitted the ugly episode from his autobiography, he was angered by Johnson's action and disappointed by the failure of the Baseball Writers Association to support him enthusiastically. The incident apparently moved Kieran to seek an early exit from the job of baseball beat writer.

In 1925, like many other journalists of his generation, Kieran let himself be hired away from the *Tribune* (which had merged with the *New York Herald* to become the *New York Herald Tribune*) by Hearst. Hearst was known to be ready to outbid anyone for a writer he wanted, and he sent Damon Runyon and Gene Fowler, two of his journalistic stars, to court Kieran. In addition to offering an increase in salary, Hearst promised Kieran that he could have his own sports column at the morning *New York American,* the Hearst syndicate's flagship paper. Kieran called his column at the *American* "Wild Oats and Chaff," and he confessed that he enjoyed the assignment partly because he was permitted to indulge his taste for composing light verse. He had not been at the *American* long before he learned that Hearst could be a fickle master. No one was safe from "the Chief's" changes in mood. When Hearst, apparently without cause, fired the widely admired sports editor Bill Slocum, Kieran decided that it was time to move.

In 1926 Kieran approached his former editor at *The New York Times* to ask whether the paper might have a place for him. The *Times* responded by offering to let him create the paper's first signed column, and "Sports of the Times" was born. In his first year writing the column, Kieran hit a memorable high note by celebrating Babe Ruth's record-setting sixtieth home run in verse. The piece on 2 October 1927 concluded:

> My voice may be loud above the crowd and my words
> just a bit uncouth,
> But I'll stand and shout till the last man's out:
> There never was a guy like Ruth!

Years later Kieran remarked that since he was conscious of his limitations as a versifier, he was careful to keep his efforts light. Nevertheless, the verse column became a favorite among *Times* readers, and thereafter Kieran rarely hesitated to invoke the minor muse if the event seemed to warrant it. Nor was he afraid to tackle verse forms more complex than ballad measure – sonnet, ballade, villanelle – though he usually did so in parody of a well-known poem. Describing one of his own frustrating outings on the golf links, Kieran plundered John Masefield's "Tomorrow":

> Oh, yesterday, flushed high with hope, I stood upon the
> tee,
> My drive I hooked behind a rock; my second hit a tree,
> And all the dreadful afternoon I flubbed in misery,
> But tomorrow, by the gods of golf, I'll try the game
> again.

Kieran could, of course, be serious when events demanded. In the spring of 1931 he was moved to eulogize Notre Dame football coach Knute Rockne, who had been killed in the crash of a private airplane.

*Caricature of Kieran by Willard Mullin in John Lardner's* It Beats Working *(1947)*

Having reviewed Rockne's brilliant career as a coach, Kieran concluded:

> But these are mere football technicalities, little incidents that come rushing to mind at the mere mention of Knute Rockne's name. But Rock was far bigger than all that. He was a national figure. He appealed to the imagination of the country. He was a leader and a builder. He was a clear thinker and a fine speaker. He had the salt of keen wit. He was an inspiring force. Even with the sudden tragedy, football has not lost Rockne entirely. They will remember him on countless gridirons. And somehow, in thinking of Rockne and his boys, there come to mind the words of Shelley:
>
> "The soul of Adonais, like a star,
> Beacons from the adobe where the Eternal are."

Given so much latitude in writing his column, Kieran gained the reputation among many readers as perhaps the best and most literate sportswriter of his generation. Because his column was particularly popular among academics and literary persons, anecdotes about Kieran's erudition abounded. A favorite involved his visit to an elite prep school to address the students. According to the popular version of the story, as Kieran writhed in embarrassment on the dais, one of the schoolmasters offered a patronizing introduction, half apologizing for the school's inviting such a lowly sort as a sportswriter to address them. The schoolmaster assured the boys, however, that in this instance they were getting the best of the breed. The schoolman ended his introduction with a quotation in Latin, which he gratuitously translated for the benefit of his guest. Thoroughly irritated by the condescending introduction, Kieran, the story goes, rose and thanked his host for the kind remarks, begged to correct a couple of small errors in the host's quotation, and then launched into an address to the students entirely in Latin. According to Kieran, the episode unfolded as told except that he did not give his address in Latin. Instead he tossed off a dozen or so Latin quotations that he knew by heart before delivering in English the speech he had come to make. Still, the performance was enough to delight the students and give rise to the legend.

Like several famous sportswriters of an older generation, among them Red Smith, Kieran was never able to generate enthusiasm for basketball. In an unpublished interview Roger Kahn, whose father, Gordon Kahn, was the cofounder of *Information Please,* told a story that illustrates Kieran's lack of interest in the sport. Sometime in the mid 1940s Kahn, then a student at New York University, was on his way to New York's Madison Square Garden to watch highly ranked New York University play equally strong North Carolina in one of the most widely publicized collegiate basketball games of the year. Kahn stopped by the offices of *Information Please* to pick up his father and take him to the game. Kieran happened to be at the office, and the excited undergraduate asked, "Who do you like in the game tonight, Mr. Kieran?" To Kahn's astonishment Kieran asked, "Who's playing?" In fairness to Kieran, it must be noted that by this time he had retired from newspaper sports reporting. In subsequent conversations with young Kahn about basketball, Kieran offered as reason for his lack of interest in basketball that "It doesn't write."

In spite of his painful experience with Ban Johnson during the Ty Cobb episode in 1922, Kieran's favorite sport remained baseball, and each spring, while covering the baseball training camps, he looked forward to "suiting up" and working out with the major leaguers. He took advantage of this privilege until he was well into his forties, and it is said that he never embarrassed himself in the field. In a classic bit of whimsy, Kieran used "Sports of the Times" to help create a unique public image for Morris "Moe" Berg, a utility catcher with the Chicago White Sox, the Boston Red Sox, and several other teams. Like Kieran, Berg was a polymath. A graduate of Princeton University with a degree in philology, Berg had a reading knowledge of more than half a dozen languages and was actively interested in a variety of scholarly subjects. Perhaps it was the incongruity of such an erudite young man making his living by playing major league baseball that amused Kieran. For many years the columnist kept his readers informed about "Professor" Berg's progress through the nation's research libraries as well as about his activities behind the plate, making Berg one of the most widely recognized players of his era, despite his lowly position as third-string catcher.

Beginning in 1938 and continuing for a decade, Kieran answered correctly thousands of questions submitted by listeners to *Information Please* – questions on sports, ornithology, literature, musicology, botany, architecture, language, archaeology, law, and other subjects. There seemed to be no field to which he was a stranger, and he became a national celebrity.

Early in the history of *Information Please,* producer Dan Golenpaul made a special effort to interest college students in the program and urged them to send in questions. John Kieran Jr., then an undergraduate at Yale University, asked the panel to recite the second stanza of each of three famous English poems. Poetry specialist Kieran Sr. gave it a try, answered part of the question, and then stumbled. He nevertheless got in a last word. Parodying a line from William Shakespeare's *King Lear,* he observed, "How sharper than a thankless tooth it is to have a serpent child."

Although as a star of radio Kieran had generally been able to preserve his privacy when in public places, the situation changed during World War II when the company of *Information Please* traveled overseas to entertain U.S. forces. On the tours he became a familiar figure to tens of thousands of men and women in the services who had previously known only his voice. *Information Please* also made a series of short features for the movies, and Kieran and the other regulars from the program – Franklin P. Adams, Oscar Levant, and moderator Clifton Fadiman – became instantly recognizable in public.

A common question among listeners to *Information Please* who lived far from New York City was how could so learned a man as Kieran, who spoke precise and unfailingly correct English, do so in what was to their ears the accent of Hollywood gangsters and New York taxi drivers. The answer was that Kieran was a man totally without pretense; to the end of his life he made no conscious effort to soften or smooth over the normal pronunciation of his native West Bronx. When Kieran rose to na-

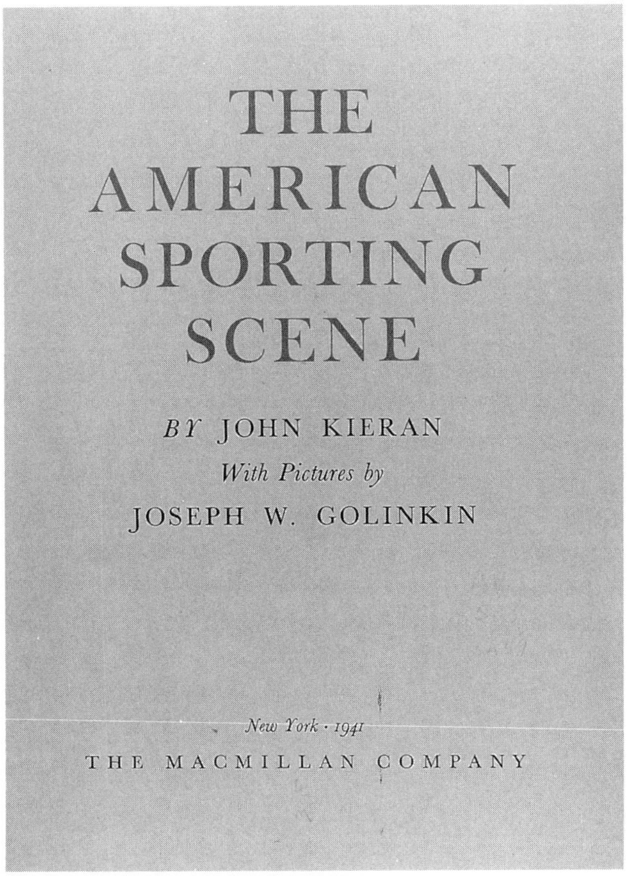

THE AMERICAN SPORTING SCENE

*BY* JOHN KIERAN

*With Pictures by*

JOSEPH W. GOLINKIN

*New York · 1941*

THE MACMILLAN COMPANY

*Title page for Kieran's 1941 book about American sports*

tional prominence on the heels of his extraordinary performance on *Information Please,* he dismissed as absurd terms such as *genius* or *walking encyclopedia.* He did not consider himself erudite in the conventional sense. His explanation for the extraordinary breadth of knowledge was simple: "I read a lot and I am interested in many things." He was fond of quoting an idea from Terence that was used by Cicero in his *De Officialis:* "Homo sum, humani nil a me alienum puto" (I am a man and nothing human can be alien to me).

At the end of 1941, convinced that his burgeoning activities made it impossible for him to do a thorough job of covering sports, Kieran left *The New York Times* and accepted a job writing a less demanding column for the *New York Sun.* The new column, called "One Small Voice," was often devoted to natural history, one of his major interests. Widowed in 1944 and already experiencing symptoms of heart disease, Kieran further cut back on his work schedule by resigning from the *Sun* at the beginning of 1945. His resignation ended his thirty-year career as a newspaperman and reduced his work commitment to writing freelance articles – principally on subjects related to natural history and conservation – and to serving as a panelist on *Information Please.* In 1947 Golenpaul persuaded Kieran to accept the editorship of the newly established *Information Please Almanac.* Feeling that he was really only window dressing for the new annual, Kieran withdrew after editing the second edition. When in the following year *Information Please* ended its spectacular run on radio, Kieran was free at last to concentrate on the activity he enjoyed most, writing on natural history.

In 1941 Doubleday, Doran had published Kieran's first book on nature, *John Kieran's Nature Notes.* He was by the time the book appeared already a director of the National Audubon Society. Immediately following World War II, Kieran wrote *An Introduction to Birds* (1946), the first book of what became a series on natural history. It was soon followed by *Footnotes on Nature* (1947), largely a compilation of his Monday evening nature columns from the *New York Sun.* The columns were in part a product of weekend nature walks, often in New York's Van Cortlandt Park, near Kieran's home, with friends Clyde Fisher, Fred Nagler, and Herman Forster. Kieran's nature books were well received by the reading public and helped launch him on a late-life career as a serious though self-trained naturalist. He brought to his writing on natural history the same charm and attention to detail that had won him many readers during the years he wrote about sports. In 1948 Doubleday published the second volume of the nature trilogy, *An Introduction to Wildflowers.* A new edition of *An Introduction to Birds* followed in 1950, and in 1954 Kieran completed the project with *An Introduction to Trees.*

In 1947 Kieran married Margaret Ford, a feature writer for the *Boston Herald.* With Kieran's new flexible work schedule, the couple was able to continue their separate writing careers. As with his first marriage, Kieran's union with Ford was by every account a happy one. In dedicating his autobiography, *Not Under Oath,* Kieran wrote of his second wife that she "has given me the best years of her life – and mine." In 1952 Kieran realized a lifelong dream of living away from the city when he and his wife took up full-time residence at Rockport, on Massachusetts's scenic Cape Ann. Rockport was to remain their home until his death twenty-nine years later.

In 1959 Houghton Mifflin published Kieran's most ambitious work in the field of natural history, *A Natural History of New York City.* As if to justify his undertaking a work whose very title seemed a paradox, Kieran wrote – not defensively but with characteristic grace:

> If you cover with buildings and pavement 999 former meadows and leave one meadow untouched, the chances are that you will find all the different kinds of meadow life in the region of the one tract left untouched. The larger animals may be driven off, but the Meadow Mice will remain. On the outskirts of the city there are still areas of greenery that offer hospitality to birds, insects, and small animals. There are more than 28,000 acres of public park land within the city limit. New York's rivers and bays and ocean front are the season haunt or permanent home of many different forms of wildlife. Let the population of the area increase and multiply as it may, let men build and pave to their heart's content, there will always be many kinds and untold numbers of wild things in the great city.

Kieran's *A Natural History of New York City* was awarded the John Burroughs Medal for 1959.

According to an unpublished interview with Kieran's son James, even with advancing years Kieran demonstrated on his occasional speaking engagements that he had lost none of the playful wit that had won him so many friends and admirers. Following a Kieran lecture at Harvard University, an undergraduate asked the guest why he had chosen to send his two sons to Yale rather than to a superior institution, one on the Charles River. "That's simple," Kieran replied. "The train fare from New York to Boston was four dollars and sixty cents, to New Haven it was only a dollar thirty-five."

In 1981, when he was eighty-nine, Kieran was awarded the President's Medal of Fordham University. Alluding to his success on *Information Please,* a portion of the citation read, "The stumping of the scholarly John Kieran was a singular occurrence." The President's Medal was the last honor Kieran received in his lifetime. He died at Rockport on 10 December 1981.

In matters of language Kieran was for the most part a nominalist and a conservative. This was hardly surprising, since throughout his boyhood English accidence, syntax, and diction were commonly subjects of dinner-table conversation in the Kieran home, and any deviation from standard American speech was promptly corrected by his schoolmaster father. Perhaps his poet mother was more flexible in such matters. From this family background Kieran emerged to become a journalist distinguished above all by his clarity of expression and pleasantness of style.

# Sam Lacy

*(23 October 1903 – )*

J. Douglas English
*Johns Hopkins University*

MAJOR POSITIONS HELD: Reporter (1918–1920, 1923–1930), managing editor, sportswriter (1930–1934), sports editor, and columnist (1934–1939), *Washington Tribune;* sportswriter and columnist, *The Baltimore Afro-American* (1939–1940); national editor, *Chicago Defender* (1940–1943); sports editor and columnist, *The Baltimore Afro-American* (1943– ).

Many writers spend their careers fighting injustice; most fail. Sam Lacy's tenacity and skill, however, enabled him successfully to aid the desegregation of American sports, especially baseball. Lacy was one of the most outspoken advocates of integration and was Jackie Robinson's close companion through the turbulent first three years of Robinson's career in major league baseball. His columns provide a window into the debate over integration in baseball and, more broadly, in the country as a whole. Still a sports editor today, in his nineties, Lacy continues to speak out against the exploitation of black athletes and the residual elements of Jim Crow in the sports world. He has earned a place in history beside Robinson as a pioneer who challenged the American conscience through the sports media.

Sam Lacy was born on 23 October 1903 in Mystic, Connecticut, to Samuel Erskine Lacy and his wife, a Shinnecock Indian. In 1905 Lacy's father moved the family to Washington, D.C., where his son would live all but four years of his life. The family's migration to Washington to find a labor market more favorable for blacks ended prosperously. After working as a legal researcher for the burgeoning legal industry of the city, Lacy's father became the first African American detective in the District of Columbia's police force. Young Lacy had big shoes to fill. But he also had a role model for the pioneering path he would walk later in life.

His father's passion for baseball accounted for Lacy's early exposure to the sport. Wearing a wool suit jacket no matter what the weather, the elder Lacy regularly sat in the Negro section of the right-field bleachers to watch the Washington Nationals (who would not be known as the Senators until 1919). Often his son was by his side. Lacy's father became friends with Nationals owner Clark Griffith, although their relationship deteriorated as the decades passed.

*Sam Lacy (courtesy of Sam Lacy)*

His father's association with Griffith landed the younger Lacy a job shagging fly balls in batting practice and acting as a general gofer for players, coaches, and staff before each game. The official compensation for this work was a position as a vendor for that day's game; the real reward for the teenage Lacy was the opportunity to meet, mix with, and even befriend ballplayers, including future Hall of Famers Walter Johnson and Goose Goslin.

At Armstrong High Lacy's athleticism inspired his permanent attachment to sports. Although short and slim, Lacy's wiry physique, mental acuity, and determination earned him three letters each in football, basketball, and baseball. His baseball prowess earned him an opportunity to play in the local semipro Negro leagues against future Negro league Hall of Famers Martin Dihigo, John Henry Lloyd, and Oscar Charleston. The experience he gained working for the Nationals and playing against black stars gave Lacy the opportunity to compare the relative abilities of the best black and white players. The early conclusions he drew from this experience formed the attitudes that would shape a professional purpose determined by his nonathletic talents. His academic merits earned him, while still in high school, a job with the city's largest black newspaper, the *Washington Tribune*. Under sports editor Lewis Lautier, Lacy earned twenty-five cents an inch writing as a stringer on high school sports.

In 1920 the seventeen-year-old Lacy angered his parents when he left the *Tribune* job, turned down admission to prestigious Howard University, and embarked on a career in the Negro leagues. After two years playing for the Atlantic City Bachrachs and the Baltimore Black Sox, however, Lacy left his baseball career behind and matriculated at Howard. While statistics for the leagues and years he played are incomplete, the record shows that Lacy played infrequently, with rather mediocre results. Lacy's reason for leaving baseball as a player, however, lay in other concerns. "When the games ended all of the players would go back to the boarding houses and smoke and drink, and have women in their company. I went once and said this is not my life," he recalled in Peter Sheingold's "In Black and White" (1992). While Lacy earned a degree in physical education at Howard in 1923, his experience in the Negro leagues left him believing what few others, black or white, believed at the time: that black players had enough talent to play in the major leagues. Lacy dedicated the rest of his life to overcoming the problems generated by segregation in sports.

After graduating from Howard Lacy spent the next seven years as a part-time sportswriter at his old newspaper, the *Washington Tribune*, and as a radio commentator in Baltimore and Washington. Following his marriage to Barbara Robinson in 1928, he spent more and more time on the struggling *Tribune*, finally taking a full-time job in the sports department in 1930. At the time the sports department consisted of three reporters, two desks, and a typewriter. The paper's financial stress demanded that sportswriters be flexible. In an interview with Jim Reisler, Lacy recalls doing "a little of everything. By working for small papers with small staffs, we also covered other beats that took us beyond just sports or baseball. We had to be well-rounded writers." Lacy's gift, according to his contemporaries, was the ability to write detailed reports concisely.

In 1934 Lacy took on more responsibility to save the financially faltering paper, becoming the sports and managing editor. His editorial byline appeared under the title "Looking 'em Over," which began with general observations on the Negro leagues and black athletics. While he always reported the action on the field, as time went by he turned his attention increasingly to the trials and successes of black athletes in a segregated society. He editorialized adamantly against the injustice of segregation in sports.

Motivated by his teenage experiences with Washington Nationals and the Negro leagues, Lacy knew black players deserved more respect and better treatment than they received. He saw the segregation of baseball as a tragedy in two ways. As he told Ron Fimrite for the latter's October 1990 article in *Sports Illustrated*, "I felt that not only were blacks being deprived of the opportunity to make some money but that whites were being deprived of an opportunity to see these fellows perform. I could see that both were being cheated. And so, with a certain amount of ego, I took it upon my self to act as a wedge." Along with others, including fellow black sportswriters Wendell Smith, Joe Bostic, and Fay Young, and, later, white writers and editors such as Westbrook Pegler, Jimmy Powers, and Shirley Povich, Lacy would prod the American conscience.

Lacy's first salvo against segregation in sports concerned college football. In October 1937 an undefeated Syracuse University football team traveled to take on the University of Maryland. The Orangemen's star, as reported in the press, was the first East Indian to play American football. Lacy, however, learned that Wilmeth Sidat-Singh, despite having been raised by parents of East Indian descent, had been adopted from a black family at the age of three months. Lacy boldly, and to much criticism, revealed Sidat-Singh's ethnic heritage in his *Tribune* column. After Maryland refused to allow

Sidat-Singh to take the field, Syracuse lost 13–0. Lacy pointed to the hypocrisy and discrimination his actions exposed rather than responding to his critics. "An undefeated football record went by the boards here today as racial bigotry substituted for sportsmanship," he wrote.

Lacy also took direct action to press the case for integration. In 1938 he met with his former employer, Clark Griffith, who still owned and operated the Washington baseball club, now known as the Senators. Since Griffith's team had landed in the American League's second division on a regular basis during the past decade, Lacy hoped that Griffith would consider taking on black players to improve the quality of the team. But while Griffith, whose team contained a remarkably high number of Latin players, claimed to be tolerant and liberal, he had only excuses when it came to hiring black athletes.

Griffith argued that the poor administrative quality of the Negro leagues prevented black players from attaining the requisite character and training necessary to make it in the big leagues. Players in the Negro leagues often jumped contracts, especially around playoff time when teams padded their rosters in anticipation of the post-season. League offices were not ruled impartially; frequently team owners served as executives with the power to punish their opponents and advance their own team's interests. Most important, racketeering rings owned several teams. In the wake of the Black Sox scandal of 1919, Lacy could understand that major league baseball would refuse any players tainted by gambling, even if only by association. Despite being privately outraged (even his father boycotted Senators' games from this point on), Lacy apparently took Griffith at his word. In his *Tribune* editorials he began to censure the misbehavior and mismanagement in the Negro leagues, hoping thereby to raise their quality.

In 1939 *The Afro-American,* headquartered in Baltimore, bought out and absorbed the *Tribune,* keeping Lacy on as a sportswriter. The change boosted the weekly national circulation of Lacy's articles to nearly one hundred thousand. Lacy now began to write with the almost exclusive intention of integrating the major leagues. Unfortunately, his editorial campaign during the next two years, mounted along the lines Griffith had suggested, would fail to advance his cause. He not only spurned but attacked forms of social protest that would become central to the broader civil rights movement two decades later. Perhaps afraid to antagonize Griffith, he condemned and eventually succeeded in defusing a black-organized boycott of white baseball in Washington.

Lacy explained the purpose of his early columns in *The Afro-American* in "In Black and White": "I lent my pen to the crusade which had as its purpose the breaking down of the barriers that stand between the dark-skinned diamonder of merit, and a place on the payroll of . . . one of the sixteen major-league clubs." The target of his attack shifted from the executive offices of the major leagues to the Negro leagues. He even suggested to his readership that "perhaps there was something to the contention that we are keeping ourselves down."

Lacy blasted legendary players such as Josh Gibson for jumping contracts, a charge often hinted at but never stated explicitly by others in the black press to that date. He harshly criticized Negro league stars who, he felt, lacked the mental discipline to represent favorably their race in the major leagues. The didactic style of these columns was brutal and unforgiving. "Whatever the reason or reasons, I find myself far more painful from the display of mental dullness by our players than I am joyful over the exhibitions of mechanical stardom," he wrote in the 10 June 1939 *Afro-American.* He quoted at length several players, including stars Dick Lundy and Jud Wilson, who seemed to put all the blame on prominent Negro league players and owners for the lack of progress in integration.

Lacy limited the focus of his analysis to baseball and took care to remind his readers that he did not blame the perceived cultural inadequacy of black society or its often regressive leadership for the lack of racial progress. But despite Lacy's laudable intentions, his columns did more harm than good. As Jules Tygiel points out in *Baseball's Great Experiment* (1983), Lee MacPhail, president of the American League, cited Lacy's negative evaluations of the Negro leagues as evidence to support the continued exclusion of their teams and players from the majors. Quoting Lacy's statement that "I am reluctant to say that we haven't a single man in the ranks of colored baseball who could step into the major leagues and disport himself after the fashion of a big leaguer," MacPhail argued that "Mr. Lacy's opinions are shared by almost everyone, Negro or White, competent to appraise the qualifications of Negro players." Talk of integration, ironically, was undermined through the words of its most ardent supporter.

A disappointed and somewhat embarrassed Lacy eventually revised his strategy. Before, he had envisioned an integrationist-minded association between the Negro leagues and the major leagues that would organize the former as a sort of minor league, or feeder system, for the major leagues, much like the successful, highly competitive American Association or Pacific Coast League. Negro league teams, then, would remain segregated, but the major league clubs would buy up the best players and provide enough cash for the Negro leagues to remain a viable financial enterprise and cultural centerpiece for black America. By arguing for this restructuring, Lacy had sought the survival of the Negro leagues even after integration at the major league level. McPhail's co-optation of Lacy's exhortations to the Negro leagues, however, forced him to change his approach.

As long as it seemed that integration of the major leagues might come about through the improvement of the administration of the Negro leagues, Lacy's pleas for internal reform were reasonable. The top-level resistance by McPhail, however, convinced Lacy that the conservatism of major league baseball lay primarily in its commissioner, Judge Kenesaw Mountain Landis. "Pickets at Griffith stadium would be misplaced," he wrote in the 1 April 1939 *Afro-American,* "but not so much so in front of the Michigan Building in Chicago, where Judge Landis holds forth." Lacy quit *The Afro-American* in 1940 to take a job as national editor for the *Chicago Defender.* Although his new position would put his sportswriting on hold for the only time in his adult life, it would allow him to carry the fight to Landis in his own city. Unfortunately, this fight would also end in frustration.

Upon his arrival in Chicago, Lacy intensified an already aggressive and voluminous letter campaign directed at major league owners and primarily Commissioner Landis. In November 1943 Landis finally granted Lacy an interview. Lacy asked Landis to permit a three-person delegation to address the baseball owners' meeting the following month. To Lacy's surprise and delight, Landis agreed, stipulating that the delegation had to be headed by a prominent black organization. Lacy immediately recommended the Negro Press Association, conveniently headed by the publisher of Lacy's *Chicago Defender,* John Sengestacke.

Lacy planned to base his argument on behalf of the committee along the lines of his appeal to Griffith, but with a crucial revision. While he would continue to lobby for the integration of the major leagues on grounds of merit, he shifted his emphasis from the Negro leagues to individual black players who deserved inclusion in baseball at its highest level. As he told Sheingold in "In Black and White," "I would have simply made the point that I had observed both operations at work and I had seen and played against some of the black ballplayers. . . . My point would have been that they should have taken some of the black players in a pool."

The effectiveness of Lacy's argument cannot be judged since he was not allowed to make it. Sengestacke did not name Lacy to the committee; instead, he brought along another black publisher, Ira Lewis of the *Pittsburgh Courier,* and the actor Paul Robeson. All three argued for the inclusion of black players through an appeal to patriotism. How could black players, they asked, face equal danger on the battlefields of Europe and the Pacific yet not get an equal chance on the playing fields at home? In Lacy's view, Robeson's well-known status as an intellectual communist destroyed the credibility of this flag-waving appeal. The meeting resulted in nothing. Lacy, crushed, returned a month later to a position as the sports editor of *The Afro-American.* The next four years saw Lacy at his most confrontational and determined – and, finally, at his most successful.

Despite the failure of the special commission's audience before the patriarchs of major league baseball, several factors, including the service of black Americans in the war, rendered black activism increasingly effective. Lacy became part of a growing wave that demanded equal treatment for equal service and ability. The death of Landis in 1944 also removed the most powerful obstacle to baseball's integration. After World War II "Looking 'em Over" grew to reach more than two hundred thousand readers in a week in *The Afro-American* and in syndication. Lacy's voice would be heard.

On his return to the Baltimore-Washington area in 1943 Lacy, who in the six years since he had listened respectfully to Griffith and achieved no results, showed an unconcealed antipathy toward intolerance in his writing and wasted no time in blasting the Senators owner. "The white-haired owner who spends four-fifths of the time he gives colored reporters extolling his virtues as a liberal has just made public that he will go into spring training with the unprecedented total of 39 would-be big leaguers – the percentage of colored among them exactly zero," he wrote in the 26 February 1944 *Afro-American.* Lacy had taken off the gloves and turned "Looking 'em Over" into a bully pulpit. He then invoked what earlier he had headed off, serving notice that a boycott of the Senators' games was in the works to protest Griffith's failure to advance the cause of integration: "Uncle Clarkie always could look out there in the right field stands and find it generously filled with loyal colored fans. Well, many of those loyal black rooters are puzzled by the Senators' steadfast refusal to consider the use of qualified colored players. . . . Many of these same fans will take this column and mail it to Uncle Clarkie."

Although Lacy's use of his column to further his political aim provoked an injunction from the attorneys for *The Afro-American,* who feared costly lawsuits if Lacy continued to advocate boycotts or strikes in print, he continued to pound on the door of the major leagues. Lacy knew that Happy Chandler, Landis's replacement, had made off-the-record comments indicating that he would not block the integration of baseball if the point was pressed. Quickly, Lacy petitioned for a second special panel to address the issue, and in 1945 the Major League Commission on Baseball Integration was born. Lacy held a position on this committee, and his optimism soared when Chandler named sympathetic Branch Rickey as its second member. But the enterprise was crushed by its third member, Lacy's old nemesis Larry McPhail, who stalled any progress the commission may have made by simply refusing to meet.

"Baseball has given employment to epileptics, kleptomaniacs, and a generous helping of saints and sinners," an exasperated Lacy wrote compellingly in the 10 November 1945 *Afro-American.* "A man who is totally lacking in character has turned out to be a star in baseball. A man whose skin is white or red or yellow has been acceptable. But a man whose character may be of the highest and whose ability may be Ruthian has been barred completely from the sport because he is colored." Two weeks later, on Sam Lacy's forty-second birthday, the breakthrough came. Rickey signed Jackie Robinson to a minor league contract with the Brooklyn Dodgers organization. Lacy then received one of the biggest assignments in American sportswriting history when *The Afro-American* assigned him to follow and cover Robinson's trials and travails for its nationally syndicated audience.

Perhaps surprisingly, Lacy's reaction was not immediately one of ecstasy. Although he no longer complained in print about the lack of discipline exhibited by Negro league players, Lacy still fretted about failure. While he recalls suggesting Robinson to Rickey in 1945 as a prime candidate for breaking the segregation of major league baseball (it should be noted that there is no evidence that Rickey heard or took Lacy's recommendation into account), Lacy feared that a poor showing, either on grounds of character or ability, might result in an even more powerful barrier to integration. "With us," he lectured his readership, "the first man to break down the bars must be suited in every sense of the word. We can't afford to have any misfits pioneering for us, and for obvious reasons. Unwilling as they are to employ Negro players, they will be quick to draw the old cry: 'We gave 'em a chance and look what we got.'" Rickey had made a superb choice, for Robinson was exemplary both on and off the field.

Integration, however, did not put an end to bigotry. Jim Crow laws and social segregation tested Robinson's moral and psychological limits even as he excelled on the field. Lacy reported that truth to the black American reader in the 1 April 1950 *Afro-American:* "Regardless of any impressions you have gotten to the contrary, the South has not accepted interracialism in baseball. It is merely tolerating it." Robinson and the black players that followed him frequently endured the humiliation of eating on the team bus or in hotel rooms rather than in segregated restaurants and hotel dining halls. They often experienced the impossibility of flagging down a taxi when late for the ballpark and the discomfiture of sleeping four to a room in black boardinghouses rather than in whites-only team hotels. Through this period Lacy and Wendell Smith of the *Pittsburgh Courier* not only brought Robinson's story to hundreds of thousands of black readers but also were pressed into service as Robinson's companions and psychologists and wrote as social critics assailing a racist society.

Although Lacy did not experience the slurs, spikes, and spit that Robinson endured on the playing field, he suffered and overcame all the indignities heaped upon a professional black man in white America. As a sportswriter Lacy often found himself banished to the roof of the press box (where several white writers, including Dick Young, joined him in protest) or reporting from the Brooklyn dugout (at Rickey's orders) when denied entrance to the ballpark. Through it all, Lacy got his stories in before deadline, and his achievements finally began to receive the national attention they deserved. Robinson, with his characteristic unselfishness, cited sportswriters as central to baseball's integration: "I know that my position was obtained only through the constant pressure of my people and their press. It's a press victory you could say." Lacy became the central figure in the integration of another aspect of baseball when the Baseball Writers Association of America admitted him as its first black member in 1948. For the first time in his sportswriting career, Lacy said, he was elated.

After covering Jackie Robinson's turbulent first three years in the major leagues, Lacy resumed his post at *The Afro-American,* retitling his (now thrice-weekly) column "A to Z" to symbolize the diversity of his focus. This broadening was necessary as a media flood saturated the Negro leagues after Robinson's major league success. White writers now aggressively covered the top black teams and players, breaking the monopoly black presses and writers had held. To compete, Lacy argued, black sportswriters had to come up with stories that had more "bite" and all too often invaded a player's privacy. But ultimately and inevitably, the collapse of the Negro leagues following the loss of its best players to the majors and minors eclipsed this problem. By 1955 the Negro leagues (which were unstable even in their heyday) had ceased to exist.

Some held Lacy and others who had pressed for integration responsible for the passing of this central element in black culture. While Lacy acknowledged the loss and his role in bringing it about, he had no regrets. "After Jackie, the Negro Leagues was a symbol I couldn't live with anymore," he told Reisler. But even as he shifted his attention to golf, college basketball, and the Olympics, he had one more service to render to the Negro leagues. In 1972 Commissioner Bowie Kuhn honored Lacy by naming him as one of only two sportswriters on the Hall of Fame's Special Commission on the Negro Leagues. During the next five years the commission named eight former Negro leaguers to the Hall of Fame before dissolving itself.

Lacy's sportswriting remained opinionated and polemical even after his major battle had been won. Lacy saw that the black press, more than ever before, was obligated to offer political and social criticism rather than the more reportorial sports coverage the general presses provided more efficiently. "The role of the African-American press is to report the positive things, and police the major papers," he said in the 14 September 1992 *Los Angeles Times.*

Lacy excelled at castigating the inequities of the sports world. Harshly but accurately, he chastised college basketball for its exploitation of amateur black athletes; boxing, a sport dominated in America by blacks, for its apathetic stance toward severe injuries; and professional golf for its residual segregationist ties. His literary style might best be described as sardonic honesty. "The fact that the PGA of America scheduled its championship at a site where black membership is excluded actually came as no surprise. To blacks who follow golf, the surprise was that the PGA had the audacity to claim it was not aware of the Shoal Creek policy of discrimination. Or that professional golf has a history of fielding its events in an atmosphere of exclusivity," he wrote in the 5 August 1990 *New York Times.* At eighty-six, Lacy was still writing the good fight.

Lacy's outsider status and critical attitude certainly led to a rather muted reception of his work by those other than his colleagues at the major black presses. While he had won four National Newspaper Awards at *The Afro-American,* the distinguished quality of his career was not widely acknowledged until late in his life. Although a complete listing of the honors accorded him would be impractical (the plaques themselves take up most of his office floor), the major awards that involve firsts alone tell the story. Lacy became the first black in the Maryland Media Hall of Fame (1984), the first black sportswriter recognized for lifetime achievement by *Sports Illustrated* (1989), and the first sportswriter in the Black Journalists' Hall of Fame (1991). His nomination for election to the sportswriters' wing of the Baseball Hall of Fame is only a partial honor. Lacy, few could dispute, merits the realization of the nomination as the second black sportswriter ever enshrined at Cooperstown.

Sam Lacy, because of his circumstances and personal character, brought to sportswriting a different philosophical purpose than many others. "Anyone can be a reporter. But a journalist is a restless reporter. No matter what the obstacles, he never gives up. He's never satisfied," he said in the 4 August 1991 *Newsday.* Lacy never gave up. His goal was to bring the talent of black players to a white audience and the rewards of that audience to the players. In so doing he also broke the color line in the field of sportswriting, and brought to a broader audience some of the best sports journalism: his own and that of those black writers who would follow him.

**References:**


Ron Fimrite, "Sam Lacy: Black Crusader," *Sports Illustrated,* 79 (29 October 1990): 90–94;

Mark Judge, "Writing the Good Fight," *Washington City Paper,* 17–23 May 1991, pp. 28–32;

Jim Reisler, *Black Writers/Black Baseball* (Jefferson, N.C.: McFarland, 1994), pp. 9–31;

Peter Sheingold, "In Black and White," bachelor's thesis Hampshire College, 1992;

Jules Tygiel, *Baseball's Great Experiment: Jackie Robinson and His Legacy* (New York: Oxford University Press, 1983).

# Al Laney

*(11 January 1896 – 31 January 1988)*

Joel Sternberg
*Saint Xavier University*

See also the entry on Laney in *DLB 4: American Writers in Paris, 1920–1939.*

MAJOR POSITIONS HELD: Reporter, *Dallas Dispatch* and *Minneapolis News* (pre–World War I), *New York Evening Mail* (post–World War I); night editor, reporter, and rewrite man, *Paris Herald* (European edition of *New York Herald,* 1925–1934); reporter and columnist, *New York Herald Tribune* (1934–1966).

BOOKS: *Paris Herald: The Incredible Newspaper* (New York & London: Appleton-Century, 1947);

*Golfing America,* edited by Edward A. Hamilton and Charles Preston (Garden City, N.Y.: Doubleday, 1958);

*Prep Schools: Profiles of More Than Fifty American Schools* (Garden City, N.Y.: Doubleday, 1961);

*Covering the Court: A Fifty-Year Love Affair with the Game of Tennis* (New York: Simon & Schuster, 1968).

OTHER: "A Dark Man Laughs," in *Best Sports Stories of 1944,* edited by Irving T. Marsh and Edward Ehre (New York: Dutton, 1945), pp. 15–19;

"Regardless of Race, Color or – " in *Best Sports Stories 1945,* edited by Marsh and Ehre (New York: Dutton, 1946), pp. 270–276;

Bob Cooke, ed., *Wake Up the Echoes: From the Sports Pages of the New York Herald Tribune,* includes contributions by Laney (Garden City, N.Y.: Hanover House, 1956): 49–52, 127–130, 211–214, 223–225;

"Ching, Ching, Chinaman," in *Best Sports Stories 1960,* edited by Marsh and Ehre (New York: Dutton), pp. 224–227;

"Wills vs. Lenglen," in *The Twentieth Century Treasury of Sports,* edited by Al Silverman and Brian Silverman (New York: Viking, 1992), pp. 367–380.

SELECTED PERIODICAL PUBLICATIONS – UNCOLLECTED: "The Tennis World Watches Wimbledon," *Literary Digest,* 117 (23 June 1934): 32;

"England, Davis Cup Host Once More," *Literary Digest,* 118 (14 July 1934): 34;

"Two Million Louisville Sluggers," *Scholastic,* 86 (6 May 1940): 44;

"How to Play Table Tennis," *Popular Science,* 141 (July 1942): 82–87;

"Stained Glass Making," *Popular Science,* 141 (October 1942): 129–133;

"Blind Gladiator," by Laney and Bertram Baynes Fowler, *Baltimore Sun,* 4 June 1944, A3;

"The Mystery of Wimbledon Tennis," *Saturday Evening Post,* 219 (21 June 1947): 23, 135–136.

Al Laney was one of the premier golf and tennis writers in the United States from the mid 1930s to the mid 1960s. During a career of more than fifty years he developed a reputation as a sensitive, graceful writer capable of beautifully crafted prose. An insightful reporter and master interviewer, Laney had an unerring ability to capture the style and character of the sports he covered and the athletes who played them.

Born on 11 January 1896 in Pensacola, Florida, Al Laney was the son of an attorney and one of six children. Mostly self-educated, he began his journalistic career with the *Pensacola Journal* in an after-school job that became full-time by the time Laney was fourteen years old. Jerome Holtzman writes that Laney emphasized, "there is no such thing as a born reporter. You have to learn, just like you learn to play the piano." Laney learned by working in the profession, by trial and error and by his association with editors who always demanded his best.

Laney's love for sports began early in the century. In *Covering the Court: A Fifty-Year Love Affair with the Game of Tennis* (1968) he writes about the newsreel days when "boys took their sports heroes more on faith as truly heroic figures." The newspaper was the chief source of information about sporting events, and while Laney was working at the *Pensacola Journal,* he was fortunate to have access to many New York newspapers that came into the shop. Consequently, he knew much about baseball – more, he was to claim, than those who were to view it on television and "adopt as their own the endlessly repeated clichés of the announcers." His enthusiasm for the game was fueled when the Cleveland Naps held spring training in his hometown. Laney later would recall the thrill of speaking with Cleveland stars Napolean Lajoie, for whom the team was named, and Shoeless Joe Jackson. However, as exciting as those experiences may have been, the two players were not New York Giants, and in those days, Laney recalled, "Christy Mathewson was a shining knight, and if the Giants won, then were the gods just and life a lyric."

Added to these interests in newspapers and baseball came one in tennis. On vacation in New York in August 1914, Laney took time from the Polo Grounds and his beloved Giants to sit in the stands of the West Side Tennis Club in Forest Hills to watch a new hero – Maurice McLoughlin, the "flame thatched" California Comet. McLoughlin defeated the Australian Norman Brookes in a historic match that Laney felt "caught the imagination of the whole nation" and raised the stature of American tennis to new levels. Red Smith reports that Laney was so taken with the event that he "conceived an ambition to become a tennis writer."

After dropping out of high school, Laney began to pursue this ambition. Prior to World War I he worked for the *Dallas Dispatch* and the *Minneapolis News.* He considered himself a tennis player in those days, but he gave up any dreams that he may have had of becoming a professional athlete after he was wounded in the Argonne Forest while serving as an officer in the 308th Infantry of the 77th Division.

After the war Laney returned to New York with no plan or prospects – just the confidence that he would find work with a newspaper. He eventually found a position with the *Evening Mail,* which had a small sports staff and sometimes called on volunteers to assist in covering sports events. Sports editor Warren Brown was sympathetic to Laney's request to cover tennis and offered him some hope by acknowledging that because none of the New York newspapers had tennis writing specialists, it might prove worthwhile for Laney to work toward such a goal. He advised Laney to write about tennis star Bill Tilden, and, if the *Mail* were to use Laney's work, the newspaper would pay space rates. When the *Evening Mail* merged with another New York newspaper in 1924, Laney found himself without work. Using his two weeks' severance pay and $500 in savings, he sailed for France with a bag of newspaper and magazine clippings and notes for possible story ideas.

Captivated by the ambience of Paris, Laney was one of many aspiring writers who spent winter afternoons at Sylvia Beach's bookshop, Shakespeare and Company. Learning that he had a typewriter, Beach quickly recruited him to type business letters for her friend, James Joyce, to whom Laney began to read. "He always wanted me to read contemporary stuff," Laney told *Newsweek* in 1960. "He was blind then and he wanted me to read everything." Roger Kahn writes that Laney recounted how Joyce would hand him a novel and say, "Begin," and Laney would read. "Sometimes after just one paragraph he would order me to stop," said Laney. "Sometimes I would read the whole book. Some were in English and some were in French."

Editor Herbert Warren Wind has speculated that Laney's time with Joyce had no apparent effect on the young journalist, but Laney's ability to distinguish individual sound and speech patterns among people he interviewed seems partly derived from Joyce's discourses on the sounds of words, the influence that the sound of one word has on another, the sounds of silences between words, and the creative power of the word. "Without the right sound," Joyce told his young reader, "writing is worthless." Three decades later Laney would emphasize how important it was for writers to hear differences in ways people talk and in what people have to say. As he told *Newsweek* in 1960, "When they write, everybody's speech is the same. You have to hear. You have to help your subject. In two or three hours, how many thousands of words are said? You have to hear and select the best."

Laney's time with Joyce was brief, as the novelist fired him when he realized that Laney was not capable of reading in Italian. Fortunately the nearly impoverished Laney found work with the Paris edition of the *New York Herald,* more commonly known as the *Paris Herald* – the newspaper of Right Bank Americans. In his 1947 memoir about that newspaper Laney recalled that being "on a newspaper pub-

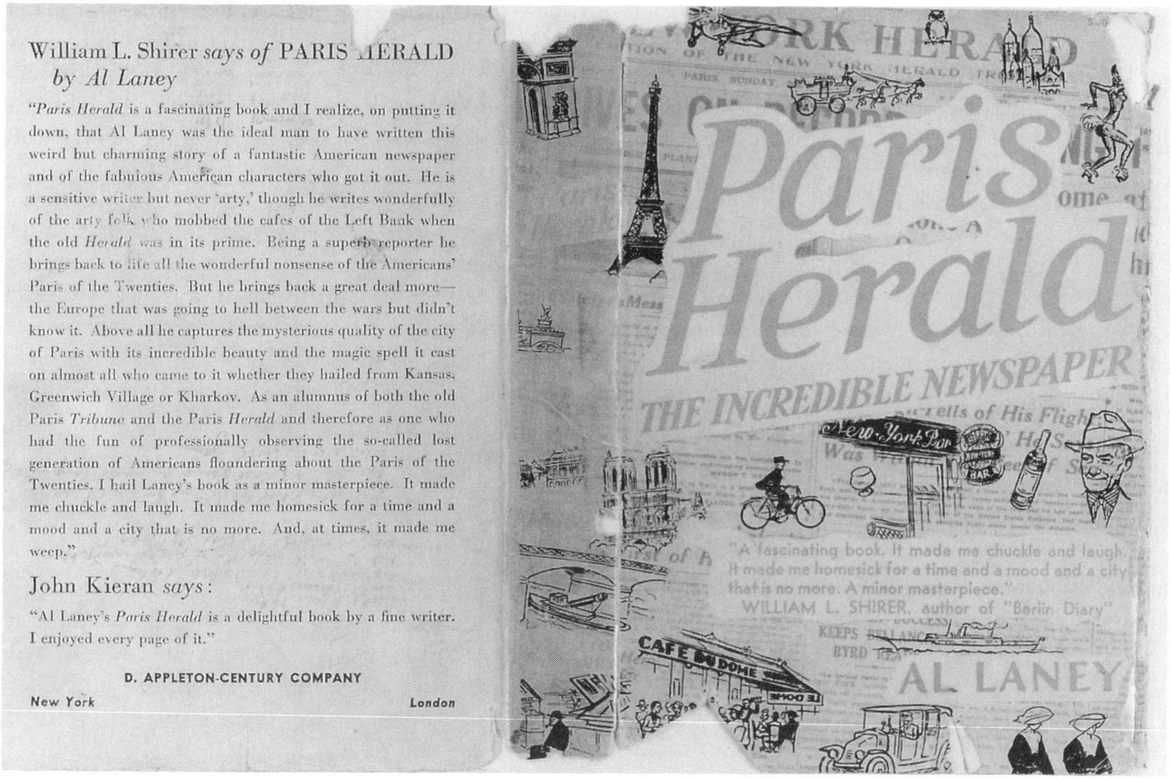


*Dust jacket for Al Laney's memoir of his years as a writer, reporter, and editor in Paris following World War I*

lished in Paris in times such as these, and to be young and in revolt was all that the heart could wish. This was for the moment the summit of desire."

From 1925 to 1934 Laney served as night editor, reporter, copyreader, headline writer, and rewrite man for the *Paris Herald.* Eric Hawkins, the managing editor of the paper, has written that Laney was a "superb news handler and an excellent writer," as well as quite a character. "His battered hat never left his head," Hawkins recalls, "and he conducted the business of putting the paper together in an uncommonly quiet manner, giving his direction and suggestions always in a half-whisper." Smith remembers Laney as a very gentle, extremely neat man who, in covering golf tournaments, could usually be seen walking alone and wearing a snapbrim hat of Confederate gray, a jacket, and a tie – even in the most stifling weather. "He spoke in a voice so gentle," Smith has written, "it was often barely audible. 'Dammit!' his sports editor, Stanley Woodward, shouted during one effort at conversation. 'I'm going to have you wired for sound!'"

In Paris Laney continued to rub shoulders with literary luminaries such as Ernest Hemingway, Ezra Pound, and Gertrude Stein. He also tried his hand at writing a novel, but nothing came of it and he never submitted the manuscript. Journalism was clearly his forte, and his work on the *Herald* coincided with the growth of that newspaper. American tourism was increasing, as was the size of the American community in Paris, and in 1924 readership of the newspaper began to grow so rapidly that within two years its circulation was to double. As Laney later recalled, "this rather forlorn newspaper . . . would blossom into a money-maker and a paper of prestige."

During these early days Laney spent time studying various sportswriters. "I read the papers," he told Holtzman, "read everything I could, and every time I read a good piece I would say, 'Gee, I wish I'd written that.' And then I'd look at it closer and say, 'And if he had done it this way it would have been even better.'" Continuing to educate himself in this manner, Laney began to concentrate on the work of Bernard Darwin, golf writer for the London *Times,* and of Neville Cardus, cricket writer and music critic for the Manchester *Guardian.*

To Laney, sports essayist Darwin, the grandson of Charles Darwin, was the finest sportswriter of all. Reading him for the first time in Paris in 1925, Laney found Darwin's literary, anecdotal style to be somewhat unusual, and he concluded that he would never write stories in quite that fashion. "But when I read [Darwin's stories] a second and third time," Laney told Holtzman, "it occurred to me that this fellow had a lot of pleasure in writing these stories and they were giving me pleasure just to read them." Laney clipped and studied everything that Darwin wrote. He realized that he could never match Darwin's style and that it would be harmful to try, but comparisons that always proved embarrassing to Laney were always to be made between the two writers.

During his time with the *Paris Herald* Hawkins relied heavily on Laney and used him principally in covering general news. Yet Laney's true love was sportswriting, and especially tennis writing, and this became as readily apparent to Hawkins as it had to Brown. Laney began filing sport stories for the *New York Herald Tribune,* stories with multiple columns on important matches. In summers he traveled between Paris and England to cover the Davis Cup, Wimbledon, and other British tournaments. When Bobby Jones scored his unprecedented grand slam golf victories in 1930, Laney and O. B. Keeler of Atlanta were the only two American reporters who covered the British Amateur and British Open tournaments.

In covering tennis Laney continually demonstrated his eloquence and his eye for detail as he captured the color, the mood, and the intensity of the crowd for his readers. As his colleagues noted, he had a fondness for describing sports with exceedingly long, graceful sentences. He used what Richard Kluger has called "an almost conversational casual vocabulary," and Kluger writes that Laney was fond of saying, "A couple of felicitous phrases go a long way in a sports story." This style is evident in his 23 June 1934 *Literary Digest* feature on Wimbledon, in which Laney reported,

> There is nothing like the great roaring noise made by thousands of feet and thousands of voices echoing and reechoing around the corridors and rebounding from the roof as these tennis-mad people pass and repass down the passages of the great gray cement building which surrounds the center court, ever hurrying and scurrying in fear of missing something.

Laney spent nearly a decade working in Paris at the *Herald,* and in 1934 this unusual time for Americans in France was clearly changing. One indication was in the decline of the Parisian competitor to the *Herald* – the Paris *Tribune,* the European edition of the *Chicago Tribune.* While the *Herald* had chronicled the Right Bank activities of Americans, the *Tribune* had reported the activities of artists and advertisers on the Left Bank. In November 1934 the competition ended when the *Tribune,* a victim of the Depression, was sold to the *Herald.* As Laney was to recall in his memoir of the *Herald,* "This act of amalgamation brought a pitiful false hope to the Herald direction, but for Herald men and Americans generally, it was one of the saddest things that could have happened. For an epoch in the lives of Americans in Paris had ended and it was impossible any longer to ignore the fact." Laney believed that there would always be nostalgia for Paris and especially for the *Herald,* "For never in the history of journalism have so many men had such a wonderful time on so little money."

Laney had occasionally traveled home from Paris for periods of approximately a month to work at the city desk of the *New York Herald Tribune.* In 1934 he returned to New York permanently and was assigned to the newspaper's sports staff. He still continued to travel to Europe, but his principal responsibilities were thereafter in the United States, where initially he covered all sports. He even covered baseball, as Kahn remembers, until the "belligerent vulgarity of Leo Durocher drove him to golf."

Golf and tennis during the mid 1930s were not considered to be rigorous sports but rather sources of recreation for members of country clubs – and both sports were relegated to the status of minor subjects on the sports page. This attitude remained, writes Kluger, until after World War II. Yet despite this general regard for golf, Laney indicated that he preferred covering the game because it was the only competitive sport not played on a regulation field, and reporting on it took him to attractive places. He later wrote in *Golfing America* (1958) of these scenic views in which golf is played, "No time is wasted in these pages trying to pin down with words the fascination of golf. . . . Rather, we have pictured a series of beautiful spots in which this congenial game is played. How pleasant to wander in the mind from one picturesque setting to another!"

Early in his *New York Herald Tribune* career Laney felt that writing a column was the ideal job for the sportswriter, and Stanley Woodward, sports editor of the *Herald,* gave him the opportunity. Within six months, however, Laney lost his enthusiasm. "There was more money in it," Laney told Holtzman, "but not enough to compensate for having to come up with something good five or six days a week." He realized that being a reporter would allow him to work his observational style around an event without having to take the more editorial position of a columnist.

In the mid 1940s Laney urged Woodward to hire Red Smith, a reporter and columnist with the *Philadelphia Record.* Woodward had been interested in hiring John Lardner, but in 1945 he hired Smith for $125 a week – but not as a columnist. Instead, Smith served as a reporter until Laney took a sabbatical from his column in order to write *Paris Herald: The Incredible Newspaper* (1947). Smith filled in for him and never relinquished the job.

Laney's reputation for reporting, for conducting interviews and writing special stories, was unsurpassed, and Woodward, who realized the value of his work, gave him free rein at the *Herald Tribune.* Laney had no regular assignments except in the fall, when he joined his colleagues in covering football. When Laney was not covering hockey in the winter, Woodward generally devised a special assignment for him and let him spend all his time working at it – and Woodward recalled that the results of this arrangement were often spectacular. One example is that of Laney's poignant story in February 1942 about a rapidly aging Babe Ruth on his forty-seventh birthday. In "Babe Ruth, Forty-seven and Ailing Goes through Usual Birthday Chore" Laney paints a sad portrait of an unhappy American hero gone from the limelight but still engaging and worthy of his fans' adoration. On 25 and 26 October 1945 Laney's story "'Regardless of Race, Color or – '" examined the strength, courage, and character of Jackie Robinson, who had just signed a minor league contract with Montreal. Helping to strike a blow against racial prejudice, Laney reported that the prospect of Negroes playing baseball was not something that was relegated to the future. "It is a fact," wrote Laney of Dodger general manager Branch Rickey, "Brother Rickey has signed a Negro to a contract and from all accounts a player who is very likely to make the grade."

In the winter of 1943–1944 perhaps the best-known series of Laney features, "Forgotten Men," appeared. Woodward had established two requirements: public attention to the subjects must have lapsed for at least ten or twenty years, but the subjects were to be known well enough to ensure that everyone who read the stories could instantly recognize them. Laney began the series with the 1908 Olympic marathon winner, Johnny Hayes, whom he found operating a tea and coffee shop in lower Manhattan. He tracked down Joe Notter, who had ridden Regret, the winner of the 1915 Kentucky Derby and the first filly ever to win the prestigious race. He found Smokey Joe Wood, a hard-throwing pitcher who had played for the Boston Red Sox and Cleveland Indians between 1908 and 1920.

In time Laney would write fifty or sixty features, perhaps the most memorable of which is "A Dark Man Laughs," winner of the first Dutton Best Sports Story Prize in 1944. The story concerned pugilist Sam Langford, the "Boston tar baby." When Laney began his search for Langford, everyone told him that the old fighter was dead. Yet Laney stubbornly continued, and in two weeks he found Langford, blind and poverty stricken, living in a one-room flat in Harlem. For Woodward, Laney's perseverance led to "one of the finest special stories I have ever read."

"Don't nobody need to feel sorry for old Sam," Langford told Laney in the story that appeared on 10 January 1944. "I had plenty good times. I been all over the world. I fought maybe three, four hundred fights and every one was a pleasure. If I just had me a little change in my pocket I'd get along fine." Letters poured in from fans who remembered Langford and from people all over the United States who had never seen him but were affected by Laney's words. Practically everyone sent a check or cash. A committee was formed to handle the contributions, which totaled more than $10,000, and to arrange an annuity designed to remove Langford from the relief rolls.

In December 1944 Laney returned to visit Langford and write a sequel that was published on Christmas morning. Laney's second story reported that Langford was pleased and that Laney's readers had made it possible. Langford was no longer alone, and Christmas decorations had brightened his dingy room. "His belly will be stuffed with turkey and fixin's today," wrote Laney, "and he will play his guitar and sing and he will laugh. To hear Sam laugh and sing is one of the most profound Christmas experiences a man can have."

Laney pointed out that the Langford stories did not just report the straight facts but were interpretive and editorial, and, aside from the quality of the writing, they effectively drew public attention. Despite the success of the stories, Laney felt he was guilty of a major oversight in having inaccurately quoted Langford, for Laney later recalled that Langford habitually addressed everyone as "Chief." In the first story, however, Laney had neglected to transcribe this, and it disturbed him. Damon Runyon, another New York journalist, had also noticed this omission and cited it in one of his columns as an example of what makes a faulty story. Laney was angry, and as he told a *Newsweek* reporter in 1960, "The piece has been in anthologies and reprinted quite a bit, but every time I see it now, I want to sit down and rewrite it."

When Woodward was fired in 1948, Laney returned to covering beat assignments – some baseball, some golf, tennis, and hockey. After the *Herald Tribune* rehired Woodward in 1959, Laney again was free to cover stories of his choosing as the winter began. "You can't push Al," Woodward explained to *Newsweek* in 1960. "Maybe he'll go a week without getting anything. But if you leave him alone, let him work at his own pace, the stories he does get are worth it." One of Laney's later triumphs was "Ching, Ching, Chinaman," detailing the life of Ching Johnson, New York Ranger hockey star. The story has been included in a Dutton anthology, *Thirty Years of Best Sports Stories* (1975), and the Dutton judge, John Hutchens, was moved to comment that the story "is a beauty of its kind, full of heart and unsentimental nostalgia. It is one of the relatively few stories I have read that gives you a real idea of why and how a great athlete can be not only admired but held in true affection."

The sources of Laney's success lie not just in the technical aspects of his writing but also in his understanding that any writing about a particular sport should have its unique style, for, as Laney told Holtzman, the pace and mood of each sport are different. Baseball, for example, has "marvelous climactic action, something that doesn't exist in any other game," and the baseball writer, said Laney, must always be looking for an angle and trying to weave threads into a coherent story. On the other hand, to cover golf the reporter has to remember that the golf match is an entire day's affair with a constantly evolving story. Competitors are not necessarily playing in close proximity, and challenges may occur at different places and times.

"Of all sports," Laney told Holtzman, "golf lends itself to writing, to literature." In the mid 1970s, however, Laney felt, golf reporting was overemphasizing meaningless quotations. He complained that any time a golfer finished near par, he or she would be promptly escorted into a press tent and there be subjected to a routine question-and-answer session. "And," said Laney to Holtzman, "these people aren't articulate to begin with. Very few. Palmer, yes, but not many others." Laney found nothing worse than opening a story with a quotation followed by dull, wooden quotations ploddingly strung together to clutter the story and blur the narrative of the event. "If my paper wants quotes," Laney asserted, "they should send a second man. I think this applies to all sports."

Laney stayed with the *Herald Tribune* until 1966, when the newspaper went out of business. For a time he continued to write, and in 1968 he published *Covering the Court.* John Quinn, an enthusiastic reviewer of this book, pointed out that tennis was the most exacting game people could play "and the most difficult to describe journalistically without sacrificing technical accuracy." As for Laney, "no one has ever better explained the subtleties and marvels of the game."

During his last years, Laney divided his time between upper Manhattan with his wife, Irene, a pianist, and two cottages located in close proximity on a two-acre plot near Redding Ridge, Connecticut, in early springtime. Holtzman reported that until Thanksgiving Laney stayed there "enjoying the solitude and the freedom from the social ramble" in almost total seclusion, with his wife visiting on weekends.

Through the years Laney won his share of awards for golf and tennis coverage. On 29 August 1965 he received the Danzig Trophy at the national doubles tennis championships at the Longwood Cricket Club. Named for its first recipient, Allison Danzig of *The New York Times,* the prize had been established in 1963 by Longwood in collaboration with the National Art Museum of Sport "for the reporting of tennis with knowledge, accuracy and literary grace and for unfailing devotion to sportsmanship and to the game itself."

On 4 September 1977 Laney was honored by the U.S. Tennis Writers Association with a gold writers' pass and plaque during ceremonies at the United States Tennis Open championships. Two years later on 20 June 1979, at the age of eighty-three, Laney was elected to the International Tennis Hall of Fame.

In 1978 the Metropolitan Golf Association chose Laney to receive that organization's Distinguished Service Award. As Smith wrote of the occasion, "The choice speaks well for the literary taste and for the memory of the selection committee, for more than a decade has passed since Al's sensitive reporting and beautifully crafted prose last appeared in the daily press." In appreciation, the Metropolitan Golf Association noted that, despite Laney's international reputation (or because of it), he had worked just as hard at covering local events as he had worked in covering the more renowned tournaments. "No one knew more local golfers and no one made more visits to local clubs than he. He made his readers aware of golf in their area and thus set a standard that forced other newspapers to re-evaluate their local coverage." The association concluded that Laney's forty years of covering the sport "form[s] a body of work that stands as a major contribution to the game of golf."

On 31 January 1988 Al Laney died at his retirement home, the Fellowship Community, in Spring Valley, New York. He was survived by his son, Michael. Having enjoyed more than four decades of sportswriting in the United States and abroad, Laney felt that he had been fortunate. Never having finished high school, he never had a job other than on a newspaper, and he wrote about the games and the people he loved. As Herbert Warren Wind has noted, "Many of his colleagues consider Laney as good a sportswriter as ever lived, and they are probably right."

**References:**

John Webster Curley, "No. 38, Rue de Louvre," *New York Times Book Review,* 2 November 1947, section 7, p. 31;

"Four on Their Jobs," *Newsweek,* 30 (27 October 1947): 68, 70;

Eric Hawkins with Robert N. Sturdevant, *Hawkins of the Paris Herald* (New York: Simon & Schuster, 1963);

Jerome Holtzman, *No Cheering in the Press Box* (New York: Holt, Rinehart & Winston, 1974);

Roger Kahn, *The Boys of Summer* (New York: Harper & Row, 1972);

Richard Kluger, *The Paper: The Life and Death of the New York Herald Tribune* (New York: Vantage, 1986);

"Old Heroes Never Die," *Newsweek,* 55 (4 April 1960): 106, 108;

John Quinn, "Tilden, Budge & Co.," *Chicago Tribune Book World,* 15 September 1968, section 9, p. 12;

Red Smith, "Al Laney's Prose," *New York Times,* 3 December 1978, section 5, p. 3;

Herbert Warren Wind, ed., *The Realm of Sport* (New York: Simon & Schuster, 1966);

Stanley Woodward, *Sports Page* (New York: Simon & Schuster, 1949).

# John Lardner

*(4 May 1912 - 24 March 1960)*

Terence Malley
*Long Island University*

MAJOR POSITIONS HELD: Reporter, *New York Herald Tribune* (1931–1933); sports columnist, North American Newspaper Alliance (1934–1948), *Newsweek* (1939–1960); war correspondent, *Newsweek,* North American Newspaper Alliance, *The Saturday Evening Post* (1942–1945); television and radio critic, *New Yorker* (1957–1960).

BOOKS: *The Crowning of Technocracy,* by Lardner and Thomas Sugrue (New York: Robert M. McBride, 1933);

*Southwest Passage: The Yanks in the Pacific* (Philadelphia: Lippincott, 1943);

*It Beats Working* (Philadelphia: Lippincott, 1947);

*White Hopes and Other Tigers* (Philadelphia: Lippincott, 1951);

*Strong Cigars and Lovely Women* (New York: Funk & Wagnalls, 1951);

*The World of John Lardner,* edited by Roger Kahn, preface by Walt Kelly (New York: Simon & Schuster, 1961).

OTHER: Ring Lardner, *You Know Me Al: A Busher's Letters,* introduction by John Lardner (New York: Scribners, 1960);

"Remember the Black Sox?" in *The Armchair Book of Baseball II,* edited by John Thorn (New York: Scribners, 1987), pp. 231–243.

*John Lardner (photograph by Ed Wergeles)*

SELECTED PERIODICAL PUBLICATIONS – UNCOLLECTED: "Reese and Robinson: Team Within a Team," *New York Times Magazine,* 18 September 1949, pp. 17, 19–20;

"Will They Bring Back the Spitter?," *Saturday Evening Post,* 222 (17 June 1950): 31, 167–168, 170;

"Touchdown by Slide Rule," *Nation's Business,* 38 (October 1950): 52, 54, 86–87;

"Trouble with Basketball," *New York Times Magazine,* 7 January 1951, pp. 16, 48–49;

"DiMaggio and Other Old Men," *New York Times Magazine,* 13 May 1951, pp. 22, 79;

"Tallulah Goes to the Ball Game," *Woman's Home Companion,* 78 (September 1951): 26–27, 80–81;

"My Case Against Sport," *American Magazine,* 152 (October 1951): 24–25, 111–113;

"So You Think You See the Fights on TV," *Saturday Evening Post,* 225 (2 May 1953): 25, 144–146;

"Young Stay Young at Dartmouth," *Look,* 17 (2 June 1953): 58–60, 62–63;

"Everybody Hates the Yankees," *American Magazine,* 156 (August 1953): 20–21, 122–124;

"Which Sport Has the Greatest Appeal?," *New York Times Magazine,* 24 October 1954, pp. 25, 32, 34, 36;

"Baseball Fan's Favorites," *New York Times Magazine,* 10 April 1955, pp. 27, 64, 66;

"Mighty Robot at the Bat," *New York Times Magazine,* 19 June 1955, pp. 13, 15;

"Perilous Plight of the Pitcher," *New York Times Magazine,* 18 September 1955, pp. 25, 38, 42;

"Sports on TV: A Critical Survey," *New York Times Magazine,* 25 December 1955, pp. 10–11, 27;

"Would It Still Be Brooklyn?," *New York Times Magazine,* 26 February 1956, pp. 19, 78;

"That Was Baseball," *New Yorker,* 32 (12 May 1956): 136, 138–154, 157–158;

"Big Man with the Biggest Punch," *New York Times Magazine,* 26 August 1956, pp. 16, 34, 38;

"Letter from the Olympics," *New Yorker,* 32 (8 December 1956): 153–157; (15 December 1956): 59–74;

"Lightweights in the Heavyweight Ranks," *New York Times Magazine,* 21 June 1959, pp. 18–20.

Being the son of a famous man is never easy. Since John Lardner was a humorist who wrote principally (though by no means exclusively) about sports, comparisons between him and his father, Ring Lardner, would be inevitable. Not the least of John Lardner's accomplishments was the way he established himself as a much-admired writer in his own way, not as one of the legion of Ring Lardner imitators but rather, as Grantland Rice says in *The Tumult and the Shouting* (1954), as "one of our finer writers, a worthy son of a brilliant father." When John Lardner died, A. J. Liebling wrote, in an unsigned *New Yorker* tribute, "John grew up in the shadow of a father who was a great writer. This is a handicap shared by only an infestimal portion of any generation, but it did not intimidate him. He made his own way."

When John Abbott Lardner, the eldest of Ring and Ellis Lardner's four sons, was born on 4 May 1912, his father was having the most difficult time of his early career. After leaving a Boston newspaper in November 1911, Ring returned to Chicago. At first the only job he could find was as a mere copyreader. At the time of John's birth the Lardners lived in a small apartment. It was not until the second son, James, came along, two years later, that they moved into a house, in the Chicago suburb of Riverside.

It must have been perplexing for Ellis Lardner to begin motherhood in a cramped apartment. Her family, the Abbotts, were people of wealth and social prominence. She grew up with every comfort and advantage. When she married Ring in 1911, she was a graduate of Smith College. Ellis was twenty-four, her husband twenty-six. The Abbotts could also boast of their American pedigree. George Abbott came to Massachusetts from England in 1642. More than two hundred years later, in 1862, William Abbott (John's great-grandfather) moved the family to Goshen, Indiana.

Although the Lardners' roots in American soil were not as deep as the Abbotts', they too had a long and distinguished history in the United States. Lynford Lardner left England for Pennsylvania in 1740. His son, the first American John Lardner, served under George Washington in the Revolutionary War; a later Lardner was an admiral during the Civil War. Like the Abbotts, the Lardners moved to the Midwest in the nineteenth century. In 1836 John's other great-grandfather, Henry Lardner, settled in Niles, Michigan. Unlike the Abbotts, however, the Lardners' prosperity did not continue into the next century. Although Ring (the youngest of nine children) grew up in luxury, by the time he was sixteen or so his father had lost most of his money. As Ring Lardner Jr. says in his family memoir, *The Lardners* (1976), "[Ring] went to high school from one of the area's wealthiest families and graduated in 1901 into genteel poverty."

However challenging it must have been for Ellis Lardner to handle straitened domestic circumstances, they did not affect her long. After covering the Cubs and White Sox for the *Chicago Examiner* for a year or so, Ring took over a popular daily column, "In the Wake of the News," for the *Chicago Tribune.* He frequently found items for this column in the activities of baby John (and the other boys as they came along). In general, these baby-inspired pieces were humorous, though some were appallingly sentimental (at least to later sensibilities). Poems about John and infant James were collected in *Bib Ballads* (1915).

John Lardner would sometimes try, for comic effect, to pass himself off as a former tough kid from the mean streets. For example, in a *Newsweek* column about Jim Norris's ludicrous offer to J. Edgar Hoover of a million dollars to run the malodorous International Boxing Club, Lardner said, "When I was a kid in Chicago, my mob, the Lake Rats, tried to get President Wilson to resign his job to front for us in the marbles racket" (19 January 1953). But Riverside, the suburb he lived in from age two through five, was an idyllic place to raise a family. In 1917, when Ring Lardner returned from a brief stint as a war correspondent in France, the family

*Lardner (second from left) at his favorite drinking spot, Bleeck's, a former Manhattan speakeasy known after Prohibition as the "Artist and Writers Restaurant" (courtesy of Susan Lardner)*

moved to the university town of Evanston, then moved again in the fall of 1918 to an apartment near fashionable Lake Shore Drive.

By all accounts, Ellis Lardner was a devoted and supportive mother. Having grown up with four sisters, she nevertheless seemed to enjoy presiding over an otherwise all-male household. Ring Lardner, perhaps surprisingly, was a doting papa. Indeed, one of his biographers, Donald Elder, calls him "an almost insufferably proud father." Every biographer agrees, however, that Ring was awkward in relating to his sons. Jonathan Yardley asserts in *Ring* (1977) that "They [Ring's sons] all inherited the Lardner reticence, and there was little in their upbringing to release them from it."

In 1919 Ring Lardner signed on with John N. Wheeler's Bell Syndicate to write a weekly column and gave up the daily newspaper grind. That year he moved his family east, to Greenwich, Connecticut. The move meant the Lardner boys would grow up almost literally surrounded by famous people from sports, journalism, literature, theater, and show business. After a year in Greenwich and some months in a hotel in Garden City, Long Island, the Lardners moved in the spring of 1921 into a large house in Great Neck, Long Island (the East Egg of F. Scott Fitzgerald's *The Great Gatsby,* 1925). Among their neighbors were Ring Lardner's brother Rex, a writer and editor; Herbert Bayard Swope, executive editor of the *New York World;* Franklin P. Adams (F.P.A.), the enormously influential columnist; and for a couple of years Scott and Zelda Fitzgerald. Fitzgerald, who engaged in some formidable drinking bouts with Ring, also found time for his sons. According to Ring Jr., "[Fitzgerald] told us stories and did card tricks for us and we felt he genuinely cared." Guests at the house in Great Neck when John Lardner was growing up included the wit and author Dorothy Parker, the composers Jerome Kern and George Gershwin, the theatrical impresarios George M. Cohan and Flo Ziegfeld, and the comedians Ed Wynn and Groucho Marx. As a result, Ring Lardner's sons grew up accustomed to the presence of celebrities in their lives.

When the Lardners moved farther out on Long Island, to East Hampton, in 1928, they had as neighbors Grantland Rice (Ring's closest friend) and his boss, John Wheeler. John Lardner would later write a syndicated column for Wheeler, too. Harold Ross, the legendary editor of *The New Yorker* for whom John would also write, was a regular member of Rice's "Sporting Club." The heavyweight boxer Gene Tunney, who abandoned his championship for more-cultivated endeavors the year the Lardners came to East Hampton, lived nearby. John met other celebrities from the sports world. Ring Jr. recalls how the Lardner boys astonished neighborhood kids when they produced a baseball for use in a sandlot game containing the signatures of the entire New York Yankees team. In a recent reminiscence, Ring Jr. says, in his foreword to *Ring around the Bases* (1992), edited by Matthew J. Bruccoli, "Sometimes one or more of my brothers would sit with [their father] in a box in the newly built Yankee Stadium and be introduced to the players, including Ruth and Gehrig, who came over to pay their respects."

The 1920s were an exciting decade for a future sportswriter moving into his adolescence: Babe Ruth and Lou Gehrig, Jack Dempsey and Gene Tunney, Earl Sande, Bill Tilden, Red Grange, Bobby Jones, and John Lardner's favorite golfer, Walter Hagen, subject of one of his best profiles. In "The Haig," published in *True* in 1959 and reprinted in *Press Box: Red Smith's Favorite Sports Stories* (1993), Lardner evoked the American sports world of the 1920s: "By land and sea, in airplanes and in Wall Street, the age of Walter Hagen was the age of gorgeous individualism and golden soloists." By his teen years John had seen many famous athletic contests. Perhaps his passion for football developed earliest. Along with baseball, boxing, and horse racing, football continued to be a favorite sport.

John and his brothers had plenty of opportunity to develop their own athletic skills. They had their own diamond and a tennis court that doubled as a hockey rink in winter on the spacious Great Neck property. When John was about thirteen, Ring hired a football coach to instruct his sons and neighborhood boys. But John was not particularly athletic. A lanky six feet two inches in adulthood, he was a chunky youngster, and his eyesight was very poor. Although all four brothers liked tennis and swimming, Ring Jr. depicts them as essentially sedentary children; reading seems to have been their favorite outdoor activity.

Both Ring and Ellis Lardner had prudish literary standards, but they left the boys to roam freely in their reading. In his introduction to a 1960 republication of Ring Lardner's 1916 novel *You Know Me Al,* John recalled how his parents fretted when he read a lurid novel called *The Plastic Age.* "But my father was even more worried when he caught me reading a baseball novel called *Won in the Ninth* . . . he warned me not to let my mind be soiled by corrupt observation of baseball procedures."

Even more than his brothers, John was a precocious writer. His first publication appeared in 1922, when he was ten. In his "Conning Tower" column Franklin P. Adams quoted a quatrain John had written:

> Babe Ruth and old Jack Dempsey,
> Both sultans of the swat.
> One hits where other people are,
> The other where they're not.

F.P.A. praised John's youthful literary efforts several other times in his columns. Raised in such a richly supportive writing environment, the Lardner boys blossomed early. As Ring Jr. says in *The Lardners,* "John, and to varying degrees the rest of us, thought of ourselves as writers at a very early age, set high standards and worked to live up to them."

John early showed a facility for light verse (a skill he would employ occasionally and effectively in his *Newsweek* columns), and when he was fourteen he wrote *U.S.S. Skinafore,* which Ring Jr. called in his memoir "an American version of *H.M.S. Pinafore.*" This show was performed by the four Lardner brothers and a female cousin for an appreciative audience consisting of Ring and Ellis and a few friends. Like his father, John Lardner had a lifelong interest in the theater: Ring was a playwright and songwriter; John wrote a fairly large amount of theater criticism.

In many ways John Lardner's childhood and adolescence appear ideal: loving, indulgent, supportive (if occasionally absent) parents; all the material comforts (by 1926 Ring was doing well enough as a freelance writer to give up journalism); the companionship of three brothers (all four of whom seem to have gotten on together uncommonly well); frequent trips to sporting events and Broadway shows; early acquaintance with some of the most interesting people in America. But in 1926, the same year he abandoned newspaper work, Ring Lardner was diagnosed with tuberculosis. His health declined steadily from that point to his death in 1933. Ring Lardner Jr. says that John predicted his own death would arrive at about the same age as his father's had. Ring was halfway to his forty-ninth birthday, John about six weeks short of his forty-eighth.

The year 1926 was also when John Lardner went off to Phillips Academy in Andover, Massachusetts, the first of the four Lardner boys to attend that institution. Since, like his brothers, John did

*Dust jacket for one of the 1951 collections of Lardner's* Newsweek *sports columns*

not complete college, Andover was a very significant part of his education. His record was good but hardly spectacular. He made the editorial board of the school literary magazine, *The Phillipian,* in his second year, and he was manager of the golf team. Following his graduation from Andover, John spent a desultory year at Harvard (1929–1930), then another year at the Sorbonne. By 1930 the Lardners, like nearly everyone else in the United States, were financially strapped, and for the first time Ellis's letters to her sons were full of admonitions to "Make the money go as far as possible." Perhaps the chief benefit of John's year at the Sorbonne was the opportunity to travel in Europe. During part of this period, Ring Lardner was in a sanatorium. Unable to work and beset by boredom and insomnia, he wrote John amusing letters, including one in rhyming doggerel beginning, "I thought you very deftly and wittily / Described your visit to [sunny] Italy."

When the 1931 school year was over, Lardner took a job at the Paris branch of the *Herald Tribune* for a few weeks. Returning to New York, he was hired by the *New York Herald Tribune.* At nineteen Lardner was about to begin his career. He moved into the first of a series of Manhattan apartments, sharing with various roommates at different times, including his brother Jim, his cousin Richard Tobin, and Thomas Sugrue, a young writer who collaborated with Lardner on his first book, *The Crowning of Technocracy* (1933).

At the *Herald Tribune* John Lardner worked under city editor Stanley Walker, writing general news stories. He had as instructive an apprenticeship as any young journalist could have. According to Ring Lardner Jr., Walker, an extremely demanding editor, "eventually reached the verdict that John 'came close to being the perfect all-around journalist.'"

John Lardner quickly showed some of the qualities he would develop in his later writing career – a gift for mixing fantasy into his reporting (for example, an account of an imaginary fox hunt through the streets of Brooklyn) and a delight in outrageous schemes. Ring Jr. claims that while at the *Herald Tribune* John was a participant in the transformation of an unknown Brooklynite into a celebrated Russian aristocrat. Years later John attempted to promote a boxing match between the eminent circus gorilla Gargantua and the fighter "Two Ton" Tony Galento.

Ring Lardner followed his oldest son's early career very closely and, adds Yardley, "with enormous pride." In a letter to his nephew Richard Tobin, who would soon join John at the *Herald Tribune,* Ring reported, "John has had five stories in the paper in the last three days and one of them each day carried a by-line. We are all swollen up like my ankles." In *The Lardners* Ring Jr. says that probably his father's last written communication was a telegram congratulating John for scooping a rival newspaper. Indeed, Ring and John Lardner may have achieved greater intimacy in Ring's last days than ever before. In her introduction to *Shut Up, He Explained* (1960), a collection of Ring Lardner's work, Babette Rosmond asserts that Ring, usually as reticent about his writing as about everything else, discussed it in detail with John shortly before he died. F. Scott Fitzgerald once consulted John about putting together a new collection of Ring Lardner's writing. Enthusiastic about the project, John wrote to Fitzgerald, "I'm a sucker for the guy's work and always have been," but he never edited such a collection. He did, however, write an appreciative introduction to the 1960 reissue of *You Know Me Al.*

spitter—2

than the spitball rule. And the Volstead Act is dead. The dry law in baseball may last forever.

In recent months, several things have been written and said about bringing back the spitball. There's an outburst of this kind of humanitarianism every couple of years—stimulated by the pitiful plight of the modern pitcher, as he bleeds and dies for his paycheck in a world he never made. No truly tender soul can watch his tortures unmoved. In the last fifteen years alone, your correspondent has heard pro-spitball sentiments uttered by Ford Frick, Branch Rickey, Billy Evans, Bucky Walters, Eddie Collins (though Eddie changed his mind back and forth), and fine old spitballers emeritus, like Frank Shellenback and Red Faber.

But the right people, the rulemakers, are not listening—even when the commissioner of baseball drops a not unmanly tear for the pitcher. Baseball rule-makers are like Society, in the rigid days that Edith Wharton wrote about: They never revise their judgments. They can't remember today exactly what was wrong with the spitball, that led to the 1920 law. But they're sure that something must have been wrong.

It wasn't a question of delicacy. The best memories agree that the public didn't give a damn when a pitcher juiced up his "wet one", as Jack Quinn used to call it, with tincture of slippery elm, sea moss, or tobacco. (Quinn himself, an austere type, used chewing-gum to spark his sputum; sometimes he wet the ball with perspiration only.)

"Who cared?" said Frick, talking about the hygiene of the spitter once. "The pitcher was out there on the field by himself. He just moistened the ball some."

The only voter who might care was the catcher—in line of duty. Branch Rickey caught a couple of spitballers, Smiling Al Orth and Harry Howell, when he backstopped briefly for the old N.Y. Highlanders in the 1900's. Orth threw a light spitter, like Quinn's, but Howell was what the boys courteously called a slobberer. Rickey says Howell's spitball was heavy to handle and hard for the catcher to throw. (The Brain holds the American League catchers' record for bases stolen off

more

*Page from Lardner's corrected typescript for his story "It Could Turn Baseball Upside Down"*

Walker was impressed by Lardner's work, too, and over the years his admiration grew. Reviewing *It Beats Working* (1947) in the *Herald Tribune* (21 September 1947), Lardner's first book about sports, Walker wrote, "Lardner is one sportswriter who can write. He has taste, judgment, knowledge, intelligence and style – an equipment which sets him above and beyond the great mass of his sportswriting colleagues."

During John Lardner's time at the *Herald Tribune,* in addition to his city room duties, he reviewed some sports books passed on to him by Stanley Woodward and made a good impression on him as well. They met a dozen years later in the South Pacific when both men were war correspondents. A few months after that, as he relates in his autobiography, *Paper Tiger* (1964), Woodward tried to hire Lardner to write a sports column for the *Tribune.* Among other prominent (or later to be prominent) writers at the *Tribune* during or close to John Lardner's tenure were Lucius Beebe, Walter Lippmann, Robert Fitzgerald, Homer Bigart (a mere copyboy in the early 1930s), and Joseph Alsop. At the *Tribune* John Lardner also met Hazel Hairston, whom he would later marry. She had been a reporter for several newspapers, but when Lardner met her she was Walker's secretary.

Every account of the last forty years of the *New York Herald Tribune* gives prominence to Bleeck's (pronounced "Blake's"), a speakeasy (later restaurant) located next door to the Tribune building. For decades it was at the center of activities at the *Tribune.* Ring Lardner Jr. emphasizes the hard-drinking environment in which his brother served his journalistic apprenticeship. "He took readily to a routine in which almost as many hours were spent in the speakeasy downstairs as in the city room or out on assignments." John Lardner and other regulars would play (and bet drinks on) a match game that involved outthinking (or outguessing) one's opponents. James Thurber was among the avid match game regulars. Others included the novelist John O'Hara and the drama critics Richard Watts and Wolcott Gibbs. Lardner's best friend, the cartoonist Walt Kelly, asserted in the preface to *The World of John Lardner* (1961) that Lardner liked the match game so much that one of his friends slipped three matches into his hand as he lay in his coffin. (Another put two dollars in his pocket.)

The year John Lardner left the *Herald Tribune,* 1933, was significant for him. He signed with John Wheeler's North American Newspaper Alliance (NANA) to write a daily syndicated sports column, beginning in 1934 and lasting until 1948; Lardner's local newspaper outlet was the *New York Post.* His salary began at $100 a week, a substantial amount in the depth of the Great Depression for a twenty-two-year-old with only two years of journalism experience. (Woodward lists Lardner among the excellent newspapermen who left the *Herald Tribune* "because they could not convince the top brass that they were worth paltry raises of five and ten dollars.")

Earlier in the year John, not yet twenty-one, published his first book, written in collaboration with Thomas Sugrue, *The Crowning of Technocracy* (or, as Ring Lardner fancifully called it, "The Crooning of Telegraphy"). Some of the characteristics of *Crowning* are evident throughout Lardner's work: a skeptical view of history, a derisive attitude toward self-righteous ideologues, scorn for manglers of the English language, a love of parody and burlesque, a flair for humorous light verse. It would be absurd to invest this forgotten little book about a forgotten movement with great significance. Reading it, however, one may be reminded of Red Smith's remark in his obituary of John Lardner: "I think the only people he disliked were the self-important and the bad writers." The Technocrats – a pompous band of scientists and engineers who believed that they had the answers to all America's social and economic ills – qualified for Lardner's dislike on both counts.

Another interesting aspect of *The Crowning of Technocracy* is the way the authors made incongruous references to sports in contexts having nothing to do with athletic contests. For example, they wrote that, thanks to Technocracy, "The Eskimo, sitting before his primitive hut within the Arctic Circle, was able to hear how they finished at Saratoga [Racetrack]." In their list of "Who's Who in Technocracy," they included "COBB, TYRUS R." In his mature work Lardner delighted in treating non-sporting events as if they were athletic contests. For instance, he wrote a hilarious account of the marriage of Grace Kelly and Prince Rainier, "Docker the Knocker," in which the various female journalists covering the event were described as if they were engaged in a fierce free-for-all.

The final memorable event of the year 1933 for Lardner was the repeal of Prohibition in December. Lardner was both a serious and humorous student of drinking customs in the United States. Ring Lardner Jr. reports that his brother wrote an unpublished history of Prohibition. This may have been a draft of a chapter in the unfinished "Drinking in America," on which Lardner was working at the time of his death. He had completed seven chapters, plus a brief introduction, bringing his chronicle only up to 1862.

While writing his daily column for Wheeler's syndicate, Lardner began dating Hazel Hairston, a Texan who had come to New York after a divorce in 1930. According to Ring Lardner Jr., she had dated him and Jim Lardner before she started dating John. In fact, Ring Jr. says, by 1937 people thought of her as Jim's girlfriend. Nevertheless, she and John married in September 1938. Their first child, Susan, was born in 1939; two more children, Mary Jane and John Nicholas, followed at two-year intervals. The year 1938 was important for John Lardner in other ways, too. He made his first significant freelance magazine sale in April, an article about the 1919 World Series, to *The Saturday Evening Post.* One week after Lardner's marriage, his brother Jim was killed fighting in the Spanish civil war, one of the last members of the International Brigade to die in combat; he was twenty-four.

Later that year John Lardner was hired by *Newsweek* to write a weekly sports column beginning in March 1939. For nearly the next decade he would turn out both a daily piece for the NANA and a weekly one for the magazine. His *Newsweek* columns, under the title "Sport Week," composed the single main body of his sports writing. For the next twenty-two years, with approximately two years out as a war correspondent, he wrote weekly short essays of seven hundred to eight hundred words on a wide variety of sports, on various aspects of sports, and on topics beyond the conventional boundaries of sports. He wrote frequently on the economics of baseball, on corruption in boxing, on the effects of television on sports. His broad conception of the sporting life included gambling.

Like the lives of so many millions of other Americans, Lardner's was disrupted by World War II. In March 1942 he suspended his regular columns to write war dispatches for *Newsweek* and NANA. ("Sport Week" temporarily became "Under Fire.") For about two of the next three years he wrote about the war from Australia, North Africa, Italy, and the South Pacific. Besides his *Newsweek* and syndicate copy, he wrote longer articles for other magazines, primarily *The Saturday Evening Post* and *The New Yorker.* In effect, Lardner was preparing for the next phase of his career, as a freelance writer for magazines.

In his 28 June 1954 tribute to Lardner, his *Newsweek* publisher said, "the effective and telling manner in which he portrayed the life of the soldier overseas brought home the human side of the conflict." Far more impressive was Lardner's ability to handle straight war reporting, as shown in two articles reprinted in *The World of John Lardner.* "D Day, Iwo Jima," published in *The New Yorker* in 1945, shows his mastery of clear narration. So does "The Rangers' Beachhead Alamo" (*Newsweek,* 1944), a grim, concise account of a disastrous engagement in which two battalions of Rangers were wiped out. His friend, fellow war correspondent, and colleague-to-be at *The New Yorker,* A. J. Liebling, compares Lardner's combat reporting to the battlefield descriptions in John William De Forest's Civil War novel of 1867, *Miss Ravenel's Conversion from Secession to Loyalty,* but readers of Lardner's war dispatches may also be reminded of Stephen Crane. Of Lardner's conduct under fire, Liebling says, "John was naturally brave; when he saw blinding bomb flashes by night, he used to walk *toward* them to see better" (in part, presumably, because his eyesight was so poor). Lardner experienced both German V-1 rockets at Anzio and Japanese kamikaze attacks. On Iwo Jima, Lardner was wounded slightly in the groin.

In between war assignments Lardner resumed his "Sport Week" column. In 1943 he collected the articles he had written for *The Saturday Evening Post* from Australia and New Guinea to form *Southwest Passage: The Yanks in the Pacific,* one of the first books by an American World War II correspondent. The next year, besides "Sport Week," Lardner wrote theater reviews for *The New Yorker,* filling in for his youngest brother, David, who went to Europe as a correspondent.

In the three years he had been at *The New Yorker,* writing about sports, movies, and the theater, David had shown himself to be yet another talented Lardner. William Shawn is quoted in *The Lardners* comparing John and David as writers: both had "discipline, tautness, control and a general similarity of humor." By the early 1940s the two brothers had become quite close, playing poker together, drinking at Bleeck's, visiting their mother in Connecticut. Ring Lardner Jr. notes that although John and David might not have been as much alike as some people thought, they adjusted to the semi-itinerant, catch-as-catch-can life of the working journalist much more easily than the other two brothers. In October 1944, while John Lardner was preparing to depart for the Pacific, David was killed in Germany. He was twenty-five, the second Lardner brother to die young.

John Lardner sailed to the South Pacific in December 1944. For the next four months he island-hopped with the advancing American forces. He landed on Iwo Jima within two hours of the first as-

*Dust jacket for the 1951 collection of Lardner's articles on heavyweight boxing in the United States*

sault. He came under fierce mortar fire in an area of extremely high casualties. By mid April 1945, Lardner was heading home. "Sport Week" began again in June of that year. But in August Lardner went out to California to share an apartment with his recently divorced, sole-surviving brother. Ring Jr. had already established himself as a major Hollywood screenwriter, an Academy Award winner by age twenty-seven. He and John collaborated on a screenplay that was never produced. Although, according to Ring Jr., neither brother was able to break through the family reserve completely, their two months of working together on the script brought them "closer than we had ever been before." In November 1945 John Lardner returned to New York City and resumed his "Sport Week" column at *Newsweek.*

For years Lardner's column included seasonal "Fearless Football Forecasts." Ring Lardner Jr. says that as a teenager John had teamed up with his father to handicap and bet on college football. In his *Newsweek* predictions he consistently averaged over 70 percent correct. For the 1952 and 1953 bowl games he made his picks in humorous quatrains. In "A Child's Garden of Bowls" he began by stating, "A couple of years ago, we noticed that while our forecasting was good, our children were very bad" (*Newsweek,* 29 December 1952). Thus, in imitation of Robert Louis Stevenson's *A Child's Garden of Verses* (1885), he not only named the winners-to-be but also passed on "some words of firm but friendly advice to children." For example,

> U.S.C. has too much power
> For Wisconsin's weary forces.
> (Kids who carry tales to teacher
> Often wind up touting horses.)

(The Trojans did win the 1953 Rose Bowl.)

The *Newsweek* column underwent a title change – from "Sport Week" to "Lardner's Week" – on 14 December 1953. Liebling offers perhaps the best description of the face in the column's photograph: "handsome, grave, and equable, only the corners of his mouth, and the eyes behind the thick lenses, betraying . . . his private amusement with what he thought about." On 28 June 1954, about six months after the column became "Lardner's Week," the publisher of *Newsweek,* Theodore F. Mueller, devoted his "For Your Information" column to Lardner. Rather belatedly, Mueller announced that for some time Lardner's "candid commentary has been broadening in scope. And now that he has decided to point his pen in all directions, readers can expect an even wider range of originality and humor."

The "Lardner's Week" columns show plenty of the same originality and humor as the pieces written for "Sport Week," but the earlier columns were equally broad in their scope. The first installment of "Lardner's Week" was an amusing account of a former fighter named Al Ettore who was suing over the edited version of a film showing him knocked out by Joe Louis seventeen years earlier. Both the legal aspects of sports and the relationship between sports and the broadcast media were central concerns throughout Lardner's career. The 28 June 1954 column described a recent heavyweight championship match in which Ezzard Charles managed to last fifteen rounds against the great Rocky Marciano (and earned $123,000, "or roughly $1 for each new lump on his face").

Mueller quoted Lardner as saying that he favored the new column title because it might help to convince people that there was no "fence between sports and the rest of the world." Two years earlier, in a humorous ad for *Strong Cigars and Lovely Women* (1951), a selection of his *Newsweek* columns, Lardner wrote that the book was certainly about sports, "though it seems to have boiled over a bit and gotten into such subjects as television's Mr. Kefauver and Hopalong Cassidy" (*New Yorker,* 10 September 1951). Lardner reminds one again and again, directly and indirectly, that sporting events are part of the real world and that the real world (the world of television producers, owners of baseball clubs, congressional committees, gamblers, and fixers) intrudes on and affects sports all the time.

As Woodward observes in *Sports Page* (1949), the seven-hundred-to-one-thousand-word column is a very difficult literary form. He singles out Lardner as one sports columnist who could handle that form, calling him a writer "who specializes in humor which is based largely on incongruity." Humor and sports, sports and humor – the two cannot be separated in any discussion of John Lardner's work. Was he a humorist who usually wrote about sports or a sportswriter who was habitually very funny? Clearly, he was both. And, like all successful humorists, Lardner made it look easy. However, Roger Kahn, his editor at *Newsweek* from 1957 on, says in *The World of John Lardner* that Lardner did not compose easily. "It took him a long time to bring each story up to his usual standard and a longer time to build that standard up to what it was." Another of his editors, Al Silverman, says of the work that Lardner submitted to *True* magazine in the 1950s, "When his pieces came in there was nothing to be done with them except perhaps double-check a fact or two." As Ira Berkow notes, the title of Lardner's first collection of sports articles, *It Beats Working,* was comically inappropriate. Lardner was legendary for his dedication; no sportswriter worked harder at his craft.

Although the work was consistently of high quality, Kahn is surely right in his assertion that Lardner's writing after World War II was generally better than before the war. One could even argue that Lardner just kept getting better and better. Some of his very best articles appeared in 1959 and 1960. His last *Newsweek* column (21 March 1960) was a joyful and entertaining piece looking forward to the baseball season, gently mocking Casey Stengel's insistence that every American League team had improved, less gently taking Phil Rizzuto to task for criticizing Mickey Mantle for presuming to hold out. From first to last, in his low-keyed way, Lardner disliked yes-men and toadies; in this final piece he said that "Rizzuto works for the Yankee office more completely than he did as a player."

The last fifteen years of Lardner's life, from 1945 to 1960, were richly productive. Besides the syndicated column for NANA, which he maintained until 1948, and the *Newsweek* column, Lardner began writing regularly for *The New Yorker,* sometimes filling in as drama critic. During his last couple of years Lardner's freelance output slowed, but he wrote a more or less regular television and radio column for *The New Yorker,* "The Air." Once more he paralleled his father's career: Ring had written a radio column for that magazine for the last fourteen months of his life.

For six months or so in 1952, Lardner wrote a weekly column for *Look* magazine, "John Lardner's New York," mostly about the theater. He was a frequent contributor to *Sport* magazine in the late 1940s and early 1950s, and for several years in the 1950s he contributed a column, "It Happened in Sports," to another monthly, *True.* In addition, he wrote articles for the Sunday *New York Times Magazine* from the late 1940s on. Lardner also published in *The Saturday Evening Post,* where he had placed his first significant freelance piece, sporadically throughout the postwar era. Altogether, Ring Lardner Jr. estimates that his brother's work appeared in more than twenty magazines, including such improbable publications as *Nation's Business* and *Woman's Home Companion* (an account of actress

Tallulah Bankhead's visit to the Polo Grounds to see the New York Giants play).

About the only popular magazines of consequence for which Lardner did *not* write were those published by Henry Luce (*Time, Life, Sports Illustrated*). This might have been because he was so closely associated with *Newsweek*, the archrival of *Time*. But Kahn is probably correct in detecting a protest against Luce's concept of "group journalism." Kahn notes in *The World of John Lardner* that when *Sports Illustrated* started in 1954 – a magazine tailor-made for Lardner's talents – "the protest became costly. Lardner stuck to it as a matter of integrity; he guarded his integrity in a fierce, uncompromising way." Silverman, in the introduction to *The Twentieth Century Treasury of Sports* (1992), adds that Lardner was "an individualist who was the best answer to the group journalism the newsmagazines practiced then, and still mostly do today."

In 1948 the "Press" section of *Time* announced that Lardner had taken on yet another job, drama critic for the *New York Star*, the short-lived successor to the liberal daily *PM*. Although it was written with obvious Red-baiting malice (yoking John to his brother, who was in serious trouble for his left-wing political views and already blacklisted in Hollywood), the author of the piece acknowledged not only John Lardner's prolificacy but also his writing ability. By the mid 1950s, says Ring Lardner Jr., his brother was "one of the best-known and respected journalists in America." By the late 1950s, Kahn adds, "Lardner occasionally had to beat off editors with bottles of club soda lest he find himself writing stories he didn't like, or didn't have time for, or couldn't afford." Lardner was a modest, reticent man, but he did not lack confidence in his abilities as a writer. As early as 1941 he said in a letter to his NANA boss, John Wheeler, that he believed his syndicated sports column was "the best of its kind in the country."

Lardner's magazine articles won places in anthologies of sports stories year after year. He was a regular in the *Best Sports Stories* annuals edited by Irving T. Marsh and Edward Ehre for Dutton. The collection of 1960 stories in that series was dedicated to his memory, and one of his articles was chosen for the best-of-best collection, *Thirty Years of Best Sports Stories* (1975). Lardner is also represented in such collections as the first three volumes of Charles Einstein's *Fireside Book of Baseball* (1956, 1958, 1968), *Press Box: Red Smith's Favorite Sports Stories* (1976), Al and Brian Silverman's *Twentieth Century Treasury of Sports,* and Dick Schaap and Mort Greenberg's *Joy in Mudville* (1992).

Many readers will prefer Lardner's twenty-five-hundred-to-six-thousand-word features to his much shorter columns. In these longer pieces he had more room to pursue odd twists and angles, to add amusing anecdote to amusing anecdote. "Mr. Henderson and the Cooperstown Myth," first published in *True* in 1950 and reprinted in *Thirty Years of Best Sports Stories* in 1975, celebrated a "spare, gray, quiet fellow in steel-rimmed glasses," head of the main reading room in the New York City Public Library. Mr. Henderson had taken on the baseball establishment, gently but firmly – and with mountains of evidence – refuting the myth of Abner Doubleday's invention of baseball in a single day. In his persistence, said Lardner, Mr. Henderson was "like a bloodhound in spectacles." The essay is very funny, but it is also central to Lardner's overall point of view; he was always against the bureaucratic bullies and, without sentimentalizing, in favor of the spunky little guys.

"Babe Herman" (*Sport,* 1952) exemplifies another major Lardner theme – joy in the individual, however zany or downright eccentric. "It is said," Lardner wrote, "that [Herman] broke the hearts of 45 per cent of [his managers]. The rest avoided cardiac trouble by getting rid of the Babe as fast as they could." Again Lardner avoided sentimentality in his account of the heavy-hitting, woeful-fielding, erratic-baserunning Herman, this time by subverting various baseball clichés. For example, after mentioning Herman's comeback during World War II at age forty-two, Lardner added, "Right to the end of his playing days the Babe retained his fresh young affection for cash money." Every aspiring sportswriter should read, reread, and memorize Lardner's admirable two-paragraph account of Herman's most egregious moment – when he hit the double that he (with some help from Dazzy Vance) turned into a double play:

> The bases were full of Brooklyns, with one out, when the Babe strode to the plate on that occasion, swinging his bat like a cane in his right hand. Physically, he was a phenomenon, a left-handed hitter with most of his power in his right arm. Scattered around the landscape before him were Hank DeBerry, the Brooklyn catcher, on third base; Dazzy Vance, the Dodger fireball pitcher, on second; and Chick Fewster, an outfielder, on first. Mr. Herman swung ferociously and the ball hit the right-field wall on a line. DeBerry scored. Vance, however, being a man who did not care to use his large dogs unnecessarily, hovered between second and third for a moment on the theory that the ball might be caught. When it rebounded off the wall, he set sail again, lumbered to third base, and made a tentative turn toward home. Then, deciding he couldn't score, he stepped

> back to third. This move confounded Fewster, who was hard on Vance's heels. Fewster started back toward second base. At that moment, a new character, with blond hair and flapping ears, came into their lives.
>
> Mr. Herman has described himself as a great runner. What he meant was, he was a hard runner. He forgot to mention that he ran with blinkers on, as they say at the race track. He concentrated on running and ignored the human and animal life around and ahead of him. Passing Fewster like the Limited passing a whistle stop, the Babe slid into third just as Vance returned there from the opposite direction. Herman was automatically out for passing Fewster on the baseline, though nobody realized it at once but the umpire, who made an "out" sign. The third baseman, not knowing who was out, began frantically to tag Herman, who was already dead, and Vance, who stood perfectly safe on third base.

Not all of Lardner's magazine articles focused on a single individual; some were concerned with an issue. Kahn reprinted "The Great Spring Training Nonsense" (*Saturday Evening Post,* 1953), a detailed analysis of the economic realities of spring training, which ended with the confident (and quite incorrect) prediction that baseball teams would shorten their training period soon.

In one of Lardner's funniest pieces, "Never Trust an Amateur" (*Sport,* 1957), he took the platitudes about sportsmanlike amateurs and win-at-all-cost professionals and turned them topsy-turvy. Taking off happily from a report of an investigation of cheating at golf courses in the New York City area, Lardner asserted that both amateurs and pros cheated. However, "being pros, these boys do not sing just for exercise, like barroom tenors. They repent at $1 a word in national magazines." On the other hand, "no amateur, from Athenian days to the time of Wes Santee, has ever confessed." This article also contains excellent specimens of Lardner's versatility – a parody of Grantland Rice's "Alumnus Football" (it's "not that you won or lost – but how you played the Game"); some apt quotations from Burt L. Standish's novels, showing that Frank Merriwell knew a thing or two about cheating; a dialogue between a suspected golf cheater and a judge, worthy of comparison to one of Ring Lardner's celebrated nonsense plays.

John Lardner's deteriorating health weirdly mirrored his father's decline. Both men were diagnosed as having tuberculosis when they were about forty. Both suffered from heart disease and died as the result of heart attacks at around forty-eight. John bore the additional burden of multiple sclerosis, which limited his mobility during his last few years. Walt Kelly, in his preface to *The World of John Lardner,* also hints ominously of "personal troubles too private for anyone else to reveal" in Lardner's life. Whereas Ring Lardner's failing health made it increasingly difficult for him to write, however, John's work was seemingly unaffected by his afflictions. Certainly, his customary wryly cheerful tone was not darkened by his health problems. Like his father, John Lardner was primarily a newspaper and magazine writer rather than a book writer, a journalist, not an "author." But in that fertile period following the World War II three collections of his work were published; a fourth appeared the year after he died.

*It Beats Working* contains a foreword by historian Carl Van Doren and illustrations by a famous sports cartoonist, Willard Mullin (best remembered for his depictions of the Brooklyn Dodgers "Bum"). Lardner himself was frequently literary, without ever being bookish, and like a good cartoonist he exaggerated aspects of his subjects, but never grotesquely. *It Beats Working* consists of material from "Sport Week," revised and edited for continuity. Lardner's third book of reprinted sports articles was *Strong Cigars and Lovely Women.* This time the illustrations were by his friend Kelly, famous for the comic strip *Pogo.* Like *It Beats Working, Strong Cigars and Lovely Women* is made up of *Newsweek* columns revised and edited. The in-between collection, *White Hopes and Other Tigers* (1951), makes a more coherent volume than the other two. For one thing, it has a single focus – heavyweight boxing in the United States from 1910 to 1930, an era in the "sweet science" that Lardner regarded fondly. The book collects a series of articles Lardner published in *The New Yorker* from December 1947 to March 1951.

In 1947 Lardner bought a summer home on Fire Island, less than two hours from his Manhattan apartment. For the next five years or so, he spent most of the summers there with his wife and three children. From the summer of 1952, however, illness and perhaps his increasingly busy work schedule kept him in the city most of the time. He died in his apartment on West Twelfth Street in Manhattan's Greenwich Village on 24 March 1960, after yet another heart attack. According to Ring Lardner Jr. in his family memoir, as two doctors worked on John Lardner, one, Dr. Louis Siltzbach, begged him to hang on. "You're a noble human being," the doctor said. Low-keyed to the last, Lardner replied, "Oh, Lou, that sounds like a quotation." Soon after, he lost consciousness and died.

The tributes following Lardner's death were so obviously heartfelt that they are still moving decades later. Liebling, who had known him well, re-

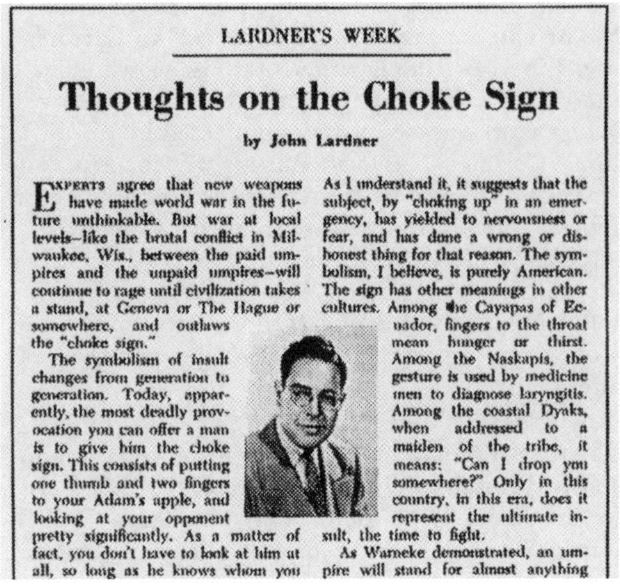

LARDNER'S WEEK

Thoughts on the Choke Sign

by John Lardner

EXPERTS agree that new weapons have made world war in the future unthinkable. But war at local levels—like the brutal conflict in Milwaukee, Wis., between the paid umpires and the unpaid umpires—will continue to rage until civilization takes a stand, at Geneva or The Hague or somewhere, and outlaws the "choke sign."

The symbolism of insult changes from generation to generation. Today, apparently, the most deadly provocation you can offer a man is to give him the choke sign. This consists of putting one thumb and two fingers to your Adam's apple, and looking at your opponent pretty significantly. As a matter of fact, you don't have to look at him at all, so long as he knows whom you As I understand it, it suggests that the subject, by "choking up" in an emergency, has yielded to nervousness or fear, and has done a wrong or dishonest thing for that reason. The symbolism, I believe, is purely American. The sign has other meanings in other cultures. Among the Cayapas of Ecuador, fingers to the throat mean hunger or thirst. Among the Naskapis, the gesture is used by medicine men to diagnose laryngitis. Among the coastal Dyaks, when addressed to a maiden of the tribe, it means: "Can I drop you somewhere?" Only in this country, in this era, does it represent the ultimate insult, the time to fight.

As Warneke demonstrated, an umpire will stand for almost anything

*Column head for Lardner's* Newsweek *column (15 August 1955)*

ferred to Lardner's reserve and graceful understatement, both as a man and as a writer, and to his courage and integrity. Liebling also provided perhaps as good an epitaph as Lardner could have wanted: "humorist, reporter, sportswriter, critic." Acknowledging that he had not known Lardner as well as some of his peers, Red Smith nevertheless said, "I never encountered him without a thrill of pleasure and I rejoiced in his company." Smith called Lardner's death "a loss to the living, to everyone with a feeling for written English handled with respect and taste and grace." The finest tribute of all appeared some sixteen months after Lardner's death, in *The World of John Lardner.* In between a loving preface by Kelly and a perceptive (and loving) epilogue, Kahn provided a rich selection of the best of Lardner's work, emphasizing his versatility as a writer. In fact, Kahn deemphasized Lardner's sportswriting to display his wide range of interests and his formidable storehouse of knowledge. The editor also selected nonhumorous pieces, ranging from a fairly long profile of Charles Lindbergh to the grim account of an American military disaster.

The book, called by an anonymous *Newsweek* reviewer a "solid gold anthology" (7 August 1961), is divided into five sections of varying lengths: "People" (the longest part), "TV and Other Phenomena," "A Variety of Sports," "War Correspondence" (the shortest part), and "Drinking in America." Placing the sports articles squarely in the middle of the volume enabled Kahn to underscore the central position of sportswriting in Lardner's career; however, allotting only about forty-five pages (one-fifth of the total) to these pieces suggested that Lardner was more than just a sportswriter. As Kahn puts it in his epilogue, "This book illustrates some of the range and sweep of a writer looking about with wit, taste, subtlety, and sometimes with horror."

Just as sporting events intrude on other aspects of life and life most certainly intrudes on sports, so Kahn's grouping of material emphasized Lardner's refusal to separate sports from "real life." In the first section, along with profiles of a legendary gambler and a Nazi war criminal, Kahn placed accounts of two boxers who led turbulent lives ending in early violent deaths – Stanley Ketchel ("Down Great Purple Valleys," *True,* 1954) and Battling Siki ("Battling Siki," *The New Yorker,* 1949, reprinted in an expanded version in *White Hopes and Other Tigers*). The latter sketch illustrated Lardner's ability to be literary without being pretentious. Noting the various patronizing labels attached to the African fighter ("jungle child," "natural man"), Lardner seized on "the black Candide" and proceeded to draw a brief but brilliant comparison between Voltaire's hapless hero and the wild man gunned down in Hell's Kitchen in 1925. Whereas Candide survived the Inquisition, cannibals, and other perils to wind up safely cultivating his garden, "Siki's story is perhaps more realistic. He failed to last out the course." The second section of *The World of John Lardner* consists of six brief articles, four from Lardner's "The Air" column in *The New Yorker* and two from "Lardner's Week." Besides "Docker the Knocker," Kahn reprinted a trifle called "Dixie Deposit" (*Newsweek,* 1957), in which Lardner treated southern speedtraps as a kind of competitive sport pitting rural communities against Yankee motorists trying to get to Florida.

Obviously, sports fans will be most interested in "A Variety of Sports." This central section includes "No Scar, No Memory" (*Newsweek,* 1957), a textbook model account of a prizefight in which the great Sugar Ray Robinson knocked out Gene Fulmer with a classic left. But Kahn also placed "Death of a Simian and Scholar" (*Newsweek,* 1949), Lardner's humorous yet moving tribute to Gargantua, the famous circus gorilla, in this part of the book. In this extraordinary obituary Lardner used a couple of his own traits to liken himself to the "genial gorilla": "We had little to say to each other, both being of a reserved, introspective turn of mind." Of the longer pieces, in addition to "The Great Spring Training Nonsense" and "Babe Herman," Kahn chose "The Roller Derby" (*Sport,* 1949), "Cockfight-

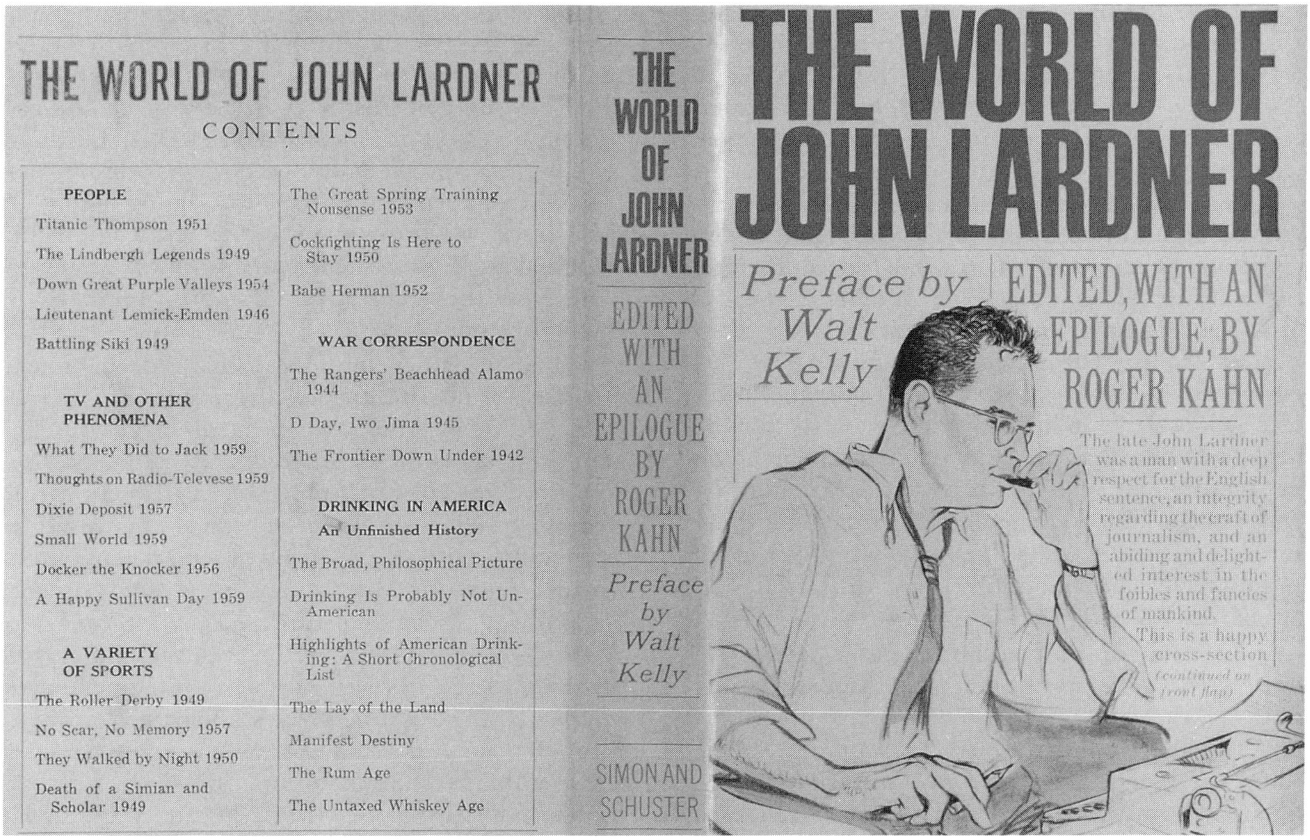


*Dust jacket for the posthumously published 1961 collection of Lardner's writing about sports and other subjects*

ing Is Here to Stay" (*Sport,* 1950), and "They Walked by Night" (*Sport,* 1950). The first two indicate the vast range of Lardner's knowledge of sports. "The Roller Derby" (also reprinted in the Silvermans' *Twentieth Century Treasury of Sports*) is an account of the growing popularity of that dubious activity, focusing on its inventor and promoter, Leo A. Seltzer (the kind of hustler always dear to Lardner), but giving attention to the diminutive, bellicose superstar of Roller Derby, Miss Marjorie Clair (Toughie) Brashun, too. With his usual interest in the relationship between sports and television, Lardner also brought out the way in which TV in effect created such sports as Roller Derby and professional wrestling (of the Gorgeous George – not Strangler Lewis – variety).

"Cockfighting Is Here to Stay" exemplifies Smith's point that Lardner was not only a good writer but also a "meticulous and thorough" researcher. In fewer than three thousand words he provided a basic education for readers who knew nothing about this unsavory pastime – breeding and training gamecocks, equipment, rules, and conditions under which cockfights were staged. Reaching out to another area he had some interest in, Lardner observed, "The status of cockfighting today is roughly what the status of liquor was during the Volstead Act: You could get it (they tell me) if you looked for it." Lardner treated the frenzied birds with respect, admiration, and amusement, rather as if they were feathery Toughie Brashuns.

"They Walked by Night" is Lardner's affectionate chronicle of those baseball players who sought to slake their thirst nightly. Beginning with Shufflin' Phil Douglas, the Giants pitcher who was banned from baseball in 1922, Lardner brought in several other players who shared Douglas's "single-minded thirst" and had "feet that [pointed] away from home." Readers in later, more straitlaced times may be perplexed by Lardner's fondness for "The era of the night-walking ballplayer." Clearly, he would have been amused by and disdainful of current sportswriters who deplore athletes taking the odd drink (or two).

Lardner was always interested in the subject of drinking, in practice as well as theory. In addition to his unfinished "Drinking in America," which echoed material from "They Walked by Night," Lardner frequently drew on drinking for sports analogies (as in "Cockfighting Is Here to Stay") and even treated drinking as if it *were* a sport, or at least a competitive game. "Drinking cheap whiskey," he wrote in "Drinking in America," "was . . . treated in some circles as an extension of athletics." Lardner once picked an all-time alcoholic all-star baseball team, starting with Babe Ruth in right field.

*The World of John Lardner* was more widely reviewed than Lardner's previous books, and the reviews were unanimously favorable. Indeed, the chorus of praise was like a second round of death tributes. Once again, his old *Tribune* editor Stanley Walker led the cheers, summing up in *New York Herald Tribune Books,* (6 August 1961) Lardner's virtues as a writer: "In his stuff there was no straining for bogus theatrical effect, no phony excitement, no nauseating breeziness, and no tricks – that is, no tricks except the wonderful trick of telling a thing simply and clearly with just enough wry overtone to supply a pleasant and unobtrusive touch of flavor, or character, or style."

Certainly Kahn's disclosure, in the epilogue to *The World of John Lardner,* that Robert Frost admired Lardner's humor is worth noting, not merely because of Frost's prestige but more importantly because the great poet and John Lardner had some things in common. Besides a wry sense of humor, they shared a tendency toward understatement, a dislike of pretentiousness, a self-effacing manner, and, in their different fields of literary endeavor, each set a high standard of excellence. In his headnote to the Lardner selection in *The Baseball Reader* (1983), Charles Einstein compares the difficulty of choosing among Lardner's articles to a pitcher having to decide which of his offerings to throw Jimmy Foxx: "Lardner, like Foxx is good for extra bases any time."

Perhaps Kahn describes Lardner's craftsmanship best: "After the painstaking business of reportage his craft was purely writing: writing the English sentence, fusing sound and meaning, matching the precision of the word with the rhythm of the phrase." About a former Brooklyn Dodgers pitcher who confessed in print that he had bent the rules of baseball during his career, Lardner wrote: "Preacher Roe, after consulting his conscience and his checkbook, made it known that he had distributed more illegal dew in his time than Al Capone."

Lardner's craftsmanship included a passion for language precision. "He considered bad writing a hanging offense," writes Smith. From the early *Crowning of Technocracy* to the end of his life, Lardner seldom passed an opportunity to chide or mock those who misused the English language. In "No Scar, No Memory," his account of the 1957 Robinson-Fulmer fight, he spent an entire precious paragraph clarifying just what a left hook is; sportswriters, he asserted, tended to call every left-hand punch a hook. In one of his late *New Yorker* columns, "Thoughts on Radio-Televese" (1959), Lardner took various commentators and sports announcers (even the revered Red Barber) to task for sloppy or inaccurate word usage. In a *Newsweek* column he criticized *Bartlett's Familiar Quotations* for incorrectly identifying the origin of the immortal sentence "I should of stood in bed."

Like his father, John Lardner was deeply interested in the possibilities of distinctly *American* English. Although John Lardner hardly ever wrote in anything that could be called dialect, he knew a lot about the varieties of American English. In the 1 December 1951 *New Yorker* he "edited" (ghostwrote or co-authored) a "Dictionary of American Underworld Lingo" for Morris Lipsius, a shady figure best known for having informed on the gangster Waxey Gordon. Lipsius belongs to that gallery of thirsty athletes, elusive gamblers, pitchmen, and conmen of all sorts who always fascinated Lardner. (Probably no American author has written more about dubious characters and the shady side of sports than Lardner.) In the 23 February 1953 *Saturday Evening Post* Lardner collaborated with Lipsius again, to write "I Put the Finger on Waxey Gordon." (In his next-to-last *Newsweek* column [4 March 1960], Lardner wrote a memoir of Gordon.)

Lardner's affection for rogues and rascals never caused him to sentimentalize them. Kahn mentions in his *The Era: 1947–1957* (1993) that near the end of his life Lardner became annoyed by all the fulsome praise heaped on Casey Stengel – long a favorite of his – particularly "the enraptured accounts of a saintly Casey" by Arthur Daley of *The New York Times.* He never urged his readers to weep for Phil Douglas, his career in ruins, or for Stanley Ketchel, murdered at twenty-four. As Kahn says in the epilogue to *The World of John Lardner,* there are no Iagos in Lardner's work. But there are many Falstaffs, and he treated them with the same kind of sympathetic amusement that Shakespeare brought to Sir John. Lardner's tolerance for human frailty (his own included) was vast; it covered just about everyone but the self-righteous, the pompous, the bureaucratic, and, of course, the intolerant. Among his favorite characters was Bill Veeck. Lardner frequently celebrated Veeck's generosity, expansiveness, and flamboyance, contrasting him to such miserly lords of baseball as Connie Mack, Clark Griffith, and Branch Rickey, men, as Lardner said, who always looked for ballplayers who "can't count higher than $5,000" (*Newsweek,* 14 February 1949). In their preface to *Thirty Years of Best Sports Stories,* Marsh and Ehre pointed out how much sportswriting had changed in the three decades spanned by their anthologies (1944–1974). They saw a shift from "the hero-worship type of writing to the critical and even cynical type." Lardner was certainly critical but never cynical, generally amused and never awestruck.

Franklin P. Adams died one day before Lardner. The "Talk of the Town" column in the 2 April 1960 issue of *The New Yorker,* recalling F.P.A.'s benign influence on younger writers, said Adams's heyday "was a fine era [whose] likes are not known today." The same issue carried Liebling's obituary of Lardner. Lardner, along with Red Smith, surely came as close as anyone of his era to fulfilling the high standards of the days of F.P.A., H. L. Mencken, and Ring Lardner, when a column or article was not just journalism but also literature.

It seems appropriate that Lardner was working on a tribute to F.P.A. when he died. In his final days, in addition to working on "Drinking in America," Lardner confided to Kahn that he was considering a novel set around the time of the 1919 championship fight between Jess Willard and Jack Dempsey. According to Kahn in *The Era,* though, "as a gifted essayist, John resented America's worship of the novel."

Although his books are out of print, Lardner, is still highly regarded by those who appreciate journalistic excellence. Lardner persists in another way, too: through his name. Among the various Lardners of the next generation (and the next) who have become journalists is his oldest child, Susan, a versatile freelance writer who has contributed film and book reviews, as well as feature articles, to *The New Yorker* – the third generation of Lardners to publish in that magazine.

**References:**


Ira Berkow, *Pitchers Do Get Lonely* (New York: Atheneum, 1988);

Donald Elder, *Ring Lardner: A Biography* (Garden City, N.Y.: Doubleday, 1956);

"John, 21, Adds New York to His Public," *Newsweek,* 3 (24 February 1934): 30–31;

Roger Kahn, Epilogue to *The World of John Lardner* (New York: Simon & Schuster, 1961), pp. 219–224; reprinted in slightly different form in Kahn's *Games We Used to Play* (New York: Ticknor & Fields, 1992);

Kahn, *The Era: 1947–1957, When the Yankees, the Giants and the Dodgers Ruled the World* (New York: Ticknor & Fields, 1993);

Ring Lardner, *Shut Up, He Explained,* edited by Babette Rosmond and Henry Morgan (New York: Scribners, 1962);

Ring Lardner Jr., Foreword to *Ring around the Bases,* edited by Matthew J. Bruccoli (New York: Scribners, 1992);

Lardner Jr., *The Lardners: My Family Remembered* (New York: Harper & Row, 1976);

Lardner Jr., "Ring Lardner & Sons," *Esquire,* 77 (March 1972): 99–103, 169–170;

A. J. Liebling, Unsigned obituary, *The New Yorker,* 36 (2 April 1960): 163–164;

Grantland Rice, *The Tumult and the Shouting: My Life in Sport* (New York: Barnes, 1954);

"Ring's Boy," *Time,* 52 (20 September 1948): 66–67;

Red Smith, "John Lardner," *New York Herald Tribune,* 30 March 1960; reprinted in his *To Absent Friends* (New York: Atheneum, 1982), pp. 365–366;

Stanley Woodward, *Paper Tiger* (New York: Atheneum, 1964);

Woodward, *Sports Page* (New York: Simon & Schuster, 1949);

Jonathan Yardley, *Ring: A Biography of Ring Lardner* (New York: Random House, 1977).

# Ring Lardner

*(6 March 1885 – 25 September 1933)*

Pete Cava

*International Sports Associates*

See also the Lardner entries in *DLB 11: American Humorists, 1800–1950; DLB 25: American Newspaper Journalists, 1901–1925; DLB 86: American Short-Story Writers, 1910–1945, First Series;* and *DLB Yearbook: 1985.*

*Ring Lardner*

MAJOR POSITIONS HELD: Reporter, drama critic, sports editor, *South Bend Times* (1905–1907); sports reporter, *Chicago Inter-Ocean* (1907), *Chicago Examiner* (1908), *Chicago Tribune* (1908–1910); managing editor and columnist, *The Sporting News* (1911); sports editor, *Boston American* (1911); sports reporter, *Chicago Examiner* (1912–1914); daily columnist, *Chicago Tribune* (1914–1919); weekly columnist, the Bell Syndicate (1919–1927); sports columnist, *Morning Telegraph* (1928–1929).

BOOKS: *Bib Ballads* (Chicago, New York & Toronto: Volland, 1915);

*You Know Me Al* (New York: Doran, 1916);

*Gullible's Travels, Etc.* (Indianapolis: Bobbs-Merrill, 1917; London: Chatto & Windus, 1926);

*My Four Weeks in France* (Indianapolis: Bobbs-Merrill, 1918);

*Treat 'Em Rough* (Indianapolis: Bobbs-Merrill, 1918);

*The Real Dope* (Indianapolis: Bobbs-Merrill, 1919);

*Own Your Own Home* (Indianapolis: Bobbs-Merrill, 1919);

*Regular Fellows I Have Met* (Chicago: Wilmot, 1919);

*The Young Immigrunts* (Indianapolis: Bobbs-Merrill, 1920);

*Symptoms of Being 35* (Indianapolis: Bobbs-Merrill, 1921);

*The Big Town* (Indianapolis: Bobbs-Merrill, 1921);

*Say It with Oil,* published with *Say It with Bricks,* by Nina Wilcox Putnam (New York: Doran, 1923);

*How to Write Short Stories* (New York & London: Scribners, 1924; London: Chatto & Windus, 1926);

*What of It?* (New York: Scribners, 1925);

*The Love Nest and Other Stories* (New York: Scribners, 1926; London: Philip Allan, 1928);

*The Story of a Wonder Man: Being the Autobiography of Ring Lardner* (New York: Scribners, 1927);

*Round Up: The Stories of Ring Lardner* (New York: Scribners, 1929; London: Williams & Norgate, 1935);

*June Moon,* by Lardner and George S. Kaufman (New York: Scribners, 1930);

*Lose With a Smile* (New York: Scribners, 1933);

*First and Last,* edited by Gilbert Seldes (New York & London: Scribners, 1934);

*Shut Up, He Explained,* edited by Babette Rosmond and Henry Morgan (New York: Scribners, 1962);
*Some Champions: Sketches and Fiction from a Humorist's Career,* edited by Matthew J. Bruccoli and Richard Layman (New York: Scribners, 1976);
*Ring Lardner's You Know Me Al: The Comic Strip Adventures of Jack Keefe* (New York: Harcourt Brace Jovanovich/Bruccoli Clark, 1979).

**Collections:** *The Portable Ring Lardner,* edited by Gilbert Seldes (New York: Viking, 1946);
*The Ring Lardner Reader,* edited by Maxwell Geismar (New York: Scribners, 1963);
*The Best Short Stories: 25 Stories from America's Foremost Humorist, Ring Lardner* (New York: Collier, 1988);
*Haircut and Other Stories by Ring Lardner* (New York: Collier, 1991);
*Ring Around the Bases,* edited by Matthew J. Bruccoli (New York: Scribners, 1992).

PLAY PRODUCTIONS: *Elmer the Great,* by Lardner and George M. Cohan, New York, Lyceum Theatre, 24 September 1928;
*June Moon,* by Lardner and George S. Kaufman, New York, Broadhurst Theatre, 9 October 1929.

Ring Lardner's place in literature as one of America's most important and influential short-story writers and humorists is secure. He was also a successful playwright, although his ambition was to be a songwriter. Lardner's original fame, however, came as a newspaperman. In the early part of the twentieth century he was one of the nation's best-known sportswriters.

In the era before the advent of radio and television – when newspapers and magazines were the country's main sources of information and entertainment – Ring Lardner was a print media superstar. To advertise his syndicated coverage of the 1925 World Series, the *Indianapolis Star* ran a photo of Lardner under the heading, "No Need to Introduce This Guy." The ad promised readers "a barrel of laughs a day," and described Lardner as "baseball's poet laureate."

Others described Lardner as solemn, noble, and dignified; long-legged, lean, and consumptive; and enveloped in an odd, caustic sympathy that was frequently mistaken for misanthropy. A columnist of the 1920s, remarking on Lardner's large brown eyes, aquiline nose, and regal bearing, described him as having "a look of eagles." Lardner would inspire other American writers, including Ernest Hemingway, J. D. Salinger, and playwright Neil Simon, whose alter ego in *Brighton Beach Memoirs* (1984) hoped "to be a writer like Ring Lardner."

Ringgold Wilmer Lardner was born on 6 March 1885 in the southwest Michigan town of Niles, about a dozen miles from South Bend, Indiana, where his writing career would begin. His first name was the name of his father's recently deceased first cousin (who had been the namesake of Cadwallader Ringgold, a Union admiral during the Civil War), and his middle name was the maiden name of the deceased cousin's mother. Lardner was the last of Henry and Lena Phillips Lardner's six surviving children. Three others died in childhood.

The first three children, William, Henry Jr., and Lena, were considerably older than Ring. Rex (born 1881) and Anne (born in 1883) were closer in age. Henry Lardner had inherited considerable wealth, and had expanded the family fortune through real estate and investments. His estate included a tennis court, a private baseball diamond, a coachman, and nursemaids. The Lardners, writes biographer Jonathan Yardley, "came as close to genuine aristocracy as a small Midwestern town could hope to see."

At an early age Lardner underwent corrective surgery for a deformed foot and wore a leg brace until he was eleven. Nevertheless, his boyhood was active, although sheltered – until he was eight years old he was forbidden to leave the family estate unless he was accompanied by an adult. Along with Rex and Anne, young Ring was educated by private tutors and by his whimsical and personable mother, who had been a schoolteacher. Henry Lardner, already in middle age when his youngest child was born, appears to have been a loving father but was content to remain in the background; it was Lena who left a lasting impression on all the Lardner children.

Lena Lardner provided her offspring with an unusually good library (she was the author of two books of sentimental verse), a solid background in English composition, an appreciation for classical music, and a sense of noblesse oblige. She instilled in them a love of music. Lardner inherited her perfect ear, a talent that may have foreshadowed his precision at re-creating dialect. In "dramas" presented by the three youngest Lardner children for the benefit of the family members and household staff, Ring usually played a leading part – or hoped to attract everyone's attention in some clever way.

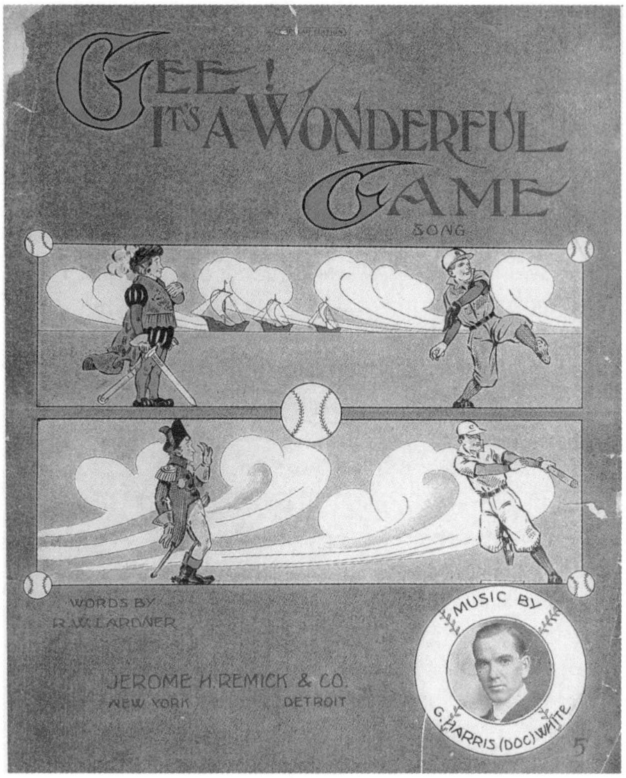


*Cover for Lardner's 1911 song that compares the achievement of pitcher Christy Mathewson, a future Hall of Famer, to that of Christopher Columbus*

When he was twelve Lardner entered Niles High School. He was a popular student who achieved honors and played football. Four years later, in 1901, the year Lardner graduated from high school, poor investments had depleted the Lardner family wealth. (Once he was established, Lardner sent money to his parents regularly; he also made frequent donations to needy friends throughout his life.) After rejecting a scholarship from Olivet College in Michigan, Lardner moved to Chicago. There he held several menial jobs – none of them for long. He returned to Niles and worked for a time as a railroad freight hustler. He and his brother Rex briefly attended Chicago's Armour Institute, where they studied mechanical engineering. "I can't think of no walk in life," Lardner later reflected in his characteristic style, "for which I had more of a natural bent unless it would be hostess at a roller rink."

Lardner spent 1903 unemployed, occupying himself with amateur theatrical productions of the Niles American Minstrel players (long after his fervor for baseball subsided, Lardner would maintain a passion for the stage). He wrote the music and most of the lyrics for a two-act musical comedy fashioned along the lines of a Gilbert and Sullivan opera. Titled *Zanzibar,* the fourteen-page booklet was the first of Lardner's published works.

In 1904 Lardner went to work for the Niles Gas Company. After two years as a meter reader, accountant, and debt collector, he began his career in journalism – entirely by accident. Rex Lardner had become a reporter for the *Niles Daily Sun* while also serving as a correspondent for several other newspapers. Edgar Stoll, whose family owned the *South Bend Times,* came to Niles in 1905 to offer Rex, who was not in town at the time, a staff position. Finding Ring instead, Stoll learned that Rex was already under contract. When Lardner found out that the salary at the *South Bend Times* was twelve dollars a week – four dollars more than his pay at the gas company – he asked Stoll for the job. His claim to journalistic experience – that he often helped his brother – Lardner later recalled, "was very far from the truth." Stoll hired Lardner as a "society reporter, court-house man, dramatic critic and sporting editor." At the age of twenty Ring Lardner entered the newspaper field.

Although he received no bylines, Lardner's work in the sports pages of the *South Bend Times* can often be identified by his incipient style. Starting with the 1906 season, Lardner covered South Bend's minor league baseball club in the Central League (as sports editor he was also expected "to read the sport pages of papers printed in other Central League towns"). Articles about the team are sprinkled with the same droll humor that would soon appear in Lardner's reporting on major league baseball.

In 1907 Lardner met two people who would profoundly affect his life. In July he was introduced to Ellis Abbott, a Smith College student from Goshen, Indiana, whom he would marry four years later. During the World Series he encountered Hugh Fullerton, a baseball writer for the *Chicago Examiner.* Impressed with Lardner's knowledge of the game, Fullerton recommended the twenty-two-year-old reporter to the *Chicago Inter-Ocean.* The newspaper hired Lardner, and he began covering sporting events alongside Fullerton, Hugh E. Keough, I. E. Sanborn, Walter Eckersall, and Charles Dryden, well-known reporters from rival dailies. Newspapers had been hiring full-time sports reporters for less than two decades, and writers in New York, Boston, and Chicago were defining the craft. Jack Lang notes in *Total Baseball* (1989) that the eastern writers emphasized expert knowledge of the game, while humor and cynicism characterized the Chicago style. The irreverent, witty Dryden, twenty-five years older than Lardner and the highest-paid sportswriter in the country, greatly influenced Lardner's style. Donald Elder, Lardner's first biographer, recounts how ballplayers nicknamed Lardner "Charlie's Hat," because he seemed to go everywhere with Dryden.

Fullerton left the *Chicago Examiner* in 1908 and recommended Lardner to replace him. Lardner began traveling with the White Sox and earned immediate popularity with the players, who called him "Lardner," "Niles," or "Old Owl Eyes," despite his having maintained an aristocratic aloofness (Fullerton described it as a "pretense of indifference") during the early phases of his career. If he was moody at times, prone to staring out of train windows and not speaking unless spoken to, he could, according to Elder, just as easily join in on the ballplayers' fun, delivering "impromptu verses and parodies of song." Lardner recalled in the 20 February 1932 *Saturday Evening Post* that his winter assignment usually entailed "call[ing], every afternoon, on President Murphy, of the Cubs; President Comiskey, of the White Sox; President Johnson, of the American League; President Hickey – or whoever it happened to be – of the American Association, and visit[ing] the haunts of Chicago ball players or ball players from out of town, pick[ing] up what news there was, and writ[ing] something, whether there was news or not."

Lardner befriended one illiterate White Sox player, who valiantly tried to conceal his deficiency. Lardner never revealed the player's true identity, referring to him only as "Jake Gibbs, a regular infielder," from Brooklyn. Out of kindness, Lardner feigned ignorance of the player's inability to read. In his presence Lardner began reading aloud, as if to himself, from menus and newspapers, so that "Gibbs" could overhear him. Soon Lardner began taking care of the player's correspondence with his wife, "Myrtle," back in Brooklyn. This charitable act gave Lardner an invaluable glimpse into the domestic life of a ballplayer. "Gibbs" would, in part, contribute to the character of Jack Keefe in Lardner's classic work *You Know Me Al* (1916).

Later in 1908 Charles Dryden retired, and Lardner replaced him at the *Chicago Tribune*. At twenty-three Lardner was near the pinnacle of Chicago sportswriting. In the spring of 1909 the *Tribune* was the city's top newspaper, and Lardner's beat was the world champion Chicago Cubs, famous for their infield combination of shortstop Joe Tinker, second baseman Johnny Evers, and first baseman Frank Chance. Alternating between the two Chicago ball clubs that season, Lardner became good friends with White Sox pitcher G. Harris "Doc" White, who collaborated with Lardner to write successful songs (words by Lardner). Their song "Gee! It's a Wonderful Game" had a clever twist on Christopher Columbus and Christy Mathewson, the great pitcher for the New York Giants and manager John McGraw:

> Who discovered the land of the brave and the free?
> I don't know, I don't know
> 'Twas Christy Columbus is what they tell me;
> May be so, I don't know
> There's only one Christy that I know at all,
> One Christy that I ever saw,
> He's the one who discovered the fade away ball,
> And he pitches for Mugsy McGraw.

At the *Chicago Tribune* between 1908 and 1910, observes Elder, Lardner's "turns of phrase became sharper, his style more fluent and his reports were colorful, dramatic and funny." He was also by this time, notes David Quentin Voigt, "familiar with human foibles in general and those of ballplayers in particular." Though much of Lardner's early work reveals the difficulties of his having to write under deadlines, Matthew J. Bruccoli and Richard Layman are convinced that "a considerable body of first-rate Lardner is entombed in . . . disintegrating newspapers."

Engaged to marry Ellis Abbott (whose father frowned on "sporting people"), Lardner tried to break away from the peripatetic life of a baseball reporter. He first agreed to accept a post in Louisville, Kentucky, as business manager of an American Association baseball club, but decided against it in favor of a position with *The Sporting News* in Saint Louis. Later known as "the Bible of Baseball," the newspaper was, at the time, an eight-page weekly that provided information from every level of professional baseball. Lardner left Chicago in 1910 to become the paper's managing editor. He regretted it immediately, writing to Ellis after his first day on the job that he had begun "heartily hating my employers." But he wrote a popular column called "Pullman Pastimes" about his travels with the Chicago teams and the players he had known. While using euphemism, hyperbole, and well-established stereotypes, he began to explore the more realistic side of the players' lives, from their obvious limitations as uneducated men with few abilities other than playing baseball to their practical jokes on rookies and their time spent in barrooms and at racetracks. It was Lardner's style more than his content that attracted readers, for as Lee Allen observes, "The stupidity of certain players was acknowledged long before Ring Lardner painted his classic pictures of bushers, alibi artists, boneheads, and bores."

After three months at *The Sporting News* Lardner had had enough. He left after telling owner Charles Spink that he would "rather work for Jesse James." He went to work for the *Boston American* as sports editor, a position that had first been offered to Fullerton, who turned it down and once again recommended Lardner. At the *Boston American* he wrote "Oddities of Bleacher 'Bugs,'" an article that was both a denunciation and apologia of the fan "in the bleachers, where the real 'bugs' sit." The "real article," wrote Lardner on 23 July 1911, "doesn't attempt to score at all, disputes every statement made by his neighbors in the bleachers whether he knows anything about said statement or not, heaps imprecations on the umpire and manager, [and] thinks something is a bonehead play when it really is good, clever baseball." The "bugs," in Lardner's view, did not know half of what they claimed to know and would be even less enthralled with baseball if they really did find themselves in the roles of coaches or reporters, who enjoy the game on an entirely different level. But in the end he admitted the bugs "are even happier" than the coaches and scribes; they "go out there to have a good time and they have it."

On 28 June 1911 Lardner and Ellis were married. His job in Boston required little travel, and Lardner had hired for his staff his brother Rex and an old Chicago crony named Frank Smith. But he soon found the management at the *Boston American* as insufferable as his employers in Saint Louis. His brother and Smith were dismissed in a cost-cutting move while Lardner was away covering the Philadelphia Athletics–New York Giants World Series, and Lardner resigned in anger. After borrowing money from the owners of both Boston baseball teams, Lardner returned to the Midwest and took a job as copyreader for the *Chicago American.*

In his first magazine article, "The Cost of Baseball," published in the 2 March 1912 *Collier's,* Lardner showed both an understanding of and an appreciation for baseball magnates, known almost archetypally in baseball lore for their notorious penury. Upon accounting for every dollar of an owner's expenses in any given year (such as payment for an overrated minor league player for whom the owner "would have been just as well off if he had spent the money for walnut shucks"), Lardner asked the disgruntled fan to "feel a touch of sympathy" for "Mr. Magnate's life": "He's not going to the poorhouse. No, but the asylum is not so far away."

In February 1912 Lardner rejoined the *Chicago Examiner,* and in May Ellis gave birth to their first child, John Abbott. Lardner took over the White Sox beat and covered the Boston Red Sox–New York Giants World Series in October, writing of the famous deciding game, in which the Giants, who had taken a lead in the top of the tenth inning, let victory slip away after Fred Snodgrass dropped a fly ball and Fred Merkle failed to run for an easy foul ball:

> [T]here was seen one of the saddest sights in the history of a sport that is a strange and wonderful mixture of joy and gloom. It was the spectacle of a man, old as baseball players are reckoned, walking from the middle of the field to the New York players' bench with bowed head and drooping shoulders, with tears streaming from his eyes, a man on whom his team's fortune had been staked and lost, and a man who would have proved his clear title to the trust reposed in him if his mates had stood by him in the supreme test. The man was Christy Mathewson.

Hugh E. Keough died in 1913, and by the spring of the following year the *Chicago Tribune* was still eager for someone to take over his popular column, "In the Wake of the News." Fullerton had tried his hand at it, but soon decided he was better suited for conventional reporting and recommended Lardner for the job. Taking over for the popular and erudite Keough, Lardner, though influenced by the bucolic midwestern characters created by George Ade and the Irish dialect humor of Finley Peter Dunne, immediately made the column his own and was an instant success. He decorated the "Wake" column with verse and doggerel and frequently wrote about his growing family. Three more sons were born during Lardner's days as a columnist: James Phillips in 1914; Ringgold Wilmer Jr., called "Bill," in 1915; and David Ellis in 1919. Lardner's Monday columns almost always centered on his children. When a reader complained that these pieces had nothing to do with sports, Lardner replied: "If kids ain't sport, what is?"

It is with the "Wake" pieces that Lardner began to hone his vernacular writing. Lardner listened carefully to the spoken language and tried to replicate its rhythms and sounds in print. (He once explained to H. L. Mencken how ballplayers who invariably dropped the *g* sound on the ends of words *nothin* and *somethin* almost always enunciated the words *everything* and *anything.*) He often wrote ungrammatical, first-person narratives that reflected the clumsy speech of an unlettered midwesterner. The voice was often that of "a athalete," but "R. W. L."

frequently adopted such a voice himself: "I wont be there to get a look at the braves and the cardinals which I would like to see the both of them and see how they get a way with it but may be if you send Mr. Sanborn out to the games you will get pretty fare reports on them but I will be over to Niles Sunday and see there game with Allegan and I hope you will print a big write up of that game and menshun that I was there and seen it and that will make the people that dont go wisht they had of went" (15 August 1914).

Lardner experimented with prototypes of Jack Keefe in his "Wake" columns. "Perhaps unconsciously," writes Tristram P. Coffin in *The Old Ballgame: Baseball in Folklore and Fiction* (1971), "he began to use these *essais* as a means of rebelling against the well-established custom of dispensing poorly spelled, cracker-barrel philosophy out of well-adjusted, small-town brains." Lardner also used the vernacular to write seriously about ballplayers he admired. Of Mathewson, he remarked,"They's a flock o' pitchers that knows a batter's weakness and works accordin'. But they ain't nobody else in the world that can stick a ball as near where they want to stick it as he can." He captured the temper of Ty Cobb in the May 1915 *American Magazine:* "Cobb wouldn't stand for nothin'. If somebody poured ketchup in his coffee, he was liable to pick up the cup and throw it the guy nearest to him. If you'd stepped on his shine, he'd of probably took the other foot and aimed it at you like he does now the ball when it's lyin' loose on the ground. If you'd called him some name on the field, he'd of walloped you with the bat, even if you was his pal."

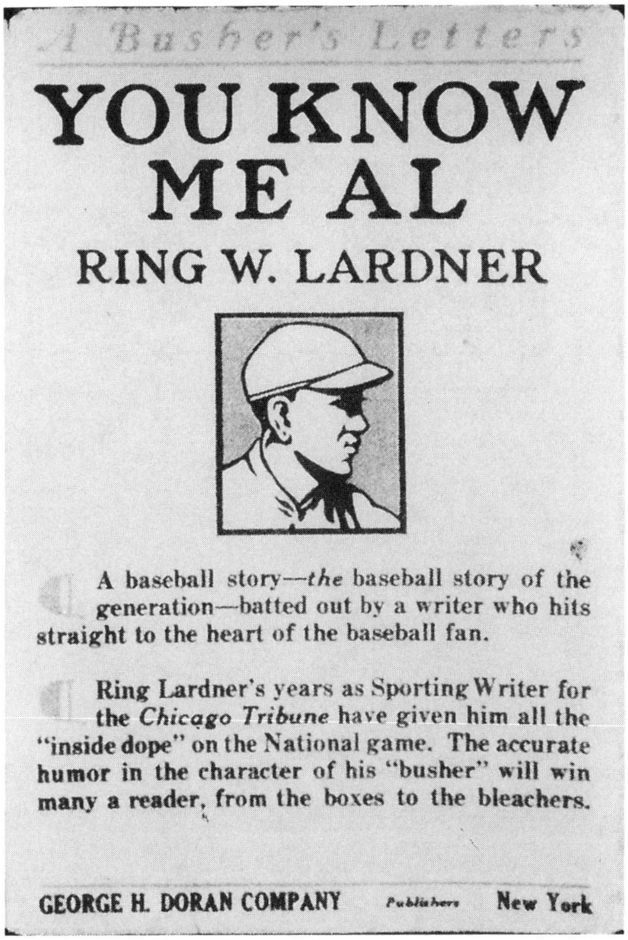


*Dust jacket for Lardner's 1916 book, in which a rookie pitcher for the Chicago White Sox, Jack Keefe, recounts his experiences to a hometown friend, Al Blanchard*

Lardner's dialect was so true to life that his colleagues encouraged him to write more. The narrative developed into "A Busher's Letters Home," a series of epistles written by Jack Keefe, a rookie pitcher with the White Sox, to a buddy back in his hometown of Bedford, Indiana, named Al Blanchard. Keefe was probably a composite of several players, including not only the illiterate "Jake Gibbs" but also Joe Benz and Ping Bodie. The boastful Benz, who pitched for the White Sox from 1911 to 1919, was, like Keefe, an Indiana native; Lardner allegedly once saw Benz wearing hip boots in an exhibition game for the White Sox. The brash and garrulous Bodie, born Francesco Pezzolo, was an outfielder for the Sox from 1911 to 1914. Keefe is fresh from from the minor leagues – "the bushes" in baseball parlance – but is destined by his nature to remain a bush leaguer for life. Although vain and often cruel, Keefe is a yokel, an innocent, and occasionally a sympathetic character, especially because of his relationship with his overbearing and rapacious wife, Florrie. The first busher story appeared in the 7 March 1914 issue of *The Saturday Evening Post.* In one letter collected in *You Know Me Al* Keefe describes a spring-training rail trip:

> FRIEND AL: Coming out of Amarillo last night I and Lord and Weaver was sitting at a table in the dining car with a old lady. None of us were talking to her but she looked me over pretty careful and seemed to kind of like my looks. Finally she says Are you boys with some football club? Lord nor Weaver didn't say nothing so I thought it was up to me and I says No mam this is the Chicago White Sox Ball Club. She says I knew you were athatletes. I says Yes I guess you could spot us for athatletes. She says Yes indeed and specially you. You certainly look healthy. I says You ought to see me stripped. I didn't see nothing funny about that but I thought Lord and Weaver would die laughing.

*The first "reels" of Lardner's syndicated comic strip that ran from 1922 to 1925*

Until the appearance of Jack Keefe, notes Lardner scholar Bruccoli, baseball fiction was hero-worship stuff written for boys. Lardner's Busher stories, Bruccoli points out, had an "inside quality" that was bolstered by the inclusion of well-known players such as Harry Lord, Buck Weaver, Kid Gleason, and others. When Keefe's body weight balloons (he writes to Al about a dinner order of "some stake and some eggs and some pie and some ice cream and some coffee and a glass of milk"), he runs afoul of coach Gleason:

> He says Well it is a good thing that they did not start bunting because if you had of went to stoop over and pick up the ball you would of busted wide open. I says Why? and he says because you are hog fat and if you don't let up on the stable and fancy groceries we will have to pay 2 fairs to get you back to Chi. I don't remember now what I says to him but I says something you can bet on that. You know me Al.

Lardner continued to write "letters" by other athletes in his "In the Wake of the News" column. A 17 April 1915 piece, for example, has "Bill" writing to "Steve" that "the old seasons oppened up and they still got me setting on the bench yet but as soon as this here Williams looses his horse shoes I will be there every day and then watch out." But Jack Keefe provided him with an ongoing character, a man whose talent and incipient moral sense are thwarted because of his own ignorance, conceit, and hypocrisy and the circumstances of fate (a recurrent theme in Lardner's later fiction). Though Keefe is surrounded by individuals who often try to tell him the right thing to do, he is never willing to take their advice or – in the case of his coaches – follow their orders:

> Lajoie was up next and Callahan sends out word for me to walk him. I thought that was rotten manageing because Lajoie or no one else can hit me when I cut loose. So after I give him two bad balls I tried to slip over a strike on him but the lucky stiff hit on a line to Weaver. Anyway the game was over and I felt pretty good. But Callahan don't appresiate good work Al. He give me a call in the clubhouse and said if I ever disobeyed his orders again he would suspend me without no pay and lick me too. Honest Al it was all I could do to keep from wrapping his jaw but Gleason winks at me to do nothing.

Enthusiastic critics dubbed Lardner's accurately re-created slang "Lardner's Ringlish." Hemingway and James T. Farrell, both still in their teens, imitated the Lardner style. Hemingway's articles appeared under the byline "Ring Lardner Junior" in his high-school newspaper. Michael Reynolds explains in *The Young Hemingway* (1987) that this debt to Ring Lardner, the sportswriter, extended into Hemingway's early career as an author.

In 1916 the first six Busher installments from *The Saturday Evening Post* (they had been published in the magazine upon the recommendation of Charles E. Van Loan, whose own sports fictions Lardner greatly admired) were published in book form as *You Know Me Al.* The book had only one printing until it was republished by Scribners in 1925. One other collection of Lardner's work had been previously published, *Bib Ballads* (1915), which consisted of verses about Lardner's family. Lardner published five more collections while he was writing the "Wake" column: *Gullible's Travels, Etc.* (1917), *My Four Weeks in France* (1918), *Treat 'Em Rough* (1918), *The Real Dope* (1919), and *Own Your Own Home* (1919). It was, however, the publication of *You Know Me Al* that transformed Lardner, as Wilfrid Sheed declares in his introduction to the 1984 Vintage Books edition, "from a local sports stiff to a national treasure."

Lardner wrote other baseball stories (a total of thirty-four, not counting the Busher abroad stories), many of them about eccentric ballplayers and most of them for *The Saturday Evening Post.* In "Alibi Ike" (31 July 1915) the title character has an excuse for everything: "His right name was Frank X. Farrell, and I guess the X stood for 'Excuse me.' " He even has an excuse when things turn out all right: "If he cracked one for three bases he ought to had a home run, only the ball wasn't lively, or the wind brought it back." "Harmony" (*McClure's,* August 1915) concerns a player who is more interested in warbling (Lardner wrote songs and formed quartets with Cubs and White Sox players) than winning. "Sick 'Em" (25 July 1914) shows how a manager deliberately pits two jealous pitchers against each other without their knowledge in order to get the best out of both of them: "On the road we'd get the telephone girls in the hotels to call up Smitty and ask him if he was Mr. Fogarty. When he'd say no she'd ring off; but she'd call him up again in about ten minutes and ask him the same question. We worked this on Fogarty, too, and both o' them pretty near went nuts 'cause the other was gettin' so many calls."

Two *Saturday Evening Post* stories of this period are masterpieces in which Lardner deals with psychologically troubled ballplayers. In "Horseshoes" (15 August 1914) Dick Grimes of the Athletics hates his hometown and boyhood rival, Speed Parker of the Giants, so much that he is determined to destroy him even if it means wrecking the Athletics chance of winning the World Series from the Giants, though ironically Grimes ends up as the series hero. As in "Harmony," the principal narrator is a sportswriter ("Even a baseball writer must sometimes work"). Intent on getting the inside "dope" on the winning game, he interviews Grimes in a dining car shortly after the series has ended and finds that all of Grimes's great plays were accidental, the result of his blind hatred for his rival. When after the game Parker alluded to Grimes's obvious luck by calling him "Horseshoes," Grimes injured Parker so badly that "no one is allowed to see" him in the hospital.

The central character of "My Roomy" (9 May 1914), Buster Elliot, is a borderline psychotic. Although he could "be the greatest ballplayer in the world if he could just play ball" (Elliot can hit but cannot field), no one can bear to room with him. "He always insisted on havin' the water run in the bathtub all night, because he said it reminded him of the sound of the dam near his home" and shaved in the middle of the night with all the lights on, stopping "every little while" to "turn round and stare at the guy who was makin' a failure o' tryin' to sleep. Then he'd wave his razor round in the air and laugh, and begin shavin' agin."

Lardner portrays a sick man sinking further into despair. The narrator, who has managed to last most of the season rooming with Elliot, cannot fully understand the process of Buster's psychological disintegration. He knows Buster is not a drunk, a dopehead, or "a lefthand pitcher"; he can only suppose that "he must of been just a plain nut and li'ble to break out any time." The story ends with Buster's crying at the news of his release. Without a check for the city series he won't be able to marry his girl so he smashes a mirror and "cut[s] his hand somethin' awful." Nonetheless the narrator, who has come to show empathy for Buster, convinces his teammates to award Buster a full share of the series money. The money, he learns, will not be spent on a wedding since the girlfriend has married a man "who runs a soda fountain" but to see that Buster "is made comf'table" in an "asylum at Kalamazoo."

"My Roomy" signals a significant shift in the overall quality of baseball literature, primarily because it plumbed more deeply into psychological waters, as it explored the consequential actions of a man whose "business" was "bustin' things," including his own fragile psyche. Writes Coffin, "when Buster Elliot broke that mirror way back in 1914, baseball writing was transformed, qualified by the buffet to aspire toward art." Though Lardner would write few baseball stories as good as "My Roomy" or even "Horseshoes," both stand with the works that would make Lardner's mark in American fiction, including "Haircut," "The Golden Honeymoon," "The Love Nest," and "There Are Smiles." The "busher" stories continued to appear in *The Saturday Evening Post* until 1919.

The favorite watering hole of the Chicago newspaper crowd was a saloon called Stillson's, just across from the *Chicago Tribune* office, and Lardner was a frequent visitor. The demands of the daily column kept him sober during most of his *Chicago Tribune* days, but later he would alternate between periods of sustained abstinence and heavy consumption. Alcohol may have numbed his insecurities, one of which was a lack of belief in his own talents. At the *Chicago Tribune* he spent long hours on the seven columns he wrote each week and was never completely satisfied. The self-doubts and self-criticism would continue throughout the rest of Lardner's life.

In 1917 Lardner turned thirty-two years old. He summed up his employment record in the form of a pitcher's lifetime statistics in the 28 April *Saturday Evening Post:*

| Jobs | W. | L. | Pct. |
|---|---|---|---|
| 13 | 8 | 5 | .616 |

The United States entered World War I in April 1917, and Lardner went to France as a correspondent, temporarily renaming his column "In the Wake of the War." Further "busher" tales, though not about baseball, began appearing in *Collier's* between 1918 and 1919, with Jack Keefe attempting to dodge the draft, but ultimately entering the army and going overseas.

When Lardner returned from Europe, he decided to move his family to New York so he could be closer to the theater world. His last column for the *Tribune* appeared on 20 June 1919. He continued to write about baseball and boxing (in addition to political conventions) in "Ring Lardner's Weekly Letter," published by John Wheeler's Bell Syndicate, from 1919 to 1927. With his work appearing in newspapers from Boston to Seattle, his reputation continued to grow. He soon published another collection, *Regular Fellows I Have Met* (1919), and *The Young Immigrunts* (1920), purportedly the story of the Lardners' move east. The latter contained several memorable exchanges between a four-year-old Ring Jr., the putative narrator, and his father, including one that became famous: "Are you lost daddy I arsked tenderly. Shut up he explained."

*Lardner in a White Sox uniform*

Lardner had not yet left Chicago when he covered the 1919 World Series for the Bell Syndicate. What ultimately became known as the Black Sox Scandal (in which eight White Sox players conspired with gamblers to throw the series to the Cincinnati Reds) was arguably the biggest scandal in sports history. Although the sordid episode did not break until a year later, Lardner's coverage of the games indicates that he knew about the fix: "Well as for the rest of the ball it looks soiled on the northwest side and I will worry my life away wondering who put a dirty finger on that ball which I have got and my children will still have it after me," he wrote on 8 October, after the sixth game of the best-of-nine series. During the series he helped compose the caustic parody of "I'm Forever Blowing Bubbles" (dedicated to Lefty Williams, the pitcher who threw the second game), which began, "I'm forever blowing ball games / Pretty ball games in the air."

Lardner's ardor for baseball had cooled somewhat, but the scandal was not the only reason. The brand of baseball Lardner most enjoyed was the "inside game," in which a team played for a run at a time through walks, steals, and sacrifice bunts. The busher story "Along Came Ruth" (1919) indicates the change that was coming. Though Babe Ruth was still a great left-handed pitcher for the Boston Red Sox in 1919, he was on his way to hitting more home runs (twenty-nine) than the aggregate totals of ten of the sixteen major league clubs. Jack Keefe's pitch to Ruth, who sends the ball "sailing into the right field bleachers where the black birds sets," announces the dawning of the era of the lively ball, in which, as Lardner disdainfully opined, "Ball players which used to specialize in hump back line drives to the pitcher is now our leading sluggers." Nonetheless, Lardner did not abandon baseball writing and continued to attend ball games with his sons and to write about baseball into the 1920s with enough firsthand knowledge to criticize it competently. When long after his days of reporting on the sport he appeared unexpectedly at a meeting of the Baseball Writers Association of America, Fred Lieb recalled in a 1963 article for *The Sporting News,* Lardner said to the surprised crowd, "What's the matter with you guys? I belong here. I am still a baseball writer and always will be."

Although Lardner had little regard for some sports, particularly basketball, he did write about boxing and football. When he covered the University of Chicago football team, the legendary coach Amos Alonzo Stagg commented: "He listens to the others' questions while those big eyes are roaming around. Then he goes to his office and writes a story that contains more real football information than any of the others have gathered." Grantland Rice, whom Lardner called "Grandland Rice," wrote in *The Tumult and the Shouting* (1954) that his friend "had two passions, Notre Dame and [Jack] Dempsey."

Lardner, who Rice believed was "equally good at football" writing as at baseball writing, had covered Notre Dame football games for the *South Bend Times.* In the glory days of coach Knute Rockne and running back George Gipp, Lardner described Notre Dame's best and oft-used offensive play: "Line up, pass the ball to Gipp, and let him use his own judgment." He referred to Gipp as a "quadruple-threat" man, who "could pass, run, point-kick and punt." Lardner also followed another midwesterner, University of Michigan football coach "Hurry Up" Fielding Yost. In his "In the Wake of the News" column he annually picked the best collegiate football teams.

Lardner covered the Jack Dempsey–Jess Willard bout in 1919 and the next year witnessed the Dempsey–Georges Carpentier fight, the first boxing match to have a "million-dollar gate." The latter fight became the inspiration for Lardner's thinly disguised story "The Battle of the Century," published in the 29 October 1921 *Saturday Evening Post,* in which hype, greed, deceit, and phoniness take precedent over the notion of sport. "The thing's got to be nursed along and worked up, and that's my job," the heavyweight champion's manager, Larry Moon, tells the narrator concerning how to promote the fight. "It'll take time, but it'll be worth it. The tough problem ain't getting the fans steamed up. They'll take care of themselves. What I've got to do is convince some guy with money and a lot of nerve that it would be a fight, not a murder." A war hero and a "romantic," Goulet (the Carpentier counterpart) has recently knocked out the English champion in the first round. He is the ideal opponent for Jimmy Dugan, the Dempsey character. But "the plain facts was . . . a good big man was going to fight a little man that nobody knew if he was good or not, and the good big man was bound to win and win easy unless he has sunstroke." Dugan does win easily, and Moon, who has helped to make all the players $200,000 richer, including the Frenchman and his manager, announces in a classic line of understatement, "Well, . . . the way I've got it figured out, he wasn't big enough."

Lardner also kept a wary eye on much ballyhooed pugilists in his daily columns. He once described in his customary acerbic tone the training regimen of a heavyweight prospect: "From 6 to 7 in the A.M. he does road work for a paving concern. From 7 to 12 he eats breakfast, usually a wolf or bear and the inevitable sweet potatoes. From 12 to 1 he reads a sentence. From 1 to 3 he manicures himself all over. From 3 to 4 he practices expression, and from 4 to 5 cheek breathing. A good many boxers pick out the wrong features to breathe through, features that may get stopped up."

Lardner's October 1916 story in *Metropolitan,* "Champion" (made into a 1949 film starring Kirk Douglas), is less a boxing story than it is a critique on the ludicrousness of "Gee Whiz" sportswriting. The brutal boxer Midge Kelly is a mordant and malevolent human being, whose first knock-

*President Warren G. Harding, Grantland Rice, Ring Lardner, and Undersecretary of State Henry R. Fletcher at the White House before playing a round of golf in 1921*

out at seventeen was his crippled younger brother and whose wedding gift to his wife is "a crushing blow on the bride's pale cheek." There is nothing endearing or redeeming about Kelly, who would throw a fight or let his wife and baby starve while "the wine flowed freely for him," yet the press accounts paint the spurious picture of a devoted family man and hero, "honest and straight as a die." No reporter interviews any of the people Kelly has abused – "the comments they made were not for publication." Readers "don't want to see [Kelly] knocked. He's champion."

After the first controversial Dempsey–Gene Tunney fight on 23 September 1926, Lardner posited his own theory as to why Dempsey lost a fight he should have won in a letter to his friend F. Scott Fitzgerald: "The thing was a very well done fake. The championship wasn't worth a dime to Jack; there was nobody else for him to fight and he had made all there was to be out of vaudeville and pictures." Rice claimed that Benny Leonard, the former lightweight champ, had given Lardner the tip about the fix and that Lardner "blistered the hide off both Dempsey and Tunney" in an overnight he filed for an ailing Rice. "Neither [boxer] spoke to me for several months," Rice wrote in *The Tumult and the Shouting*. "I couldn't blame them. I had a ghost."

Lardner also contributed regularly to *American Golfer,* which Rice started editing in 1921, but "only seldom, and only peripherally," noted Charles Fountain, "on the subject of golf. Lardner would write of duck hunting, and of travels in Egypt that never happened, with his peculiar sense of exaggeration and nonsense." In "Mr. Frisbie," a story originally published in the June 1928 issue of *Hearst's International-Cosmopolitan,* the title character is a millionaire who becomes "a slave of the game" on his private eighteen-hole golf course. Told by Frisbie's personal chauffeur and caddy (another of Lardner's uneducated and not always reliable narrators), the story shows how the narrator manages to gain the beneficence of his employer by taking advantage of him on the days he golfs well, even if it means helping along with his score or giving his "ball a little kick."

The chauffeur also manages to secure for Frisbie's wife and children the things they want most from him. Frisbie even consents to his daughter's marrying her ne'er-do-well boyfriend when the narrator sets a trap for the rich family friend Mr. Frisbie would like her to wed instead. Though the narrator would "just as leaf go up and tickle [Mr. Frisbie] as tell him that his stance is wrong," he gets the young suitor to correct Mr. Frisbie whenever he does anything incorrectly. At one point he even shows Mr. Frisbie, whose face turns "a kind of purple black color," how to swing by driving "the ball 250 yards straight down the middle of the course." Despite Lardner's comic approach to the game, "Mr. Frisbie" does contain an excellent description of playing what is known in golf as a blind hole:

> It is 350 yards long in lenth and . . . you can not see the green from the tee as you drive from the tee up over a hill with a direction flag as the only guide and down at the bottom of the hill there is a brook a little over 225 yards from the tee which is the same brook which you come to again on the last hole and in all the times Mr. Frisbie has played around the course he has seldom never made this 16th hole in less than 7 strokes or more as his tee shot just barely skins the top of the hill giving him a downhill lie which upsets him so that he will miss the 2d. shot entirely or top it and go in to the brook.

More than any other sport, however, baseball as metaphor, inspiration, allusion, and touchstone pervades Lardner's work. In the comic *The Story of a Wonder Man: Being the Autobiography of Ring Lardner* (1927) Lardner uses a baseball metaphor in an account of his service as a thirteen-year-old general for the U.S. forces in Cuba during the Spanish-American War: "We all had two sets of uniforms, white for home and gray for traveling, because there was no telling if the battles were to take place at our park or on the road." According to Allen, Lardner's creation of Jim Kendall, the notorious protagonist in "Haircut," was inspired by William A. Phelon, a noted baseball writer and poet for the *Cincinnati Times-Star*. The title of Lardner's 1925 book *What of It?* purportedly came from his overhearing a conversation between Cubs manager Hank O'Day and Cubs owner Charlie Murphy in a hotel lobby. Lardner alludes to baseball to suggest his character's inner turmoil in "Insomnia," a stark story published in the May 1931 *Hearst's International-Cosmopolitan:* "Sheckard, Evers, Schulte, Chance, Steinfeldt, Hofman, Tinker, Kling, Brown. Twenty years or so ago I was a baseball nut and could recite any club's batting order. Nowadays I hardly know which Waner bats second and which third." Voigt considers Lardner a "Voltaire among sportswriters" and speculates that his "devotion to baseball writing rested on his faint hope that the sport might offer a refuge from life's phoniness."

Lardner's salary from the Bell Syndicate for his "Weekly Letter" enabled his family to live comfortably, though he never bothered to save or invest, except to buy life insurance. In 1921 he published *The Big Town* and *Symptoms of Being 35.* Jack Keefe resurfaced in a Bell Syndicate daily comic strip beginning in 1922. "Every word in the 'Balloons' is Lardner's," exclaimed an advertisement for the new *You Know Me Al.* "The Plot, the Episodes, the Letters . . . all are ORIGINAL, and all are WRITTEN BY LARDNER EXPRESSLY FOR THIS STRIP." Lardner continued to write the strip (with illustrations by Will B. Johnstone and later Dick Dargan), albeit reluctantly, until 1925; it was earning him about $17,000 a year.

In 1922 Lardner's sketch "The Bull Pen" was performed in the Ziegfeld Follies of 1922, with Will Rogers in one of the three roles. The published text of the piece (in *Judge,* July 1922) seems to omit a scene that became the source of a well-established baseball anecdote. Lardner, according to *Detroit News* sportswriter H. G. Salsinger, who recalled the sketch, has a bullpen pitcher eating a sandwich as the pitcher is called upon to go to the mound. "Who's up? he asks the catcher. "Cobb, Crawford, and Veach," the catcher says. "Well, hold this, then," the pitcher says, handing him the sandwich. "I'll be right back."

Lardner moved his family in 1921 to Great Neck, Long Island, where from 1922 to 1924 his neighbors included Scott and Zelda Fitzgerald. Fitzgerald's roots were also in the Midwest, and Fitzgerald and Lardner, both heavy drinkers, loved to talk through the night over a bottle of Prohibition liquor. Both were amused by the nouveau riche displays of their neighbors (Lardner referred to one oversized living room as "the Yale Bowl – with lamps").

Lardner, according to his son Ring Jr., "found so much to like in [the Fitzgeralds] . . . that he waived all his normal criteria and displayed more affection for them than for anyone else outside the family." Even after the Fitzgeralds moved to Europe, Fitzgerald maintained an interest in Lardner's work and tried to wring a novel out of Lardner, whom Fitzgerald thought was squandering his talent by writing for newspapers and magazines. Most important, Fitzgerald helped move Lardner to the prestigious Scribners publishers, whose writers included Fitzgerald, Hemingway, and Thomas Wolfe. Scribners' literary editor, Maxwell Perkins, with

Fitzgerald's encouragement, worked with Lardner to bring together a collection of ten of his previously published short stories.

*How to Write Short Stories* (1924) – the title suggested by Fitzgerald – includes seven sports-related stories, among them "Champion" and another boxing story, "A Frame-Up," about a slow-witted boy whose handlers have to get him enraged before he can punch effectively. "A Caddy's Diary" depicts the venality of the country club set – its inept golfers who toss around nickel tips and cheat on their scores at every turn. "Alibi Ike," "Harmony," "My Roomy," and "Horseshoes" rounded out the sports stories in the collection.

In 1925, the year Scribners published *What of It?* and reprinted *Gullible's Travels, The Big Town,* and *You Know Me Al,* Lardner wrote his farewell tribute to Christy Mathewson, dead at forty-five from tuberculosis:

> My eyes are ever misty
> As I pen these lines to Christy
> Oh, my heart is full of heaviness today.
> May the flowers ne'er wither, Matty
> On your grave in Cincinnati
> Which you've chosen for your final fadeaway.

The next year Lardner himself was diagnosed with tuberculosis. He also published *The Love Nest and Other Stories,* which included "Haircut," his most famous story.

Lardner covered five more of "The World's Serious" for the Bell Syndicate. His account of the 1923 series between the Yankees and Giants ("even more important . . . than the famous world series of 1919 which was win by the Cincinnati Reds greatly to their surprise") was one of his most digressive. Reprinted in the posthumously published *First and Last* (1934), Lardner's narrative focused less on the action on the field than on his three cats at his "heavily mortgaged home in Great Neck" and his attempts to get his wife a fur coat (gambling also figures into the narrative as a motif, as it did in other Lardner sports prose and the sportswriting of the era in general). Still, he could observe "a few incidence" of the game from his unique perspective and take pot shots at two of his nemeses, Babe Ruth and bleacher "bugs":

> The way Jess [Barnes] pitched to Ruth did not seem to rouse unanimous enthusiasm amongst the bugs in the grandstand. Slow balls is what Jess feeds the Babe and the reason for the same is because Babe dont hit slow balls out of the park. If Jess did not feed the Babe slow balls when he knows he cant hit slow balls so good, why that would make Jess a 1/2 wit and when he does feed the Babe slow balls, why it shows he is thinking. That is why the crowd hoots him for pitching slow balls, because the average baseball bug hates to see anybody think. It makes them jealous.

Lardner's account of the 1927 Pirates-Yankees series, almost as circuitous as his 1923 account, focused on the colorful people who crossed his path during his coverage of the games, for instance the woman with a mouthful of candy:

> On one occasion, when Pipgras and Bengough stopped the game for a conference, she wanted to know what they was talking about and I told her they were trying to find out how to pronounce each other's name. . . .
>
> "Can you pronounce them yourself? " she asked me.
>
> "No," I said. "But you certainly ought to be able to with all that candy in your mouth."
>
> Well, she tried, and the less said about the result the better. Luckily I am going where I have got a clean suit of clothes.

From 1927 onward Lardner hoped to concentrate on writing short fiction for magazines such as *Cosmopolitan, Liberty,* and *Redbook* and on writing for the stage. In 1928 his *Elmer the Great,* which was based on his story "Hurry Kane" and written in collaboration with George M. Cohan, played to tepid reviews on Broadway but was made into a movie in 1933. In the story Kane is a pitcher who "could pitch every third day and be at his best" but also was a "bone-headed sap." He is a windbag "too dumb" to "half believe" that "the boys pulled all the old gags on him that's been around in baseball since the days when you couldn't get on a club unless you had a walrus mustache." Although Kane accepts a bribe to throw a World Series, he finally comes to his senses and plays to win, while still trying to keep the check the gambler gave to him. A year later Lardner collaborated with George S. Kaufman to produce the more successful Broadway play *June Moon.*

In 1928 the Lardners moved to East Hampton, New York, and in December Lardner returned briefly to newspaper work, writing a sports column called "Ring's Side" for the *Morning Telegraph,* which paid him a salary of $50,000. Lardner left the paper in February 1929 to work on other projects, including a March *Collier's* piece on Babe Ruth called "Pluck and Luck" and the critically acclaimed *Round Up,* a collection of sixteen new stories plus all of those from *How to Write Short Stories* and *The Love Nest. The New York Times* hailed Lardner as "our best short story writer."

*Ellis and Ring Lardner with their children (left to right): Jim, David, Bill (Ring Jr.), and John (courtesy of Ring Lardner Jr.)*

In *The New Yorker* of 13 September 1930, during a record-setting season for the most home runs and highest batting averages, Lardner published his critique of the "lively ball," which he also called "the TNT ball" or "crazy ball":

> Manufacturers of what they are using for a ball, and high officials of the big leagues, claim that the sphere contains the same ingredients, mixed in the same way, as in days of old. Those who believe them should visit their neighborhood psychiatrist at the earliest possible moment.

The essay's title, "Br'er Rabbit Ball," was a double pun, referring to the lively ball as well as type of baseball Lardner preferred, in which one could "see efficiency rewarded" and clever strategy pay off (like the resourceful Br'er Rabbit in the Uncle Remus stories).

Chronic illness in 1931 forced Lardner to return to writing for newspapers. "A Night Letter from Ring Lardner" began appearing in February for the Bell Syndicate in the *Chicago Tribune* and the *New York Daily News*. The columns were, according to Yardley, a "pathetic attempt to locate, much less recapture, a spark of the old Ring." About a month after the column's debut, Lardner, debilitated by the pain of tuberculosis, had to withdraw to a sanatorium in Tucson, Arizona. When he returned to his column in early April, Lardner found his "Night Letter" being used primarily as filler for sports pages. He then introduced into the column Willis Clough, a bush league pitcher, whose letters home concerned his attempts to land a berth on the Pittsburgh Pirates' roster. He was a pale imitation of Jack Keefe. Clough's "career" ended with the last edition of "Night Letter" on 24 April 1931.

In 1932 Lardner, while mounting a quixotic campaign against what he perceived as illiteracy and obscenity in radio in an irregular column for *The New Yorker* called "Over the Waves," was also writing a series of articles for *The Saturday Evening Post,* which included some of his best reminiscences and anecdotes about his early years on the sports beat:

> When the football and soccer seasons are over, when there is no longer anything to keep a morning-paper sports reporter outdoors in the daytime, his life becomes a little bit unhealthy. It also becomes a little bit dull, even in a town of Chicago's size – that is, after the mere sight of a club owner or ball player fails to make him swoon.

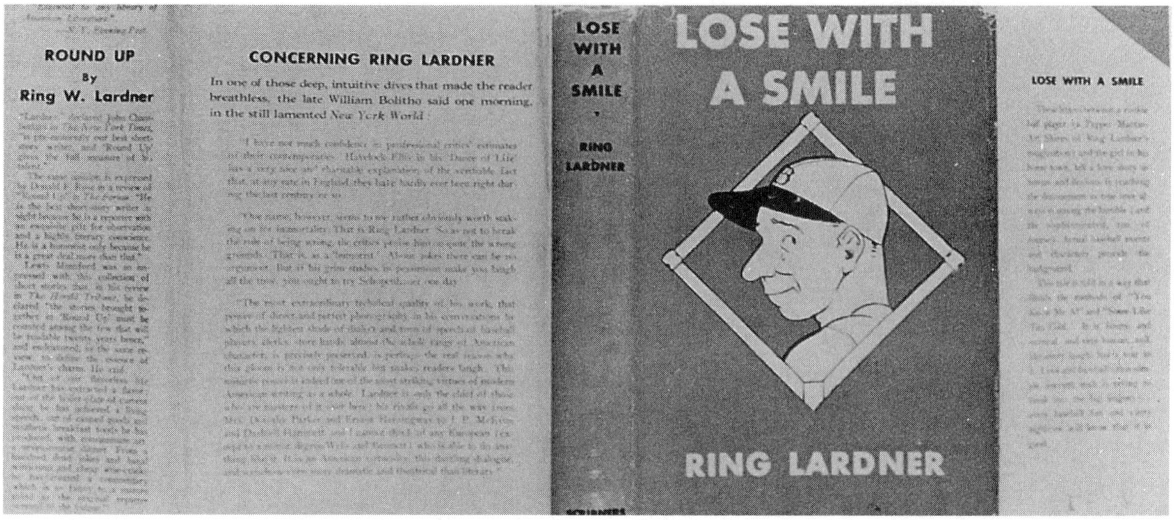


*Dust jacket for Lardner's 1933 book, in which Brooklyn Dodger rookie Danny Warner writes home to his girlfriend, Jessie Graham*

In recalling the real Dempsey-Carpentier match, which he had fictionalized in 1921, Lardner wrote in "Some Champions" (3 June 1933) that Carpentier kept his workouts down to a minimum because "he realized what a fat chance he had and that it was silly to go through a rigorous grind of preparation for inevitable defeat."

Lardner resurrected the ghost of Jack Keefe one last time in several 1933 short stories for *The Saturday Evening Post,* later collected in *Lose With a Smile.* The stories consist of a series of letters between a Brooklyn Dodger rookie named Danny Warner and Jessie Graham, his girlfriend in Centralia, Illinois. Warner's epistles contain some vintage Lardner: he mentions a contest "with a club called the House of David witch aint in no league but they are members of a religious sex in Michigan" and describes a pitcher who "throws a fire ball and a curve ball and a change of paste." Warner's roommate is Casey Stengel, whose own comic style Lardner obviously appreciated. In a rare moment of insight, Warner writes to Jesse that Stengel "is all ways sane things . . . witch dont make no sense."

The tuberculosis, along with the years of hard drinking and smoking, exhausted Lardner's health and his creativity. On the morning of 25 September 1933 he suffered a heart attack at his home in East Hampton. Later in the day he lapsed into a coma and died at the age of forty-eight. The news of his death received front-page coverage throughout the nation. His Long Island neighbor and close friend, Grantland Rice, eulogized Lardner in a long poem (addressed to the mythological Charon, the ferryman who conveyed the dead souls to Hades over the river Styx) that read in part "a friend is dead; the friend of a shattered age / Standing upon Time's brink – The friend of the printed page, / For those who could read – and think." Fitzgerald, who in his novel *Tender Is the Night* (1934) had patterned the doomed, dignified, sympathetic character Abe North after his close friend, greatly underestimated the achievement of Lardner in his eulogy titled "Ring." He concluded that Lardner "fell short of the achievement his was capable of" because he spent his formative years "in the company of a few dozen illiterates playing a boy's game," a shaping experience that had limited his scope to "the diameter of Frank Chance's diamond."

Virginia Woolf, who "didn't know from left-field," was closer to the mark. In her essay "American Fiction" published in the *Dial* in 1925, Woolf wrote that "games gave [Lardner] what society gives his English brother." Lardner observed the baseball diamond with an eye as discerning and a tongue as sharp as English novelists employed on the English drawing room, pulling the "wise boob" (to use Bruccoli's phrase) up from his folklore roots and into American literature. Clifton Fadiman suggested in *The Nation* six months before Lardner's death that he had "gone farther on hatred alone, without a mitigating soft streak in him" than anyone save for Jonathan Swift but that he had directed his hatred at venality, stupidity, mendacity, and disingenuousness. As Henry Morgan noted in an introduction to a 1962 collection of Lardner's work titled *Shut Up, He Explained,* "there are always those among us who have a fairly low opinion of the human race and are still amused by it, . . . who see clearly the folly of the whole proposition, and can be funny about it."

Thirty years after Lardner's death, the Baseball Writers Association of America proclaimed him the second winner of the J. G. Taylor Spink Award for meritorious service to baseball writing. The award is named for the son (who won the first award) of Lardner's old employer at *The Sporting News*. Critic Alfred Kazin called Lardner "undoubtedly the best literary sportswriter that America has produced." Stanley Frank concurred, "The man who gave literary stature to sportswriting . . . compressed enough distinction and genius in an all-too-brief quarter-century for three careers."

**Letters:**

*Ring Around Max: The Correspondence of Ring Lardner and Max Perkins*, edited by Clifford Caruthers (DeKalb: Northern Illinois University Press, 1973);

*Letters from Ring*, edited by Caruthers (Flint, Mich.: Walden Press, 1979).

**Bibliography:**

Matthew J. Bruccoli and Richard Layman, *Ring Lardner: A Descriptive Bibliography* (Pittsburgh: University of Pittsburgh Press, 1976).

**Biographies:**

Donald Elder, *Ring Lardner: A Biography* (Garden City, N.Y.: Doubleday, 1956);

Ring Lardner Jr., *The Lardners: My Family Remembered* (New York: Harper & Row, 1976);

Jonathan Yardley, *Ring: A Biography of Ring Lardner* (New York: Random House, 1977);

Elizabeth Evans, *Ring Lardner* (New York: Ungar, 1979).

**References:**

Lee Allen, *The Hot-Stove League* (New York: Barnes, 1955);

Sherwood Anderson, "Four American Impressions: Gertrude Stein, Paul Rosenfeld, Ring Lardner, Sinclair Lewis," *New Republic*, 32 (11 June 1922): 171–173;

Anderson, "Meeting Ring Lardner," *New Yorker* (25 November 1933);

John Berryman, "The Case of Ring Lardner," *Commentary*, 22 (November 1956): 416–423;

Tristram P. Coffin, *The Old Ballgame: Baseball in Folklore and Fiction* (New York: Herder & Herder, 1971);

Clifton Fadiman, "Ring Lardner and the Triangle of Hate," *Nation*, 136 (22 March 1933): 315–317;

F. Scott Fitzgerald, "Ring," *New Republic*, 71 (11 October 1933): 254–255;

Stanley Frank, *Sports Extra* (New York: Barnes, 1944);

Ring Lardner Jr., "Ring Lardner & Sons," *Esquire* (March 1922);

Thomas L. Masson, "Ring Lardner," in *Our American Humorists* (New York: Moffat, Yard, 1922);

Allan Nevins, "The American Moron," *Saturday Review of Literature* (8 June 1929);

Robert Obojski, "Kazin Ranks Lardner as a Top Baseball Writer," *Sports Collectors Digest* (13 January 1989);

Walton R. Patrick, *Ring Lardner* (New York: Twayne, 1963);

V. S. Pritchett, "The Talent of Ring Lardner," *New Statesman* (25 April 1919);

Leverett T. Smith Jr., "'The Diameter of Frank Chance's Diamond': Ring Lardner and Professional Sports," *Journal of Popular Culture*, 6 (Summer 1972): 133–156;

Murray Sperber, *Shake Down the Thunder: The Creation of Notre Dame Football* (New York: Holt, 1993);

Dana Tiverton, "Ring Lardner Writes a Story," *Writer* (January 1933);

David Quentin Voigt, *American Baseball, Volume II: From the Commissioners to Continental Expansion* (University Park & London: Pennsylvania State University Press, 1983);

Howard W. Webb Jr., "The Development of a Style: The Lardner Idiom," *American Quarterly*, 12 (Winter 1960): 482–492;

John Wheeler, "Ring Lardner," in *Collier's Greatest Sports Stories*, edited by Tom Meany (New York: Barnes, 1955), pp. 99–104;

Jonathan Yardley, "The Real Frank Merriwell," in *The Ultimate Baseball Book*, edited by Daniel Okrent and Harris Lewine (Boston: Houghton Mifflin/A Hilltown Book, 1981), pp. 65–80.

# Fred Lieb

*(5 March 1888 - 3 June 1980)*

J. Douglas English
*Johns Hopkins University*

MAJOR POSITIONS HELD: Sportswriter and columnist, *New York Press* (1911–1916), *New York Morning Sun* (1916–1921), *New York Evening Telegram* (1921–1927), and *New York Post* (1928–1934); correspondent (1935–1958) and columnist (1943–1947), *The Sporting News*; columnist, *St. Petersburg Times* (1971–1977).

BOOKS: *Sight Unseen: A Journalist Visits the Occult* (New York: Harper, 1939);

*Healing Mind, Body, and Purse* (Saint Petersburg, Fla.: Privately printed, 1941);

*The St. Louis Cardinals: The Story of a Great Baseball Team* (New York: Putnam, 1944; revised, 1945);

*Connie Mack: The Grand Old Man of Baseball* (New York: Putnam, 1945; revised, 1948);

*The Detroit Tigers* (New York: Putnam, 1946; revised, 1950);

*The Boston Red Sox* (New York: Putnam, 1947);

*The Pittsburgh Pirates* (New York: Putnam, 1948);

*The Story of the World Series* (New York: Putnam, 1949; revised, 1950 and 1965);

*The Baseball Story* (New York: Putnam, 1950);

*The Philadelphia Phillies,* by Lieb and Stan Baumgartner (New York: Putnam, 1953);

*The Baltimore Orioles: The History of a Great Team in Baltimore and St. Louis* (New York: Putnam, 1955);

*Comedians and Pranksters of Baseball,* by Lieb, Bob Burnes, J. G. Taylor Spink, and Les Biederman (Saint Louis: Spink, 1958);

*Baseball As I Have Known It* (New York: Coward-McCann & Geoghegan, 1977).

SELECTED PERIODICAL PUBLICATIONS – UNCOLLECTED: "Hits Are My Bread and Butter," *Saturday Evening Post,* 210 (16 April 1938): 33, 104–106;

"Strictly Screwball," *Collier's,* 105 (1 June 1940): 16, 69;

*Fred Lieb, 1913*

"The Great Exhumation," *Collier's,* 109 (16 May 1942): 22, 60–61.

As a boy Fred Lieb dreamed of being a baseball writer, and his accomplishments exceeded even his dream. His sportswriting career, which began in 1911 and lasted sixty-nine years, was recognized by the Baseball Writers Association of America, and he was inducted into the Hall of Fame in 1973. During that career Lieb worked at four New York papers and was surrounded by sportswriting legends such as Damon Runyon and Grantland Rice. At their greatest circulation his syndicated columns reached more than one hundred newspapers. He retired from daily reporting in 1934 only to publish eleven books on baseball and to become the top columnist and correspondent of *The Sporting News* for twenty-four years. Lieb's death deprived his fans of a treasured writer who was still typing at the age of ninety-two.

Frederick George Lieb was born on 5 March 1888 in Philadelphia to George and Teresa Lieb, German immigrants. As a boy in the mid 1890s, Lieb sold scorecards – complete with the previous day's results and updated standings – in the Philadelphia business district. Later recalling his days as a young player, Lieb modestly identified himself as "Mr. Strikeout," although the fact that he played on his church team and in newspaper leagues and charity games during those years belies his lack of athletic ability. He spent his time poring over the sports pages of Philadelphia's *North American* and the columns of Charles Dryden and Jimmy Isaminger, journalists whom he would later attempt to emulate. Provided at an early age with a small printing press, a Christmas gift from his parents, he soon showed his literary aptitude at the age of fifteen, when he won a five-dollar prize in his high school's short-story contest. An enthusiastic Philadelphia Athletics fan, he finally saved enough to afford the twenty-five-cent admission to Columbia Park in 1904 and saw Rube Waddell beat the Detroit Tigers, 3–0.

That same year Lieb graduated from Central Mechanic High School at the age of sixteen and began pestering Philadelphia sports editors for a job as a sportswriter. Nothing was open, but one editor acknowledged that his award-winning short story was "damn good." Lieb secured a job as a junior clerk for the Norfolk and Western Railroad, and the short hours required by this job gave him plenty of time to write. His first publications – short mysteries in pulp journals such as *Top-Notch Magazine* and *Railroad Man's Magazine* – were quite removed from the world of baseball. In 1908 he published two pieces of baseball fiction in *Short Stories.*

A breakthrough came in 1909, when Lieb convinced a Boston monthly, *Baseball Magazine,* to carry his short biographies of baseball stars. In writing these he relied entirely on statistics pillaged from other publications, his memory, and conversations with former players whom he had befriended, players who had played against the greats whose biographies Lieb wrote. Memory and conversation were tools that Lieb relied on for the rest of his life. The appearance of his first two biographies – on Honus Wagner and Ty Cobb – began a series that lasted into the 1920s.

With these successes and a letter of introduction from his railroad boss in hand, Lieb arrived for a job interview in 1910 at the office of the tyrannical Charles Barron, owner of *The Wall Street Journal.* The examination consisted of Barron's asking Lieb to draw him a bath and describe his desire to be a writer; Lieb did so and landed a job writing financial news with Barron's Philadelphia news bureau, the research office and outlet for *The Wall Street Journal* in Philadelphia.

By this time Lieb had compiled a rather impressive résumé. Early in 1911 *The Sporting News* published *Who's Who, People Worth Knowing in Baseball,* with Lieb listed as "baseball's foremost biographer." When he saw an advertisement for a sportswriting position at the *New York Press,* he ignored Barron's advice to stay in financial writing (where the money was) and applied for the job. This interview was more difficult for Lieb than the earlier one with Barron had been. As Lieb recalled, sports editor James Price handed him a two-hundred-word skeleton summary of the preseason prospects for each New York team and told him to "write . . . a column on the Giants and one on the Highlanders. And make it sound as if you are today in Marlin and Athens," the spring training sites that each team then maintained in Georgia. Two hours later Lieb returned with the columns. Having passed the test, he replaced the departing Ernie Lanigan in February 1911 as the baseball writer for the *Press.* The generally poor quality of the newspaper, Lieb reflected later, was offset by the exclusive focus of his position: he was to cover only baseball, and for the full year.

With the *Press* job came the responsibilities of being the American League official scorer for New York (Lanigan and Price reportedly had to teach Lieb how to keep score) and of writing the syndicated Associated Press reports for American League home games. Lieb's move to New York was followed shortly by marriage in 1911 to his Philadelphia sweetheart, Mary Ann Peck, and then by the birth of their daughter, Marie, in 1912. Lieb remained a New York baseball writer for the next twenty-three years.

He arrived as a sportswriter in New York at an auspicious time. Other new sportswriters included Heywood Broun, Rice, and Runyon (who was a veteran writer but new to New York) – all of whom were to receive recognition as sportswriters in baseball's Hall of Fame. The existence of at least six or seven competitive major newspapers and an

*Sportswriters at the New York Yankees spring training camp in Savannah, Georgia, 1915: Grantland Rice, Ed Curley, unidentified, Bunk MacBeth, Bill Slocum, Fred Van Ness, Harry Schumacher, and Lieb*

increasingly volatile financial market forced mergers and business failures that made sportswriting a peripatetic profession. Through the two decades following his beginning with the *Press,* Lieb would write for the *Morning Sun, Evening Telegram,* and *Post.*

Price, his first editor, quickly recognized Lieb's ability. Beginning in 1915 the *Press* sent him to Georgia with the Yankees to cover spring training. By 1919, after Lieb had joined the staff of the *Sun,* the Elias Bureau syndicated his preseason columns to more than one hundred American newspapers; soon his World Series columns achieved the same circulation. Lieb was elected to serve as a nonvoting member of baseball's Rules Committee in 1920, and he served as the president of the New York chapter of the Baseball Writers Association from 1921 until 1924.

While the rest of what Lieb referred to as the "freshman class" of 1911 (Broun, Rice, and Runyon) left to cover other sports or left sportswriting, Lieb remained. As he explained, "only I 'went the route.' I admired the others for their accomplishments; but never for a moment was I jealous of them. I preferred baseball – it had become part of my life." His daily routine was enviable:

> I scored the game, made notes on outstanding plays, sent the Associated Press lead and box score by messenger to their office downtown, and then went home for a leisurely dinner. . . . Later I took the subway down to the newspaper offices, usually reporting around 7:45, and two hours later handed my copy to the editor.

Although Lieb's deadlines and office locations would change, the essence of his life did not – at least for the next two decades.

He considered the popular writing styles in these early years of sportswriting to be vulgar. He characterized the approaches as the "Gee Whiz" style (in which the reporter played a cheerleading role for the hometown team) and the "aw shucks" style (in which the writer's cynicism obscured the details he reported of the game). Refusing to adopt such styles, Lieb reported as if the audience were composed of readers such as he – as if they wanted what he had wanted as a young connoisseur of sportswriting in Philadelphia. "I think I wrote first

*Lieb (right) with Babe Ruth in Saint Petersburg, Florida, circa 1923*

of all to satisfy Fred Lieb the reader of sports pages. As a reader I wanted honest reporting, and I wanted to know not just who had won, but how and why." Yet Lieb also aimed to communicate the emotion of the game to his readers, for, as he added, "If I could be made to feel by the writer that I had missed something exciting, so much the better."

For example, consider Lieb's reporting on the 1940 All-Star Game:

> On a hit-and-run play, Herman whistled a single past Third Baseman Travis, sending Vaughan to third. West, making his first appearance as a mid-summer All-Star hitter, swung and missed a first strike, but connected with the second pitch Ruffing threw to him and sent into the unscreened section of the right center stand for a home run as the crowd erupted, giving the Nationals a quick block of three runs.

Lieb's column let readers know what had happened and also let them know that they had missed something electrifying. Then he assessed why it had happened:

> One must question Cronin's judgment for starting Ruffing. . . . With an abundance of pitchers, including Milnar, one of the standouts of the first half, Cronin led off with a hurler who has had difficulty in winning half of his games and who was badly battered by the lowly Athletics a few days before.

Lieb used a different style in writing as a columnist. Using skills developed through the short stories written in his youth, he painted pictures of players, scenes, and ballparks in his feature stories. Jimmie Foxx became "a broad-shouldered Maryland plowboy." Burleigh Grimes's spitball was an "anointed globule," and, in another sobriquet Lieb bequeathed to baseball vocabulary, Yankee Stadium became "The House that Ruth Built." His prose thickened, and some of his formulations (Lieb called the Black Sox scandal of 1919 "the big blotch on baseball's escutcheon") attained a kind of abstract melody perhaps traceable to the German-speaking home of his boyhood. Ascending from the lowly *Press,* where it had taken him only a year to get a byline, to the more highly regarded *Post,* Lieb climbed rapidly, in both salary and prestige, in the ranks of New York sportswriters. His election as president of the New York chapter of the Baseball Writers Association of America indicated that his peers as well as his readers regarded him highly.

Eventually the Yankees, in order to promote fan interest by having extensive preseason coverage, began to pay for Lieb's annual trek to the team's spring training site. Lieb became quite close to the Yankee players, especially Babe Ruth and Lou Gehrig, and this naturally gave Lieb some scoops, although he refrained as much as possible from dragging the private lives of players into the public eye. Ruth and Lieb, for instance, became frequent poker and golf companions, and such friendship enabled Lieb to mediate in disputes between Ruth and Yankee ownership, Commissioner Kenesaw Mountain Landis, and much of the rest of the press. His relationship to Ruth also may have allowed Lieb, as he later boasted, to dictate Bob Considine's *The Babe Ruth Story* (1948) in a little more than two weeks – although it seems more likely that Lieb simply provided voluminous information to Considine and perhaps assisted in the research.

As close as Lieb and Ruth were, Lou Gehrig became Lieb's closest friend. The sportswriter's fluency in German made such a favorable impression with Gehrig's mother that Lieb was the one who talked her into assenting to Lou's marriage and also, on the day of the event, into attending. Lieb was also with Gehrig's wife, Eleanor, when Lou broke the news of his fatal disease to her by phone. Yet while these intimacies accelerated his climb in both pay and prestige among New York sportswriters, Lieb's friendship with Landis changed his life forever.

As president of the Baseball Writers Association, Lieb met Landis almost as soon as the latter took office in 1921. Their relationship remained one of mutual respect rather than personal intimacy, and Landis's trust in Lieb's judgment and character paid off for Lieb in 1931. After the World Series that year entrepreneur Herbert Hunter, who had conducted a barnstorming tour in 1922, approached Landis to propose another such tour of Japan by major league players. Landis balked. He knew that Hunter had financial difficulties and believed that he had mismanaged the previous barnstorming team – perhaps even had thrown a game to a team of Korean all-stars. But when Lieb, who had invested wisely and was rather secure during the Depression, approached Hunter and offered to help underwrite the proposal financially, Landis approved it. Lieb was unable to coax Ruth into coming along, but he did land Gehrig and five other future Hall of Famers to play on the team. Splitting the profits evenly with Hunter, Lieb returned as a rather rich man. This financial independence led him to entertain thoughts of retirement, and the financial situation of New York newspaper publishing indirectly hastened his departure.

Lieb had become dissatisfied with changes the Depression was causing in the life of the New York baseball writer. He had achieved a top position at the *Post,* but unfortunately in 1930 the six papers competing for the evening market were seeking to save expenses and reached an agreement that produced only one play-by-play wire from each New York ballpark. This meant that Lieb could write only second-day game stories and feature articles, many of which dropped his byline. Because fans were not interested in recountings of yesterday's game, baseball writers had to turn to other subjects, and, as Lieb wrote, writers "gave less and less attention to the game and more space to the personal lives, habits, and whims of the players," a focus that he found journalistically repugnant. For Lieb, something of a golden age of baseball writing had passed. Following the 1934 season, twenty-three years after entering New York and its newspaper world, he left them both behind. He and his wife moved to Saint Petersburg, where he had gone every spring since 1915 to cover the Yankees' training camp and where Lieb was to begin the most prolific period of his career.

Lieb moved to Saint Petersburg intending to freelance full-time and even termed his migration a "retirement," but J. G. Taylor Spink, publisher of *The Sporting News,* prevailed on him to return to newspaper sportswriting. Lieb and Spink had met in the press box of the Polo Grounds during the downpours that had plagued the 1911 World Series, and Spink had occasionally solicited from Lieb sketches of New York players and expansions of New York feature articles. Spink had also published one of Lieb's pieces of baseball fiction.

Lieb became one of the two national correspondents on the staff of *The Sporting News,* and from 1935 until 1958 Lieb was to cover every World Series and All-Star game for Spink. Lieb also began to publish in major periodicals. Although he had sold a few pieces to *Liberty* and *Eagle Scout* earlier in his career, he first reached a large national audience in *The Saturday Evening Post* in 1938 with "Hits Are My Bread and Butter," a story that detailed the often stormy relations between players and official scorers. Lieb himself had experienced many of these moments and had built a reputation as a stubborn, respected, and influential scorekeeper.

In 1922 he refused to reverse a controversial first-inning call that prevented Howard Ehmke

*The team that Lieb and promoter Herbert Hunter formed for a barnstorming tour of Japan following the 1931 World Series. With seven future Hall of Fame members, the touring party included: (back row) Larry French, Mickey Cochrane, Frank O'Doul, Lieb, Lefty Grove, Hunter, Bruce Cunningham, Tom Oliver, George Kelly, and Lou Gehrig; (front row) Willie Kamm, Rabbit Maranville, Frank Frisch, Muddy Ruel, Al Simmons, Ralph Shinners, and Doc Knolls, the trainer.*

from pitching two consecutive no-hitters. In the same year he was also party to the reversal of official scorer John Kieran's ruling on a error, and this reversed call credited Ty Cobb with a hit that boosted Cobb's final batting average from .399 to .401 for the season. He had shouting matches with Babe Ruth and Oscar Vitt, whose argumentative refrain served as the title for "Hits Are My Bread and Butter," but, as Lieb told his readers, he never reversed a call that he had made. He also placed two pieces in *Collier's* in the early 1940s – "Strictly Screwball," a tragic biography of alcoholic gambler-ballplayer Hal Chase, and "The Great Exhumation," an account of the influx of older players returning to the major leagues during World War II.

His first book appeared in 1939. Strangely enough, it had virtually nothing to do with baseball. *Sight Unseen: A Journalist Visits the Occult* treated instead the mystic experiences in which he and his wife regularly engaged. Apparently Lieb genuinely believed that his wife spoke to deceased acquaintances through her Ouija board. She also supposedly divined, through palm reading and transcendental meditation, the onset of Lou Gehrig's fatal illness, the 1932 world championship for the Yankees, and other events.

In 1940 Spink began serializing Lieb's accounts of past World Series in *The Baseball Register*. From 1935 to 1942 Spink also persuaded Lieb to come to Saint Louis during the summer vacation months when *The Sporting News* was understaffed. Although Lieb kept Saint Petersburg as his permanent address, his stays in Saint Louis sometimes lasted as long as six months.

The outbreak of war further depleted the ranks of sportswriters, and in 1943 Lieb agreed to take a full-time position at the office of *The Sporting News* in Saint Louis. During this period he regularly covered the hometown Browns and Cardinals, and he supplied weekly features and editorials to *The Sporting News* and other Spink publications such as the *Register* and the *Baseball Guide and Record Book*.

During this period Lieb embarked on his most famous contribution to sports literature, the Putnam

Team History series. Beginning in 1944 with *The St. Louis Cardinals: The Story of a Great Baseball Team,* Lieb produced five similar volumes during the next eleven years. Each history provided an exhaustive and embellished narrative of professional baseball in a particular city. What makes his achievement all the more remarkable is that he wrote about teams with which he had little experience – Detroit (1946), Boston (1947), Pittsburgh (1948) – and about the Browns and their successors, the Baltimore Orioles (1955). *Connie Mack: The Grand Old Man of Baseball* (1945) served as the official history of the Philadelphia Athletics, and Lieb collaborated with Stan Baumgartner on *The Philadelphia Phillies* (1953). Each volume went through several printings (the Cardinals book alone went through eight), and many were updated and republished in the late 1940s and early 1950s, especially if a team enjoyed a banner season. In addition, three works were republished and distributed to American servicemen in Europe and Korea by Editions for the Armed Services.

This series made Lieb more widely known as a baseball historian. This should not be taken as a denigration of his sportswriting talent but rather as a credit to one of the most imposing writers of baseball history. Given the breadth and scope of Lieb's familiarity with baseball's past, it is hard to imagine anyone in a better position to recount the old days to the current readers. Lieb's histories still serve as foundational references for baseball research, and it is difficult to find any historical work on baseball that does not contain Lieb's name either in its bibliography or index.

Unfortunately the team histories were to be marketed to hometown audiences, and their writing style too often slips into that of the Gee Whiz rooter that Lieb had earlier disparaged. Often his straining to put everything in a positive light smacks of effrontery – for example, Detroit, Saint Louis, and Philadelphia are each characterized as "the greatest ball town in the nation." But the historical value of these works remains invaluable, as in many cases they provide the only remaining records of baseball voices that antedate the turn of the century.

In the midst of this productive period Lieb left Saint Louis in 1948 and returned to Saint Petersburg for good. Lieb wrote later, "I felt I could do the same work in St. Petersburg as I did in St. Louis and get in an occasional game of golf as well." He kept himself busy by completing the team history set and collecting pieces for an anthology, *The Story of the World Series* (1949). This book was reprinted several times and was revised and expanded in 1950 and 1965 editions.

His vignettes in *The Story of the World Series* retain the heavily figurative, tightly packed style of his early newspaper features, making Lieb's experienced eye one with the reader's. "[Eddie] Collins doubled down the left-field foul line to start the sixth, and then came the denouement," Lieb describes the 1911 World Series; "[Frank] Baker, the succeeding batsman, caught one of [Rube] Marquard's fast shoots on the end of his bat and sent it hurtling over the right-field fence for a home run, the gleeful Collins prancing in ahead of him." Reaching back decades, Lieb captured the moment for the reader. As Grantland Rice put it, "Freddy Lieb could write a story on official fielding averages and make it sound interesting."

In 1950 he published *The Baseball Story,* a nondescript genealogy that discounted the roles of Abner Doubleday and English cricket in the origins of baseball, which Lieb attributed to ancient Egypt. While completing these projects Lieb continued to write for *The Sporting News,* frequently as a features author, a World Series and All-Star Game correspondent, and an obituary writer.

At the age of seventy Lieb finally discontinued his correspondent position and – except for collaborating with Bob Burnes, Les Biederman, and Spink in publishing *Comedians and Pranksters of Baseball* (1958) – slowed down considerably. While he still contributed an occasional feature article or obituary to *The Sporting News,* his effort in *Comedians and Pranksters* amounted to little more than an editing job, filling in spaces between sections by the other authors. His claim that he ghostwrote Johnson Spink's *Judge Landis and 25 Years of Baseball* (1947) during this period is generally accepted, although the workmanlike prose of the book is impossible to identify as Lieb's.

In 1965 he was elected to a seat on the Veterans Committee of the national Baseball Hall of Fame, which was responsible for selecting for membership players who had not been selected by the younger members of the Baseball Writers Association of America (BBWAA). At the age of seventy-seven Lieb had seen more than eight thousand games, forty-seven World Series, and thirty All-Star Games. Given his astonishing memory, Lieb was an obvious choice for the position.

He also resumed some conventional sportswriting with some occasional columns for the *St. Petersburg Times.* By 1971 his "Hot Stove League" column, which was syndicated through the Elias Bu-

*Lieb with Joe Cronin, attending the annual Hall of Fame induction ceremonies in Cooperstown, New York, August 1976*

reau, became a regular weekly feature that appeared from October until Opening Day.

In 1972 the BBWAA bestowed on Lieb the J. G. Taylor Spink Award, which mandated his induction into the Baseball Hall of Fame in 1973. Lieb, Daniel, and Stockton were to be the tenth, eleventh, and twelfth sportswriters so honored. Lieb and Daniel were the first two living writers ever elected, and for many years Lieb was the only writer with his own exhibit in the Hall of Fame Museum.

While Lieb was in Cooperstown he decided to write a final book, and upon returning to Saint Petersburg, where he lived alone in a rooming house, he began this last labor. In 1977 he published *Baseball As I Have Known It,* a work which, by his own assessment, made him at age eighty-nine "the oldest professional baseball writer to complete a book on the subject."

Generally autobiographical but frequently incorporating chapters on baseball's tragic or heroic figures and events from 1911 to the 1930s, the book offers observations about the present and future status of baseball. Lieb tended to exalt the past at times, but he recognized that, while the form of the game remained constant, its content was constantly changing. He therefore praised expansion of the two major leagues, proclaimed the long-term financial health of the game, and admitted that today's players were superior to those of the past – sentiments rare among writers of his generation.

Lieb discontinued his regular winter column in 1977, but even as his health began to fail he occasionally wrote a column for the *St. Petersburg Times* or a feature for *The Sporting News.* In February 1980 Lieb's only child, Marie Lieb Pearsall, admitted him to a nursing home in Houston. On 3 June 1980, exactly one month after his final byline in *The Sporting News,* Lieb died in the company of his daughter and her only child, Mary Pearsall Holloway.

It is doubtful that Lieb could have been anything but contented. In reflecting on his life, Lieb told Jerry Holtzman, "there's nothing else I would have rather done. When I walked into the New York press box for the first time, I couldn't have been happier, not if I'd made it to the Oval Office of the White House." A critical appraisal of Lieb's work suggests that, for the most part, his contentment was well deserved. He had strong opinions, but he was dignified and soft-spoken and cultivated friends at every level of baseball, from owners and managers to players and fellow writers.

What made him a great sportswriter was his relationship with his audience – baseball fans. Through a sixty-nine-year career Lieb may have earned the supreme professional honor in his Hall of Fame induction, but the most telling award is one he received informally late in life. As Lieb walked through Saint Petersburg, an old gentleman approached him and introduced himself. "You've brought a lot of joy into my life with your baseball writing," he said. "I never miss your columns. Gosh, when I was a boy in New York, I used to enjoy your father's writing in the *Press* and later in the *Sun*. He used to write under the name of Frederick G. Lieb." To top it off, he added, "I tell you truthfully, you're a better writer than your father was!"

Beneath the irony occasioned by this old gentleman's words, his sincere appreciation shows the quality and scope of Lieb's gift, a gift so great that it must seem to have required two men to embody it. For Lieb, baseball writing was not a job, but a life. He once remarked, "I love baseball. I could watch it every day, every year. And to think I get paid for watching it." Baseball fans everywhere are grateful that he was paid to write about it.

**References:**

Jerry Holtzman, ed., *No Cheering in the Press Box* (New York: Holt, Rinehart & Winston, 1973);

Jonathon Yardley, "Reading of the Game," *Washington Post,* 11 December 1977, p. E7.

# A. J. Liebling

*(18 October 1904 - 28 December 1963)*

Sanford Pinsker
*Franklin and Marshall College*

See also the Liebling entry in *DLB 4: American Writers in Paris, 1920-1939.*

MAJOR POSITIONS HELD: Sportswriter, *The New York Times,* (1925); reporter, feature writer, *Providence Journal and Evening Bulletin* (1927-1939); feature writer, *World Telegram* (1931-1935); sportswriter and columnist, *The New Yorker* (1935-1963).

BOOKS: *They All Sang, From Tony Peter to Rudy Vallée,* as told to Liebling by Edward B. Marks (New York: Viking, 1934);

*Back Where I Came From* (New York: Sheridan House, 1938);

*The Telephone Booth Indian* (Garden City, N.Y.: Doubleday, Doran, 1942);

*The Road Back to Paris* (Garden City, N.Y.: Doubleday, Doran, 1944; London: M. Joseph, 1944);

*The Republic of Silence* (New York: Harcourt Brace, 1947);

*The Wayward Pressman* (Garden City, N.Y.: Doubleday, 1948);

*Mink and the Red Herring: The Wayward Pressman's Casebook* (Garden City, N.Y.: Doubleday, 1949);

*The Honest Rainmaker: The Life and Times of Colonel John R. Stingo* (Garden City, N.Y.: Doubleday, 1953);

*Chicago: The Second City* (New York: Knopf, 1952);

*The Sweet Science* (New York: Viking, 1956; London: Gollancz, 1956);

*Normandy Revisited* (New York: Simon & Schuster, 1958; London: Gollancz, 1958);

*The Press* (New York: Ballantine, 1961; revised editions, 1964, 1975);

*The Earl of Louisiana* (New York: Simon & Schuster, 1961; London: W. H. Allen, 1962);

*Between Meals: An Appetite for Paris* (New York: Simon & Schuster, 1962);

*The Jollity Building* (New York: Ballantine, 1962);

*The Most of A. J. Liebling,* selected by William Cole (New York: Simon & Schuster, 1963);

*A. J. Liebling*

*Mollie and Other War Pieces* (New York: Ballantine, 1964);

*The Best of A. J. Liebling,* selected by William Cole (London: Methuen, 1965);

*Liebling Abroad* (New York: Playboy Press, 1981);

*Liebling at Home* (New York: Wideview Books, 1982);

*A Neutral Corner: Boxing Essays,* edited by Fred Warner and James Barbour (San Francisco: North Point Press, 1990).

OTHER: *The Republic of Silence,* edited by Liebling (New York: Harcourt Brace, 1947).

In "Tradition and the Individual Talent" (1919) T. S. Eliot argues that if tradition consisted of little more than following the way of a previous generation it should be positively discouraged – if for no other reason than novelty is better than mere repetition. He goes on, however, to add that tradition, properly understood, is of much wider significance, for it embodies "the historical sense," not merely in the sense of the pastness of the past, but more important, in the recognition of its abiding presences. Eliot was talking, of course, about modern poets in general and of himself in particular, but he was well aware that "no artist of any art has his complete meaning alone. His significance, his appreciation is the appreciation of his relation to the dead poets and artists. You cannot value him alone; you must set him, for contrast and comparison, among the dead."

That an essay about A. J. Liebling, sportswriter par excellence, should be prefaced by quotations from a mandarin such as Eliot may seem surprising, but its central point about the tradition and an individual writer is not limited to poets; indeed, Liebling would be the very first to insist that his reflections on boxing were both formed by, and indebted to, Pierce Egan, late-eighteenth-century London's consummate chronicler of what he called "milling" and we know today as boxing. Indeed, the very title of Liebling's first collection of essays about boxing, *The Sweet Science* (1956), is yet another phrase first popularized by Egan. Liebling, however, borrowed much more than chunks of Egan's colorful lingo, for it was Egan's *Boxiana,* a collection of boxing pieces published in 1813, that first demonstrated how such writing might go well beyond a mere compilation of synopses of fights. As Liebling explains the phenomenon:

> Egan's round-by-round stories, with ringside sidelights and betting fluctuations, are masterpieces of technical reportage, but he also saw the ring as a juicy chunk of English life, in no way separable from the rest. His accounts of the extra-annular lives of the Heroes, coalheavers, watermen, and butchers' boys, are a panorama of low, dirty, happy, brutal, sentimental Regency English that you'll never get from Jane Austen. . . . Egan's pageant scenes of trulls and lushes, toffs and toddlers, all setting off for some great public, illegal prizefight, are written Rowlandson, just as Rowlandson's print of the great second fight between Cribb and Molineaux is graphic Egan.

Not only does Liebling trot out Egan's name (usually followed by a half-playful apostrophe linking him to a classical historian such as Thucydides), but he also reduplicates Egan's working assumption that a prizefight happens within a series of concentric circles. There is, of course, the ring in which the contestants meet, but as Liebling's essays make clear, there is also the circle of trainer-seconds who surround their fighter in the weeks, often months, of preparation and, finally, the circle of the crowd that makes its way to the arena on fight night. Taken together, these disparate levels, or "rings," of involvement constitute the colorful life that Liebling's essays explore.

Egan himself felt that boxing deserved precisely this manner of treatment, and in his introduction to the first volume of *Boxiana,* he answered potential critics this way:

> To those, Sir, who prefer *effeminacy* to hardihood – assumed *refinement* to rough *Nature* – and whom a *shower of rain* can terrify, under the alarm of their polite frames, suffering from the unruly elements – or would not mind Pugilism, if BOXING was not so shockingly vulgar – the following work can create no interest whatsoever; but to those persons who feel that Englishmen are not automatons. . . . Boxiana will convey amusement, if not information . . .

Egan wrote about boxing as if it were life – "low life," to be sure, but life nonetheless. By contrast, Hazlitt wrote but a single boxing story, as, in our age, writers such as Norman Mailer or Joyce Carol Oates would each produce a small volume chronicling a single, paradigmatic fight; but for Egan – as for Liebling – boxing deserved continuing treatment. As such, the sweet science was neither an occasion for amateur psychoanalysis nor an extended metaphor for something else. Rather, "milling" and the "coves" (fans) attracted to it participated in an exercise that demanded discipline, skill, and perhaps most of all, heart. Thus, if Egan represents the best of a noble "tradition," Liebling is the "individual [American] artist" most formed by its suppositions. He is, at the same time, also the beneficiary of other significant influences.

The first is New York City itself. Abbott Joseph Liebling (known to his friends as Joe) was born there in 1904; he was also educated there – not only in its public schools, but also on the Manhattan streets he walked with his father, a well-to-do furrier, as well as at the racecourses and ballparks they frequented. According to Joseph Epstein in "A. J. Liebling – The Minnesota Fats of American Prose" (1971), Liebling was "a true New Yorker when that meant something other than somebody who is put upon by high rents, rudeness, human squalor, and an urban scene that is not so much distracting in its variety as it is jarring in its tumult." Liebling savored the variety because he loved life – whether it took the form of a gigantic meal (he was a legendary eater, with the girth to prove it) or the chance to meet interesting city types. Manhattan, one might say, made Liebling receptive.

When in 1917 a bachelor uncle arrived from California bearing vivid tales about boxers who made their home in the Golden State (James J. Corbett, Joe Choynski, James J. Jeffries, Tom Sharkey, Abe Attell, and Jimmy Britt), his thirteen-year-old nephew listened with rapt attention. Listening deeply and recording what he heard with both accuracy and panache would remain Liebling trademarks. But it is one thing to learn about boxing secondhand and quite another to experience it fist-in-glove. To his credit, the young Liebling knew the difference. After Uncle Mike's indoctrination – what might be called an oral tradition as opposed to the writerly one he acquired later from Egan – Liebling tells us in his introduction to *The Sweet Science* that:

> I boxed for fun whenever I had the chance until I was twenty-six and earning sixty-three dollars a week as a reporter on the *Providence Journal and Evening Bulletin*. I continued to box occasionally for many years more, generally just enough to show I knew what was all about it, as the boys say. I went shorter rounds every time. The last was in 1946, and the fellow I was working with said he could not knock me out unless I consented to rounds longer than nine seconds.

Liebling's self-deprecating humor bears more than a few resemblances to *New Yorker* writers such as Robert Benchley, E. B. White, and James Thurber. He was, after all, writing a series of essays on boxing for the magazine's pages during the days when columns like "Talk of the Town" and the two-thousand-word sketches were redefining the traditions of American humor; furthermore, Liebling's peculiar joie de vivre often took comic turns. The result is probably one of emphasis rather than appropriation, a way for Liebling to incorporate aspects of Benchley's "little man," a character frustrated by modern technology and increasingly anxious about urban life, into some of the turns his own style took. Both Benchley and Liebling were rotund figures, hardly "little men" in the physical sense, but that was part of the joke; so too were the blows they concocted for themselves and then delighted in reporting. Given their conservative temperaments, much of modern life must have struck them as a diminution, a pale reflection of a manlier past. No doubt this spirit partly accounts for Liebling's attraction to Egan and the "sweet science of bruising" (the operative phrase he condensed into his own title) it chronicled.

Mostly, though, Liebling lacked the floating anxieties, nervous tics, and neurotic compulsions that are often associated with *The New Yorker*'s humor. Perhaps he simply was not sad enough to require the mask modern humorists uneasily wear; but if so, he was every bit as independent of spirit and prone to mock pretentiousness as they were. Like his voracious appetite for the city's sights and sounds, Liebling foreshadowed early a lifelong habit of bristling in the face of authority. He enrolled in Dartmouth College in 1920, but was expelled during his junior year for cutting too many required chapel services (he later described himself as "an agnostic of Jewish origin"); and after graduating from Columbia University's School of Journalism with a B. Litt. in 1925, he lasted a scant eight months with the sports department of *The New York Times* before being fired for christening a basketball referee "Ignoto" because he did not know what his name was.

After a year in Paris, where he developed a passion for French cuisine, Liebling worked on various newspapers, including the *Providence Journal and Evening Bulletin,* and New York City's *World Telegram,* before he and his style found a congenial home at *The New Yorker* magazine in 1935. He remained there for the rest of his life, writing about what interested him (his profiles included jockey Eddie Arcaro, Father Divine, Gen. George Marshall, and of course, a wide assortment of boxers), and always in a style he thought made sense: unpretentious, irreverent, passionately involved.

Liebling wrote well about a wide variety of subjects, including food, travel, and popular amusements, but his best nonsports book is arguably *The Earl of Louisiana* (1961), a profile of Earl Long and the 1959 gubernatorial race in Louisiana. The Louisiana State University Press republished the book in a paperback edition (1970) not only because of its concentration on Louisiana politics but also because the book itself holds up extraordinarily well. Like Liebling's best writing, it has quietly moved from the realm of journalism to that of literature. Much the same thing can be said about *Chicago: The Second City* (1952).

Editors who preferred timid journalism often did not know what to do when faced with one of Liebling's paragraphs; however, Harold Ross, *The*

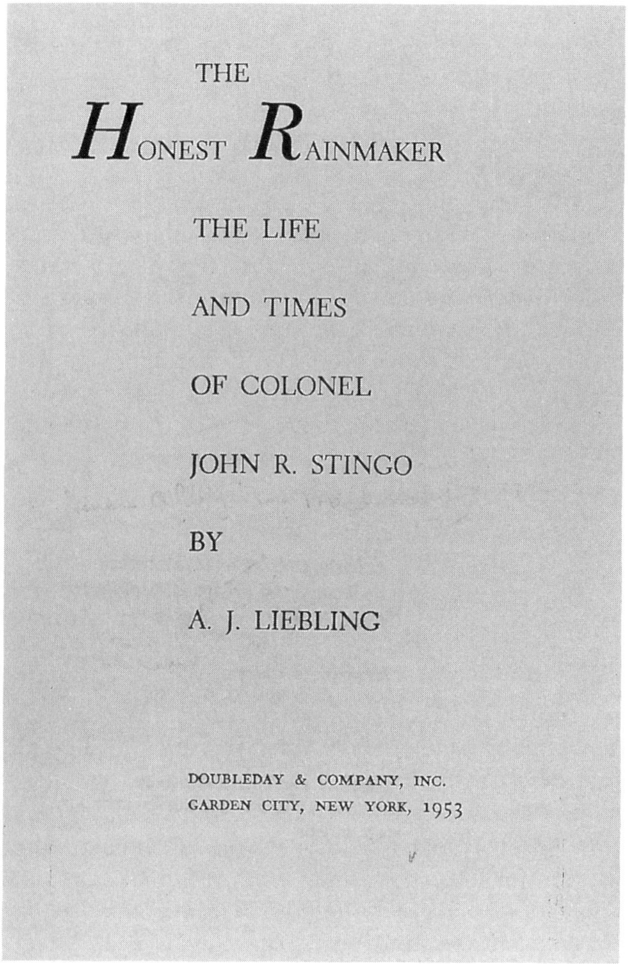

THE

HONEST RAINMAKER

THE LIFE

AND TIMES

OF COLONEL

JOHN R. STINGO

BY

A. J. LIEBLING

DOUBLEDAY & COMPANY, INC.
GARDEN CITY, NEW YORK, 1953

*Title page for Liebling's 1953 biography of the sports columnist John R. Stingo*

*New Yorker*'s editor, did: he simply gave Liebling the freedom and space that distinguished writing requires. The result was a style as difficult to describe as it is impossible to duplicate. Nonetheless, many writers have tried to describe that style, not because there was any disagreement about the glories of Liebling's prose (there was not), but because each wanted to emphasize the aspects that seemed to matter most. In his foreword to *A Neutral Corner* (1990) Bill Barish suggests that Liebling's pieces "operated on the principle of inclusion, not exclusion, and unfolded at the leisurely pace of an evening stroll," while Red Smith in *To Absent Friends* (1982) called attention to Liebling as the "witty writer with a remarkable ear for the speech of a Whitey Bimstein or Al Weill"; but most critics agree that Epstein probably comes closest to what makes a Liebling sentence both characteristic and memorable in "A. J. Liebling – The Minnesota Fats of American Prose." Epstein compares what Liebling does with words to the incredible pool shots Minnesota Fats regularly pulled off:

> Like Minnesota Fats with a cue ball, so Liebling was with an English sentence – there was nothing he couldn't make it do, including the work of three normal paragraphs. He could make a single sentence ride the crest of a classical allusion, dip down to capture a passing irony, pick up a startling analogy along the way, curl over to the side for a comic touch, kiss off a serious observation, and, *plunk,* drop neatly into the pocket of a period to score a delicious original point.

Liebling worked several beats and wrote a wide variety of essays, casuals, and books, but it was as a writer about boxers and boxing that his permanent reputation is likely to rest. As Epstein put it, "No American before or since Liebling's time has written so well about boxing. He wrote about the sport knowingly, feelingly, and absolutely without crap."

Ross, however, had his doubts – not about Liebling's prose style (he recognized good writing when he saw it), but, rather, about running boxing essays in his magazine's pages. As such, he was decidedly cool when it came to approving such assignments or indeed for anything that smacked of the low life; but Liebling was the exception that proved the rule. There was an indefinable *something* about his accounts that added up to more than who fought whom and which man emerged victorious.

Liebling could have followed his low life bent wherever it might have led him, but in 1939 he left both boxing and *The New Yorker* behind to become a war correspondent. *The Sweet Science* represented his return to material that had engaged him before, but that now he polished to a perfection not seen since the days of Egan. The result was a new, distinctly American "tradition" of sportswriting, one fashioned from what might be called the trope of the fist. For it was the laying on of hands balled into fists that connected him, first, to Jack O'Brien, and then backward, in a succession of blows, to the likes of Jim Corbett and John L. Sullivan:

> He [Jack O'Brien] hit me, for pedagogic example, and he had been hit by the great Bob Fitzsimmons, from whom he won the light-heavyweight title in 1906. Jack had a scar to show for it. Fitzsimmons had been hit by Corbet, Corbet by John L. Sullivan, he by Paddy Ryan, with the bare knuckles, and Ryan by Joe Goss, his predecessor, who as a young man had felt the fist of the great Jem Mace. It is a great thrill to feel that all that separates you from the early Victorians is a series of punches on the nose.

At this point one is far from what Eliot had in mind when he reshuffled the major players of English literature so that John Milton was "out" and John Donne "in." At the same time, however, it is precisely this strong sense of historical connection that distinguishes Liebling's reportage from sportswriters unfamiliar with the fists that came before them. As Liebling himself put it, in phrasing chosen to resemble a well-oiled machine: "The Sweet Science is joined onto the past like a man's arms to his shoulder." Such a writer sees context-as-continuity, the prizefight before one's eyes as refracted through tradition's prism.

Liebling, in short, was a clear-eyed, decidedly unsentimental reporter. Deeply committed to the sweet science, he knew that boxing's long history of crosses and double-crosses (once again, terms he borrowed from Egan) – not to mention its longer history of pummeled bodies and battered brains – made outrage about some recent black-eye the sport presumably suffered so much nonsense. He also knew that boxing had long ago run out of black eyes – and this *before* the likes of promoter Don King or boxers who might be as widely known for their indictments as for their boxing skills. But for Liebling the seeds of boxing's demise had been sown much earlier – in sociocultural conditions such as "full employment and late school-leaving age" (Egan, in 1829, had pointed out that "drummers and boxers, to acquire excellence, must begin young," and Liebling agreed.) Liebling also believed that the more immediate crisis in the sport of boxing was brought on by what he roundly dismissed as "the ridiculous gadget called television." The following screed, taken from his introduction to *The Sweet Science,* leaves little doubt about Liebling's credentials as a purist:

> [Television] is utilized in the sale of beer and razor blades. The clients of the television companies, by putting on a free boxing show almost every night of the week, have knocked out of business the hundreds of small-city and neighborhood boxing clubs where youngsters had a chance to learn their trade and journeymen to mature their skills. Consequently the number of good new prospects diminishes with every year, and the peddlers' public is already being asked to believe that a boy with perhaps ten or fifteen fights behind him is a topnotch performer. Neither the advertising agencies nor brewers, and least of all the networks, give a hoot if they push the Sweet Science back into a period of genre painting. When it is in coma they will find some other way to peddle their peanuts.

Given Liebling's crankiness about the cultural forces that, taken together, put boxing into such a sour pickle, it is worth noting that *The Sweet Science* records the "Heroic transactions" (Liebling's phrase) that occurred between June 1951 and September 1955. Liebling understood that the history of boxing is the story of periodic doldrums punctuated by stretches of delight. Thus, the dull space between the defeat of John Broughton in 1750 and the rise of Daniel Mendoza in 1789 (recorded in *Boxiana*), or what Liebling called "the Dark Age" between the retirement of Tunney in 1928 and the ascension of Joe Louis in the middle 1930s, were periods in which "champions of little worth succeeded each other with the rapidity of the emperors who followed Nero, leaving the public scarce time to learn their names." If Liebling's sense of "tradition" was an expansive one, so, too, was his sense of history. Or put a slightly different way, he brought equal measures of learning and wit to his enterprise.

For a while Liebling stopped writing about boxing, and during this period he bashed away at the American press in collections such as *The Wayward Pressman* (1948) and *Mink and the Red Herring: The Wayward Pressman's Casebook* (1949). In 1949 his marriage to his first wife, Anne Beatrice McGinn, ended, and he married Lucille Hille Spectorsky. Eventually he returned to boxing as a subject, partly because the press proved less competitive than the ring ("Faced with a rival," Liebling observed, "an American newspaper will usually offer to buy it.") and partly because June 1955 seemed a propitious time for his return. Liebling's instincts proved to be as on-target as a well-aimed punch. Not only was Rocky Marciano about to make his charge toward the heavyweight championship, but Joe Louis, now entering his eighteenth year as a competitor, was preparing to make his last, gallant stands. The result was a particularly fruitful combination for Liebling's interest and his prose. In the new cycle he recorded, "Marciano knocked out three world's heavyweight champions, Louis, [Joe] Walcott, Ezzard Charles, and wound up beating Moore, the heavyweight-light-heavyweight, who challenged for the title at thirty-nine." Moreover, there were intriguing subplots as well: the graph of Sugar Ray Robinson's comebacks and downfalls, as well as accounts of such minor heroes as Sandy Saddler.

In effect, *The Sweet Science* has the porous unity of its model, Egan's *Boxiana.* What gives the collection heft, however, is Liebling's ability to capture the ethos of a time and place, long after the fighters and trainer he writes about have been replaced by the likes of Muhammad Ali (a fighter Liebling knew, and did not especially care for, as Cassius

Clay), the machinations of promoter Don King, or the terrifying spectacle of Mike Tyson in his prime. The collection's dominant motif is the rise of Marciano (heavyweight boxing's only undefeated champion) and, in Liebling's pithy, no-nonsense words, "the fall of everybody who fought him," along with such interthreading subplots as "the comeback of Sugar Ray [Robinson] after his downfall before Turpin, and his re-downfall before Maxim" and, as in Egan's model, "the characters who hold the book, and the whole fabric of the Sweet Science together" – namely, the trainer-seconds.

*The Sweet Science* begins with an account of the fight between Louis and Lee Savold that took place in the Polo Grounds on 15 June 1951. The occasion not only afforded Liebling the opportunity to go on record with the obvious – namely that "there hadn't been anybody remotely like Louis since Dempsey in the early thirties" – but also to make it abundantly clear how unsatisfactory, indeed, inferior, televised fights are to the real event:

> Watching a fight on television has always seemed to me a poor substitute for being there. For one thing, you can't tell the fighters what to do. When I watch a fight, I like to study one boxer's problem, solve it, and then communicate my solution vocally. On occasion my advice is disregarded, as when I tell a man to stay away from the other fellow's left and he doesn't, but in such cases I assume that he hasn't heard my counsel, or that his opponent has, and has acted on it. Some fighters hear better and are more suggestible than others – for example, the pre-television Joe Louis. "Let him have it, Joe!" I would yell whenever I saw him fight, and sooner or later he would let the other fellow have it. Another fighter like that was the late Marcel Cerdan, whom I would coach in his own language, to prevent opposition seconds from picking up our signals. *"Vas-y, Marcel!"* ["Go for it, Marcel!"] I used to shout, and Marcel always *y allait* [went]. I get the feeling of participation that way that I don't in front of a television screen. I could yell, of course, but I would know that if my suggestion was adopted, it would be by the merest coincidence.

That Liebling personalized the boxing essay by giving it a voice and distinctive character is true enough; but he also injected measures of wit and wide cultural referencing that were noticeably absent before. Even more important, Liebling was a sportswriter who rigorously eschewed "sports English" – that is, the accumulated clichés of the trade, trotted out without a first thought, much less a second one – preferring, instead, to rack his brain for the precise word.

Liebling made writing about boxing look easy. His engagement and delight probably had something to do with the quality of his writing, but one also suspects that *The New Yorker*'s high standards played a significant role. Liebling knew that fighters are so focused on the job in front of them that they take neither heart nor counsel from the crowd noise – including the presumably savvy advice directed their way. Thus, the examples of Joe Louis "giving" it to an opponent (something he did regularly, if one merely waited long enough) or of Cerdan *allaying* because he took Liebling's French to heart are instances of his antic humor.

But Liebling was dead serious when he railed against televised fights, partly because they changed the old conditions with the flick of a knob, and partly because fighters no longer had a venue in which to learn the rudiments, much less the finer points, of the Sweet Science. Televised fights were, in Liebling's construction, "like the potato, which is only a succedaneum for something decent to eat but which, once introduced into Ireland, proved so cheap that the peasants gave up their grain-and-meat diet in favor of it." *Succedaneum,* meaning "substitute," is not a term one often encounters on the sports page, but for a writer who enjoyed tweeking the noses of highbrows who wrote for what he called "the quarterlies," Liebling was not above raising the level of his prose to the Latinate; nor was he shy about elaborating his analogy of the televised fight and the potato with an impromptu history lesson. Liebling's point is that after Irish peasants developed an unfortunate taste for the potato:

> the landlords let them keep just enough money to buy potatoes. William Corbett, a great Englishman, said that he would sack any workman of his he caught eating one of the cursed things, because as soon as potatoes appeared anywhere they brought down the standard of eating. I sometimes think of Corbett on my way home from the races, looking at the television aerials of all the little houses between here and Belmont Park.

No doubt some will read Liebling's fulminations about television and conclude that he was an elitist snob. They would be wrong, and one need only cite passages from his essay about Stillman's Gym to make the case. Titled "The University of Eighth Avenue" (included in *The Neutral Corner* [1940]), it is less a sketch about a famous New York City landmark, the place where countless fighters sweated and grunted their way to boxing championships, than it is an extended metaphor that cobbles the like and unlike in ways the poet William Wordsworth scarcely had in mind when he wrote his preface to the *Lyrical Ballads* of 1798. In this case, the *like* is Stillman's Gym as verbal portraiture would have it: "There are posters for the coming week's metropolitan fight shows . . . a wide wooden staircase leading up to the gym . . . [and] the joint always smells wrong." By contrast, the *unlike,* one that keeps winding itself around the ramshackle building at 919 Eighth Avenue like a tight fist, is decidedly more academic. Thus, turf around Stillman's Gym is, metaphorically at least, akin to the ivy-covered blocks that surround Columbia University, and Dr. Stillman himself likened to its president:

> Dr. Stillman, like so many college presidents nowadays, is not himself a teacher but rather an administrator, and the smell in the hall makes him feel there are limits to academic freedom. . . . In earlier stages of his academic career he used to speak exclusively in shouts, but now that he is in his latter sixties, his voice has mellowed to a confident rasp. The great educator has never, so far as is known, himself put on the gloves; it might prove a psychological mistake. Stillman excels in insulting matriculants so casually that they don't know whether to get sore or not.

Long before practitioners of the New Journalism crowed about pressing fictional methods into the service of nonfictional needs, Liebling knew the value of yoking a rarefied academic vocabulary with the rougher idioms of boxing, then sustaining the metaphor until the insistence seemed natural; or, by contrast, of taking a stray phrase and capitalizing on its potential:

> When I asked Weill he looked happy. "He is a nice boy," he said. "The dollar is his God. That is to say, he is a poor Italian boy from a large, poor family, and he appreciates the buck more than almost anybody else. Them guys is hard to get outa there. You want to look out for them young broken fighters," By "broken fighter," Weill, who is a purist, meant a fighter who is broke.

If Egan represents the tradition of popularizing boxing's colorful language, Liebling, the individual talent, made it his business to keep faith with its contemporary possibilities. Other sportswriters hung around training camps and schmoozed with the entourage, but Liebling kept his ears and eyes open far wider than most; and in the process he collected gems of verbal indirection that he later wrinkled into his essays, rather like a fighter who feints with his right and then clobbers you with a short left hook. Thus, a "broken fighter" is not one suffering from a fractured rib or concussion, but, rather, a man on an economic mission. Such fighters were dangerous in ways that the tale of the tape cannot always predict.

When Liebling's essays reached the main event, he could describe a decisive punch in ways that combined realism (he insisted on accuracy above all else) with similes that ran rings around the competition at ringside. Most fight reporters were lucky if they could keep the left hooks and right crosses straight. Liebling, often sitting rows from the action, had an eye as sharp as his wit. Here, for example, is how he describes the punch that provided Jersey Joe Walcott with a moment of truth and Marciano with a championship:

> Rocky had knocked Walcott out with a right that traveled at most twelve inches, straight across his chest to the champion's jaw. . . . The sports writer of the Philadelphia *Inquirer,* sitting ringside, wrote that Rocky had hit Joe with a "roundhouse right, swung from his hip and his heart." The punch was the antithesis of a roundhouse; it was a model of pugilistic concision. The newsreel film of the fight shows that both men started right leads for the head at the same moment. Walcott, the sharp, fast puncher, figured to get there first in such an exchange. But Marciano hit sharper, faster, and, according to old-timers, about as hard as anybody ever hit anybody. Walcott, the film shows, flowed down like flour out of a chute. He didn't seem to have a bone in his body. And so, after old Jersey Joe had piled up a lead by fighting the way he wasn't supposed to, Rocky knocked him out with the kind of punch he wasn't supposed to know how to use. "In other words," Charles Goldman said to me at Jimmy Tomato's party in the Hotel Warwick after the fight, "he equalized."

Liebling's interest in a fighter depended on a wide range of factors: his skills as a sweet scientist, his dramatic potential, and perhaps most of all, his place in boxing history. Taken together, *The Sweet Science* is a chronicle of the new cycle Marciano brought into being by knocking out fighters of genuine stature, a time when the sport produced fighters who generated wide interest and who more than justified Liebling's investment of time and sweat as his well-wrought paragraphs formed into essays.

By contrast, the essays edited by Fred Warner and James Barbour as *A Neutral Corner,* which gathers the previously uncollected *New Yorker* pieces Liebling wrote between 1951 and 1963, seem arbitrary. These essays emphasize Liebling's unflappability, the enduring soft spot he had for the Sweet Science, and the hard journalistic truths of trying to square his gourmet taste with a diner-sized wallet; in fact, they have little overall shape beyond that given by the ending of the boxing essays in 1963, when Liebling died. Still, the record assembled in *A*

*Liebling and his wife, novelist Jean Stafford, in 1961 (photograph by Therese Mitchell, courtesy of Joseph Mitchell)*

*Neutral Corner* is impressive, because Liebling was arguably the best journalist of his generation. Priding himself on his ability to write faster than the best writers and better than the fast writers, he must nevertheless have wondered at times about the gap between his talents and the uses to which they were put. If he had such misgivings, however, they did not appear in his work. *The New Yorker* pieces Liebling wrote in the last dozen years of his life retain an enthusiasm and freshness that lesser writers failed to achieve as they entered Madison Square Garden to cover another championship fight.

Liebling's success may be attributed to the fact that he was a better, purer writer than most of his contemporaries. And while Liebling continued to rail against that which corrupted the sport he loved, he refused to see the Sweet Science through psychoanalytic eyes or as symptomatic of large sociocultural conditions. He charted the rise and fall of various ethnic fighters because he knew that poverty was the spur that brought many would-be fighters to the gym in the first place, but Liebling left sociology to the sociologists and intellectualizing to the eggheads. What Liebling brought to the fight game was expertise and curiosity, and it was the last item that continued to energize his paragraphs until the end. He knew, for example, that greatness and its opposite could coexist in the same fighter; when a champion thumbed his opponent's eye, Liebling did not feel obligated to launch into a tirade about sportsmanship any more than he felt compelled to point out that the boxer who ended up on the canvas received what he deserved. Only the uninitiated rattled on – usually in shocked indignation – about bloodied eyes and noses; Liebling took such barbarism in stride, considering it an integral part of the Sweet Science.

Long before notions of political correctness became commonplace, Liebling's cheery descriptions of the Sweet Science's darker side were decidedly incorrect. One suspects he would have cared little about such criticism, arguing that those who turn away from boxing are also likely to turn away from life. Moreover, his style often served to transmogrify the seamy into the exotic, not only because his elaborate metaphors joined low life with high culture (one will look in vain for another sportswriter who could appeal to Ibn Khaldun, the Tunisian historian, as a way of understanding why Floyd Patterson was felled by Ingemar Johansson's mighty right hand); but also because, in the words of Barish's foreword to *A Neutral Corner,* "he loved to inflate a sentence, then release it like a helium balloon and watch it skitter across a page." Here, for example, is Liebling's description of the rematch between Patterson (a disciplined fighter he much admired) and Johansson, whose cavalier training methods earned him the moniker of "hedonist":

> In the third and fourth rounds, Patterson, full of energy, continued to hustle and cuff, pushing his man and spinning him. As Johansson tired, or grew discouraged, the challenger could afford to get set and punch more solidly. He was hurting Johansson with body punches in the fourth round, and although Whitey [Bimstein] whom I have since seen several times, insists that Johansson wasn't exhausted then, I thought he was. In the fifth, Patterson, very gay, as Pierce Egan, the Cholly Knockerbocker of the London Prize Ring, would have said, went about his man and nobbed him merrily. With the Swede, it was a case of bellows to mend, in Pierce's idiom. He was gasping. I saw him pull the right back for a punch; he was moving it so slowly that I could watch it from the lower stands. There was no snap left in his muscles, and he probably wished he had never seen an ice-cream soda. Patterson had no time to think about the rewards of virtue – he was too busy reaping them. He hit the hedonist with a left hook to the body and then switched it to the head. The Swede was in the same place when the punch landed as when it started. He went down like a double portion of Swedish pancakes with lingonberries and sour cream.

Few sportswriters have had a better eye for the shape and ring of individual sentences and the way that rhythms build to become coherent paragraphs.

Finally, though, what makes Liebling's boxing essays important can be encapsulated in a single word: *voice.* His phrases had the uniqueness of a thumbprint, and one would hardly mistake what he had to say about an aging Archie Moore (nearly always approving) or the flamboyant Cassius Clay (at best skeptical) with the sentences of anyone else. If places go through our best travel writers, rather than the other way around, one could say that boxing went through Liebling – which simply means that he was peculiarly attentive to boxing's ethos, as opposed to merely its more celebrated matches. Individual fighters were important to him, but so too was everything that contributed to their making: the trainer-seconds and prizefight bureaucrats, his tribe of fellow scribblers, and, of course, the fans who made their way to the staging area by subway and limousine. Like Egan, he saw prizefighting as emblematic of a larger social context, and his eye took in the human comedy that resulted from an Olympian viewpoint; and like Egan, he honored the best traditions of reporting about the Sweet Science while also advancing them.

Some may object that a talent as large as Liebling's might have produced short stories, novels – serious literature. No doubt Liebling must have had moments when he wondered much the same thing, especially during the years between 1959 and his death in 1963, when he was married to the novelist Jean Stafford. But given Eliot's pronouncements about the affinities between tradition and individual talents, apologies on Liebling's behalf are unnecessary. His best journalistic pieces were always close to literature.

In 1963, at the age of fifty-nine, Liebling died from complications of viral pneumonia. His three marriages produced no children, but his life at the writing desk created a legacy of a different sort. Speaking at the funeral, Joseph Mitchell, a fellow journalist and Liebling's closest friend at *The New Yorker,* recalled a conversation he had with the owner of a secondhand bookstore. The man told him that every few days a customer would inquire about one or another of Liebling's collections, and that, the proprietor insisted, was a certain indicator a writer's books would last. "Literary critics don't know what books will last," he exclaimed, "and literary historians don't know. *We* are the ones who know."

Epstein, a literary critic himself, would, no doubt, demur – not because of what the bookseller said about Liebling, but because he sees Liebling as belonging to a category reserved for "special writers," those who fall somewhere between major and minor. They are special, Epstein insisted, because "we feel a kinship between their imaginations and our own. A. J. Liebling was a special writer for a great many people. I, for one, have missed his prose more than that of any other writer who has died in my lifetime." It is doubtful that one could name a later essayist on boxing who reaches Liebling's level of excellence. Liebling maintained an unflinching gaze at the worst that boxing could be and combined it with an enthusiast's joy in boxing at its best.

**Reference:**

Joseph Epstein, "A. J. Liebling – The Minnesota Fats of American Prose," *Book World* (10 October 1971).

# Albert Jules McClane

*(26 January 1922 – 21 December 1991)*

Maggie Nichols
Field & Stream

MAJOR POSITIONS HELD: Fishing editor (1947–1977), executive editor (1972–1977), contributing editor, *Field & Stream* (1985–1991).

BOOKS: *100 of the World's Best Fishing Spots* (East Hartford, Conn.: United Aircraft Corporation, 1952);

*Spinning for Fresh and Salt Water Fish of North America* (New York: Prentice-Hall, 1952);

*The Practical Fly Fisherman* (New York: Prentice-Hall, 1953);

*The American Angler* (New York: Holt, 1954);

*Field & Stream International Fishing Guide* (New York: Holt, Rinehart & Winston, 1971); revised as *The Field and Stream International Fishing Guide: The First Fishing Guide Ever Published for Globe-Trotting Anglers* (New York: Scribners, 1973);

*Fishing With McClane: 30 Years of Angling with America's Foremost Fisherman,* edited by George Reiger (Englewood Cliffs, N.J.: Prentice-Hall, 1975);

*The Encyclopedia of Fish Cookery* (New York: Holt, Rinehart & Winston, 1977);

*Secrets of Successful Fishing* (New York: Holt, Rinehart & Winston, 1979); republished as *McClane's Secrets of Successful Fishing* (New York: Holt, Rinehart & Winston, 1980);

*McClane's North American Fish Cookery* (New York: Holt, Rinehart & Winston, 1980);

*McClane's Game Fish of North American* (New York: Times Books, 1984);

*McClane's Great Fishing and Hunting Lodges of North America* (New York: Holt, Rinehart & Winston, 1984);

*McClane's Angling World* (New York: Truman Talley Books, 1986);

*The Compleat McClane,* edited by John Merwin (New York: Dutton, 1988);

*McClane's Fish Buyer's Guide* (New York: Holt, 1990);

*A Taste of the Wild* (New York: NAL/Dutton, 1991).

A. J. McClane

OTHER: *McClane's Standard Fishing Encyclopedia and International Angling Guide,* edited by McClane (New York: Holt, Rinehart & Winston, 1965); revised and enlarged as *McClane's New Standard Fishing Encyclopedia and International Angling Guide* (New York: Holt, Rinehart & Winston, 1974);

*McClane's Field Guide to Freshwater Fishes of North America,* edited by McClane (New York: Holt, Rinehart & Winston, 1978);

*McClane's Field Guide to Saltwater Fishes of North America: A Project of the Gamefish Research Association,* edited by McClane (New York: Holt, Rinehart & Winston, 1978).

Albert Jules (A. J.) McClane is certainly America's best-known fishing writer in the twentieth century. From 1947, when his byline first appeared in *Field & Stream,* till his death in 1991 at his home in Palm Beach, Florida, he educated and entertained readers throughout the world with his eloquent mix of anecdote and expertise. Although he considered himself more a reporter than an expert fisherman, few anglers in the history of the sport have gathered such a wide range of experience in so many different places. Whether he was writing about fishing for bluegills in farm ponds or for marlins off the Florida coast, he brought the same enthusiasm and care to his work. Experts and beginners alike could read a piece by McClane and feel that it had been written expressly for them.

McClane began collecting fish stories and experiences from his earliest years. Though his family lived in Queens, New York, his grandfather, a retired railroad engineer, had a home at Hampton Bay on Long Island, and he spent much of his young life on the docks there or riding buses and trains to various parts of Long Island to explore fishing spots. An often told anecdote illustrates McClane's lifelong enthusiasm – and luck – at fishing. At age nine he was riding on a Long Island train when it stopped unexpectedly on a bridge outside the Babylon station. Since McClane had his steel casting rod with him, he quickly rigged up and, leaning out the window, dropped a bass bug into the water, where it was taken by a big bass. According to one version of this story, he hauled the fish through the window to the cheers of fellow passengers. Another version has the train starting up again and breaking off the fish before he could haul it in. In either case, the act of taking advantage of any fishing opportunity, no matter how unorthodox, and the hooking of a fish so quickly typify the way McClane's life as an angler and a storyteller evolved.

During the difficult days of the Great Depression, the teenage McClane searched for work in the Catskill Mountains. He settled in Margaretville, near the East Branch of the Delaware River, and found jobs on dairy farms and digging postholes for the power company. He also found some great Delaware anglers. McClane fished with the likes of Dan Todd, Mike Lorenz, and John Alden Knight, storing away knowledge and anecdotes that he would later call on to illustrate his articles.

In "The 'Ferocious' Commuter," a September 1990 article in *Field & Stream,* McClane drew on one of his experiences as a youth as he presented the life history of the smallmouth bass in America, combining stirring descriptions of the East Branch of the Delaware with good how-to-catch-it advice. "I remember the river in spring with its rosy-fingered dawns mirrored on hemlock-shaded pools when bankside cobwebs were spangled with luminous dew," he begins. Next McClane introduces Todd, "nut fly fisherman" and stationmaster, which smoothly leads to a short disquisition on the spreading of the smallmouth across America via the railroads where the bass "was at one time a regular commuter with a finfull of one-way-tickets." After packing all one needs to know about the fish into less than two magazine pages, McClane concludes with an unforgettable image: "I like the spring season best, fiddlehead and apple blossom time, when a bronzeback in the shallows will spring pantherlike at a surface lure, then make a rocketing jump; such a simple act but always a startling experience no matter how often it is repeated. It is about as explicable as the electric pulse of sound through silent space."

In 1939 McClane entered Cornell University, becoming a fisheries major and working at the state hatchery in Ithaca. He also began his writing career around this time, selling his first article in 1941 to *Outdoor Life.* McClane's budding career was put on hold in 1942 when he enlisted in the U.S. Army. He was sent overseas with the 398th Self-Propelled Automatic Weapons Battalion and later was transferred to Gen. George Patton's Third Army. In 1944 Sergeant McClane received a Purple Heart for wounds received in France, and he also received a letter from *Field & Stream* suggesting he come for a job interview. It was not until 1947 that he was able to accept the offer, but through a series of fortunate circumstances – the sort that followed McClane throughout his life – he was hired as the fishing editor. (He was fired by the publisher that same morning because he had taken a coffee break but was rehired in the afternoon.)

Although he was young and virtually unknown when he began what would be a thirty-year stint as *Field & Stream* fishing editor, McClane quickly proved his abilities both as a fisherman and as a writer, sprinkling even his earliest pieces with the spice of well-turned phrases, colorful descriptions, telling anecdotes, and fascinating historical references – a style he would hone and perfect throughout his life. A good example of this mix can be found in the April 1965 *Field & Stream* article "Angling in Austria":

> In the Austria of 1945 there was no *Lageltrager* on the rivers. He is the faithful man who wades behind his angler with a floating barrel to keep the trout alive. Each time a fish is caught he salutes the fisherman, then drops the trout into his keg. . . . Nearly two decades later I went back to Austria and still didn't have the pleasure of observing this centuries-old custom. Because of refrigeration and a booming economy, the *Lageltrager* has virtually disappeared from Alpine rivers. Being the kind of nut who has to go to the top of the Empire State Build-

> ing . . . I had to have a *Lageltrager*. On a dozen streams I could find only one guide who was willing to bring a barrel along, and he thought the idea pretty silly. It was. The barrel leaked and sank.

Although the bulk of McClane's work dealt with practical matters such as how and where to fish for virtually every freshwater and saltwater species, his range of interest was wide and he regaled readers with features on such disparate subjects as poisonous sea snakes ("Cobras of the Sea," February 1965); a caviar dealer ("The Caliph of Caviar," December 1965); an account of a scientific expedition to collect specimens of South American fish ("Amazon . . . Wonder World of Fishes," May 1972); and a day in the life of an angler in the New York City of 1776 ("The Saturday of Marten Van Kleek," February 1976).

In 1950 McClane married Patti Murphy, who was herself an expert angler and shared many of her husband's adventures throughout the world. McClane and Patti moved from the New York area to Palm Beach, Florida, in 1953, having persuaded *Field & Stream* publisher E. T. Warner that there was no need for the fishing editor to maintain an office in New York. In that same year their daughter, Susan, was born. She too would become a world-traveling angler.

Florida and the Bahamas played a major role in McClane's development as a fisherman and writer. He not only honed his fly-fishing skills on the bonefish flats in the Bahamas and Florida Keys, but he became involved in important conservation issues, such as opposing the dumping of raw sewage into the ocean just off Palm Beach. But though he eventually became disheartened by the overdevelopment of Florida, he retained his love for the state's potent beauty, as he eloquently expressed in "The Mouth That Swims," written for *Field & Stream* in February 1989:

> When I first moved to Florida many years ago, the palmetto and cypress landscape west of our town was pocked with shallow, crystal-clear, sand-bottomed ponds where I could wade wet and take bass for hours on end. . . . When the morning mist gave way to a sky laced with the vast skeins of mauve and orange and the bass started working noisily on potgut minnows and those elusive dragonflies that played in the bells of yellow lilies, I felt like a trespasser in paradise. Sometimes I stayed until the air turned spicy and the water looked like cooling copper at the advent of dusk – supposedly the magic hour for a double-digit fish. . . . It was the best of all times. Today, bulldozers roam our land like alien monsters obscuring the past. Fortunately, the largemouth is a survivor adaptable to even marginal habitat.

Patti McClane played a major role in a hilarious episode that took place during a fishing trip that they undertook at the invitation of the Venezuelan government in 1957. They were fishing near the border of Colombia, and their assigned guide, Guillermo, went shopping across the border every day while McClane and his wife fished. "Some guide," McClane commented in an article titled "The Field & Stream Hat Trick," written for the magazine's seventy-fifth-anniversary issue in June 1970. "He must be a fashion editor for Consumer Reports."

The description of what happened next demonstrates not only McClane's great talent for storytelling, but the way stories literally followed him throughout his life. After Patti remarked that Guillermo sometimes looked strangely fat, McClane began to take notice:

> That evening I saw Guillermo waddle into the lobby. He was a veritable Latin Sidney Greenstreet. Yet when he came to the table, he was skinny as a rail. . . . My funny bone told me that something would happen soon.
>
> We were nailed at a roadblock on the way back to Caracas . . . the main route for smuggling contraband into Venezuela. . . . There was an endless clutter of cars ahead of us and soldiers waving tommyguns were stalking the road shoulder making a random check. I had sweaty armpits. Guillermo was shaking like a hound dog.
>
> The military didn't say anything. They just yanked the doors open and waved us out. . . . I produced some papers. . . . No dice. One of the soldiers barked something at Guillermo, who was vibrating so bad now I thought he'd come unbolted. He proceeded to strip.
>
> To this day I've never figured out how any man, even a skinny one, can wear eight suits. But eight suits he took off before he got to the buff. When they opened his duffel, what appeared to be another fifty suits were tossed on the road. A suit smuggler! Guillermo stood naked as a jaybird in a rag nest waiting to hear the gun go off.
>
> "I have more papers in my bag," announced Patti. "Maybe we have a permit for smuggling." She proceeded to empty her purse, which is always a fascinating operation. Even the tommygunners shook their heads in disbelief.

Patti found a bunch of papers and passed them around "like a Salvation Army worker handing out Biblical tracts." One of them did the trick. "Everybody saluted everybody," McClane reported. "They even put a suit back on Guillermo, who looked stunned." The paper – which was indeed a magic permit – bore the signature of the head of the Venezuelan secret police.

*Dust jacket for McClane's 1986 book, a characteristic blend of storytelling and fishing expertise*

Although McClane's initial fame as a fisherman and writer was developed through his column for *Field & Stream,* it was his monumental work *McClane's Standard Fishing Encyclopedia and International Angling Guide* (1965) that made him a true celebrity. Years of work went into the making of this impressive volume, which weighed about ten pounds and ran more than one thousand pages, with contributions from 141 different authors totaling more than a million words. In 1974 a revised edition was published, *McClane's New Standard Fishing Encyclopedia and International Angling Guide.* It cost considerably more: $75.00 as opposed to the $19.95 price for the first edition. Exact sales figures are not available, but the encyclopedia in all its editions sold well over a million copies, making McClane a millionaire.

Besides his expertise at catching fish, McClane was equally knowledgeable about cooking them, partly as a result of his long friendship with Charles Ritz, owner of the Ritz Hotel in Paris and another world-class angler. Renowned as a gourmet (he served as *Grand Officier* of the *Confrerie des Chevaliers du Testavin*), McClane often wrote about the art of cooking fish and game. His masterworks on food were *The Encyclopedia of Fish Cookery* (1977) and *A Taste of the Wild* (1991). He was a consultant to restaurants, hotels, and airlines on matters of fish and seafood. He was also a regular guy who simply liked to fish and eat. According to Duncan Barnes, editor of *Field & Stream,* "Al was a master angler, an ichthyologist, a world traveler, the author of more than 1,000 magazine articles and twenty-one books, a patient and philosophic man with a real zest for life. He was also a highly knowledgeable gourmet who would gladly settle for baloney on white, as long as he was breaking bread with friends."

In all McClane's writing and in the many stories about his life, one fact stands out: strange and funny things happened to this man, things a writer would pray to have come his way – except no writer would have been able to dream up such experiences. Once in Alaska, after the pilot of the float trip he was traveling on had to make an emergency landing, he was greeted by a woman who appeared out of the fog and offered him tea. It turned out that he and the pilot had stumbled onto a pair of Disney photographers who were lost, out of food, and decidedly not in their right minds. In his travels (he visited and fished in more than 140 countries), he walked away unscathed from plane crashes, formed friendships with the rich and famous, got stranded in the Amazon, costarred in a *Field & Stream* movie with a bad-smelling fox, and generally lived a life that would have made an exciting, though improbable, movie or television series.

One little known event seems particularly typical of the quirky charm that hung over his life. Several years before his death McClane participated and won a celebrity fishing tournament in Woods Hole, Massachusetts. As the winner, he was the first person interviewed by the television reporters waiting on shore to cover the event, and when they were done with him, he walked off the dock. Moments later he heard a loud crashing sound and turned to see the television crews and all the other anglers floundering in the water. The dock had collapsed – but not until McClane was safely on shore.

In his later years McClane specialized more and more in bonefish, a fascination that culminated with his catching the fish on a dry fly, something generally thought to be impossible. But he had no pretensions about this choice of sport; all fishing was worthwhile. He was a celebrity in the truest sense, yet he thought of himself simply as a man who loved to fish and as a reporter. McClane's sense of himself is clear from a piece titled "Shore Lunches" that he wrote for *Field & Stream* in March 1990. After describing some of the terrible meals he had eaten at streamside, most notably a daily lunch at a salmon camp that consisted of a hot dog – "not the digestible kind, but one of those red-dyed weenies that taste like excelsior" – he went on to speak of his enjoyment of a pair of tiny brook trout cooked in butter over a fire built on a gravel bar next to a deep pool. "The greatest halls of gastronomy could never duplicate that simple ritual," he wrote. And he added, "A famous epicure of centuries past, Brillat-Savarin, who, incidentally was an avid sportsman, supplied an aphorism: 'Tell me what you eat: I will tell you what you are.' Well," McClane signed off, "I know that I'm not a red-dyed hot dog."

Many of McClane's works remain as current, important, and widely read as they were when they first appeared. Published in 1988 and edited by angling expert John Merwin, *The Compleat McClane* (the title is an allusion to Izaak Walton's *The Compleat Angler,* published in 1653) serves as the best single guide to McClane's writing. It gathers many of his finest uncollected pieces, which originally appeared in *Field & Stream* as well as in such diverse magazines as *Fishing World, Fly Fisherman, Food and Wine, Gourmet, Sports Afield, Venture,* and *Esquire.*

Merwin divided the book into four sections that roughly can be categorized as techniques ("Squatting, crawling, hiding and casting in situations that would give a squirrel claustrophobia are not without compensations"), "gamefish behavior" ("Sunfish, for example, usually study a surface lure for some time, finning toward it and backing off as though waiting for it to explode"), cookery ("The discipline of fish cookery does not end in an iron skillet. It begins"), and recollections. McClane was a frequent fishing companion of *Esquire* founder and publisher Arnold Gingrich as well as hotelier Charles Ritz. When both men passed away within days of each other in July 1977, McClane recalled in *Fly Fisherman* how "their presence [was] still as tangible as the nearest river." He remembered the elegant Ritz as "impeccably dressed, even on the river," while Gingrich, otherwise the epitome of American male fashion in the 1950s, was the opposite: "Even the wrinkles in his waders had wrinkles. He was a walking tackle shop. With extra reel spools and fly boxes stuffed in every pocket, he clanked down the brook like a Sherman tank." "As a writer," Merwin observed in his introduction, McClane was "one of a small handful in the angling field who qualify as genuine stylists."

In 1987 Patti McClane died after a long illness, and in 1989 A. J. married Diane O'Hara, who shared his final years. When McClane died of congestive heart failure on 21 December 1991, *The New York Times* acclaimed him as "the most familiar and respected byline in angling literature." "[H]is features," the obituary continued, "roamed a vast canvas of destination, species, historical lore, technical issues, adventures, and observations" (25 December 1991). There is no reason to doubt, as Barnes intimated in a March 1992 tribute to his late colleague, that McClane's writing will continue to "inspire . . . and influence . . . millions of readers over the years."

# W. O. McGeehan

*(22 November 1879 – 29 November 1933)*

Dennis Gildea
*Springfield College*

See also the McGeehan entry in *DLB 25: American Newspaper Journalists, 1901–1925.*

MAJOR POSITIONS HELD: Sports reporter, *San Francisco Call, San Francisco Chronicle, San Francisco Examiner* (1900–?); city editor and managing editor, *San Francisco Evening Post* (?–1914); sports reporter, *New York Evening Journal* (1914–1916); sports editor, *New York Tribune* (1916–1920); managing editor, *New York Tribune* (1920–1922); sports columnist, *New York Herald* (1922–1924); sports editor, *New York Herald Tribune* (1924–1933).

BOOK: *Trouble in the Balkans* (New York: Dial, 1931).

SELECTED PERIODICAL PUBLICATIONS – UNCOLLECTED: "Barber of Broadway," *Collier's,* 74 (1 November 1924): 14–15;

"Why Pick on Dayton?," *Harper,* 151 (October 1925): 623–627;

"Baseball: Business as Usual," *North American,* 224 (March 1927): 119–124;

"Ivory Industry," *Saturday Evening Post,* 200 (11 February 1928): 12–13;

"Glorified Tomboys," *Ladies' Home Journal,* 45 (July 1928): 25, 72;

"Our Changing Sports Page," *Scribner's Magazine,* 84 (July 1928): 55–60;

"The Last Gladiator," *Saturday Evening Post,* 202 (28 September 1929): 37, 149–150, 153;

"But They Do Come Back," *American Magazine,* 111 (April 1931): 52–54.

Regarded as the founding father and one of the leading practitioners of what came to be known as the "Aw Nuts" school of sports journalism, William O'Connell McGeehan was reviled by sports promoters and admired by colleagues and successors for being an honest, authoritative, and imaginative sportswriter. Paul Gallico, who was covering sports for the *New York Daily News* while McGeehan was writing and editing for the *New York Herald Tribune,* called him the "brilliant McGeehan, the greatest sports-writer who ever lived." Gene Tunney, the heavyweight boxing champion about whom McGeehan wrote both favorably and unfavorably, referred to him as "that prince of sports writers." Randall Poe notes that in an era when the sports hero was "spiced and fattened" on the sports pages, McGeehan "refused to serve the Hero." Poe ranks McGeehan and his successors, John Lardner and Red Smith, as "perhaps the best three sportswriters of this century" and observes that they "always went for critical revelation, presaging the sports writing to come."

During the zenith of McGeehan's career his columns and stories were fueled by a habit of mind opposed to that of reporters in the "Gee Whiz" school of sports journalism that sought to mythologize both the athletes and the games they played. Ironically, Grantland Rice, McGeehan's friend and colleague on the *New York Herald Tribune,* is generally regarded as the foremost writer of the "Gee Whiz" school. Unlike Rice, though, McGeehan punctured the illusion of sport as a redemptive resource that was akin to a secular religion in American culture. Yet together they contributed to making the sports pages of the *Herald Tribune* among the best written of any American paper at the time. Stanley Walker noted that McGeehan and Rice, along with Ring Lardner, Hughey Fullerton, and Damon Runyon, "brought to the business a grown-up judgment, some sense of proportion, a gentlemanly taste and even some literary quality."

While Rice's influence on contemporary sportswriters may have been greater than McGeehan's, the latter achieved prominence and lasting importance by treating sport not as something sacred and separate, but as an integral part of society. Aiming at colleagues who made athletes larger than life, McGeehan wrote critically of those who "compare a fat young man swinging a club at a little ball to Thor of the Hammer." Working in what many sports historians consider the "Golden Age of Sport" in the 1920s, McGeehan's intellectual resistance to the overwhelming trends of sportswriting distinguished his work. He was one of the first sportswriters to recognize and treat sports as another embodiment of corporate America. In an age of athletic boosterism in the nation's sports pages, McGeehan refused to cheer, preferring instead to examine, dissect, and ruminate on sports events, heroes, and issues of the day. Walker writes, "In his exposure of papier-mâché gods, and in his allusions to the intricate financial and political system controlling sports, McGeehan relied mostly upon a gentle sarcasm." The most telling feature of McGeehan's writing was his language, which was seldom overblown and trite, always fresh and lively. When he died at age fifty-four on 29 November 1933, *The New York Times* noted that, writing in a field rife with clichés, McGeehan had been a "coiner of new phrases."

McGeehan grew up in San Francisco, one of the most competitive newspaper cities in the nation at the time. Cartoonists Rube Goldberg and Tad Dorgan, boxing writer Herbert "Hype" Igo, and reporters Jimmy Swinnerton, Robert Edgren, and Will Irwin all worked in San Francisco at the same time and all were to achieve varying degrees of fame in New York. As the eldest of six children of Hugh and Theresa O'Connell McGeehan, W. O. McGeehan attended public schools in San Francisco and then entered Stanford University. When the Spanish-American War started, McGeehan left school to enlist with the First California Volunteers and eventually served with that unit during the war in 1898 and during the Philippine insurrection in 1899. Following his release from the service, he returned to San Francisco and took his first newspaper job as a reporter with the *San Francisco Call*. He was soon assigned to the boxing beat and covered that sport in a city and region that had been a hotbed of prizefighting since the Gold Rush days and where, in contrast to most states in the union, the sport was legal.

McGeehan built on the public's interest in boxing with a writing style that earned him a reputation as a promising reporter. With the *Call* he covered a famous bout in Goldfield, Nevada, on 3 September 1906, when lightweight champion Joe Gans defeated Battling Nelson in forty-two rounds, a fight that marked the debut of Tex Rickard as a boxing promoter. Years later when Rickard was running Madison Square Garden, where he was credited with bringing a press agent's promotional instincts to boxing, he admitted that he was both puzzled and frustrated by his inability to buy McGeehan's favor. Rickard told his publicity man, Gene Fowler, who recounts the story in *Beau James* (1949), his biography of New York City mayor James J. Walker,

> I can't understand some of them there sportswriters like Dan Parker and Bill McGeehan. McGeehan was at Goldfield with Rube Goldberg. I put thirty thousand in twenty-dollar gold pieces on show in the bank window, the purse for the fight. . . . Well, I passed around some twenty-dollar gold pieces just to advertise myself and the fight. When I tried to give one of 'em to Bill McGeehan, do you know what he told me to do with it?

Fowler, a former newspaperman who knew McGeehan well, allowed that Rickard "had a good idea, or a bad one, however you want to look at it." Rickard's anecdote served as both a grudging tribute to McGeehan's integrity and as a puzzled plea to Fowler to try to get McGeehan to stop referring to Rickard in print with the nickname "Dink," by which McGeehan heard that Rickard had been known during his gambling days in Alaska. "I wish [McGeehan] hadn't dug up that 'Dink' label," Rickard said. "Down in Texas it means somethin' private." McGeehan, an expert at needling whomever he felt needed to be deflated, never ceased.

During his San Francisco newspaper days McGeehan earned his own nickname, "Sheriff," which colleagues had given him after McGeehan joined a posse chasing escaped convicts, and the name stuck throughout his life. In New York McGeehan became part of an informal group of sportswriters including Irvin S. Cobb, Bozeman Bulger, and Runyon, a group that Cobb called "The Four Horsemen." Together they took hunting trips, ranging from nearby Long Island Sound to as far away as the Canadian woods and the Georgia backcountry.

While he gained valuable journalistic experience during his years in San Francisco, McGeehan never looked fondly on it; instead, he regarded that experience as the product of a time and place in which good newsmen were exploited. "There was no money and no future for newspaper men," McGeehan later said of those early years. "Today reporters and editors are better paid, and the type of men who are entering the business is higher." Nevertheless, McGeehan worked for fourteen years in San Francisco, where he moved from the

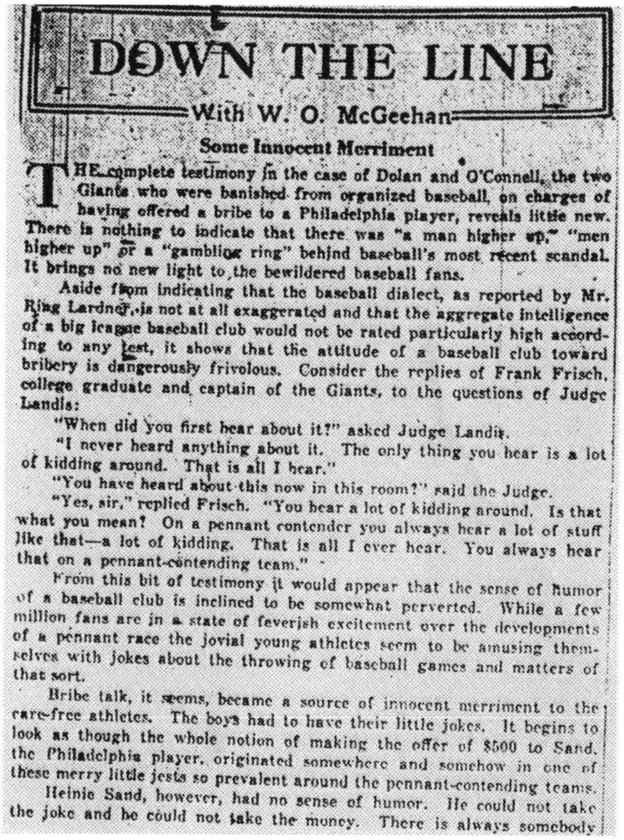

DOWN THE LINE

With W. O. McGeehan

Some Innocent Merriment

THE complete testimony in the case of Dolan and O'Connell, the two Giants who were banished from organized baseball, on charges of having offered a bribe to a Philadelphia player, reveals little new. There is nothing to indicate that there was "a man higher up," "men higher up" or a "gambling ring" behind baseball's most recent scandal. It brings no new light to the bewildered baseball fans.

Aside from indicating that the baseball dialect, as reported by Mr. Ring Lardner, is not at all exaggerated and that the aggregate intelligence of a big league baseball club would not be rated particularly high according to any test, it shows that the attitude of a baseball club toward bribery is dangerously frivolous. Consider the replies of Frank Frisch, college graduate and captain of the Giants, to the questions of Judge Landis:

"When did you first hear about it?" asked Judge Landis.

"I never heard anything about it. The only thing you hear is a lot of kidding around. That is all I hear."

"You have heard about this now in this room?" said the Judge.

"Yes, sir," replied Frisch. "You hear a lot of kidding around. Is that what you mean? On a pennant contender you always hear a lot of stuff like that—a lot of kidding. That is all I ever hear. You always hear that on a pennant-contending team."

From this bit of testimony it would appear that the sense of humor of a baseball club is inclined to be somewhat perverted. While a few million fans are in a state of feverish excitement over the developments of a pennant race the jovial young athletes seem to be amusing themselves with jokes about the throwing of baseball games and matters of that sort.

Bribe talk, it seems, became a source of innocent merriment to the care-free athletes. The boys had to have their little jokes. It begins to look as though the whole notion of making the offer of $500 to Sand, the Philadelphia player, originated somewhere and somehow in one of these merry little jests so prevalent around the pennant-contending teams.

Heinie Sand, however, had no sense of humor. He could not take the joke and he could not take the money. There is always somebody

*Part of McGeehan's 12 January 1925 prizefighting column for the* New York Herald Tribune

*Call* to the *Chronicle* and then to William Randolph Hearst's *Examiner* before taking his last western position with the *Evening Post.* During his professional climb in San Francisco he switched from sports to news coverage, and he eventually served as both city editor and managing editor at the *Evening Post,* a pattern he also was to follow in New York, although there he was ultimately to settle in as the sports editor of the *Herald Tribune.*

While in San Francisco, McGeehan married newspaper reporter and playwright Sophie Treadwell on 27 January 1910. They had no children. They worked together on at least one memorable story: the Jack Dempsey bout with Georges Carpentier for the heavyweight championship at Boyle's Thirty Acres in Jersey City on 2 July 1921. The fight, elaborately promoted by Rickard, became the first million-dollar gate for boxing and the first fight to be broadcast.

In 1914 McGeehan responded to the tug of ambition and moved to New York City. There, despite his accomplishments on the West Coast, he found himself starting anew by returning to his origins as a sportswriter who specialized in boxing. Writing a column that he signed "Right Cross," McGeehan landed a job on Hearst's *Evening Journal,* and it did not take long for him to begin to make his mark in New York City journalism. In 1916 McGeehan joined the sports staff at the *Tribune,* where he had been lured by the assurances of sports editor Heywood Broun that his stories would appear under his rightful name rather than a pseudonym. Soon afterward Broun was named *Tribune* drama critic, and McGeehan began covering various sports – major league baseball, tennis, intercollegiate football, and golf. He was with the paper a relatively short time before being named its sports editor.

In many ways McGeehan was an ideal sports editor for the "Golden Age of Sport," because he had matured journalistically in that previous era and knew the nineteenth-century history of sport in America. Relying on such knowledge, McGeehan often contrasted the events and people he had known and covered during those days with the more sensationally promoted events and stars of the

1920s. In 1925, for example, one of his "Down the Line" columns for the *Herald Tribune* recalled "the most fiercely contested battle ever fought since the bareknuckle days" of prizefighting – a 22 February 1910 bout in Port Richmond, a San Francisco suburb, where Ad Wolgast had ended Battling Nelson's reign as lightweight champion in a fight that the referee had finally stopped after forty rounds.

The occasion for the column was the retirement of Wolgast, "now reported to be suffering from a mental and physical breakdown as the result of his meteoric ring career," McGeehan wrote. But McGeehan used the story of the fight not so much to reminisce or to rail against the brutality of boxing, but to reiterate one of his favorite sports themes – the emergence of the indomitable human spirit. In this fight McGeehan had seen such spirit displayed by both men. Wolgast, knocked to the canvas and seemingly beaten in the seventeenth round, "leaped to his feet furious and . . . drove Nelson back to the ropes." Nelson, "The Durable Dane," as McGeehan had dubbed him, was a "mass of red bruises," one eye battered shut by Wolgast's blows. Yet after two hours of "sheer viciousness" Nelson was still on his feet when the referee stopped the fight. McGeehan wrote that "Battling Nelson had lost his championship, but he was not knocked out. He lost it standing on his feet and fighting to the last."

McGeehan's rise to prominence in New York City sportswriting was interrupted by the entry of the United States into World War I. He obtained a commission as a captain in the infantry in 1917, but he was not sent overseas and was later appointed a major in the U.S. Infantry Reserve Corps. McGeehan's belief in American involvement in the war contributed partly to his dislike of Jack Dempsey, who by working in the shipyards of Chester, Pennsylvania, had elected to avoid military service under the "work-or-fight" option offered by the Selective Service. McGeehan was one of many prominent newsmen who argued that Dempsey's choice constituted a moral failing. Randy Roberts writes that "McGeehan did not quarrel with Dempsey's alibi, but he did question the champion's actions. Did not Dempsey, as the world's greatest fighter, have an ethical obligation to enlist?"

Dempsey won the title on 4 July 1919 when he battered the bigger Jess Willard in Toledo, Ohio. The lead for McGeehan's story of the fight provides an example of the almost playful rhythm of his language in contrast to the brutal image he evokes. McGeehan writes: "A towel tossed into the ring from the corner where Jess Willard floundered on a stool with his right eye closed and a torrent of blood gushing from his gaping mouth was the signal that a new heavyweight champion had arrived here this afternoon."

Dempsey exploded on the boxing scene at an opportune time for fame and fortune, because in 1920 New York State senator James J. Walker pushed past Gov. Al Smith and the state legislature a bill to legalize prizefighting. McGeehan applauded Walker's legislation, but, because McGeehan's knowledge of the kinds of people professionally involved in the sport enabled him to recognize phonies and fixers, he continued to criticize how the sport was governed. In 1927 McGeehan's sarcastic touch appears, for example, in his persistent questions about the integrity of certain bouts:

> Strange as it may seem when you consider the splendid type of businessmen associated with the business, many of whom would not climb a porch to get at the safe when there was any chance of jimmying the back door, many of the customers are beginning to suspect all has not been up and up in the business. . . . These suspicions seem to have reached the inner councils of the New York State Boxing Commission, and I understand there is to be an informal hearing shortly to consider a few of the fights that have caused the customers to feel these unwarranted suspicions. This is in line with the boxing commission's policy in taking every precaution after the safes have been cracked. It is what is known as locking the barn door after the horse has been stolen.

Typical of McGeehan's "Down the Line" column, this passage bears the essentially irreverent McGeehan stamp. His columns insisted on viewing sport as a business played and run by mere mortals with human failings rather than as heroic encounters that somehow transcended reality. McGeehan was especially tough on Bill Carey, Rickard's successor as top fight promoter at Madison Square Garden, a job for which he was ill suited. Carey knew as much about fights, Grantland Rice observed, as "an elephant knows about contract bridge." Carey promoted the heavyweight bout between Jack Sharkey and Young Stribling in Miami Beach in 1929, a fight that McGeehan regarded as a sensationalized pugilistic fraud. Nevertheless, he went to Florida to judge for himself, and Rice tells of McGeehan's meeting with Carey in Miami Beach:

> Bill had been roasting Carey daily in his column. When he arrived, Bill expected Carey to have him tossed out. I introduced them, "Mr. Carey, Mr. McGeehan."
>
> Carey beamed his welcome through his milk-bottle lenses. "And by the way, Mr. McGeehan," he said,

"what business are you in?" That was the only time I ever saw McGeehan stopped cold. He fled the camp. He refused to write a word.

No doubt because of men such as Carey, McGeehan came to have little regard for fight promoters. Of boxing, which he termed "the cauliflower industry" and "the manly art of modified murder," McGeehan wrote:

> Prize fighting is not a sport. It is a business. As a business it is probably as honest as American politics or the average business, and probably a little more honest than the stock market.
>
> Prize fights today are merely entertainments staged for the box office. The sporting element went out when the business became respectable and legal. In the old days, when the fighters had to hop across the country ahead of the Sheriff and stage a bout on the quiet, prize fighting was more of a sport than it is today.

These views appear in "The Last Gladiator," another magazine piece that typically blends cynicism with an almost sentimental respect for the individual fighting a good fight. In this article McGeehan rails against the hypocrisy and shady dealings of professional boxing while he praises New York State athletic commissioner William A. Muldoon, the "last gladiator" of the title. Muldoon was a political appointee with whom McGeehan had disagreed, but McGeehan respected him for the various roles Muldoon had played during the outlaw years of boxing and for Muldoon's lifelong devotion to physical fitness and health. McGeehan's description of a typical group of fight fans (or "customers," as he liked to call them) gathered in the lobby of Madison Square Garden on a fight night in the 1920s contrasts with his description of Muldoon, a former boxer and wrestler who at the time of McGeehan's article was eighty-four years old.

Of the fight fans, McGeehan wrote:

> on fight nights, groups of rat-eyed, furtive young men gather and whisper. These are the fight fans or, as they used to call them in the pre-Queensbury days, the "gentlemen of the fancy." They are not the romantic characters of [Sir Arthur] Conan Doyle's Rodney Stone. Most of them look as if they might be classed in the category of the racketeers, if you do not care to use the softer term of "sporting characters."

McGeehan was never one to resort to the softer term, even when referring to the people who largely comprised his readers. He describes the crowd in Madison Square Garden as growing "a little more shifty and uneasy as an elderly gentleman walks through the lobby with a firm tread and looking around him with a piercing blue eye," and he characterizes Muldoon as the "Solid Man," a contemporary gladiator who defies the whims of the masses.

> The hair of the living gladiator is white and his closely cropped mustache is white, but the frame of the elder Muldoon is quite as erect as that of the gladiator of half a century ago, and the same light of defiance blazes from his eyes.
>
> He sniffs the smoke-laden air angrily and his expression indicates contempt for the hangers-on in the lobby. Of all human weakness, Muldoon despises most the addiction to tobacco. He would hold no conversation with a man smoking a cigarette. Inside, on fight nights, they may not smoke. It is not the law, but it is Muldoon's will.

McGeehan's admiration for Muldoon stems not so much from the latter's role as a physical culturist, but from his rugged individualism, his iconoclastic "defiance of the mob," as McGeehan writes, and Muldoon's manner in many ways accords with McGeehan's iconoclastic style of journalism.

Yet McGeehan did not share Muldoon's spartan regimen of exercise and diet. McGeehan writes that Muldoon "preaches fiercely against the tendencies that are softening up the race" – the "race" of white Americans – and Muldoon's "great aversion and his great fear is the gin drinking, cigarette smoking potential mother of the coming generation" of American manhood. Bob Cooke has noted that McGeehan's "favorite companion was the cigarette which invariably dangled from a corner of his mouth." Arriving in Shelby, Montana, for the 4 July 1923 fight between Dempsey and Tommy Gibbons, McGeehan and Runyon, who was reporting for the *New York American* and the Hearst syndicate, discovered that Bill Corum, a cub reporter for the *American,* had secured the only hotel room in town with running water. McGeehan coveted not the running water in Corum's room but the bathtub, which he soon appropriated to use as his beer cooler. In New York City McGeehan was a member of the sporting, theater, and newspaper crowds at Billy LaHiff's tavern on West Forty-eighth Street. Throughout his adult life McGeehan struggled against alcohol, a struggle he consistently lost.

His drinking so harmed his work that at least once it cost him his job. Impressed by the sophistication and depth of McGeehan's sports pieces, president and editor Ogden Mills Reid had appointed him managing editor of the *Tribune* in 1920. Reid, who also had bouts with alcohol, should have known better. Drinking and column writing, in which the columnist is free to choose his topics and determine his working hours, may not have been incompatible, but drinking and running a busy newsroom were, at least in McGeehan's case. He went off on occasional binges, which resulted in his being fired in 1922.

McGeehan became a sports columnist for the *Herald,* however, that same year, and the paper, delighted to have acquired him, announced his arrival with the boast that "the fighting pen of McGeehan has been secured in the interest of clean sport." With the *Herald* McGeehan argued for and won an expanded sports section, which grew in 1922 from two to three pages daily. Nevertheless, McGeehan was to argue six years later in "Our Changing Sports Page," this expansion did not signal the paper's attempt to change public attitudes toward sports but to respond to the increasing interest of the public in sports. Not being swept away by professional euphoria with the growing presence of the sports page, McGeehan struck a characteristically cautionary note in his article:

> as to the overwhelming general increase of interest in sports in the United States there are many indications of an increasing pessimism. The Association of American College Professors last year deplored the false sense of values in the colleges illustrated by the ever-increasing interest in intercollegiate football. There has been much discussion of the evil of exalting the gladiatorial spirit at the expense of the intellectual side.

McGeehan continued writing his "Down the Line" column for the *Herald,* and in 1924 the owners of the *Tribune,* Reid and his wife, Helen, bought out Frank A. Munsey, owner of the *Herald.* Combining these two papers in what would become one of the country's most respected newspapers, the Reids welcomed McGeehan back as their sports editor and columnist.

In 1923 McGeehan earned a niche in radio history when he assisted Graham MacNamee with the first live World Series broadcast. The opening game between the Yankees and the Giants was played at Yankee Stadium, and with the score tied 4–4 and two out in the ninth, Casey Stengel, later to become the celebrated manager of the Yankees, hit an inside-the-park home run for the Giants.

McGeehan flourished at the *Herald Tribune,* where he produced some brilliantly biting columns and pieces on sports, business, and politics. In examining the correlation between sport and big business, he was well ahead of his time, as some of his comments presaged those of sportswriters by more than half a century. Most noteworthy, especially in light of his World Series radio experience, is his column written in 1925 to warn of the potentially powerful cultural role of what McGeehan called the "radio-cinema," a technology then being tinkered with and that was to become known as television. McGeehan accurately predicts that the influence of the invention would be widespread: politicians such as the mayor of New York would have access to national audiences. He was not far wrong in predicting that this radio-cinema would mean the end of "preliminary prize fighters and semipro baseball players." The reception on his crystal ball was a bit fuzzy, however, when he foresaw that television would "spell the doom of professional sports," which would not survive without gate receipts.

Another McGeehan piece, "The Ivory Industry," in which he likened the major league baseball scouting operations of 1928 to ivory hunting in Africa, was also perceptive. "It is a big business, organized baseball of today, and highly specialized," he wrote. "In fact, on thorough analysis, organized baseball is what would have been called a grasping monopoly a few years back." Both this article and its predecessor, "Baseball: Business as Usual," were written in the wake of the 1922 Supreme Court ruling that designated major league baseball to be primarily a sport that was not subject to antitrust laws and interstate commerce regulations.

In many of his pieces and columns McGeehan used analogies, such as that of baseball scouts and ivory hunters, to satirize the evils or foibles of a prominent character or issue. In "Macbroth, or the Fatal Sausage" McGeehan used William Shakespeare's *Macbeth* (1623) to ridicule Harry M. Stevens, the man who had cornered the hot dog concessions at New York City sports stadiums and arenas. In "Hylanus Geezer," another Shakespearean parallel, here to *Julius Caesar* (1623), McGeehan spoofed New York City mayor John Hylan, considered to be Hearst's handpicked man in city hall.

In 1925 the paper sent Forrest Davis and McGeehan to join more than one hundred other reporters – including such luminaries as H. L. Mencken, Westbrook Pegler, and Nunnally Johnson – to cover the John T. Scopes monkey trial that ran throughout a hot July in Dayton, Tennessee. Yet this was an assignment that did not entirely break from McGeehan's past experience and cultural interests: he had served as managing editor of the *Tribune,* and he had held the same position while covering city news for San Francisco papers. At the

Scopes trial, though, he treated the proceedings as a sporting event, with the forces of reason competing against the fundamentalist Christian home team of rural Dayton. The athletic tropes that McGeehan used in at least one of his stories were particularly those of boxing, which provided appropriate metaphors because the trial became an oratorical slugfest between lawyers William Jennings Bryan and Clarence Darrow.

In his report of 17 July 1925 McGeehan presented Bryan's climactic address to the court in an image familiar to all sportswriters and fans – that of an aging athlete attempting to beat the odds in making a comeback:

> They never come back. The Great Moment in the case of the people of the State of Tennessee versus John Thomas Scopes has come and gone. William Jennings Bryan, the Plumed Knight of the Fundamentalists, the Prince of Pacifism, the Champion of Chautauqua, was given the floor and in exactly one hour and ten minutes he was lying on it horizontally in a figurative sense.

McGeehan, the veteran boxing writer, summed up the trial in a good sports-page lead – Darrow by a knockout in one hour and ten minutes. Even his allusions to Bryan as "the Plumed Knight of the Fundamentalists, the Prince of Pacifism, the Champion of Chautauqua" echo the kinds of nicknames that fighters took with them to the prize ring. In fact, in a "Down the Line" sports column that had appeared just a few days before he left for Tennessee, McGeehan had referred with similarly caustic alliteration to Harry Wills, a challenger for Dempsey's heavyweight championship, as the "Senegambian Stevedore." Moreover, he had referred to the men who formed the New York State Boxing Commission board as the "Three Dukes of Cauliflowerdom." As was his coverage of the Scopes trial, McGeehan's language in his boxing column was irreverent and humorous in combining information and commentary with tongue-in-cheek jabs. Explaining to his readers in the same column that a title fight between Dempsey and Wills could not be arranged that summer because both were in Europe, McGeehan presented the sporting belligerence of the fighters in a context of their records as draft dodgers: "The restlessness of the Senegambian at this particular time has a very strange look. Like Mr. Dempsey, Wills had little or no interest in the capitals of Europe during the World War."

Readers of the *Herald Tribune,* then, had every reason to expect the same acerbic humor in McGeehan's coverage of the monkey trial, and he did not disappoint them. After giving the result of the oratorical bout of the attorneys in his lead, McGeehan resorts to a round-by-round narrative in which he refers disparagingly to Bryan as both the Great Realtor and the Florida Realtor:

> There was a tense hush in the sweltering room as Judge John Raulston made the most important announcement as far as this congregation was concerned, "Mr. Bryan will speak."
>
> The Great Realtor arose, revealed stripped to the white starched shirt with the winged collar and the little black bow tie of the true believer. First he reached for his three-gallon graniteware jug and drank deeply. He walked to the center of the enclosure where the lawyers were huddled. Then, waving in one hand the palm leaf fan of the Florida Realtor and in the other the iniquitous Foster's "Civic Biology," he began.

McGeehan's portrayal is again reminiscent of fight coverage, in which the reporter spends a few lines describing the physical condition of the combatant as he takes the center of the ring to answer the opening bell. When Bryan begins his attempt at a comeback, McGeehan reports that it falls considerably short, as the lawyer's performance in the Dayton courtroom does not live up to his famous "Cross of Gold" speech delivered during his unsuccessful bid for the presidency in 1896. His voice

> was only the ghost of the voice of the Cross of Gold. The corrosion of nearly three decades became apparent at once. Once the voice had in it the qualities of brazen trumpets, but the resonance had gone from the brass. Once it had at times the qualities of a stringed instrument, but to-day the strings were loose and discordant. Once it had the booming note of the drum, but to-day the drum showed that it had been punctured in many places and its hollowness was evident.

McGeehan's paragraph has a powerful, insistent rhythm relying on the repetition of the word *once* to draw attention to the passage of time and the figure of the failing fighter's comeback that governs the piece. Once Bryan may have been able to achieve a great triumph, but not today. The paragraph also relies on the metaphor of a musical instrument slightly out of tune and unable to be played as it once was. McGeehan uses the same techniques in his story of an actual heavyweight championship fight, that between champion Jack Dempsey and challenger Luis Angel Firpo in 1923 in New York City at the Polo Grounds. Of that fight, McGeehan writes:

> Seven times Dempsey dropped the Argentine giant last night, and seven times the giant pulled himself up to his feet. He came up fighting, and lashing savagely but

> blindly at the champion. It was a right to the jaw that dropped Firpo the first time. The other six times the giant was felled by volleys of blows, on the head, on the body, all over his huge hulk. As he dropped for the fourth time it seemed he could not rise.

In this passage McGeehan repeats *times* and *time* to emphasize the beating that Firpo takes and to emphasize his courage in getting up for more – a curious emphasis, perhaps, for a writer who regarded boxing as "viciousness [that] could not be equalled in the jungle." The focus in both passages is not on the winner of the contest but on the loser, and McGeehan seems to treat the beaten Firpo with more respect than he does Bryan, whom he depicts as the loser in the trial – although in the actual outcome of the trial Bryan was to be victorious.

When Bryan concludes his address, McGeehan notes that he is greeted with applause, "but it was not inspired applause." Bryan, he writes, "seemed to feel that he had swung and missed," and McGeehan builds on his boxing trope by noting that the applause for Bryan's oration was "the applause that is given to an ex-champion." McGeehan sums up his figurative fight coverage by observing:

> William Jennings Bryan lies horizontal on the field, mortally wounded in his vanity. He crossed jaws with a younger and better man. Oratorically, like many a champion of a less intellectual game, he went into the ring once too often. Once more, as the boys say, youth was served.

McGeehan thus covered one of the most famous trials of the twentieth century as if it had been a championship fight in which the former champ suffers the fate of so many veteran boxers who lack the sense to know that their time has passed, a judgment that McGeehan implied was generally true of the fundamentalist Christians of Dayton, Tennessee.

The responses of readers to McGeehan's coverage of the Scopes trial were mixed. While some readers wrote letters to the editor to congratulate the newspaper and McGeehan for the coverage, others were less enthusiastic. One man wrote:

> It may be true that W. O. McGeehan is well qualified to write on sports, but I am very sure he possesses no qualifications for the work you would have him do in Dayton, Tenn. Your readers are entitled to an unprejudiced version of the events down there, but the established prejudice of Mr. McGeehan is very evident.

Several months after the trial McGeehan made it clear in "Why Pick on Dayton?" that any prejudice that he had exhibited was directed at Bryan, not at the people of Dayton, and he chastised newspapers for rushing to judge small-town America as "the home of the moron." Instead, McGeehan wrote, he had found Dayton to be a small town full of "beauty, courtesy, and intelligence." While some *Herald Tribune* readers may have chafed at his apparent prejudice in covering the Scopes trial, it would be difficult to argue that he possessed inadequate qualifications for any newspaper work he chose to pursue.

As a newspaperman in America in the 1920s, McGeehan was an anomaly. In a decade when athletic heroes seemed to appear in droves, he refused to do the fluff pieces calculated to manufacture heroes. While he never genuflected before Dempsey, Babe Ruth, Red Grange, or golfer Bobby Jones, however, McGeehan had his heroes, and invariably they seemed to be independent, intelligent spirits whose accomplishments soared beyond the boundaries of mere games. In a *Herald Tribune* column on the death of baseball player Christy Mathewson in 1925, McGeehan saluted the values that Mathewson had brought to the game:

> In this sporting game more than in any other walk of life they raise false idols. They make heroes of brutes, they make demigods of inconsequential young men without character or true courage, they make saintly characters out of the vicious.
>
> But when they made a national idol of Christy Mathewson of Bucknell College, they made one out of true metal.... He played for all that was in him, he fought the good fight and the clean fight. He was the incarnation of all of those virtues with which we endow the ideal American.

McGeehan's tribute to his friend Mathewson was written while the reporter was covering the 1925 World Series between the Pirates and the Washington Senators in Pittsburgh, where he heard the news of Mathewson's death. Tom Clark writes that McGeehan, seeking to drown his sorrows at the death of his friend, "went to his room with a bottle of whiskey and composed 'The passing of Matty,' his tearjerking obit for the *Herald Tribune*."

McGeehan, who spent his career chronicling the misdeeds of the less than ideal, lavishly praised any who rose above the pack, as he lauded one great American hero of the age – Charles A. Lindbergh. McGeehan could sound like a "Gee Whiz" writer about Lindbergh: following the aviator's solo trip across the Atlantic in 1927 McGeehan devoted to Lindbergh a "Down the Line" column stylistically characterized by the mythological resonance of Grantland Rice. McGeehan wrote:

> Hermes, of the winged foot, alighting on a "heaven kissed hill," never carried a more important message, even from Olympus, than did his successor, young Lindbergh. And not the least significant part of the message is that it took the clear eyes of youth to see the way and the indomitable spirit of youth to keep on it to the end.

In many ways this passage is untypical of McGeehan, although it reverberates with themes important to him: the power and promise of youth and the courage of the individual in seeking his own way among the masses. McGeehan saw the same spirit Lindbergh exhibited in America's women athletes, who at the time were certainly struggling to earn places in the nation's sports pantheon. When Gertrude Ederle became the first woman to swim the English Channel in the summer of 1926 and broke the men's record in doing so, McGeehan hailed her achievement as "the greatest sports story of the year." Ederle's achievement was "more than a great melodramatic sporting event," because, among other reasons, McGeehan quoted Ederle in writing that it would provide a reason "for all the women in the world to celebrate."

In a piece for the *Ladies' Home Journal* in July 1928, he offers praise and prophetic remarks about the future of women's sports in a preview of the 1928 Amsterdam Olympic Games and of the first American women's track and field team ever to compete in the Olympics. Having cited some statistics of past performances, McGeehan reflects:

> Some day the figures I am giving may be interesting in light of what I believe will be a rapid development in the progress of women in the track and field events. I feel that the girl athletes of today in this branch will be excelled by their successors of twenty-five years from now as Helen Wills and Mrs. Molla Bjurstedt Mallory outshine the women tennis players of a quarter of a century ago.

McGeehan's remarks about the participation of women in the 1928 Games came during a controversy about the advisability of allowing women to compete in track and field at the highest level, and he predicted that by the 1932 Games in Los Angeles American women athletes would make tremendous progress. During those Games Babe Didrikson was to set three Olympic records.

McGeehan's only book, *Trouble in the Balkans* (1931), was a tongue-in-cheek account of his travels. Reviewer Lisle Bell, writing in the 6 September 1931 *Books,* found this book to be vintage McGeehan – "a continuous chuckle – human, spontaneous, unpretentious. It is packed with laughs that hit you below the belt."

McGeehan and Rice must have shared a similar laugh at McGeehan's role in a 1926 Sportlight motion picture, one of a series of short motion pictures on sporting topics that Rice had started making in 1920. This particular production, *The Call of the Wild,* had been made in New Brunswick, Canada, and Rice had cast McGeehan in the role of a hunter tracking a giant moose through the woods. Rice recalls that the "combination of Bill's bad heart and non-cooperative moose made it tough sledding," and his production crew saved the day – and much hiking for McGeehan – when it "rented a particularly noble, if dead, moose from a native guide and propped him up in the densest bush with stilts, baling wire and ropes. Then Bill banged away with his muzzle loader at the defunct critter." Rice concludes, "That moose should have shared equal billing with McGeehan."

Largely because of his drinking, McGeehan's health was to fail rather early. His drinking had affected his work for some years, and the *Herald Tribune* found that it could not always count on him to produce his column on schedule. Ultimately the paper arranged for McGeehan to be paid on delivery, at a sum of $100 per column. Often his colleagues sought to cover for him by doing their best to turn out a column in what they hoped would pass for the McGeehan style. Stanley Walker writes, "There were days, in the last years, when even McGeehan, whose 'Down the Line' was one of the soundest of all columns, was actually hard to read."

In autumn 1933 McGeehan and his wife went to Brunswick, Georgia, to visit Wilbert Robinson, an old friend and the former manager of the Brooklyn Dodgers. At Robinson's home McGeehan suffered a massive heart attack and was hospitalized for a week. Yet even in the hospital McGeehan sought to maintain his reputation as a newsman, as he insisted that a typewriter be brought to his hospital room so that he could turn out his column. When doctors allowed McGeehan to return to Robinson's home to convalesce after a week of confinement, he weakened and was forced to return to the hospital. Again his condition improved, and he was allowed to go to a cottage at Sea Island Beach, although doctors warned him that he could die suddenly. On 29 November 1933, a little more than a month after his first heart attack, McGeehan suffered a fatal attack. Following a funeral ceremony attended only by his wife and a few close friends,

McGeehan was buried in Frederica Cemetery on Saint Simons Island.

McGeehan's place in the history of American sports journalism is lofty and secure. His writing set a standard for creativity and honesty in an age when overwriting and puffery dominated the sports pages. Above all else, McGeehan was a good reporter who happened to devote most of his career to sportswriting. One can view his work as continuing the reformative aims of the muckrakers during the first decade of the twentieth century, and his "Aw Nuts" style prefigures much of the crusading sports journalism that distinguishes the final third of the twentieth century.

Stanley Woodward, one of McGeehan's successors as sports editor of the *Herald Tribune,* has praised sportswriters who founded modern sportswriting. Rather than belonging to either the "Gee Whiz" or "Aw Nuts" schools, writers such as McGeehan, Rice, and Westbrook Pegler were "giants of our craft" who founded what he calls the "On-the-Button school which attempts straight reporting." Writing of McGeehan and other colleagues, Woodward insists that "There are moderns who . . . outdo the precursors as columnists. There are many who outdo them as reporters, but everyone in the craft owes them something, for they and others of their caliber were the pioneers of modern sports writing."

**References:**

Tom Clark, *The World of Damon Runyon* (New York: Harper & Row, 1978);

Bob Cooke, ed., *Wake Up the Echoes* (Garden City, N.Y.: Hanover House, 1956);

Charles Fountain, *Sportswriter: The Life and Times of Grantland Rice* (New York: Oxford University Press, 1993);

Gene Fowler, *Beau James* (New York: Viking, 1949);

Fowler, *Skyline* (New York: Macfadden, 1962);

Richard Kluger, *The Paper: The Life and Death of the New York Herald Tribune* (New York: Knopf, 1986);

Robert Lipsyte, *SportsWorld: An American Dreamland* (New York: New York Times Books, 1975);

Grantland Rice, *The Tumult and the Shouting: My Life in Sport* (New York: Barnes, 1954);

Randy Roberts, *Jack Dempsey: The Manassa Mauler* (Baton Rouge: Louisiana State University Press, 1979);

Stanley Walker, *City Editor* (New York: Stokes, 1934);

Stanley Woodward, *Sports Page* (New York: Simon & Schuster, 1949).

# Tom Meany

*(21 September 1903 – 11 September 1964)*

Richard Orodenker
*Pennsylvania State University, Ogontz*

MAJOR POSITIONS HELD: Sports reporter, *New York Journal* (1922), *Brooklyn Daily Times* (1923–1929), *New York Telegram,* later *New York World-Telegram* (1929–1940), *PM* (1940–1948); sports editor, *New York Star* (1948); sports columnist, *New York Morning Telegraph* (1949); sports editor, *Collier's* (1950–1956).

BOOKS: *Babe Ruth: The Big Moments of the Big Fellow* (New York: Barnes, 1947);

Joe DiMaggio, *Baseball for Everyone: A Treasury of Baseball Lore and Instruction for Fans and Players,* ghostwritten by Meany (New York: Whittlesey House, 1948);

*Baseball's Greatest Teams* (New York: Barnes, 1949);

*Baseball's Greatest Hitters* (New York: Barnes, 1950);

*Theodore Samuel Williams: Hitting Unlimited* (New York: Barnes, 1951);

*Stanley Frank Musial, the Man* (New York: Barnes, 1951);

*Ralph Kiner, the Heir Apparent* (New York: Barnes, 1951);

*Joseph Paul DiMaggio, the Yankee Clipper* (New York: Barnes, 1951);

*Baseball's Greatest Pitchers* (New York: Barnes, 1951);

*The Magnificent Yankees,* by Meany and others (New York: Barnes, 1952; revised edition, New York: Grosset & Dunlap, 1957);

*Baseball's Greatest Players* (New York: Grosset & Dunlap, 1953; revised edition, New York: Dell, 1955);

*The Artful Dodgers,* by Meany and others (New York: Barnes, 1953; revised edition, New York: Grosset & Dunlap, 1958; revised, 1963);

*Milwaukee's Miracle Braves,* by Meany and others (New York: Barnes, 1954);

*The Incredible Giants,* by Meany and others (New York: Barnes, 1955);

*The Boston Red Sox,* by Meany and others (New York: Barnes, 1956);

*Mostly Baseball: A Twenty-Year Collection of the Best Magazine Articles of a Nationally Known Sportswriter* (New York: Barnes, 1958);

*The Yankee Story* (New York: Dutton, 1960);

*The Forward Pass,* by Meany and Charlie Conerly (New York: Dutton, 1960);

*There've Been Some Changes in the World of Sports* (New York: Nelson, 1962);

*How to Bat,* by Meany and Harry Walker (New York: McGraw-Hill, 1963);

*Baseball's Best: The All-Time Major League Baseball Team,* by Meany and Tommy Holmes (New York: Watts, 1964);

*Kings of the Diamond,* by Meany and Lee Allen (New York: Putnam, 1965).

OTHER: *Collier's Greatest Sports Stories,* edited by Meany (New York: Barnes, 1955);

*Major League Baseball Annual,* edited by Meany (New York: Grosset & Dunlap, 1960);

*The Best of Red Smith,* foreword by Meany (New York: Watts, 1963).

Prolific sportswriter, author of popular books, promotional director, raconteur, after-dinner speaker, and master of ceremonies, Tom Meany was known for his remarkable memory and his sharp repartee, "illumined," wrote Red Smith in the *New York Herald Tribune* (15 September 1964), "by his sometimes double-edged wit." In 1947, when Clint Hartung of the New York Giants was being given an unprecedented rookie buildup, Meany's quip was widely quoted: "Why bother to even play the season? He should go straight to Cooperstown." Having worked for a series of publications that went out of business, Meany was convinced that he had given them "the journalistic kiss of death." In baseball lore he will be remembered as the man who nicknamed Brooklyn Dodger owner Branch Rickey "the Mahatma." His most widely quoted line occurred in 1948, when Yankees owner Del Webb approached Bucky Harris at a party shortly after

Webb had abruptly fired Harris as manager. To the question "What does Del Webb want with Bucky Harris now?" Meany replied, "He came back for the knife."

Born in Brooklyn, New York, on 21 September 1903, Thomas William Meany attended St. John's Preparatory School, a great basketball school, where he played only football and baseball – which, Meany later said, was like going to Notre Dame and not playing football. He was, however, manager of the 1921–1922 basketball team, which won the New York state championship. After graduating in 1922 from St. John's, where he had been on the staff of *The Red Owl* yearbook and a student sports correspondent for the *Brooklyn Eagle,* Meany worked for the Brooklyn-Manhattan subway system as a timekeeper until Gary Schumacher, sports editor of the *New York Journal,* hired him in 1922 to cover high-school sports at space rates for the paper's new Brooklyn section. In the introduction to Meany's *There've Been Some Changes in the World of Sports* (1962) Schumacher recalled,

> I remember guaranteeing him $20 a week regardless of how much his space writing brought him. . . . Tom made $65 the first week, $89 the second and $103 the third. Whereupon the Hearst hierarchy decreed that, no matter how many columns Tom wrote he was to receive no more than $60 a week. He was the first man in America . . . to have a ceiling placed on his salary.

Rival newspapers were, however, willing to accept carbons of his *Journal* stories. By 1923 he was working full-time for thirty-six dollars a week, with a byline, covering the Dodgers for the *Brooklyn Daily Times.* Ebbets Field, where he reported the Notre Dame and Army football games in addition to the Dodgers during those early years, would be both a starting and, when it was demolished in 1960, an ending point for Meany's career. "It served," he would write in *There've Been Some Changes in the World of Sports,* "to symbolize the vast and sweeping changes which have come to baseball and, indeed, to my own profession of sports writing."

At Ebbets Field, where he was official scorer in 1926 and 1928, Meany fell victim to the wrath of the legendary Dodgers "bleacher bum" Hilda Chester. According to Jerry D. Lewis in *Great Baseball Stories* (1979), "After Brooklyn bought a clumsy outfielder for $8,500, Tom Meany observed that the $500 was for his fielding. The following day, Hilda waited outside the Press Gate and chased Meany until she ran out of gas."

In 1929 Meany moved to the *New York Telegram* (later the *World-Telegram*) to cover the Giants.

*Tom Meany (left) with Lou Gehrig and sports cartoonist Willard Mullin (photograph by William C. Greene)*

On 3 June 1932 Meany beat the other papers by breaking the story that John McGraw had resigned and Bill Terry was the new manager of the Giants. Meany's influence was far-reaching by this time. The former *Sports Illustrated* editor Robert W. Creamer says in his biography of Babe Ruth (1974) that Meany was "a fine reporter" and "a gifted writer," and he says that it was "Meany's repeated accounts of that colorful World Series" of 1932, in which Ruth supposedly hit his "called-shot home run," that "got the legend started and kept it going." On 26 December 1932 Meany married Clara M. Maxwell, who became his constant traveling companion and a knowledgeable partner in Meany's sporting world. They had no children and resided on the Park Slope in Brooklyn.

At the *New York Telegram* Meany began to hone the descriptive leads that would become his trademark: "Dimly through the dust cloud, which has swept southeastward from Nebraska and the Dakotas to envelop this town in a murky haze, the

Giants today found themselves sharing third place with the Cardinals" (11 May 1934). He covered games with the casual tone of the generation of writers before him – a generation that included Ring Lardner and Damon Runyon. He would choose a theme and build a story around it. During a series of games in which the temperature approached one hundred degrees he used the weather as a metaphor: "If the far-flung Cardinal empire maintains a branch in Arabia, now is the time for Branch Rickey to call in his Bedouin farmhands. They'd be a cinch in this temperature" (25 June 1934). Meany regularly dropped ironic or comic comments into his columns: "Durocher hit a pop fly to right, which Homer Peel lost in the wind. . . . Some slugger, that Babe Durocher!" (11 May 1934). He was also stocking up on anecdotes for off-day and rainy-day pieces, such as the one about "Lefty O'Doul's double. . . . Not a double that Lefty clouted in the pinch, but a guy who has been going around New York masquerading as Lefty and doing a pretty good job of it" (12 June 1934).

During the 1930s Meany wrote occasional freelance articles for the *Saturday Evening Post,* whose sports editor, Harry T. Paxton, became a mentor. Meany's 6 March 1937 article, "Once a Dodger," on "the days when the Dodgers were much closer to the sheriff than they were to the first division," included his well-known reflections on Brooklyn manager Wilbert "Uncle Robbie" Robertson: "Robbie was a rule-of-thumb manager, a gentle Falstaff, who could get more out of less material than any manager before or since." He also revealed that Robertson "selected Dick Cox to play right field because he couldn't spell 'Oscar Roettger,' his original choice for that position."

In a *Post* piece on football, "Dixie Plays for Blood" (18 November 1939), Meany analyzed the success of the sport below the Mason-Dixon line. He debunked the notion that teams were refighting the Civil War or that southern players were simply superior: "there is such a Northern infiltration in the South today that Dixie may no longer take bows for all its stars." Nevertheless, Meany, with lively anecdotes and quotations, demonstrated "why football is a more exalted sport in the South than in any other section of the country." Describing how University of Georgia students at Yankee Stadium "demolished the goal posts before Colonel Ruppert's house infantry could so much as raise a club" after a victory over New York University, Meany focused on a student who "had clawed off more than he could carry." The student was finally assisted uptown in a cab by a Georgia dean, on whose desk "that chunk of goal post reposed" for the rest of his tenure at that university. According to Meany, "It was a perfectly natural gesture. He had reacted to the triumph exactly as had the freshie."

From 1940 until its demise in 1948 Meany worked for the New York evening newspaper *PM;* during World War II he also wrote the sports section of a newsletter published for the large contingency of servicemen from Saint Xavier, the church he attended in Brooklyn. He continued to freelance for the *Saturday Evening Post* with "inside" articles such as "The Chorus Men of Baseball" (25 August 1945). In one of the many allusions to current events that appeared in his pieces, he compares baseball scouts to eminent statesmen of the day: the scouts "must have the affability of a [Edward] Stettinius, the stubborn craft of a [Vyacheslav] Molotov, the persuasive powers of an [Anthony] Eden and, the record proves, the guile of a [Saburo] Kurusu," all the while getting "less spinach a year than the average Philly outfielder."

By this time his voice and style had hit their stride. Meany continued to write from the perspective of the fan in the stands, even as that approach was beginning to go out in favor of the more objective style of writers such as Frank Graham of the *New York Sun.* Still, Meany's columns were a refreshing break from what he called the "numbing pedestrian mediocrity" found on other sports pages, whose writers' "idea of reporting a game was to record, chronologically, how the runs were scored." Meany talked casually and familiarly to his readers, never down to them. A 19 September 1941 *PM* piece shows his relief at the news that the pennant-chasing Dodgers, who had just lost a crucial game, "received a reprieve from the governor" when their nearest rivals, the Saint Louis Cardinals, miraculously lost to the Boston Braves: "Any time Manuel Salvo beats a pitcher like Mort Cooper it must be construed as a manifestation of divine blessing." The loss was not "a pleasant exhibition for any one who has grown enamored of the Dodgers as I have," Meany wrote. Down 4–0, "The Dodgers really rallied like lions . . . then played like hoodlums," losing 6–5 on a flagrant balk and three wide pitches to a batter's head, one of which "rolled 127 feet behind the plate." The Dodgers did not take losing lightly: "the beefing in the Dodger clubhouse was a new high for the year." "Well," he concludes, "it's all water under the bridge now, fellows. . . . Let's go get the Phillies – nice fellers, who never harmed anybody, and aren't going to stop now, I hope."

Meany's first book, *Babe Ruth: The Big Moments of the Big Fellow* (1947), demonstrates his commitment to accuracy as well as drama. In recounting the called-shot home run fifteen years later, Meany came to question the key aspect of the legend he himself had helped, unconsciously, to promulgate:

> It was then the big fellow made what many believe to be the *beau geste* of his entire career. He pointed in the direction of dead centerfield. Some say it was merely a gesture toward [pitcher Charlie] Root, others that he was just letting the Cubs bench know that he still had the big one left. Ruth himself has changed his version a couple of times. . . . Whatever the intent of the gesture, the result was, as they say in Hollywood, slightly colossal.

Meany's leads increasingly came to resemble the openings of short stories, anticipating – or perhaps influencing – the later New Journalism of writers such as Jimmy Breslin and Gay Talese. Two pieces, the first from *Sport* (April 1947) and the second from *The Saturday Evening Post* (March 1949), begin:

> Thalmann, Georgia, is a whistle stop so tiny that its name is spelled incorrectly on some maps, which leave out the second "n." Its population is listed as 100, but that must include the scrub cattle and the razor-back hogs as well as the human beings.

> It was a typical convention room, filled with enough smoke to nominate twenty Hardings. The occupants, however, were no statesmen.

Such descriptive leads soon became Meany's trademark. They, along with his interviewing and researching skills, enabled him to give the reader a firsthand sense of an event or a life, from his biography of Ruth, his old card-playing crony, to his "instant biographies" of Ted Williams, Stan Musial, Ralph Kiner, and Joe DiMaggio, all of which were published in 1951. In 1948 Meany ghostwrote the humorous *Baseball for Everyone,* which carried DiMaggio's byline though Meany's participation was revealed on page 173 of the 224-page text. "We have suspected this all along, of course," wrote Harold Kaese for *The New York Times Book Review* (30 May 1948), "because the preceding 172 pages sound exactly like nobody but Tom Meany." That same year Meany became "sports director" (he did not like the term *editor*) for the *New York Star,* and then in 1949 he became sports columnist for the *New York Morning Telegraph.*

When he wrote about players whom he had come to know only late in their careers or those he had only heard about, Meany absorbed the stories told to him by older sportswriters, making their voices part of the piece. In "The Flying Dutchman," his essay on the 1909 Pirates in his *Baseball's Greatest Teams* (1949), Meany writes: "The late Sid Mercer, one of the charter members of the Baseball Writers Association, told me that for some time after, traveling on the Pullmans, the Giants would douse all the lights and somebody would strike a match to reveal [Giants outfielder Red] Murray posed, tableaulike, in the position in which he had made the sensational catch." Meany was at his best writing these short pieces about ancient teams or old players, particularly in his series of books on the "greatest" aspects of the game: *Baseball's Greatest Teams; Baseball's Greatest Hitters* (1950); *Baseball's Greatest Pitchers* (1951); and *Baseball's Greatest Players* (1953). His essay "The Cat," on the young Roy Campanella, in *Baseball's Greatest Players,* shows how Meany could capture the essence of a contemporary player while also showing the player's connection to the past. The essay also reveals Meany's willingness to draw these comparisons without the references to Campanella's color that were so common in the early days of baseball desegregation:

> Until he dons his armor and crouches behind the plate, Campanella doesn't look like what he is – the best catcher in baseball today. He's a roly-poly sort of guy, a real butterball. He reminds old-timers a little bit of Bob O'Farrell, who caught in the National League for two decades, quitting in the mid-30's. He's chunky like O'Farrell but even more catlike behind the plate. And faster afoot, too.

Meany became sports editor for *Collier's* in January 1950; he remained with the magazine until its final issue in December 1956. His *Collier's* articles were frequently reprinted in the Dutton *Best Sports Stories* anthologies. At *Collier's* Meany wrote mainly about baseball; *Mostly Baseball* is the title of his 1958 collection of pieces, most of them from *Collier's,* from a twenty-year period. "Night Life in the Three-Eye League" (15 July 1950), about the Illinois-Indiana-Iowa League, details "the epitome of the bushes . . . the butt of baseball jokes through the years." Like most of Meany's articles, this one is full of anecdotes (including one about the Cubs farm team in Decatur, Illinois, that went by the nickname "Commies" in honor of Stephen Decatur's ship, the *Commodore*) but centers around a particular theme – in this case "the trials and tribulations of baseball's grass roots."

On 2 June 1951 Meany wrote the first national magazine piece on the young Mickey Mantle, who was already beset with some of the physical problems that would plague him throughout his career. Meany acknowledges that "no other rookie ever

The Yankee Story

By TOM MEANY

With Illustrations

E. P. DUTTON & CO., INC.
NEW YORK 1960

*Title page for Meany's second history of the New York Yankees*

started a season in the big-time with so little public notice and such enthusiastic professional fanfare." By 1957, when the inevitable comparisons were being drawn between Mantle and Willie Mays (who, Meany believed, could "do more things better than most any ballplayer you ever saw"), Meany observed, "Always within Mickey is the knowledge that any game he plays could be his last – one throw, one swing, one slide could write the end of his major-league (and money-making) career. Yet Mick plays every game to the hilt, a tremendous tribute to his courage." He never tired of writing about Mantle or DiMaggio, and he was convinced that "It wasn't the quality of the writing that got the piece past the doorman, it was the subject matter."

Each February and March, Meany would make predictions – including pennant winners, twenty-game winners, and rookies of the year – for the coming baseball season. These were not mere lists of guesses but detailed analyses based on statistics, although he considered what he was doing to be analogous to a "politician's promises" or the forecasts of a "slightly cracked crystal ball and a seer in the same condition." Meany was a forerunner of today's "sabermetricians," who report on baseball largely by making mathematical comparisons and quantifying the record book. He examined old-time players in the same light, making evaluations based on a close reading of their numbers. He suggests in *Baseball's Greatest Players,* for example, that Lou Gehrig's awesome runs-batted-in production of 1,991 is a truer measure of his worth as a player than his home run totals or batting average. Gehrig, Meany noted, often batted with no men on base, since Ruth, who hit before Gehrig, frequently cleared the bases with a home run or struck out to end the inning. With that fact in mind, Meany says, "You get some idea of the devastating power Lou packed in his bat."

When he was not writing about baseball Meany covered sports he neither liked nor knew much about; but he still wrote with ease and versatility, as in a 1949 piece for *Sport* magazine, reprinted in *Mostly Baseball,* on the original Boston Celtics, a team whose "next stop was always just around the corner" because they played wherever and whenever they could. For *Collier's* he wrote about Miami's Hialeah racetrack, which, with its botanical beauty and the cunning and foresight of owner Joseph E. Widener, brought thoroughbred racing to America during the usually idle months between November and April (October 1950); golfer Jack Burke Jr., who "has the true champion's gift of being able to concentrate on his game to the exclusion of all else" (14 June 1952); the "headaches that arrive at the trouble gate" of big football events, from ticket and concession worries to problems with the media (29 November 1952); heavyweight champion Rocky Marciano, who seemed unbeatable in the early 1950s unless a fighter came along who was "a composite of seven different champions" (17 January 1953); and hockey great Jean Beliveau, who "learned the game on the rivers and ponds of his native land Quebec, just as Mays learned his baseball on the fields and sandlots of his native Alabama" (4 January 1957). In writing on sports with which he was only vaguely familiar Meany sought the help of experts, often crediting them in the text or with a joint byline.

Meany's finest piece for *Collier's* may have been "Hard Luck Bill Keeps Pitching," his 28 July 1951 profile of Yankee pitcher Bill Bevens, who in 1947 came within one out of pitching the first-ever World Series no-hitter. The article is supposed to have helped Bevens get another chance with the

Cincinnati Reds in 1952, though he did not make the team. Meany recalled "the day of [Cookie] Lavagetto's hit," which broke up Bevens's no-hitter with two outs in the ninth and cost him the game, as "a time of near greatness and ultimate calamity." It was not a pretty game: Bevens had walked several batters, and the Dodgers had a run against him, but he was still on the verge of World Series history. Meany relates Bevens's "downfall" in major league baseball after that event: in 1948, sore-armed and dispirited, "Bevens was back in [his hometown of] Salem [Oregon], working for a beer company, hoisting cases onto trucks"; the year before he had been "a goodwill ambassador" for the firm, "standing around and letting potential customers gape at him."

Meany finds Bevens playing minor league baseball in Salem in 1951. He is a reticent man, but Meany believes that he is just like the young minor leaguers who dream of "comfortable Pullmans and chartered transport planes" and of "someday kissing the buses good-by." The article contrasts life in the minor and major leagues through the eyes of someone who has experienced both and must now start all over again. Bevens's major league career had been one "of great frustration" – of close ball games, near wins, and "almost three scoreless innings in the payoff game" of the 1947 World Series, for which he could have been credited with the victory but, because of a scorer's decision, was not.

Meany relates the desperate moments in Bevens's career: going to doctors, some of them quacks; losing his confidence and self-control; struggling through poor seasons in the low minors; and putting up with second-guessing from fans and reporters while always having to relive the momentous game "in reverse English" – whether he would pitch the same way to Lavagetto if he had it to do over again. Meany ends the piece with Bevens's answer to the question of what it was like to come so close to immortal glory and then lose it in a second: " 'I figure I was lucky to be up there in the first place.' " When the article was included in *Best Sports Stories 1952,* the judge, John Chamberlain, said that it "has a note of courage and of philosophy. It isn't just pathos."

Throughout the 1950s Meany wrote baseball books almost exclusively for the A. S. Barnes Company, one of the leading publishers of sports books in the growing adult market. In an effort to compete with Putnam's series of team histories Meany coauthored *The Magnificent Yankees* (1952), *The Artful Dodgers* (1953), *Milwaukee's Miracle Braves* (1954), *The Incredible Giants* (1955), and *The Boston Red Sox* (1956). These books, however, were not authentic histories, and some reviewers found Meany's anecdotal style too disjointed. After *Collier's* folded, Meany wrote his yearly baseball previews and other articles for *Look* magazine for several years. In 1958 he became what he called a "drum-beater," a promotions man, for the Yankees and an occasional radio sportscaster. He also continued to write books, including collaborations on football (1960) and batting (1963), and freelance magazine articles. Reviewing Meany's second book on the Yankees, *The Yankee Story* (1960), in *The New York Times Book Review* (10 July 1960), Arthur Daley said: "So ingratiating is the Meany style with its fluent, witty and anecdotal approach that each chapter seems all too brief, a literary vaudeville that follows the cardinal rule of all successful vaudevillians – stop while the audience is still begging for more."

In 1962 Meany joined the newly formed New York Mets as publicity director – his "first success in twenty-five years," he said, as the Mets were losing a record number of games but drawing big crowds. He was a popular after-dinner speaker and toastmaster, giving one of his most notable performances in 1963 on the night Toots Shor's renowned restaurant closed its doors for good.

*There've Been Some Changes in the World of Sports* was Meany's last significant book. Mostly autobiographical and, again, heavily anecdotal, the book comprises Meany's pithy observations on developments in sports and sportswriting during the preceding four decades: "Whatever vines are left in the minors, the teams are dying on them"; "Back in the preradio and pretelevision days it was about as easy to see the average college football coach as it was to have a chat with a Trappist monk"; "Sports, generally, have improved because of some promotions, and in spite of others." The book is especially fascinating for its stories about early sportswriters. In the section "Odd Ball Sports," for instance, Meany says:

> The writers who covered wrestling were frankly skeptical of it, unless they were on the payroll and I don't think many were. The late Lloyd Lewis, biographer of U.S. Grant, served a hitch as sports editor of the Chicago *News* and had all wrestling results carried on the drama page, where, he felt, they belonged. Ben Epstein, who had done some middleweight wrestling in carnivals himself, attracted the attention of Dan Parker, sports editor of the New York *Mirror,* with a column in the *Arkansas Gazette* in which he predicted the outcome of all wrestling matches in the South for the following week. Thus was Ben launched on a happy career covering the Yankees for the *Mirror*.

Though Meany dwells in the past in these pages, he is convinced that "Baseball is played better today, and written better, than it was in 1923. . . . Those who sigh for the good, old days are taking refuge behind rose-colored glasses."

After a brief illness, Meany died of an internal hemorrhage at Roosevelt Hospital on 11 September 1964. The sports and sportswriting worlds paid final tribute at a requiem mass at Saint Francis Xavier Catholic Church in Brooklyn. In 1975 he was posthumously given the J. G. Taylor Spink Award and thus entered the writers' wing at the Baseball Hall of Fame in Cooperstown, New York.

Meany's work was frequently reprinted during his lifetime but has now largely faded into obscurity. As a writer and editor Meany helped set a new standard for the light sports essay, as well as for baseball books that purport to say which player or team was the "best" – even though the writer may never have seen that player or team play. Grantland Rice writes in the foreword to Meany's *Baseball's Greatest Teams,* "I've been around covering baseball since before Tom Meany was born. I've known him as a diligent questioner and a thorough reporter. I've seen all the clubs he has selected – seen them play during regular season competition and World Series play. I think his choices are sound. . . . that there is a valid argument to be put forth on behalf of those he has selected." Ford Frick, former baseball commissioner and sportswriter from an earlier era, says in the foreword to *Baseball's Greatest Players* that in Meany's books ballplayers "come alive again, not only as players, with the conditions under which they played made plain, but also as flesh-and-blood human beings." In the introduction to *Mostly Baseball* sportswriter Bill Corum says of Meany's magazine writing, "There is more to a sports piece than the merry quip, regardless of how merry, or the happy phrase, regardless of how happy. It takes reporting and leg-work to research the facts and this is evident throughout these articles."

Meany was one of the last of a breed of sportswriters who put a positive spin on the events and persons they covered. "He managed, without trying," wrote Red Smith, "to breach the wall that separates the baseball players from the baseball writer." He represents, in that respect, a bygone era; but his prose remains crisp and clear, and his depth of knowledge and lore more than confirms Arthur Daley's opinion of Meany as one of "Baseball's Boswells." As Meany wrote in the foreword to *The Best of Red Smith* (1963), "To write well is a talent of sorts, but to be an objective reporter requires hard work, a step removed from manual labor." What he says of his friend Smith could easily apply to Meany himself: "He possessed . . . not only the gift of writing entertainingly, but also the gift of knowing his subject."

**References:**

Robert W. Creamer, *Babe: The Legend Comes to Life* (New York: Simon & Schuster, 1974), p. 365;

G. H. Fleming, *The Dizziest Season* (New York: Morrow, 1984), pp. 83–84, 115–116, 128, 130–131, 183–184;

Stanley Frank, ed., *Sports Extra* (New York: Barnes, 1944), pp. 257–260;

Jerry D. Lewis, *Great Baseball Stories* (New York: Tempo, 1974);

Red Smith, *To Absent Friends* (New York: Atheneum, 1982), pp. 426–428.

# Joe H. Palmer

*(1904 - 31 October 1952)*

Bill Hughes

MAJOR POSITIONS HELD: Associate editor and business manager, *The Blood Horse* (1935–1944); editor, *American Race Horses* (1944–1946); racing columnist and editor, *New York Herald Tribune* (1946–1952).

BOOKS: *Names in Pedigrees* (Lexington, Ky.: The Blood Horse, 1939);

*Horses in the Blue Grass,* by Palmer, Bert Clark Thayer, John Hervey, and W. Jefferson Harris (New York: Duell, Sloan & Pearce, 1940; revised edition, 1947);

*An Introduction to the Thoroughbred Horse: Origin, Distribution, Breeding, Conformation, Uses,* by Palmer and J. A. Estes (Lexington, Ky.: The Blood Horse, 1942);

*Feeding Practices on Kentucky Thoroughbred Farms* (Lexington, Ky.: The Blood Horse, 1943);

*This Was Racing,* edited by Red Smith (New York: Barnes, 1953).

*Joe H. Palmer (courtesy of* The Blood Horse)

Joe H. Palmer belonged to that rarest class of sportswriters: those whose writing can interest readers with no initial interest in the particular sport. During his six years as a columnist for the *New York Herald Tribune* his writing always appealed to a large following of non-sports fans. His sport, and indeed the chief interest of his whole life, was thoroughbred horse racing and all that it entailed, and Joe Palmer always remained consumed by the drama of the competition and by the colorful characters who owned, rode, and wagered on the magnificent animals.

Growing up in Lexington, Kentucky, the heart of the American horse breeding and racing industry in the mid twentieth century, Palmer was quickly and completely surrounded by horse racing, and this background gave him a privileged perspective on the sport. His lifelong affiliation with racing enhanced his access to it, and his reputation and expertise on the subject lent authority and credibility to his writing. The broad appeal and accessibility of his columns resulted from his combination of concise and scholarly observations with the wit and good humor of a man who obviously loved what he was doing.

Born in Lexington to Joe H. Palmer and Sarah Frances Doyle Palmer, Palmer graduated Phi Beta Kappa from the University of Kentucky in 1927 and received a master of arts degree the following year. From 1928 to 1932 he taught English there before moving to the University of Michigan, where he taught English for two years while working for a Ph.D. Palmer completed all the necessary requirements for a doctorate, but in 1935 he decided that rather than writing a dissertation he would write for *The Blood Horse,* a weekly periodical devoted to horse breeding. He left a promising academic career and never looked back.

Palmer's love of language, sense of humor, and remarkable facility for a well-turned phrase impressed and inspired much respect from his students. Red Smith recounted a story of one Palmer meeting with a former student at a racetrack in the 2 November 1952 *New York Herald Tribune.* Palmer was acting as a judge at Keeneland Racetrack in Lexington when a mob of bettors, disgruntled at the call of one race, began clambering up the walls of the judging booth to release their frustrations on the hides of the officials. Palmer recognized the leader as one of his former students.

> "Bull," he snapped.
> "Y-Yes, sir, Mr. Palmer."
> "Get down outta there."
> "Y-yes sir," Bull replied, and the riot was quelled.

But the academic world's loss was the racing world's gain, especially for those who liked to read. To those who asked Palmer why he left teaching, he explained simply that he loved horse racing and wanted to spend his life around it. Yet he balked at being called a "horse lover," for in one column he wrote,

> A horse lover is a horse that is in love with another horse. In the main, the actual racing of horses around a track is an economic necessity which makes possible the existence of a small and, I think, very pleasurable world which is populated more by men and women than by horses, just as bridge or canasta is an excuse for the assemblage of people who like each other's company but do not trust their unaided powers of conversation.
>
> It is necessary to get out from seven to nine fields every afternoon, to take money from people who would otherwise spend it foolishly, but the real fun of racing lies elsewhere. Put a crowd of racetrackers together and they will talk about horses, to be sure, but the horses will be used, as they are in most good racing fiction, merely as symbols which bring out very human rivalries and deceptions and strategies of a game which does not ever actually end.
>
> It is a game played with such extreme gravity that many parties to it ultimately forget that it is a game, and sometimes confuse themselves with normal people.

In addition to this formidable wit Palmer could draw on a broad background in literature, and he would often pepper his columns with references to great works. Though he could quote *Beowulf,* in its original Old English, from memory, his writing was never stuffy or verbose. He once told a young sportswriter that in writing a column an author should include one word that would send the average reader to the dictionary, but he should never use two such words.

As a young writer Palmer began with the fire and zeal of a reformer, determined to expose the more unscrupulous characters and activities in horse racing and thereby clean up the tarnished image of the sport that he felt the general public had. As the years passed he realized that such a tarnished image was not only well deserved but part of the mystique and charm of professional horse racing.

Thomas B. Cromwell, publisher of *The Blood Horse,* viewed Palmer's early efforts with skepticism. After reading a few of Palmer's more ambitious efforts at removing the parasites on horse racing, Cromwell gave him a piece of advice that the writer carried with him throughout his career: "There's no future in wising up suckers. They're born faster than you can get at them." With that, Palmer later acknowledged, "the kindergarten was over."

At *The Blood Horse* his position quickly advanced from that of writer to associate editor to business manager, and he remained with the periodical until 1944. He continued to contribute columns to *The Blood Horse* later that year, after he became the executive secretary of the American Trainers' Association and also began a three-year term as editor of *American Race Horses.*

In 1946 Palmer was hired by the *New York Herald Tribune* as the paper's racing editor, and his popularity began to spread widely. Stanley Woodward, the legendary sports editor of the *Herald Tribune,* went out of his way to hire Palmer, who suspected that Woodward scorned racing writers. During a first interview Woodward told Palmer, "I have discovered after taking an opinion poll that you are the man we need. So let's stop this ineffectual talk about who likes whom and get down to the matter of how much money it will take to get you to come with us."

Palmer had written three books – *Names in Pedigrees* (1939), *Horses in the Blue Grass* (1940), and *An Introduction to the Thoroughbred Horse* (1942) – but these were not widely read outside racing circles. His talent as a versatile, entertaining, and engaging writer quickly earned him a large following in New York and a nationwide reputation.

At the time Palmer joined the *Herald Tribune* the paper was strenuously competing with *The New York Times* for readers and advertisers. While the *Times* excelled in some departments, the *Herald Tri-*

*bune* was regarded as having the better sports department: Red Smith, Jesse Abramson, Al Laney, and Roger Kahn all worked with Palmer on the *Herald Tribune* sports section. When asked about those days at the paper Kahn said in an unpublished interview, "This was like Olympus. We had more talent assembled in that sports department than any paper in history, and we all looked up to Joe Palmer."

One of Kahn's early assignments on his way up the ladder from copyboy to beat reporter was to substitute for Palmer at the racetrack on the latter's day off. Kahn remembers sweating bullets during his first assignments, and Palmer responded with a reassuring ease in advising him, "Whatever you do, kid, remember, it ain't the Iliad." In his 2 November 1952 column paying tribute to Palmer, Red Smith wrote,

> It can be stated only as one man's opinion, yet unquestionably it is shared by thousands, that Joe Palmer could write better than anybody else in the world whose work appeared in newspapers. And that may be limiting the field too narrowly. In the field of racing, which he preferred, there was never another in his time or before to compare with him. He knew his subject profoundly, . . . and he wrote as he talked, with wit, urbanity and grace and sure confidence.

Palmer's column, "Views of the Turf," appeared in the *Herald Tribune* on Mondays during the racing season, when he was writing daily news stories, and three times a week when the eastern racing season was in recess. He covered racing not only at the local New York tracks of Aqueduct, Belmont, Jamaica, and Monmouth, but also at New York State tracks as far north as Saratoga and, in Florida, as far south as Hialeah, the site of his first *Herald Tribune* story. Frequently he would include items from tracks as far away as England, Argentina, or Australia.

In his first column for the *Herald Tribune,* Palmer laid out his views on horse racing, on what intrigued him about it and how he intended to cover it:

> A prejudiced witness is all right, if you know what his prejudices are. . . . The statement is frequently made that people go to the races to bet. This is a half-truth, or a three-quarter truth, or a seven-eighths truth. . . . At any rate, it is the remaining half or quarter or eighth, which this onlooker finds most interesting, and which he will try to report.
>
> Another opinion the writer has held for years, and consequently isn't prepared to change, is that racing, apart from being big business in itself and an outlet for the good-sized breeding industry, contains a fairly large element of fun, born largely of the fact that people who must daily accept hazards must joke about them or become introverts. . . . There is a humorous fatalism among professional racing people, best expressed perhaps by the riding instructions given to a steeplechase jockey some years ago by a stable foreman: "Don't be skeered of dyin'; just let him run!"
>
> Out of the contact and occasionally the clash of such personalities a good many interesting stories emerge.

Horse racing in Joe Palmer's day was indeed colorful, a sport in which gentlemen assembled in sunny parks and sipped afternoon drinks, worked angles, placed wagers, swapped stories, and then went home contented and good-natured, whether winners or losers. Palmer was enamored of the people he spent time with and wrote about. He loved the idea that in the sport of racing every man was out for himself and yet still managed to get along with everybody else. He never pretended that racing lacked a seedy side, and he frequently acknowledged some of the scandalous stories of his day; but he considered fixed races, for instance, not just as isolated aberrations but as necessary evils that lend an air of chance to every contest in a sport featuring few reliable factors to begin with. As one racing official put it, "If we could convince everybody that this game is on the level, we'd go out of business."

"The contention isn't that everything is all right in racing," Palmer wrote in his first *Herald Tribune* column. "If there is any considerable industry involving millions of dollars and thousands of men in which everything is all right, it ought to be stuffed and put on exhibition. As far as I know racing has never claimed to build character: It does what other business does – it develops what a man has, be it good or ill."

The column, "One Apostle's Creed," made it clear to a tough audience that Palmer would not take himself or his sport too seriously, and his audience should do likewise. To New York sports journalism Palmer brought a style and perspective never before or since duplicated, as Red Smith said. Into this field of often jaded northern writers and readers came Palmer, a consummate southern gentleman who opened a window through which much of the public had never seen horse racing before.

A big, burly man with a slow way of speaking that one could almost hear in his writing, Palmer made no secret of his love of comfortable living, and nothing suited him better than the life of a horse racing aficionado. He once wrote,

> This department, whatever you may hear, is no noted lover of the horse, but of a way of life of which the horse was once, and in a few favored places still is, a symbol – a way of charm and ease and grace and leisure. Grace and charm should perhaps not be tampered with at this late date but, at whatever risk of boasting, I am as good at ease and leisure as any man alive.

Some horse racing devotees liked to get up early to watch the horses go through their training exercises, but Palmer did not. "This department," he wrote, "cannot conscientiously endorse any uncivilized nonsense like getting up and watching the horses work with the morning mists swirling behind them, and while the wood smoke from under the hot water kettles may be stimulating to brighter-eyed people it always reminds me that it hasn't been long enough since last night. Still, it's worth doing once. You will remember it a long time, and that saves having to do it again." For those who rose early, he recommended finding a comfortable table with a view of the training grounds and waiters to serve breakfast, "if only for the sublime sense of sitting at ease and watching other people work."

While Palmer may have been an easygoing man, he had a fast wit. One tale about him, recounted by Smith, concerned a night when Palmer and Smith had gone with a group of men to an amateur boxing match. The discussion afterward turned to cockfighting, which Palmer defended as a fine old sport. When someone objected (simply to get Palmer riled up, according to Smith), Palmer indignantly inquired what was wrong with chicken fighting.

> "It's cruel," replied the instigator, "It's inhuman."
>
> "Inhuman?" Palmer shouted. "Of course it's inhuman. And that stuff we saw tonight – that's unchicken."

Among other pearls of Palmer's wit was his description of stable owner Clarence Buxton as "a small, white-haired gentleman, very quiet except when he is awake" and who was not known for "adding notably to any silences that prevailed." Palmer wrote of another owner that he had acquired a horse that received "more publicity than any non-winner since the late Gen. George Armstrong Custer. [The horse] did not win any races because of a slight defect in velocity or, as they would say in Lexington, because he could not beat a fat man from here to the Lafayette Hotel."

Palmer's depiction of himself as a man of leisure is questionable, both by the testimony of his colleagues and the prodigious output of columns during his brief career as racing editor of the *Herald Tribune*. He once began a column,

> There are several ways of putting a column together. One is to work. This is generally frowned on in the profession. Another is to sit opposite John B. Campbell, a steward at Hialeah and the racing secretary for the New York tracks, and spoil his lunch. This is five times better business; he is not a man in inordinate need of nutriment, so virtue is served both ways.

Yet his self-portrayal as a somewhat lazy race enthusiast who had never won a big bet on a horse since 1923 was part of the truth-stretching persona of the horse player. As Palmer once confessed,

> This department had a reputation for unswerving truthfulness until approximately the age of seven, and would no doubt have it still except for leaving Kentucky temporarily at that age. But since then various things have happened, and now a certain admiration is held for a well-told falsehood. This is wrong, of course, but there you have it.

Racing proved a superb source of material for a man who loved to recount a tall tale. One of Palmer's gems concerned

> a character around the tracks known as Lyin' Fitz, though how he got this appellation is mysterious, for he would answer honestly when asked the time of day, or whether he had a match. He had a stable cat which slept a little too long in the sun one day and was stepped on by a horse and, though the cat ultimately recovered, it lost about three inches from one foreleg. Fitz watched it with compassion as it hobbled about and finally whittled down an old harness peg to serve as an artificial limb.
>
> "After the cat got used to it," Fitz used to say, "he was just as good as ever, in fact better. It was really wonderful to watch him catching mice with one hand and blackjacking them with the other."

Other tales included an owner who, according to Palmer, "Had a great regard for the truth and used it sparingly," or one who had entered a lightning-fast eight-year-old horse in a race and, when asked why he had not raced the horse when it was three years old, replied, "We couldn't catch him till he was seven."

Palmer was always discreet in writing about dishonesty in the racing world, and he always omitted the names of the living or the vulnerable dead. In one such case the owner of a particular stable had died, and, as was the custom, all the horses from his stable were scratched from all race cards for a week. After the week had passed,

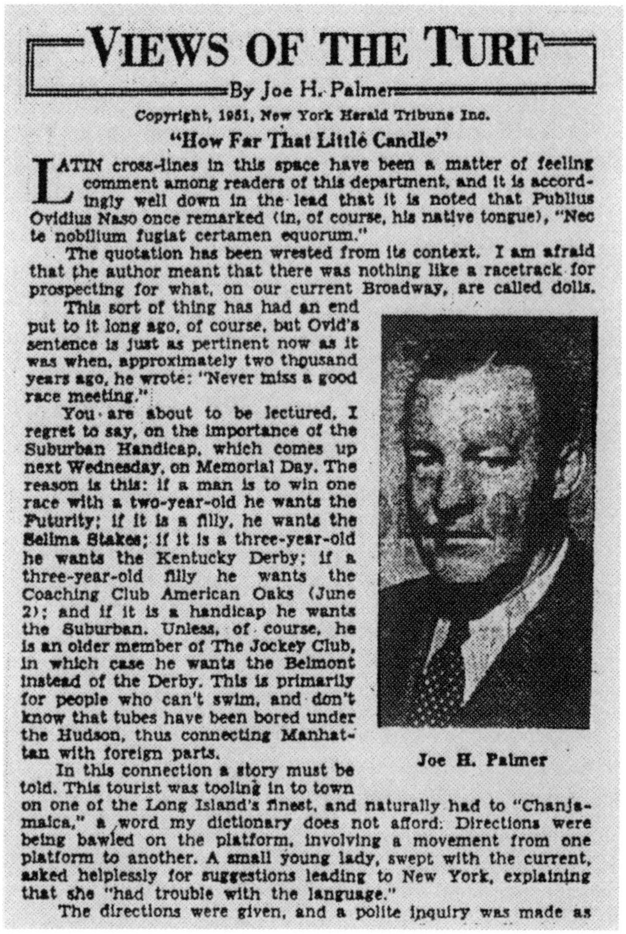

VIEWS OF THE TURF

By Joe H. Palmer

Copyright, 1951, New York Herald Tribune Inc.

"How Far That Little Candle"

LATIN cross-lines in this space have been a matter of feeling comment among readers of this department, and it is accordingly well down in the lead that it is noted that Publius Ovidius Naso once remarked (in, of course, his native tongue), "Nec te nobilium fugiat certamen equorum."

The quotation has been wrested from its context. I am afraid that the author meant that there was nothing like a racetrack for prospecting for what, on our current Broadway, are called dolls.

This sort of thing has had an end put to it long ago, of course, but Ovid's sentence is just as pertinent now as it was when, approximately two thousand years ago, he wrote: "Never miss a good race meeting."

You are about to be lectured, I regret to say, on the importance of the Suburban Handicap, which comes up next Wednesday, on Memorial Day. The reason is this: if a man is to win one race with a two-year-old he wants the Futurity; if it is a filly, he wants the Selima Stakes; if it is a three-year-old he wants the Kentucky Derby; if a three-year-old filly he wants the Coaching Club American Oaks (June 2); and if it is a handicap he wants the Suburban. Unless, of course, he is an older member of The Jockey Club, in which case he wants the Belmont instead of the Derby. This is primarily for people who can't swim, and don't know that tubes have been bored under the Hudson, thus connecting Manhattan with foreign parts.

Joe H. Palmer

In this connection a story must be told. This tourist was tooling in to town on one of the Long Island's finest, and naturally had to "Chanjamaica," a word my dictionary does not afford: Directions were being bawled on the platform, involving a movement from one platform to another. A small young lady, swept with the current, asked helplessly for suggestions leading to New York, explaining that she "had trouble with the language."

The directions were given, and a polite inquiry was made as

*Part of Palmer's 28 May 1951 column for the* New York Herald Tribune

a trainer entered one of the stable's horses in a stakes race, but he pulled the horse after he received a phone call from a man who claimed to be the son of the former stable owner and who requested that the trainer wait a little longer in honor of his father's memory. The caller turned out to be the trainer of the rival horse that won the race.

While this incident might seem a morbid transgression, racing people did not consider such underhanded tricks with much gravity. Palmer wrote of this deception,

> This will never be offered as a model of ethics in any courses on that subject, but it illustrates the interplay of personality and originality which is to me the real interest of racing. Just as a man is allowed to grunt in disgust when he picks up the fourth king in a poker game, so he is allowed to lie cheerfully to his friends about his horses. The races themselves, except for the very big ones, are just the laydown after the maneuvering is over.

What makes sportswriting among the most difficult trades to ply is that one must take an elemental occurrence and treat it as if it has happened for the first time. As a correspondent Palmer did not simply report on the current racing events, nor did he simply recount unlikely yarns and amusing anecdotes about outrageous characters. Writing more as an astute social anthropologist, Palmer portrayed the world of horse racing in his time. While his meticulous tracing of something like the genealogy of a particular thoroughbred might occasionally baffle the casual reader of some of his columns, aside from that topic there was little that Palmer could not make captivating. As Palmer often pointed out, the little people who made two-dollar bets made horse racing what it was, and, according to Smith, Palmer never referred to racing as "a Sport of Kings. A sport of ladies and gentlemen and cheerful brigands and small, skillful boys was good enough for him."

In one of his most famous columns, "Stymie – common folks," Palmer paid tribute to a horse that epitomized the elemental spirit underlying horse racing – that of the common man. As a horse of obscure breeding and average features who lost fourteen of his first fifteen races, Stymie went on to win nearly a million dollars for his owners. Through his career the horse ran more than 142 miles, a total greater than that by Man o' War, Equipose, and Whirlaway – the three greatest competitors of his time – combined. When the owners of Stymie decided to retire him, the horse was not given a grand reception at a Kentucky racetrack, as were the bluebloods. Instead, Palmer lauded the small but dignified affair given for the horse at Jamaica, for he saw this as the only fitting reception for such a horse – at a place where two-dollar bettors could come and pay tribute to a horse that they identified with, a commoner who could run against the aristocracy of horse racing and win once in a while.

"Racing is a little world, constructed in imitation to the real one," Palmer once wrote, "and paradoxes distinguish both. To own and operate a thoroughbred stable has always been an extremely expensive venture, yet the sport has a certain levelling effect and a sort of common-man humility among the owners." As Palmer put it, in horse racing, "All men are equal on the turf or under it."

While the possibility of winning big at the racetrack contributes to the popularity of the sport, Palmer saw enough to know that such an event was the exception, not the rule. "Racing is a rather expensive amusement and not an avenue to sudden and undeserved riches," he once said. He occasionally wrote columns about the folly of horseplayers, but he did not see his job as that of scaring off potential customers. "Possibly out of excessive modesty," he reflected in one column, "I have never felt myself capable of protecting mature adults who can make enough money to play the races."

Quite a few rags-to-riches stories appeared in his columns through the years, but invariably they appeared to be the results of lucky guesswork by the uninitiated, brilliant mistakes by hardcore horseplayers, or sheer accidents by members of either group. Palmer admitted that he wagered, but he insisted that he did so for fun, not money, and that he could never pick a winner after he had learned to study form. The other vice Joe Palmer cheerfully acknowledged was his enjoyment of a good stiff drink. He devoted entire columns to the history of the mint julep, the possibility of marketing jellied martinis, and the proper way to prepare a piña colada in a fresh coconut as a morning bracer at Hialeah.

Although Palmer rarely had a harsh word to say about anyone in his column, he loved to poke fun whenever he had an opportunity. Yet even the butts of his occasional humor – mainly racing secretaries, whose knowledge of horses was rumored to consist only of being able to identify which end bit and which end kicked – usually took it well. According to Smith, "the people Joe skewered didn't resent him because (a) he knew what he was talking about and (b) they appreciated his genuine enthusiasm for their sport."

Palmer truly did not like anybody who gave racing a bad name or tried to disturb the way things were done. He opposed most technological advances, such as public address systems, light boards, and even the photo finish. He most outspokenly opposed legalizing off-track betting – not from any moral grounds, but from a fear that it would lead to abuses and thus provoke public condemnation of racing.

With the possible exceptions of boxing and cockfighting, Palmer considered all sports other than horse racing to be childish, and his writing rarely refers to them other than to poke fun at them. "In other sports," he summarily concluded, "in theory, triumph is merely of the spirit. In racing it is likely to have a beautiful, sordid, financial side, and surely no money can be spent more joyously or more quickly than that accruing from a successful handling of the numbers game on the first two races."

Because horse racing was his whole life, Joe Palmer's writing never strayed far from the world of horse racing. Perhaps also because of his death of heart failure at the age of forty-eight, readers were never to be treated to anything better than his columns, but that is not to say that they do not suffice. His command of the language easily qualified him to be a first-rate novelist. He described the atmosphere at Saratoga as being "as restful as old slippers, as quiet as real joy." The New York State town of Goshen reminded him of his roots in small southern towns "partly because of the miscellaneous nature of the sidewalk paving, partly because of the sort of ruined grandeur in the domestic architecture, partly because the town had a slow pulse, and partly because people you didn't know spoke to you as they passed." He could write wryly of a morning in Aiken, South Carolina, where "The sun was squinting balefully through the magnolias. . . . The dew was glistening everywhere, a circumstance which appealed to some onlookers as beauty, but which suggested to me only that it would be an hour before anything was dry enough to sit down on."

As Red Smith said, it was better to read Palmer's work twice so that one could catch what one had missed the first time around, "because Joe slipped them in so deftly, like a dirk under the fifth rib." After Palmer's death in 1952, the *Herald Tribune* was deluged with letters inquiring about copies of Palmer's old columns. In an effort to satisfy this demand and partially to help his widow, Mary Cole Holloway Palmer, Smith and Palmer's two sons, Joseph Holloway and Stephen Noland Palmer, compiled *This Was Racing* (1953), a collection of some of Palmer's classic columns.

The day after Palmer's death, Smith's column of tribute to Palmer concluded, "Professionally there will not be another like him. Personally, he was one of the best and dearest friends a man could have. We came to New York almost together and have worked and played and wrangled together ever since. There was a friendship that could have ended only one way. As it did yesterday." Shortly before Palmer's death he had interviewed Alfred Vanderbilt for *The Saturday Evening Post,* and this article appeared posthumously. "The Riddle of Alfred Vanderbilt" was chosen for *Best Sports Stories 1954* (1954). Palmer had covered the races at Jamaica on the last day of his life, and at his Malverne, Long Island, home he left his finished column in his typewriter.

# Westbrook Pegler

*(2 August 1894 – 24 June 1969)*

Richard Orodenker
*Pennsylvania State University, Ogontz*

MAJOR POSITIONS HELD: Reporter, United Press (1912–1916); sports editor in New York City (1919–1925); sportswriter, *Chicago Tribune* (1925–1933); columnist, *New York World-Telegram* (1933–1944); syndicated columnist, *New York Journal-American* (1944–1962); freelance journalist (1962–1969).

BOOKS: *'T Ain't Right* (New York: Doubleday, Doran, 1936);

*The Dissenting Opinions of Mister Westbrook Pegler* (New York: Scribners, 1938);

*George Spelvin, American, and Fireside Chats* (New York: Scribners, 1942).

OTHER: "Guilty," in *The Omnibus of Sport,* edited by Grantland Rice and Harford Powel (New York: Harper, 1932), pp. 580–583;

"From the Woman's Angle," in *The Fireside Book of Baseball,* edited by Charles Einstein (New York: Simon & Schuster, 1956), pp. 263–264.

*Westbrook Pegler*

Westbrook Pegler had two careers, one as a well-known sportswriter and one as a famous, caustic, confrontational, Pulitzer Prize–winning political journalist. Although both careers have been essentially forgotten, Pegler achieved notable success in each. His career as a political journalist brought him fame and controversy. It ended with a sensational libel suit against him, which he lost. His eventual neglect, however, was probably less because of Pegler's vituperative style than because of his right-wing politics. Indeed, Pegler's last major job was writing for the John Birch Society's *American Opinion* magazine, which historian Clarence B. Carson, in *A Basic History of the United States* (1986), has called "almost certainly the glossiest of conservative journals, and definitely one of the hardest hitting." Even that journal severed ties with Pegler almost at the beginning.

Pegler's first career as a sportswriter was not only successful, but it influenced him at every stage of his later writing life. "It was during his hitch in sports," wrote Red Smith on the occasion of Pegler's death, "that he refined his exceptional gifts as a writer, reached his peak as a light essayist, and brought more pure delight to his readers than he ever did again" (*Daily News Record,* 27 June 1969).

Francis Westbrook Pegler was born on 2 August 1894 in Minneapolis, Minnesota. When his sister was born in 1897, she became Frances (after her mother Frances Nicholson Pegler, a Canadian of Irish-Catholic ancestry), and Pegler's name was changed to James. His nickname, common to many younger brothers of the era, was "Bud." Jack, the older brother, had been born in 1892. In Pegler's youth, his father, Arthur James Pegler, an Englishman, was sports editor of the *Minneapolis Journal.* Minneapolis in the early part of the century was a quaint town – not unlike the mythical American community of Grover's Corners, New Hampshire, which Thorton Wilder portrayed in *Our Town* (1938). Arthur Pegler was a struggling, itinerant newspaperman most of his life, working also at freelance writing jobs, including a stint as an evangelical ghostwriter and playwright. He began his working life as a farmhand but soon became a cub reporter on the *Sioux City Times,* a job that led to future newspaper positions in Minneapolis, Saint Paul, Chicago, and, finally, New York. Arthur Pegler, according to Westbrook's lifelong friend, sportswriter Marshall Hunt, "was sort of a nogoodnick. Peg never liked him. I think Peg felt he had been brought up on the wrong side of the tracks by this fellow." During Pegler's boyhood, his family moved to a better apartment in downtown Minneapolis, but there the young Pegler contracted diphtheria. The family moved again to the town of Excelsior, Minnesota, where Pegler hunted, fished, and had his first paper route.

In 1902 Arthur Pegler, having worked two newspaper jobs – one during the day, one at night, along with other writing chores – suffered a nervous breakdown. Facing severe financial hardship, he moved his family to Chicago, where he hoped to get a job with one of the Hearst papers. By 1909, working city side, he was one of the leading columnists on the *Chicago American,* and his family's circumstances improved. The young Pegler and his brothers attended the Horace Greeley District Public School, and Westbrook continued delivering papers at dawn.

Pegler's boyhood ambition was to be a cartoonist, but the great Hearst sports cartoonist Tad Dorgan, whom Pegler once met, encouraged the boy to do something else, apparently seeing no real talent in him. Yet biographer Finis Farr, in *Fair Enough: The Life of Westbrook Pegler* (1975), observes that a cartoonist "in a way, was what he turned out to be, using the medium of words rather than the drawing pencil." There is much in Pegler's writing style, especially in his depictions of the 1930s Brooklyn Dodgers as "the Daffiness Boys" and of an old-time horse player named "Old Paddlefoot," that supports Farr's contention.

Pegler began attending Lane Technical High School in 1909; but, showing no aptitude for manual skills, he dropped out and worked for a short time in the Chicago stockyards. Arthur Pegler helped his son get a job reading United Press (UP) dispatches over the telephone. Westbrook next attended the Jesuit Loyola Academy, failed to make the football varsity, and toyed with socialist ideas. A Catholic priest kept Westbrook from tilting leftward politically but not from returning to the UP in 1912 and attending his first political convention. Pegler was rejecting the pomposity and humorlessness of the Hearst reigning star Arthur Brisbane (and, indirectly, Arthur Pegler, who mimicked the Hearst staccato style), and picking up instead the rich cadences of comical writers such as George Ade and Finley Peter Dunne. In his first sports-oriented job at the *American* – compiling sports scores – Pegler was still known as "Bud." In time he would be W. J. Pegler, J. W. Pegler, and finally Westbrook Pegler.

At age twenty, still a bachelor and standing more than six feet tall, Pegler began his peripatetic journalistic career for the Scripps-Howard newspaper chain. His assignments took him to New York City, Denver, Saint Louis, Dallas, and eventually London during the outbreak of World War I, where he served as a UP foreign correspondent and rented a furnished room in the literary Bloomsbury district. As a war correspondent Pegler took on the renowned Brig. Gen. John Joseph "Black Jack" Pershing, writing critically about this American hero as he would another legend, Notre Dame football coach Knute Rockne, in the years to come. When Pershing got Pegler's press credentials revoked, Pegler returned to the UP desk in London and then enlisted in the navy in June 1918. He returned to New York after the war, hoping to become the UP sports editor for Scripps-Howard.

If politics became Pegler's pursuit in the second phase of his career, advancing his career – any sort of journalistic career – became his goal at this time, and he was willing to use the UP wires in any way that might help him. Sports, which Pegler once labeled the newspaper's "toy department," seemed the best route. He once "borrowed" another reporter's criticisms about the poor design of U.S. yachts in the America's Cup races of 1920. Pegler's version made front-page news. About this time he met Julia Harpman (known as Julie), a reporter for the *New York Daily News,* who left the business after falling ill. Her final column covered Gertrude Ederle's swimming of the English Channel in 1926. He married Julie on 29 August 1922, and she remained his great love. She was ill throughout her life, and the Peglers had no children.

In 1921 Pegler began writing a daily morning sports summary, which he sent out through the UP's United News feature syndicate. His filler material, known as "Six Short Pegs from Pegler" (colleagues later came to know him as "Peg"), which accompanied the sports column, began receiving national attention. In January 1924 Pegler published his one short story, "The Palooka," in *Liberty* magazine. It is a story about a mediocre prizefighter whose purpose in the ring (and in life) is to "fall backward," losing fight after fight just so up and coming fighters can look better at his expense.

By 1925 Pegler was making $125 a week, when the *Chicago Tribune* signed him at twice that salary to become its eastern sports editor. The *Tribune* syndicated Pegler's sports letter as "The Sporting Goods," which briefly became "The Say-So of Holleran Yell" and eventually "Speaking Out on Sports." At age thirty-one Pegler became one of the nation's most widely read sportswriters during the so-called Golden Age of Sports in the 1920s, which Pegler coined "The Era of Wonderful Nonsense." He and Julie took up residence in Larchmont, New York. Though written in New York City, Pegler's column did not appear in a New York paper until 1928, in a racing form known as *The Morning Telegraph.* The *Chicago Tribune*'s New York outlet was the *New York Daily News,* which already featured the popular Paul Gallico. Occasionally Pegler covered some other big assignment, but throughout the late 1920s and early 1930s he remained essentially a sportswriter – and that was how he was generally known.

Pegler wrote many pieces about college football during those years, concluding a series on West Point with the observation that its football players were being recruited from other lesser-known college teams and that the focus on education was secondary, if that. Such a story, like the America's Cup piece, as Farr notes in his Pegler biography, revealed "the germ of his future work . . . for this was a Pegler story in pure elemental form: something was wrong, and needed correction." He frequently damned Coach Knute Rockne as well as Notre Dame football until Rockne's untimely death in 1930, upon which Pegler expressed genuine sadness and respect. But Marshall Hunt recalled one untypical Pegler column on Rockne "filled with the clichés and corn that the other writers, admirers of Rockne, had used for years." Whether Pegler was being serious or satiric, the piece "backfired on him. . . . It wasn't in Peg, ever, to write that kind of stuff. He had no Pollyanna phrases in his vocabulary."

Pegler roundly criticized the Carnegie Report, which aimed to clean up college athletics. He found hypocritical "the pretense and concealment, couched in resolutions and agreements of the most elaborate amateur piety" (*Chicago Tribune,* 24 October 1929). He also took shots at the "mink and sable" crowd that often showed up at the most important college football matches. Still, as Murray Sperber points out in *Shake Down the Thunder: The Creation of Notre Dame Football* (1993), Pegler could catch the mood of an event perfectly, most notably in a piece concerning the bad weather that attended a 1930 Army–Notre Dame game (one of the great rivalries of the era). At this event, Sperber observes, Pegler evoked not only the weather but also the onset of the misery caused by the Great Depression:

> The damp heavy air was foul with the breath of the railroad locomotives down in the cut and with the outpourings of the city chimneys. You didn't breathe it but bit off hunks of the atmosphere on this lake front and Fletcherized [vacuumed] it in.

Although he never pursued opportunities in magazine writing, Pegler, as Marshall Smelser reveals in *The Life That Ruth Built* (1975), ghosted several autobiographical essays for Babe Ruth. Pegler was first and always a newspaperman, as his *Chicago Tribune* pieces on the significant sporting events of the late 1920s and 1930s demonstrate. But he could move beyond a factual reporting of events. A 17 December 1925 article on the ignominious death of the African boxer Battling Siki, who "had come all the way from the jungle to the haunts of civilization and chivalry to be shot in the back," becomes a commentary on the society that robbed Siki of his innocence and destroyed him in the process. For Pegler, Siki's fall from grace was "no mere matter of complexion." After all, "when blood showed, the civilized crowds came up from their chairs, roaring approbation." The boxer's death dramatized the corrupting nature of a society whose expectations of Siki were hypocritical and little more than "a puzzle and a josh to him."

Throughout his career Pegler was often accused of character assassination, and he could, at times, sound Ciceronian in his denunciations. In a 9 February 1926 *Chicago Tribune* story he accused French tennis superstar Suzanne Lenglen of being a professional in amateur guise and remarked that she would never be beaten on the courts because "she is absolutely shameless about quitting a match or defaulting a tournament." There was truth in Pegler's claims, and he corroborated some of them. Furthermore, as Lenglen prepared to meet Helen Wills in their famous match at Cannes on 16 February 1926, Pegler also criticized the so-called "Little Miss Poker-Face," Wills, for fretting so much about Lenglen's questionable financial dealings and "the Riviera scheme of business" (his term for "the ingenuous French amateur system") instead of being grateful for the opportunity finally to try to beat Lenglen (which she never would).

Readers increasingly began to see that Pegler's columns referred to matters beyond the realm of sports. Pegler touched on larger issues that were personal, moral, and controversial, and he began offering analysis rather than straightforward reporting. Sports in his view was more than the event itself; ultimately, what happened on the field of competition was shaped by the American ideals of personal liberty and responsibility; what happened in these competitions could affect people's consciences as well as their pocketbooks. Pegler believed that sports fulfilled an important moral function in American life, and he saw himself as the gadfly of the sports pages.

Pegler quickly joined the "aw nuts" school of sportswriting, the cynical, colorful bunch of writers, as opposed to the "Gee Whiz" group, which, he thought, was too poetic and romantic. In reading Pegler the sportswriter, one recognizes few heroes. Whomever he inflates, he also deflates. But there were exceptions: the golden age heroes Babe Ruth, Jack Dempsey, and Bobby Jones, all of whom Pegler knew and all of whom provided him excellent copy.

Pegler considered Ruth to be "the greatest artist the profession of sport has ever produced" (*Chicago Tribune,* 1 October 1932), although Pegler also wrote that Ruth played far past his time and was frivolous with money. Writing about Ruth also brought him occasionally to grandiloquence. Of Ruth's three home runs in the final game of the 1928 World Series, the second time in three years that Ruth performed such a feat, Pegler wrote (*Chicago Tribune,* 9 October 1928):

> Three times he laid his crutch in the way and knocked the baseball out of the St. Louis ball park and finally, in the last half of the ninth inning, shook his groaning chassis into a wild, loose-legged run to the rail of temporary seats in left field, where he leaned over and plucked a foul ball out from the feathers on a lady customer's millinery for the final put-out of the World Series of 1928.

When boxer-turned-journalist Jack Kearns began writing kiss-and-tell columns about Dempsey, prior to the latter's match with Gene Tunney, Pegler wrote that "you find yourself trying to choose between two sides of a dirty argument, between two fellows who came up from the barrelhouses, but who never did stamp all the sawdust off their shoes" (*Chicago Tribune,* 17 September 1926). But the article also questions the psychological edge the erudite and gifted Tunney believed he enjoyed in the match, considering all of Dempsey's personal problems – including various lawsuits. Pegler

*Pegler in his library at Casa Cholla, his home in Tucson, Arizona*

agreed that Dempsey, who "lives in monastic seclusion in a gray mortar bungalow" in Atlantic City, New Jersey, was preoccupied; but, he insisted, "the best minds in pugilism are still searching for a case in which the factor of psychology reached out and blocked a left hook such as Jack Dempsey's."

A week later, however, when Tunney defeated Dempsey for the world heavyweight championship, Pegler saw Dempsey as something of a caged tiger. In the 23 September 1926 *Chicago Tribune,* in one of the most descriptive accounts of his career, Pegler found the "man killer" in Dempsey gone; what was left was "a timid, pawing second rater." Pegler battered Dempsey with words almost as brutal as Tunney's fists: "The first punch raised a puff over the eye and the ensuing bombardment not only closed it tight but gashed the eyebrow and raised a gruesome purple knob as big as an egg on his cheekbone. Gore dropped from Dempsey's puffed lips as he staggered to keep his feet to the finish and his jaw dangled limp." Though Pegler could not pinpoint the reason for Dempsey's sudden collapse, he portrayed his new hero Tunney as a shrewd, punishing fighter who "must have taken the heart out of Dempsey."

Pegler also found it difficult to make negative comments about Bobby Jones, especially after Jones clinched the Grand Slam by winning the U.S. Amateur at Merion Cricket Club on 27 September 1930. According to Pegler, Jones was simply "the greatest golfer of all times" (*Chicago Tribune,* 27 September 1930). But Pegler mocked "the footsore thousands on the fairways," referring to the fawning crowds that follow Jones. Jones, whose fiercest rival contends that Jones "could give him two strokes and beat him the best day he ever lived," was a near perfect athlete, save for "a fleshy tendency denoted by a small frontal bulge in his sweater."

In Pegler's opinion, the professionals of later years were second to the purest amateurs of the 1920s and early 1930s. Pegler, for instance, was all pomp and hyperbole in his account of the first International Golden Gloves competition on 13 May 1931, a program sponsored by the *Chicago Tribune.* Even the spectacle impressed him – the music and fireworks and the military flourishes and "the sharp scent of fresh pine [that] tinged the air, an odor which is to big fight productions as the smell of tanbark is to the circus." He seemed genuinely touched when the French amateurs presented their victory medals as a token of friendship to their American opponents. When sports reached what seemed to him a level of purity and perfection, Pegler could become the sports romantic he often chastised. Moreover, as with all his pieces, he strove not only for a sense of social history and perspective but also for the right image and measure of humor needed to convey the idea that sports, after all, was great fun. Even a writer as accomplished as Red Smith admitted to learning from Pegler. Smith would later compose, perhaps more prolifically than Pegler, the compressed, witty similes that Pegler made famous, such as "They fought like a couple of husbands representing their loving spouses in a backyard row over a clothesline" and "There was about as much difference in the score as there is between a couple of dollars bills."

Pegler was preeminently concerned about injustice and hypocrisy wherever they appeared in the sporting world. When the great Finnish miler Paavo Nurmi was barred from the 1932 Olympics because he had shown a slight profit on his expense account, Pegler railed against the officials who argued that first-class accommodations were corrupting influences on the usually poor amateur athletes, while "For those officials themselves nothing is too good, because, being officials, they are used to good living" (*Chicago Tribune,* 30 July 1932). One can also see how Pegler's process of writing essays began to develop in these relatively early sports articles: contention (more often sarcastic than not) followed by support based on personal insight and experience or past encounters. The Nurmi affair, for example, reminded him of Charles Ebbets, owner of the Brooklyn baseball club, who was convinced that high salaries produced distracted ballplayers. The salary difference, of course, ended up in Ebbets's pockets. "But Mr. Ebbets felt that his strength was the strength of ten because his motives were pure and he always turned a deaf ear to any insinuation of this kind."

The essay then becomes a discourse, hilarious at times, on the nature of expense accounts, including those of foreign correspondents who write off their track losses. Pegler's conclusion delivers the coup de grace. Considering the cost of things, including "the tips for room service," Nurmi would have to "be branded as a piker" to show a profit on his expense account. The real injustice lies in the fact that the athletic clubs wind up "drawing gate receipts of majestic dimensions," while athletes like Nurmi become "convicted of making some money out of his principal business in life, a charge which most citizens of most countries would like to plead guilty of, preferably in the first degree." Pegler defended the track and field star Jim Bausch on similar grounds when Bausch was accused of holding an insurance job while playing football at the University of Kansas.

Pegler was also struck by the profound irony of Jim Thorpe's having to watch Bausch at the 1932 Summer Olympics "outdo him in the achievements that made him historic in American athletics" (*Chicago Tribune,* 7 August 1932). Given the nature of decathlon athletes, who might perform less effectively in individual events because they must spread their talents among so many, Bausch is all the more remarkable "when a man of his meaty physique goes mincing down the path with a fish pole poised and suddenly heaves all that freight up and up, more than twice his own altitude, to clear a delicately balanced crossbar and flop all of a hunk in the sand from that height without smashing something." Pegler acknowledges that the once phenomenal Thorpe would have come well behind Bausch under the modern scoring system and, in a slightly prophetic moment, wonders whether "some one may come along to prove that Babe Ruth was only a so-so sort of man with a bat."

One of Pegler's favorite subjects during the 1930s was the heavyweight boxer known as Primo Carnera. Pegler was among those who seemed to suspect something amiss in Carnera's championship victory over Jack Sharkey (later known to be a fix), although Pegler admitted that Carnera "troubled Sharkey with a left hook to the body, followed by a left jab which spun him slightly, but not out of control, to the ropes" (*Chicago Tribune,* 29 June 1933). Pegler was suspicious (Sharkey was winning the fight up until the sixth round), noting that "The punch that finished Sharkey did not seem sufficient to cause the damage which it was purported to cause" and that Sharkey "was some distance from the end of his endurance." Whether Pegler actually knew of the fix (as some sportswriters did but allegedly covered the fact up) is not known, but his dislike of Sharkey would have prejudiced him in any case. If Pegler suspected a fix, he perhaps implied it in his description of the knockout punch: "Sharkey flopped and outwaited the numerals being tolled in his ears as certain patrons without charity, tact, or any regard for the ennobling sport of prize-fighting turned to one another with thumb and forefinger clamped to the nose in an expressive gesture."

Having finally earned financial security by the early 1930s, the Peglers moved to Westchester, near the Connecticut border. But Roy Howard of Scripps-Howard soon lured Pegler back to New York with a lucrative offer that promised to net Pegler $75,000 in the Depression years.

Thus, the second phase of Pegler's career, his political writing, began on 11 December 1933, when his "Fair Enough" column appeared to the right of his friend and neighbor Heywood Broun's column, "It Seems to Me," in the *New York World-Telegram.* On leaving the sports beat, Pegler informed his readers, "there will be days over here in the Sacred Heritage of Liberty Department when I will pine for good old Primo and the Ol' Bambino." Years later Stanley Frank, in *Sports Extra* (1944), reflected, "Fight promoters, baseball owners, and football coaches gave three loud, fervent cheers when [Pegler] moved to the other side of the street." In one of his first columns, Pegler took the side of a California lynch mob that had strung up two kidnapper-murderers with the blessings of their governor after the fact. From that day forward Pegler rarely steered clear of controversy and became one of America's most beloved or despised, talked-about, and widely read newspaper columnists.

Pegler was never a sports columnist again, although a few years later, as Charles Fountain notes in his biography of Grantland Rice, *Sportswriter: The Life and Times of Grantland Rice* (1993), Pegler decried the state of sportswriting in America, pointing to the "thousand flabby counterfeits" who attempted to imitate truly original writers like his good friend, the by-then-passé Rice, who, Pegler joked, could indeed write much of "that pantywaist stuff." The new breed of writers, Pegler believed, were far more interested in "the night life of sports alone" and knew little about "sports and sportsmen . . . strictly as athletes" (*New York World-Telegram,* 18 October 1935).

Pegler still occasionally wrote about sports – the Kentucky Derby, for instance, which he savaged as a meeting place for underworld figures and snobbish society types. His accounts of the 1934 World Series between the Detroit Tigers and the Saint Louis Cardinals, featuring Dizzy Dean, demonstrated his ability still to bring incisive sports commentary into his everyday columns. Always preferring baseball to all other sports, Pegler wrote at the start of the 1934 season (*World-Telegram,* 18 April 1934) of baseball's great fortune in having survived the Depression, which, according to Pegler, was "a reminder that the country didn't quite get mad enough to sting itself to death when things were at their worst."

Pegler's best writing demonstrates fairness and balance. In many of his pieces it is difficult to determine which side of the fence Pegler sits on in regard to a particular issue because he refused to sit (until he became more reactionary) firmly on either side. Examining the business side of sports, he lashed out at the magnates for "overlook[ing] great advertising possibilities open to them" and depending on the free publicity newspapers provided. Many of these owners were sound businessmen in other fields yet unwilling to spend the requisite amount of money to keep their clubs going. Meanwhile, semipro clubs managed to operate in the black while the magnate, according to Pegler, often held his franchise "until the bank sends the sheriff around to claim it." Fans, on the other hand, would feel "morose . . . if told that the major leagues would be compelled to quit operations"; but they, too, can be unsupportive when the team is not playing well. Though the players are drastically underpaid and are "forbidden to shop around for another job with a more liberal employer than the one to whom he finds himself rather agreeably enslaved," few people would be willing "to shove his complaint ahead of that man's who is willing to grab a hoe, hod, or hammer and hire out to the CWA" [Civilian Works Administration].

By the postseason Pegler was covering the World Series in Saint Louis. Red Smith argued in his 27 June 1969 obituary of Pegler that "To Peg, a little knowledge was a useful thing but too many facts could get in the way." In his book *City Editor*

*Pegler at dinner in May 1954, after losing the libel suit brought against him by Quentin Reynolds*

(1934) Stanley Walker, former *New York Herald Tribune* editor, agrees: "His sports stuff was highly amusing, although often, unfortunately, he wrote of matters of which he knew little." Nonetheless, Pegler's columns demonstrate a sense of cultural history and background rarely found in contemporary coverage. In his 6 October 1934 story (also syndicated in the sister publication of the *World-Telegram,* the *Chicago Daily News*), Pegler described the unique excitement that accompanies a World Series played in the country's southernmost and westernmost major league city. The South, observed Pegler, is baseball country, from which many of the players (like Dizzy Dean) hail and where "there is a long spell between planting and picking when there isn't much to do but play ball or go to minor league games." The game Pegler reported seems almost incidental ("Brother Daffy didn't pitch much better than Tom Bridges, the unfortunate victim of the Tigers' indifference in the pinches"), and Pegler provided concisely and colorfully only the necessary highlights. His emphasis was on the human element and on the sound of his own prose. The Tigers, Pegler concluded "have the look of losers, whereas the Deans are overbearing even when mediocre."

On the issue of race, Pegler could be ambivalent. Initially embracing old-fashioned liberal ideals, he vehemently opposed baseball segregation (once writing about the "silly unwritten law that bars dark Babe Ruths and Deans from the fame and money they deserve"). Furthermore, as Chris Mead points out in his biography of Joe Louis, *Champion: Joe Louis Black Hero in White America* (1985), Pegler argued strongly against the racist ideas about black athletes, rejecting the notion that a superior black sprinter was "still so close to the primitive that whenever he runs a foot-race in a formal meet between schools his civilization vanishes and he becomes again for the moment an African savage." At the same time, Pegler warned that the Joe Louis–Primo Carnera fight in 1935 had the potential to incite another race riot in Harlem. In his columns, Louis was "the colored boy" and "cottonfield Negro."

Whatever specific criticisms he may have had, Pegler saw sports – baseball and prizefighting, in particular – as embodying the prized notion of individual liberty and freedom (which is one reason he admired Gilbert Patten's idealistic juvenile novels about sports). In later years Pegler chided boxing fans who wanted to soften the inherent brutality (in regard to ring deaths) of the sport, noting that "Prizefighting has changed so much in the years since teevy [*sic*] corrupted this noble sport with temptation toward insincerity that we who have its finest tradition at heart will only avert our eyes but not protest if public indignation now gives it the coup de grace with a hobnail kick behind an ear as it lies unconscious in the arena."

His crowning moment in sportswriting came in 1936 (the same year he published his first collection of writings, *'T Ain't Right*), when he traveled to Garmisch-Partenkirchen to cover the Winter Olympics in Nazi Germany. Pegler cabled articles attesting to anti-Semitic activities back to the United States. He had also been one of the few writers urging an American boycott of the games. Pegler, who detested fascism, had grilled the Nazis once before, when Max Schmeling defeated Joe Louis on 19 June 1936 and white sportswriters and boxing fans were lauding Schmeling at the expense of Louis. Pegler was quick to remind them that Schmeling had now become the "official patriot and representative of Nazi manhood" and that Schmeling's upset was being used "as a triumph for Adolph Hitler and his government."

A 17 February 1936 *World-Telegram* article detailed incidents of "strong-arm authority wholly corroborating the old tradition that the German people's favorite sport is to be shoved around by men in uniform." Pegler wrote in vivid detail about an atmosphere in which the "Olympics were of secondary importance." Swastikas and Brownshirts pervaded what was supposed to be an event "in the name of sport and human brotherhood." German tactics were less than subtle, but it was obvious that Olympic officials were turning a blind eye to them, appeasing the Nazi dictatorship. Pegler also attacked Avery Brundage (then president of the AAU [Amateur Athletic Union] and the recently elected U.S. Olympic Committee chairman), who had been wined and dined by Nazi leaders "to overpower the boycott movement in America." The athletes, Pegler suggested, were being used as political tools. Pegler's target was not so much Nazi ideology (though he was no sympathizer), but any sort of authoritarian government (he found Gov. Huey Long of Louisiana equally alarming). "This was the Dictator's day," Pegler wrote, "and it's a good thing for Americans present that this was so, because they have nothing important to learn from the athletes but much to learn about absolute authority in government."

The article resulted in a formal complaint to the U.S. State Department by German officials. The State Department relayed the complaint to Scripps-Howard, which asked Pegler to write an apology. Pegler obliged his bosses with one of his most biting, satiric essays, dissecting the sham ideals of Olympic sportsmanship in the face of an alarming German militarism. Why was a sporting event suddenly looking like a war maneuver? he asked.

His "apology" on 19 February 1936 is a classic piece of American journalism and one of Pegler's most brilliant essays. Admitting that he is "the victim of my own ignorance," Pegler, getting the inside scoop from "a kindly Bavarian cobbler in a long black overcoat," informs his critics that the soldiers wearing soldiers' uniforms were not really soldiers at all but "peace-loving German workmen in their native dress." Their bayonets are cobbling tools, and the other troops are merely gardeners; the infantry are only plasterers; aviators are actually poultry farmers, and so on. A man was not giving another man a military salute, "he was only shading his eyes." Pegler apologizes again for his mistake, reminding his readers that "When thousands of men seem to march but don't in clothing and tin hats which seem to be military uniforms but aren't, and carry harmless utensils which appear to be bayonets, any stranger is likely to make the same mistake."

Pegler took pleasure in exposing frauds and phonies, fools and dupes. He lampooned any sort of hero worship – of athletes as well as presidents – and he saved his most potent venom for men in power who let themselves be corrupted by that power. Oddly enough, friends and enemies alike often commented that the Pegler they knew personally was nothing like the persona that emerged from his prose. In "Square Peg" Heywood Broun remembers (in the *Collected Edition of Heywood Broun,* 1941) the "hardboiled, querulous, and materialistic" Pegler in the newspaper. "But the man who assumes this post morning after morning is actually shy, sensitive, and sentimental." In *City Editor* Walker counters that Pegler's "scheme of shocking the reader can be effective, but if carried too far it comes to be regarded as mere show-off."

In summarizing Pegler's later, nonsports writing, it may be noted that he went from backing Roosevelt to bashing the Roosevelts (Eleanor as well as

Franklin), from liberalism to anti-Semitism. Some of his so-called extreme views were not always wrong: he has been proved right on the subject of the threat of communism, and his antiunionism was, at least, grounded in principles rather than politics. He attacked administrative corruption and unionism's insistence that workers (even independent contractors like himself) could not operate independently of the union. His contempt extended to other areas where known racketeers associated. He called for Dodgers manager Leo Durocher's ouster from baseball because of Durocher's known associations with reputed mobsters. Pegler's smoking gun was the fact that Durocher was living in actor George Raft's house, and Raft made no pretenses about his friendships with known criminals such as Bugsy Siegel or Joey Adonis. Pegler wrote: "If the moving picture business knows that one of its stars prefers the underworld and doesn't care, that is one matter. Nobody would bet on the outcome of a picture, but a few killings on fixed ball games would degrade something that the public, perhaps without realizing as much, has come to regard as an American ideal or treasure." Corruption and moral turpitude in baseball were no different from corruption in the unions, as far as Pegler was concerned. In his autobiography *Nice Guys Finish Last* (1975), Durocher claims that Pegler's accusations were "all implication and innuendo" and, sounding a bit like Pegler, writes that Pegler was once a "great muckraking reporter" who "was beginning to lose some of his marbles." When Pegler turned to sports again during these years, he was, as Rice noted in *The Tumult and the Shouting* (1954), "usually working in the role of offering bitter protests against various people and conditions." He might take surprising sides on any issue. Thus he condemned the reserve clause in baseball but objected to collective bargaining in general. He might defend the actions of the exploited "Black Sox" players of 1919 but attack Durocher for his moral failings.

In the 1930s Pegler was still friends with many of the people who would later become his enemies, playing softball on Lowell Thomas's estate with Broun, Tunney, Bernard Gimbel (of the department store Gimbels), and Harold Ross, editor of *The New Yorker.* He was friendly with baseball statistician Al Munro Elias and *Ring* magazine editor Nat Fleisher. In the later 1930s Pegler also published his second collection of essays, *The Dissenting Opinions of Mister Westbrook Pegler* (1938). In 1941 he was awarded the Pulitzer Prize for his articles on union racketeer George Scalise, who coined the phrase "I've been Peglerized." During World War II he introduced into his columns his alter ego "George Spelvin, American" (the basis of his third book *George Spelvin, American, and Fireside Chats,* 1942). In 1944 Pegler ended (amicably) his longtime relationship with Roy Howard and the *New York World-Telegram* (after Pegler's attack on *Chicago Daily News* publisher Frank Knox). On 11 September 1944 "Fair Enough" became "As Pegler Sees It" on the front page (later page three) of Hearst's *New York Journal-American.* Ironically (in view of Pegler's early aspirations to be a cartoonist), King Features, known for its comic strips, syndicated his column.

After the war Pegler drifted further right ideologically, alienating more friends (though not necessarily his readers). The highly regarded author Gene Fowler, a former sportswriter himself, was one of the few people who remained loyal to Pegler, even when the latter offered some negative comments on Fowler's biography of former New York mayor Jimmy Walker. Pegler voiced support for Ezra Pound, who was being kept in a sanatorium in Washington, D.C., after being judged mentally incompetent to stand trial for treason because of his pro-Axis broadcasts in Italy during World War II. Pegler wrote against U.S. involvement in the Korean War, and yet, in a series of articles that brought him unprecedented support from both his supporters and his detractors in the press, helped demolish the career of Henry Wallace, a politician associated with the Communist movement. He wrote a famous column in which one line, "I must not mix wine, whisky, and gin," was repeated fifty times and a lighthearted piece on chili con carne that became one of his most popular articles.

Few people supported Pegler during the libel suit brought against him by his former friend, the author Quentin Reynolds. In 1949 Pegler, after reading a review Reynolds had written of Dale Kramer's biography of Heywood Broun, called Reynolds a coward, a war profiteer, and a moral degenerate, among other things. The case was pending for several years, but when it finally came to trial in May 1954, Reynolds performed well under the cross-examination of the Hearst lawyers. Pegler, however, virtually self-destructed under the questioning of famed attorney Louis Nizer. The result was the largest award ever granted at that time for punitive damages alone, paid with interest after Hearst and Pegler lost the case again on appeal. The Broadway play *A Case of Libel* (1963), by Henry Denker, was based upon the Pegler-Reynolds affair.

Pegler, now sixty, continued writing his columns, with a more vigilant Hearst organization peering over his shoulder, editing his prose, and publishing him deeper and deeper in the pages of the newspaper. Julie Pegler had died on 9 July 1955, and Pegler himself was often sick. He blamed her death on the effects of the Reynolds trial – and on the Jewish attorneys and the Jewish judge who presided. He continued writing – five, six columns a week. He had once said that it took him four or five hours to get out a column, and his talents were spread thin. His continued attacks on the Roosevelts (by then deified by most Americans) alienated more readers, as did his publicized conflicts with the Anti-Defamation League of B'nai B'rith over Pegler's now frequent comments concerning Jews. The comments often contained little more than innuendo at their worst, but these continued encounters with Jewish adversaries led Pegler to full-fledged anti-Semitism in later life. Those who remembered his brave dispatches about the harassment of Jews in Nazi Germany were both heartbroken and appalled. His notebooks, never published but made available after his death, reveal an even more contemptuous attitude toward Jews and even regret for his writing against the German position in the first place. His few remaining admirers, such as journalist Murray Kempton, could see this revelation as nothing short of deplorable – and it remains a blot on the Pegler legacy. On 11 May 1959 Pegler married Pear E. Doane, like Pegler an outspoken supporter of Sen. Joe McCarthy; but they divorced on 30 October 1961. On 22 November 1961 he married Maud Towart.

By 1962, almost a decade after the libel judgment, only seventy-five papers carried Pegler's column, which now attacked the Kennedys (John and Robert) and the Hearst organization itself. On 4 August 1962, the day after Pegler lambasted William Randolph Hearst Jr. and his bread-and-butter King Features at an Anti-Communist Christian Crusade in Tulsa, Oklahoma, Pegler's column, which contractually had two more years to run, was cancelled. Pegler, who began his career writing about sports and spent his whole life in the newspaper business, now had no outlet at all. At age sixty-nine he began a series of columns for the John Birch Society's *American Opinion* magazine; but Pegler disappointed even Robert Welch, founder and president of the society, with monotonous columns on familiar subjects. Welch dismissed Pegler without publishing any of his articles.

At the end Pegler was covering the 1964 Democratic convention for the *Toledo Monitor,* on the condition that he stay off the subjects of the New Deal, the Roosevelts, and Jews. Not liking censorship under any circumstances, Pegler parted ways with the *Monitor* and wrote unrestrained columns for two right-wing, parochial southern publications, the *Jacksonville Chronicle* in Florida and the *Councilor* in Shreveport, Louisiana. Ill health continued to plague him. A bitter, misanthropic Pegler now replaced the once curmudgeonly satirist who had composed such widely quoted metaphorical gems as "He could throw a lamb chop past a wolf" (on Lefty Grove); "When you hold it between your thumb or forefinger, you can hear a rabbit's pulsebeat" (on the lively ball); and "He looks like a beaten up tin can" (on Knute Rockne). Pegler kept in touch with Kempton, who remained a friend but was certainly no political ally, while a few other friends wrote to Pegler on his seventy-second birthday. He tried to fight further productions of *A Case of Libel* as an invasion of privacy but never lived to see the fight through, succumbing to complications of stomach cancer on 24 June 1969 in his home in Tucson, Arizona.

Pegler's impact on American journalism was significant and deserves further comment and study. Several of his pieces (the Olympic reports, most notably) are worthy of republication in anthologies of American essays or journalism. Pegler spent more than a decade on the sports pages (with his colleagues Rice, W. O. McGeehan, Paul Gallico, and Ring Lardner), and perhaps, in retrospect, one great pity is that he did not remain in that field or venture back to sports after his falling out with Hearst.

Broun recalled "the . . . energy which possessed him" as a beat reporter. Red Smith noted in his obituary of Pegler that "Wit and irreverence were the chief ingredients of Pegler's sports columns." "He was a fine sportswriter," wrote Rice in his posthumously published autobiography *The Tumult and the Shouting.* "He was the same in writing about sports as he is in writing of politics and politicians, including union labor leaders. He usually has a hot target and rarely misses his aim."

Pegler influenced younger sportswriters as well, Quentin Reynolds among them, who, despite his differences with Pegler, considered him, as he wrote in *By Quentin Reynolds* (1963), one of those "highly individualistic men . . . all of them craftsmen to emulate as well as real sports lovers." In *Sports Extra* Stanley Frank of the *New York Post* admired how Pegler "made his first muscles, beating the brains out of sports phonies with his trenchant typewriter." Marshall Hunt of the *Washington Daily Olympian,* who helped collect many of Pegler's later syndicated columns in their entirety, when they were often cut or relegated to the back pages, told

Jerome Holtzman, in *No Cheering in the Press Box* (1974), that Pegler "had a tremendous influence on the writing of sports and other things. He had a sharp mind. He saw things in a baseball or football game the other writers didn't see, or didn't choose to write about. His leads and descriptions of a game were so sharp."

At his best Pegler helped establish a style of writing (currently found on the pages of the finest conservative journals in America – *National Review, The American Spectator,* and others) with which, as Farr concluded in his Pegler biograpy, "he pursued the great clown [Henry] Wallace, harried crooks in unions, ridiculed Presidents, and lampooned Congress and the courts." (He once called John F. Kennedy "a low-browed donkey with the honor of a pickpocket.") Pegler honed his skill on the sports pages and created a climate for later acid-tongue sportswriters such as Dick Young, Maury Allen, Jim Bouton, and Jimmy Breslin. Left-leaning writers such as Kempton and Seymour Krim admired his courage and respected his talents and contributions to contemporary journalism. Liberals, Kempton once remarked, often wished they had had someone like Pegler on their side.

Pegler's distaste for group journalism, which came into fashion just as he was going out of fashion, speaks to his absolute insistence on the writer's right to speak his or her mind on any issue (even against his own employer). Pegler, for all his faults, believed unabashedly in the notion that blind faith in anything or anyone (even the writer himself) was a dangerous idea in a democracy. Perhaps his greatest failure was his unwillingness as he grew older to distinguish between simply wanting to offend someone rather than risking that someone might be offended by something he said. His friend and dissenter Kempton, in a review of Oliver Pilat's biography of Pegler (*Pegler: Angry Man of the Press,* 1963), wrote perhaps the best last words on Pegler's behalf:

> A cold, proud bias for the athlete as against the promoter . . . informed Pegler's work as a sportswriter in the twenties. He was not a Marxist sportswriter, but something better, an anarcho-syndicalist one. It is 44 years since the Chicago White Sox threw a World Series, but the Westbrook Pegler who would rail at a Postmaster General if he stole a five-cent stamp still thinks of Joe Jackson and Hal Chase as toilers so badly paid that taking bribes was only a way to get back a bit of their own. It was hard for him to believe that an athlete who took had violated the meaning of his life or the honor of the sport, since that meaning and honor seemed to Pegler to have been already violated by his employer (*New Republic,* 8 June 1963).

While some scholars will think it politically incorrect to resurrect Pegler (finding his spirit alive and well in the prose of combative conservatives such as Pat Buchanan and Joseph Sobran), those individuals willing to overlook ideological differences or Pegler's failings in character will discover a masterful writer who played a significant role in two distinct fields of American journalism, of which his sportswriting is the more enduring.

**Biographies:**

Oliver Pilat, *Pegler: Angry Man of the Press* (Boston: Beacon, 1963);

Finis Farr, *Fair Enough: The Life of Westbrook Pegler* (New Rochelle, N.Y.: Arlington House, 1975).

**References:**

G. H. Fleming, *The Dizziest Season: The Gashouse Gang Chases the Pennant* (New York: Morrow, 1984), pp. 63–65, 295–297;

Stanley Frank, ed., *Sports Extra* (New York: Barnes, 1944), pp. 203–212;

Murray Kempton, *Rebellions, Perversions, and Main Events* (New York: Times Books, 1993);

Red Smith, *To Absent Friends* (New York: Atheneum, 1982), pp. 185–187;

Murray Sperber, *Shake Down the Thunder: The Creation of Notre Dame Football* (New York: Holt, 1993);

Stanley Walker, *City Editor* (New York: Stokes, 1934);

Arch Ward, ed., *The Greatest Sports Stories from the Chicago Tribune* (New York: Barnes, 1953).

**Papers:**

Westbrook Pegler's collected papers are in the Herbert Hoover Presidential Library at West Branch, Iowa.

# Shirley Povich

*(15 July 1905 – )*

James C. Kaufman

MAJOR POSITIONS HELD: Reporter (1923–1925), sports editor (1926–1933, 1963), sports columnist (1926–1974), war correspondent (1945), occasional contributor (1974– ), *Washington Post.*

BOOKS: *The Washington Senators* (New York: Putnam, 1954);
*All These Mornings* (Englewood Cliffs, N.J.: Prentice-Hall, 1969).

*Shirley Povich (photograph by the* Washington Post)

Shirley Povich, who did not see a baseball game until he was out of high school, rose in just a few years from golf caddy to copyboy to sportswriter extraordinaire. In his column, "This Morning," which appeared in the *Washington Post* from 1933 to 1974, he covered the glory days of baseball with style and a sense of the dramatic. One of his best-known columns appeared on 9 October 1956, when Povich reported "a pitching job for the books": "The million-to-one shot came in. Hell froze over. A month of Sundays hit the calendar. Don Larsen today pitched a no-hit, no-run, no-man-reach-first game in a World Series."

Shirley Lewis Povich was born on 15 July 1905 to the Lithuanian immigrants Nathan and Rosa Povich, who owned a furniture store in Bar Harbor, Maine. He was one of nine surviving children; the first child, still an infant, had crawled out onto the back porch, which was built on pilings over the water, and had fallen in and drowned. In an interview with Jerome Holtzman published in *No Cheering in the Press Box* (1973) Povich discussed his unusual – for a male – first name:

> I have often been asked if I was bothered by the name of Shirley. Well, heavens, it was such a boy's name in Maine where I grew up. I went to school with three or four boys named Shirley. And the golf pro – he was named Shirley Liscomb. One day, in high school, there was this strange new girl named Shirley Johnson. We thought that was odd.
>
> I was named after my grandmother, Sarah. In the Orthodox Jewish faith, a child is named after someone deceased. I was the eighth child and they were running out of boys' names. They already had used Abe, Julius, Morris, and Bernard. My parents wanted something close to Sarah, and they took the Yiddish transliteration of Sarah, which is Sorella, and said, "Gee, that's close enough to Shirley."

Povich grew up with a close-to-home sports hero: his brother Abe, a star high school athlete in football, basketball, and baseball. He learned about national sports from the newspapers brought in on the night boat from Boston. In an unpublished interview Povich said that going from reading baseball stories to writing them was "an accident, kind of a fairy tale."

Bar Harbor was a resort for the wealthy, and, like most boys in town, Povich caddied on the golf course. He caddied for two years for Edward B. "Ned" McLean, who, as Povich puts it, "owned the Hope Diamond and the *Washington Post.*" In 1922 McLean invited Povich to attend McLean's alma mater, Georgetown University, at McLean's expense, and to work at his newspaper. McLean offered to transport Povich, who had never been outside Maine, to Washington on his private railroad car; it would have been Povich's first train ride, but Povich had to turn down the offer because the train was scheduled to leave on Rosh Hashanah, a Jewish high holiday. Traveling to Washington later on his own, Povich stopped off at New York City, climbed up Coogan's Bluff to gain a partial view of the Polo Grounds, and saw the Giants defeat the Yankees in the World Series, his first baseball game. "To my tender eyes," Povich says in his autobiography, *All These Mornings* (1969), "this was the real miracle of Coogan's Bluff. Bobby Thomson's historic homer in 1951 was distinctly a second-place event."

On Povich's first day in the nation's capital, he caddied for McLean and McLean's guest, President Warren G. Harding. Ironically, Povich made twice as much as a caddy as he earned as a copyboy for the *Washington Post,* where he began working in 1923. McLean, however, supplemented his twelve-dollar-a-week income with twenty-dollar tips for caddying on Sundays.

In his first regluar season major league game, in April 1923, Povich saw Washington's legendary pitcher Walter Johnson work. "I had never seen anything like Walter Johnson," Povich would tell sportswriter Dave Kindred in 1994, "and I never would again. I saw Walter that day in 1923, just four years before he retired. What they said then was you can't hit what you can't see – and even then they couldn't see Walter."

Povich quickly worked his way up from the copydesk to the city room, becoming a police reporter and night rewrite man before jumping at an offer in 1924 to work in the sports department. "I was attracted, also," Povich said in the unpublished interview, "by a five dollar raise." His first article described the Senators being greeted by a huge crowd when the team returned home as the new leader in the American League. His editor, Norman Baxter, liked the piece and gave Povich his first byline. "I could scarcely wait for the morning paper to see my name in print," Povich remembered. "In fact, I did not – I found my story in type and my name at the head of it, backward and upside down, as it appeared to me. Not until I ran my fingers across the metal and actually fondled my name in type was I certain of this dream come true."

Two years later, when Baxter was promoted to managing editor, the twenty-one-year-old Povich – who had seen his first baseball game only three years earlier – became the youngest sports editor at any metropolitan newspaper in America. One problem he faced was the budget restrictions imposed by the low circulation of the *Post* in comparison to its rivals, the *Washington Star* and the *Washington Herald.* It was also in 1926 that Povich began his column, "This Morning."

One of his well-known early columns concerned the famed "long count" Jack Dempsey–Gene Tunney fight on 26 September 1927, in which he sat ringside with such sportswriting greats as Grantland Rice, Damon Runyon, and Heywood Broun. "Not only was I in their company," Povich says in his autobiography, "I was expected to compete with them." After being down for longer than ten seconds, Tunney, reprieved by the referee's incompetence, went on to beat the aging Dempsey. Povich argues in his autobiography that Dempsey was not entirely hurt by the defeat: he became a hero, an underdog with whom the average person could empathize. Before the defeat, Dempsey had not been nearly as popular.

On 21 February 1932 Povich married Ethyl Friedman of Baltimore, whom he had met on a blind date in 1930. As the Great Depression continued, the financial troubles at the *Post* worsened. Returning from his honeymoon, Povich found out that his salary, like that of every other *Post* employee, had been cut 15 percent; Povich's pay dropped from sixty dollars to fifty-one dollars a week. Managing editor Baxter was fired after arguing with McLean about the pay cuts. Some workers had to sleep in the *Post* building, as they could not afford the price of lodging. In 1933 the money problems cost the *Post* the services of a promising young sportswriter when Bob Considine moved to the *Herald* for a significantly larger paycheck. That same year McLean had to auction off the *Post;* one of the first acts of the new owner, Eugene Meyer, was to rescind the pay cuts. Povich stepped down as sports editor in 1933 to concentrate on his column.

A man with whom Povich had frequent contact during the 1930s was Clark Calvin Griffith, president of the Washington Senators and a former pitching great. Povich wrote a ten-part series on Griffith; by the time he was through, he says in his autobiography, "I knew more about him than he knew about himself." During the late 1930s Povich and Ty Cobb joined forces to try to get Senators outfielder Sam Rice elected to the newly founded Hall of Fame. Cobb talked to sportswriters about his friend Rice at every opportunity, and Povich wrote several columns about Rice's outstanding ability and qualifications. They were eventually successful, but only after a generation had passed: Rice

*Povich (left) as a war correspondent in April 1945, with Capt. (later Admiral) H. B. "Min" Miller*

would enter the Hall of Fame in 1963, two years after Cobb's death.

On 5 July 1939 sixty-one thousand fans paid tribute in Yankee Stadium to a great player who had been forced into retirement by the degenerative nerve disease amyotrophic lateral sclerosis – the affliction that would come to be known as Lou Gehrig's disease. Povich opened his account of Lou Gehrig Day: "I saw strong men weep this afternoon, expressionless umpires swallow hard, and emotion pump the hearts and glaze the eyes of 60,000 baseball fans in Yankee Stadium. Yes, and hard-boiled news photographers clicked their shutters with fingers that trembled a bit."

During the 1930s the career of black heavyweight Joe Louis exposed Povich to the nature of prejudice in the sports world – and in himself. As Chris Mead observes in *Champion* (1985), Povich began with a more or less dehumanized depiction of Louis, seeing the fighter "as some machine, a methodical, mechanical destroyer, geared for destruction" – a vast improvement, nonetheless, over the "jungle" and "panther" images that other sportswriters used to describe Louis. Of the 1935 bout between Louis and Max Baer, Povich remarked: "It seemed that the crowd wanted the better man to win. Twenty-five years ago, even less, it would not have been a question of skill or conduct, but color." Povich was one of the few writers to take seriously Louis's comeback after his defeat by Max Schmeling in 1936, noting that the boxer "has been following [a] pattern of modesty, self-effacement." After Louis defeated James J. Braddock, Povich predicted, correctly, "Methinks Louis is in for a long reign as champion – the longest perhaps in the history of the heavyweight division." Povich told Mead that in retirement, Louis "educated the sports writers to be more tolerant. We became a little more, as I say, enlightened." Throughout his life Louis – whom Povich once described as "just a big, playful brown boy who happens to be deadly serious when he's in that ring" – "helped to alert people to the fact that blacks are good human beings."

Povich covered the Senators through the 1930s as they went from pennant winners to contenders to a perennial second-division team. They may not have made many runs for first place, but they did provide a group of interesting characters for Povich's columns, including Buck "Bobo" Newsom, Moe Berg, Sheriff Harriss, and Zeke Bonura. Povich recounted their antics in memorable anecdotes, such as Harriss's telling the intellectual Berg that Berg could speak seven languages but could not hit in any of them and Newsom's claiming that he won thirty-three games in the minors instead of thirty: "Who you going to believe, me or the record book?"

Povich argued early for the desegregation of baseball. After observing Negro league spring training in Florida in 1941, he wrote that the major leagues were missing out on "a couple of million dollars worth of talent." He saw players who were just as good or better than current major league stars and concluded, in his finest persuasive style:

> Only one thing is keeping them out of the big leagues – the pigmentation of their skin. They happen to be colored. That's their crime in the eyes of the big-league club owners. . . . Their talents are being wasted in the rickety parks in the Negro sections of . . . cities that comprise the major leagues of Negro baseball. They haven't a chance to get into the big leagues of the white folks. It's a tight little boycott that the majors have set up against colored players.

On 17 January 1942 Povich responded to President Franklin D. Roosevelt's statement that "it would be best for the country to keep baseball going" during the war. "That man in the White House did, indeed, earn the label of The Nation's First Fan," Povich said; nonetheless, he argued that able-bodied ballplayers, stars and journeymen alike, had a duty to serve their country and not repeat what had occurred during World War I, "when organized baseball's record was none too shining, with the shipyards a haven for scores of players."

Later in 1942 Povich learned that George Preston Marshall, president of the Washington Redskins football team, had pocketed a large share of the profits from a charity exhibition game played in Los Angeles to benefit widows and orphans of soldiers killed in World War II. Povich wrote an open letter in the *Washington Post* to Secretary of War Henry Stimson, suggesting a full investigation. Marshall sued for libel, demanding $200,000 in damages but promising Meyer that he would rescind the lawsuit if the *Post* printed a retraction; but Meyer and *Post* editor Casey Jones stood behind Povich. In court when Povich's testimony that the Redskins were paid off first was challenged, he pointed out, "That game was played eighteen months ago. It was played for the benefit of the widows and orphans of our American servicemen. The Redskins have long since been paid off. This war, however, is still going on, and these widows and orphans might not have even become widows and orphans until tomorrow, next week, next month, or next year. Consequently, they have not been paid off yet." The jury voted unanimously in favor of Povich and the *Post*.

Povich begged his editor to send him abroad as a war correspondent, but Jones continually refused. At a party where drinks were freely served, Povich and a friend, *Washington Times-Herald* columnist Vincent Flaherty, asked the air corps chief, Gen. Henry "Hap" Arnold, to issue credentials for them as war correspondents. Arnold did so, but before Povich could accept, his editor told him that Secretary of the Navy James Forrestal wanted Povich to be sent to the Pacific. Povich wound up covering the Marines. At Pearl Harbor he met with one of the commanders there, future Hall of Fame catcher Bill Dickey of the Yankees, and another soldier, Senators first baseman Mickey Vernon.

Povich flew out of Pearl Harbor only a few hours after a similar plane had gone down, killing forty-seven passengers. He spent some time with the legendary correspondent Ernie Pyle and would have accompanied him to Ie Shima but had to cancel out when severe air turbulence smashed two of his ribs; it was at Ie Shima that Pyle was killed by a sniper's bullet. Later, Povich was transferred from a hospital ship that was bombed, with heavy casualties, shortly after he left it. Povich returned to America just in time to cover the end of the 1945 baseball season.

Two years later Jackie Robinson broke baseball's color barrier when he was hired by the Brooklyn Dodgers. In a column titled "No More Shutouts" Povich revealed that Robinson was not, in fact, the first African American to don a major league baseball uniform: in 1884 Moses Fleetwood Walker had caught for Toledo in the American Association, which was then a major league. Povich's column was one of the first places that an early black player received belated recognition. By 1953 Povich would be able to report "the complete emancipation of the American Negro in America's game" when the Cleveland Indians' Larry Doby, after being decked by an Art Ditmar fastball, became the first black ballplayer to throw "the first punch in a player argument."

In January 1950 Povich published a profile of Johnson in *Sport* magazine. Using his own memories and collected anecdotes ("the most common lie in Washington" is "'I saw Johnson pitch his first game'"), Povich re-creates the baseball life of "The Big Train," who had died in 1946, from the day he "modestly allowed he could do better" than all the pitchers on his Fullerton Union High School team to his "all but invincible" years with the Washington Senators in the 1910s and 1920s to his last managerial "debacle" in Cleveland. Throughout, writes Povich, "Legend couldn't wait on the death of Walter Perry Johnson, as properly it was supposed to do. Even while he lived, he was a fabled figure of the pitching mound, for so fanciful were his feats, so all-fired fast was his fast ball, that only by overdrawing the tales of Johnson's speed could the man be put in proper perspective, even in contemporary times."

*Povich, as president of the Baseball Writers' Association, presenting Yogi Berra with the American League's Most Valuable Player Award for 1955*

In 1961 Povich found himself included in *Who's Who of American Women* between Louise Pound and Hortense Powdermaker. Several months previously he had received a letter from the *Who's Who* publisher, the Marquis Company, addressed to "Miss Shirley Povich" and asking for biographical information; without further thought he had thrown it away. He forgot about the letter until an Associated Press reporter called to tell him of his unique accomplishment. As the only man among the 19,670 entries Povich became an instant celebrity, posing with fellow inductees Elizabeth Taylor and Eleanor Roosevelt for magazines and newspapers, including *Time* and the *Los Angeles Examiner*. He appeared on the television show "I've Got a Secret" and stumped the panel. His fame reached the point where the post office correctly delivered a letter addressed only to "One of Nineteen Thousand, U.S.A." The letter was from William Friesell Jr., who had achieved a similar mark of idiosyncratic fame twenty years earlier by allowing Cornell an extra down over Dartmouth, thus earning the sobriquet "Mr. Fifth Down."

The *Who's Who* incident was neither the first nor the last time that Povich encountered confusion over his first name. After the war he had received a letter from the League of American Penwomen asking why he had not joined, and he had filled out an application on a lark. As Povich recounted in *All These Mornings,* this joke paid off years later when he addressed the group's Virginia chapter and opened his talk: "Good Evening, fellow club members." In 1963, when he again assumed the duties of sports editor at the *Post* for one year, the Associated Press sent out a story headlined "Woman Gets Post's Sports Editor Job," which actually ran in an Alaska newspaper.

In the 1960s Povich dealt in his column with Marshall's refusal to hire African American football players at a time when most other football teams were finally signing them. After Jim Brown paced the Cleveland Browns to victory over the Redskins, Povich opened his column with a typically witty and telling line: "Jim Brown, born ineligible to play for the Washington Redskins, integrated their end zone three times yesterday." The Redskins were not integrated until Stewart Udall, secretary of the interior under President John F. Kennedy, threatened to ban the team from playing in the government-owned D.C. Stadium.

Povich covered the 1972 Munich Summer Olympics, at which eleven Israeli athletes were murdered by Arab terrorists. In boxing he followed the rise to superstardom of Cassius Clay (later known as Muhammad Ali). Observing the evolution of professional basketball, he wrote in *Sports Illustrated:* "Basketball is for the birds – the gooney birds. The

*Povich with the 1959 edition of* Who's Who of American Women, *in which he was included by mistake (photograph by Edward Clark)*

game lost this particular patron years back when it went vertical and put the accent on carnival freaks who achieved upper space by growing into it." He also covered golf and horse racing – "Bobby Ussery had picked spinach and beans and cotton for a dollar a day in the acres around his native Vian, Oklahoma, and he vowed that 'they'll never get me back in those fields,' " Povich wrote of the Kentucky Derby winner; "Later he turned to race riding, and today, well. . . . " But baseball remained his favorite sport.

Povich retired in 1974, but he still writes occasionally for the *Post.* Many of his recent columns are reflections on years past. A 1991 column recalled how, fifty years earlier, the crowd at a Redskins-Eagles football game remained unaware of the bombing of Pearl Harbor. Another column in 1991 explained in its headline why he has not fully retired: "Idyllic retirement? The truth is, I missed working." Other recent columns have focused on old-time players such as Rice and Joe DiMaggio. Povich also writes the obituaries for former Washington Senators. In 1994 he was a commentator in Ken Burns's nine-part Public Broadcasting Service series *Baseball.*

Reflecting on his career, Povich said in an unpublished interview: "I'm not sure what's my favorite column. I've written thousands. My favorite players to interview were Ted Williams and Walter Johnson. I've found them to be the most fascinating of all the players I've dealt with." Povich had the opportunity to know Williams both as a brash young player and as an older and wiser manager. "He became more mature and more personable." He and DiMaggio were among the hardest players for Povich to interview before they mellowed in their later years. On the differences between sportswriting when he first started at the *Post* and today, he says:

> I think sportswriting was more basic in former days than it is now. Today, they are devotees of New Journalism. They personalize the game more than we used to. Sometimes it's good, but unfortunately, sometimes the new breed of journalists, whereas they are good writers, often write as if they're trying to avoid the subject. They're not as penetrating as we used to be.

But Povich also criticizes the sportswriters of past eras, whom he and his colleagues imitated: "Too many young sportswriters in my era had the wrong writing heroes," he told Holtzman:

> We made the mistake of imitating . . . the Associated Press writers. Many of us, in our sheer stupidity, assumed that since they were the stars of the Associated Press they had to be fine reporters. . . . They made everything cataclysmal. They couldn't give you the short action word. They were overmoderated, the adjectives were endless and the same. . . . When I see it now I want to upchuck.

Povich is also disturbed by the state of baseball today. He thinks that the games take too long and is bewildered by the astronomical salaries of the players.

Povich was president of the Baseball Writers Association of America in 1956. He won the National Headliners Award in 1947, the E. P. Dutton prize for *Best Short Story* news-coverage in 1957, the Grantland Rice Award for sportswriting in 1964, four awards from the Washington Newspaper Guild, and the Red Smith Award in 1983. He joined his own favorite sportswriter, Red Smith, in the Sportswriters Wing of the Baseball Hall of Fame in 1976 by winning the J. G. Taylor Spink Award. "It was a great pride to be elected. I'm in good company," he said in an unpublished interview. Povich has been a member since 1987 of the Veterans Committee, which decides which old-time players, executives, and officials who were passed over by the baseball writers will make it into the Hall of Fame.

Povich has also served as a mentor to other sportswriters. Two of his best-known protégés are Considine and Bill Gilbert. Considine wrote in the foreword to Povich's autobiography: "Shirley led me by my nose-ring and taught me how to sit down in the press boxes of the land, and consequently, the box seats of the world. . . . Some day I'm going to ask him a great favor, if I find him in a pleasant mood. I'm going to ask him to let me carry his typewriter." Povich hired Gilbert, an author of many sports books and now the editor of *The Diamond,* as a copyboy for the *Washington Post* and gave him advice that he never forgot. "Povich told me years ago," Gilbert said in an unpublished interview, " 'Always remember: The story has never been written that could not have been written better.' " While Povich has served as a mentor to others, he cannot single out any writer or editor who was a true mentor to him. "I picked a couple of *wrong* role models," he says. "Some were bad judgments. As I matured, I changed role models."

Povich's knack for nurturing young people with talent extended to his own children. His son David is a successful attorney in Washington; his daughter, Lynn, is the editor of *Working Woman* magazine; and his son Maury is the host of his own television talk show.

As Povich approached his ninetieth birthday, he remained active and hopeful that baseball would return to Washington. "Baseball belongs here," he told Kindred. Gilbert says, "The quintessential Povich column or article would be about baseball, his first love in sports. The one piece he would dearly love to write now undoubtedly would be to announce the return of major league baseball to Washington."

According to Gilbert, "Shirley's strengths as a writer are the same as Ted Williams's as a hitter. Both have enormous natural talent, and both worked harder than their peers. Povich is extremely hard on himself. He has often told me, 'I write in a state of perpetual discontentment.' " Povich had reflected on his writing philosophy in his interview with Holtzman: "Never think that you're writing for the people out there. Anybody who believes this is riding for a fall. You're writing for that white sheet of paper in the typewriter, for your newspaper, and for yourself. There's nobody out there."

*Povich with his sons, David and Maury, in the late 1960s (photograph by Bill Mark)*

**Interview:**

Jerome Holtzman, *No Cheering in the Press Box* (New York, Chicago & San Francisco: Holt, Rinehart & Winston, 1973), pp. 114–128.

**References:**

Alan Ebert, "Will Maury Povich Please Hug His Dad?," *New Choices for Retirement Living,* 33 (November 1993): 52;

Dave Kindred, "Yes, Virginia, Baseball Should Expand," *Sporting News,* 218 (21 November 1994): 8;

Chris Mead, *Champion* (New York: Scribners, 1985), pp. 63, 73, 116, 128, 267.

# Grantland Rice

*(1 November 1880 – 13 July 1954)*

Mark Royden Winchell
*Clemson University*

See also the Rice entry in *DLB 29: American Newspaper Journalists, 1926–1950.*

MAJOR POSITIONS HELD: Sports editor, *Atlanta Journal* (1902–1905), *Cleveland News* (1905–1907); sportswriter, *Nashville Tennessean* (1907–1910); sportswriter, *New York Evening Mail* (1910–1913); sportswriter and syndicated columnist, *New York Tribune* (1913–1924); *New York Herald Tribune* (1924–1954).

BOOKS: *Base-Ball Ballads* (Nashville: Tennessean, 1910);

*The Winning Shot,* by Rice and Jerome D. Travers (Garden City, N.Y.: Doubleday, Page, 1915);

*Songs of the Stalwart* (New York & London: Appleton, 1917);

*Sportlights of 1923* (New York & London: Putnam, 1924);

*Songs of the Open* (New York & London: Century, 1924);

*The Duffer's Handbook of Golf,* by Rice and Clare Briggs (New York: Macmillan, 1926);

*Understand Football,* by Rice and John W. Heisman (Bronxville, N.Y.: McGraw-Hill, 1929);

*Only the Brave, and Other Poems* (New York: Barnes, 1941);

*Steel and Flame: A Collection of War Poems* (New York: Barnes, 1942);

*The Bobby Jones Story,* by Rice and Oscar Bane Keeler (Atlanta: Tupper & Love, 1953);

*The Tumult and the Shouting: My Life in Sport* (New York: Barnes, 1954);

*The Final Answer, and Other Poems* (New York: Barnes, 1955);

*Curtain Call [The Lost Path],* edited by Charles C. Fleming and Martha Fleming (Richmond, Va.: Press of the Hicuppy Herring, 1956);

*The Best of Grantland Rice,* edited by Dave Camerer (New York: Franklin Watts, 1963).

OTHER: *The Boy's Book of Sports,* edited by Rice (New York: Century, 1917);

Grantland Rice

*The Omnibus of Sport,* edited by Rice and Harford Powel (New York & London: Harper, 1932);

*Reveille: War Poems by Members of Our Armed Forces,* edited by Rice, Daniel Henderson, and John Kieran (New York: Barnes, 1943).

Certain phrases and images enter the popular imagination almost from the moment they are created. Passages from Henry Wadsworth Longfellow remain in the minds of readers who know no other poetry. Millions who have no idea who George Orwell was know that "1984" and "Big Brother" are ominous phrases. The sportswriter Grantland Rice made at least two such contributions to popular mythology. As long as men and women engage in athletic competition, one can expect to hear his most famous lines of verse: "For when the One Great Scorer comes to write against your name, / He marks – not that you won or lost – but how you played the Game." These lines were first published on 16 June 1908 in the *Nashville Tennessean.* The lead paragraph for Rice's best-known newspaper article provided an image almost as enduring. Describing a gridiron encounter between Army and Notre Dame in 1924, he wrote:

> Outlined against the blue-gray October sky, the Four Horsemen rode again. In dramatic lore they are known as Famine, Pestilence, Destruction and Death. These are only aliases. Their real names are Stuhldreher, Miller, Crowley and Layden. They formed the crest of the South Bend Cyclone before which another fighting Army football team was swept over the precipice at the Polo Grounds yesterday as 55,000 spectators peered down on the bewildering panorama spread on the green plain below.

In an age when the prevailing attitude in sports is closer to Vince Lombardi's famous view that "Winning isn't everything – it's the only thing," Rice's image of the One Great Scorer giving points for sportsmanship and effort strikes many observers as an exercise in "Gee Whiz" idealism. On the other hand, some might argue that it is evidence of a far more sophisticated perspective, one greatly influenced by Rice's identity as a southerner. The analogy of the Notre Dame backfield to the Four Horsemen of the Apocalypse (or Rice's naming Red Grange the "Galloping Ghost") is symptomatic of an even more significant contribution to American culture – the mythologizing of sports and sports heroes.

As warfare in the twentieth century has become more a matter of technology than of individual valor, ritualized battles on the athletic field have assumed many of the symbolic values of ancient combat. In place of David going against Goliath or knights jousting for the medieval equivalent of bragging rights, many sports fans in contemporary society tend to project their dreams and aspirations onto the figures of sports heroes. These champions represent modern city-states, races, cultures, and – in the case of the Olympics – nations (Jesse Owens at the Berlin Games in 1936 is the most obvious example). Rice did not invent this phenomenon, and it would certainly have occurred had he not written a single one of the sixty-seven million words he published during a fifty-three-year career. But he understood the importance of sports in American life perhaps better than any other writer of his time. He spoke for the American people because he expressed what they believed in an elevated rhetoric that they could admire if not emulate.

Henry Grantland Rice was born in Murfreesboro, Tennessee, on 1 November 1880. He was named for his maternal grandfather, Henry Grantland, who served as a major in the Confederate army. Although he fought under Gen. Braxton Bragg, Henry Grantland's hero was Nathan Bedford Forrest, the yeoman farmer who became the most daring and colorful fighter for the Southern cause. Prior to the Civil War, Grantland had hidden six five-hundred-pound bales of cotton in a cave near his home. By the time the war ended, the water in the cave had swollen the rotting bales until they weighed about a thousand pounds each. Since the selling price for cotton at that time was a dollar a pound, Grantland sold his worthless stash to a gullible carpetbagger for $1,000 a bale. With that stake he established the Grantland Cotton Company in Nashville and became one of the city's leading merchants. After Rice's birth in 1880, Henry Grantland's daughter and son-in-law, Beulah and Bolling Rice, moved in 1884 from Murfreesboro to Nashville, where they lived in the Grantland home for the next four years. Then in 1888 the Rices moved into their own home on Vaughan Pike in east Nashville.

It was during his first Christmas on Vaughan Pike that young Rice found a football, a baseball, and a glove under the family tree. Sixty-six years later, in his autobiography, *The Tumult and the Shouting: My Life in Sport* (1954), Rice would recall that "those three presents were the sounding instruments that directed my life. I can still hear the echoes from far away and long ago. They were the Pied Piper in my march through life." Throughout his adolescence Rice played sports with great enthusiasm and above-average skill, even though his lean build (120 pounds on a six-foot frame) was ill suited to football. Upon graduating from the Wallace University School in 1897, he entered Vanderbilt University at the age of sixteen.

Although Rice excelled in the classroom in Greek and Latin, his primary interests lay in extracurricular activities. During his first year he was the starting left end on the freshman football team. He made the varsity team his sophomore year and earned a letter as a junior. Throughout his four

*Column heading for Rice's column in the* New York Tribune, *which ran from 1913 to 1924 and continued in the* New York Herald Tribune *from 1924–1954*

years as an undergraduate, he never weighed more than 130 pounds, and during his career suffered a broken arm, a broken collarbone, a broken shoulder blade, and the separation of four ribs from his spinal column. Rice performed more naturally on the baseball diamond, where he was a smooth-fielding shortstop who hit for power, even though his shoulder injury forced him to throw underhand during his senior year, when he was team captain.

Rice did not become involved with any of the student publications at Vanderbilt. It was only after he had graduated and was barnstorming Tennessee with a semiprofessional baseball team that his father suggested he come home and try his hand at journalism. Although he had earned a Phi Beta Kappa key, Rice had not gone into engineering, law, or medicine while in college. Since he had "done creditably well in the arts," sportswriting seemed like a natural career choice. When Rice asked for a job as sports editor of the newly formed *Nashville Daily News,* he was told that there was no such job. According to Charles Fountain, in *Sportswriter: The Life and Times of Grantland Rice* (1993), the editor told Rice, "If you want to write sports and cover the State Capitol and the Courthouse and the Custom House along with it, we'll hire you." (When assigned to cover a society ball one night, Rice copied the account of a similar function in New York, changing only the names, and published his story undetected.) As Fountain notes: "The salary was five dollars a week, and . . . his career path was set."

A few months into his new job, Rice accepted a position with *Forester* magazine in Washington, D.C., although he would later admit in *The Tumult and the Shouting,* "I didn't know a Christmas tree from a Northern Blue Horned Spruce." Immediately upon arriving in Washington, he was felled by appendicitis and returned to Nashville to recuperate. Although he went back to work for the *Daily News,* he was already looking for a more challenging job and a higher salary. In 1902 he found both when he went to work for the *Atlanta Journal* for $12.50 a week. Even as he was establishing his reputation as sports editor of the *Journal,* Rice began to get telegrams from all over the South praising a rising minor league star with the Anniston, Alabama, team of the Southeastern League. These testimonials to Tyrus Raymond Cobb were signed with a variety of names. When Rice finally checked out this phenomenal player, he discovered that Cobb was as good as his admirers had claimed. Nearly fifty years later, after Cobb and Rice had both become national institutions, Cobb admitted that he had sent all the telegrams himself. Rice relates in his autobiography that, when he asked Cobb why he had sent the telegrams, Cobb replied: "Because I was in a hurry. . . . We were both youngsters on the way up. I didn't know it then but I was trying to put you onto your first big scoop!"

Rice left the South in 1905 when the *Cleveland News* offered him a salary of fifty dollars a week. This increased financial security allowed him to marry Katherine "Kit" Hollis of Americus, Georgia, a small town 120 miles south of Atlanta. Rice and Kit Hollis became acquainted during her frequent trips to Atlanta to visit friends. On one of their first dates he hired a horse and buggy and took her for a ride. When it started to rain a clap of thunder frightened the horse, who galloped away, pitching the two young people out of the buggy. Despite this inauspicious beginning and the Hollis family's skepticism about the young sportswriter's ability to support Kit, the couple married on 11 April 1906. Their wedding night was spent aboard the Pullman coach of a train heading to Louisville, where Rice rejoined the Cleveland baseball team on its trip north for the start of the major league season. Cleveland manager Nap Lajoie presented the newlyweds with what Rice describes in *The Tumult and The Shouting* as "a huge barrel of china . . . enough chinaware to stock a hotel." During their time in Cleveland, Kit gave birth to the couple's only child, a daughter, Florence (called "Floncy"), who became a movie actress.

After spending less than two years in Cleveland, Rice was lured back to his hometown by the newly formed *Nashville Tennessean* for a salary of seventy dollars a week. By this time he had earned a reputation not just as a sports authority but as a highly literate journalist who sprinkled his columns with poetry as well as prose. All of his talents were needed for his demanding new job with the *Tennessean.* In addition to writing two pages of sports daily and four on Sunday, he had to write a daily column of verse and provide copy for both the editorial and theater pages. His typical working day began at 8 A.M., and on nights when he was covering shows it did not end until after midnight. In addition to working this punishing schedule, Rice moonlighted as a football referee and coached the Vanderbilt baseball team in the spring of 1908.

On 16 June 1908 Rice published a poem called "Alumnus Football" in the *Nashville Tennessean.* This poem was written for a gathering of alumni on the Vanderbilt University campus and contained the two lines that constituted its author's surest claim to immortality. When he wrote that the One Great Scorer "marks – not that you won or lost – but how you played the Game," Rice was referring to men who had returned to their alma mater, years after their athletic prime, to play for fun. Certainly, winning and losing would have been more important in a professional contest or even in an intercollegiate varsity game. Nevertheless, Rice genuinely believed that character finally counted far more than success. He was appalled by evidence of cheating and bad sportsmanship. Also, as a southerner, he had been brought up to admire noble losers and to revere the lost cause for which the South had fought. Forrest and the other Confederate generals who had been his boyhood heroes were paragons of defeated valor. Two of the great folk heroes of his home state – Davy Crockett and Casey Jones – were better known for how they died than for anything they did during their lives. Rice could not have written as he did about sports had he not respected the athletes he covered. That he sometimes judged these men and women to be more admirable than they may actually have been is simply a sign of Rice's basic generosity and kindliness.

In the fall of 1910, after covering the Vanderbilt-Sewanee football game, Rice traveled to New York in search of a job. Rice remembers in his autobiography that, when he presented himself to Henry Stoddard, editor of the *New York Evening Mail,* Stoddard said: "Rice, I've been reading your verse. I never met a sportswriter worth fifty dollars a week, but in your case I'll risk it!" Although the salary was considerably less than the seventy dollars a week he was making in Nashville, his workload was only about a third as great. He was simply required to write a column six days a week, with no reporting duties on the sports desk, no theater reviews, and no humor column for the editorial page. Also, working in New York provided him with exposure and contacts unavailable in Nashville. When he arrived in New York in 1911, the city had fourteen daily newspapers. The *Mail* alone employed such legendary journalists as Franklin P. Adams, O. O. McIntyre, and Rube Goldberg. When Rice and his family set up residence at 450 Riverside Drive in February 1911, they shared a building with Walter Trumbull, editor of the *New York Tribune;* the music critic Deems Taylor; the movie actor Milton Sills; and the writers Max Foster and Irvin S. Cobb. Also in the same neighborhood were Rice's colleague Adams, the *Tribune* sportswriter Heywood Broun, and young Damon Runyon, newly arrived from Denver.

At that time the most important sports franchise in New York was the Giants baseball team. The spirit of the team was epitomized by its hot-tempered manager, John J. McGraw. As Rice writes in his autobiography, "McGraw, a bantam rooster, a fiery leader, was always in active charge, on the field and off. He was the umpire's foe and the foe of every team he played. He courted dislike in various cities, for he was shrewd enough to realize that additional fans would come out to see the 'hated' Giants play." Just as competitive, but far more courtly in manner, was the Giants star pitcher Christy Mathewson. Rice enjoyed a cordial professional relationship with McGraw and became a close personal friend of Mathewson. One Sunday during the 1911 season Rice enticed Mathewson and two teammates onto the golf course. "It was a hot day," Rice recalls, "and Matty insisted on playing 36 holes. We both staggered in, exhausted. Matty had won 11 straight games at that time. But McGraw pitched him Monday and he was knocked out of the box in the fourth inning. I hid out from McGraw for four days."

The big sports story in the winter of 1913 was the discovery that Jim Thorpe, American star of the 1912 Olympics, had once taken money for appearing in a summer baseball league. An outraged Amateur Athletic Union (AAU) stripped Thorpe of his amateur status and ordered him to return his Olympic medals. Rice freely expressed his scorn for the hypocrisy and self-righteousness of the AAU. For years star college athletes had received covert payments for their services without ever being investigated, much less punished, by any regulatory body. (Rice cited the case of the college athletes who played "without salary" in the same league as Thorpe but sold their suspenders to their manager for $800 at the end of the season.) According to Rice, "the best definition of an amateur is one who

*Rice broadcasting play-by-play in 1923 during the first World Series to be carried live on radio*

can get away with it and not be nicked with the goods." More than forty years later Rice wrote: "The act that barred Thorpe could never be justified. What right did the AAU have to Thorpe's private gifts, fairly won in those 1912 Olympics? They merely robbed the Indian in a cold-blooded fashion."

In 1913 Rice left the *Evening Mail* for a position with the older and more prestigious *New York Tribune.* The move also gave him the chance to reach a national audience with a column that the *Tribune* syndicated to approximately eighty newspapers across the country. (By the end of the 1920s the column was carried in more than 250 newspapers.) It was during this time that Rice formed a close friendship with his fellow sportswriter Ring Lardner. As many observers have noted, a more unlikely pair can hardly be imagined. According to Ring Lardner Jr., his father "was a debunker of sports heroes. Granny a glorifier of them. Ring was reserved and taciturn, Granny outgoing and loquacious. Ring's views tended to be pessimistic with a sardonic flavor. Granny's optimistic and tinged with sentiment." Nevertheless, as Fountain has pointed out in his biography of Rice, "both men shared a common foundation in sports, both men enjoyed games like golf, poker and bridge, both liked to gamble and both liked to drink. Friendships are more often forged of shared interests than of shared personalities."

In 1917, at the age of thirty-seven, Rice volunteered for the U.S. Army and fought in World War I. As a southerner who grew up knowing many Confederate veterans, he believed in military duty and in heroism on the battlefield. Although he enlisted in the 115th Field Artillery as a private, he was immediately promoted to sergeant and placed in charge of training a platoon. While stationed in Greenville, South Carolina, he was made the unit's athletic officer and told to clear a baseball field from what he described as a "solid green forest." When he was shipped overseas to Paris, already a second lieutenant, Rice was assigned to the staff of the military newspaper *Stars and Stripes.* His colleagues included Harold Ross, later the founding editor of *The New Yorker,* and the drama critic Alexander Woolcott. When he was put in charge of the paper's sporting page, Rice immediately killed it off. "There is but one Big League today for this paper to cover," he wrote, "and that league winds its way among the S.O.S. stations scattered throughout France and ends at the western front."

Although he rarely criticized the athletes he covered, Rice made an exception in the case of able-bodied sports figures who avoided military service. Shortly after the war he wrote: "It may or may not interest those frenzied baseball patriots who ducked both service and the draft by jumping to some ship league that the forthcoming summer may not be entirely pleasant for them." Referring to one of the icons of the age, he called Jack Dempsey:

> the champion boxer – not the champion fighter. For it would be an insult to every doughboy that took his heavy pack through the mules' train to front line trenches to go over the top at dawn to refer to Dempsey as a fighting man. If he had been a fighting man he would have been in khaki when at twenty-two he had no other responsibilities in the world except to protect his own hide.

The decade after World War I proved to be a golden age for American sports, and Rice helped mythologize the figures who made it so. When he first met a Red Sox pitcher named Babe Ruth, Ruth told Rice that he sounded as if he had a cold. When Rice, as he relates in his autobiography, admitted that he did, Ruth produced an enormous red onion, saying, "Here gnaw on this. Raw onions are cold-killers." "While Ruth talked, I gnawed," Rice recalled, "with tears streaming from my eyes." For Ruth's guest appearance on Rice's radio program in 1928, Rice crafted a script that called for Ruth to quote the duke of Wellington's famous line that the Battle of Waterloo had been won on the playing fields of Eton. However, in Ruth's rendition it came out: "As Duke Ellington once said, the battle of Waterloo was won on the playing fields of Elkton."

Although Rice was a man of his time, his writing occasionally struck a prophetic note. When Cobb demanded the seemingly outrageous salary of $15,000 a year in 1913, Rice argued that, from a purely business standpoint, Cobb was worth every penny of it. He pointed out that, when Cobb and the Detroit Tigers came to New York to play the Yankees, the crowds would be two to three times as large as when the Yankees played any other team in the American League. If only three hundred extra fans a game came out to see Cobb, that would mean an extra forty-five thousand fans during the course of the season. Based on a fifty-cent admission, Cobb would be putting an extra $22,500 in the Detroit coffers. (The available evidence suggested that he was doing much more than that.) Thus, the team would be making a clear profit, even if it gave in to his salary demands.

At the same time, Rice was wary of a free-market approach that would let players go to the highest-bidding team. "Everyone knows," he wrote, that "the reserve clause in baseball will not stand the test of American law. And yet without the reserve clause how long could Washington hold [Walter] Johnson? How long could Brooklyn hold [Nap] Rucker? How long could Detroit hold [Ty] Cobb?" Decades before free agency finally did drive up players' salaries and threaten the competitive balance of the game, Rice wrote that, in the absence of the reserve clause, "there would simply be a private agreement among club owners not to raise the ante." More than seventy years later the courts determined that major league clubs were indeed engaged in collusion to hold down the bidding for free agents. The baseball strike of 1994 was prompted by similar efforts on the part of club owners to hold down the salaries paid to their players.

By the time his half century as a sportswriter was coming to a close, Rice had seen many changes in the game of baseball. Early in *The Tumult and the Shouting* he referred to the public vendetta waged between Cobb and Ted Williams over whether the old-time players were better than the modern ones. In his autobiography Rice agrees with Connie Mack that such arguments are pointless because the game as it evolved in his day is vastly different from the one played in earlier times. He believes that "you simply can't match two entirely different games which call for dissimilar skills." He concedes that "an outfield of Cobb, [Tris] Speaker and Ruth, even with Ruth, lacks the combined power of DiMaggio, Musial and Williams." For that reason, the outfield on Williams's all-star team might well be "more useful in today's game than the old trio, as great as they were."

The picking of all-star teams became something of a specialty for Rice. When Walter Camp, the originator of the All-American football team, died in 1925, *Collier's Weekly* turned to Rice to make the annual selections. With the advice of friends, he did so for the next twenty-seven years, until the task was taken over by the nation's coaches. During the time that he made the picks, probably no other sportswriter in America had the stature and credibility to do so. "If it was the 'eleven best' the fans wanted," he wrote, "we tried to give them just that without any particular favoring of 'geographical distribution' of players to fit the national scope of readership. In 1929 Dink Templeton, my West Coast operative, told me, 'There's nobody out here worth a damn!' That year the team went no farther west than Minnesota."

Rice's most enduring contribution to the mythos of American football came in the first paragraph of his account of the Notre Dame–Army game played in the Polo Grounds on 18 October 1924. This was the story in which Rice compared the Notre Dame backfield of Harry Stuhldreher, Don Miller, Jim Crowley, and Elmer Layden to the Four Horsemen of the Apocalypse. Over time, all but a handful of people have forgotten that Notre Dame won that game 13–7. The details of the contest have faded even further from memory. The individual careers of the Four Horsemen have been consigned to footnotes in the history of sport and only the Notre Dame coach Knute Rockne continues to live in legend. What has remained in the public imagination, more than seventy years later, is Rice's striking analogy. As Fountain observes in *Sportswriter: The Life and Times of Grantland Rice:*

> It was not even necessary to have read, or even heard of, Vicente Blasco-Ibanez's novel, *The Four Horsemen of the Apocalypse,* or to have seen the Rudolph Valentino movie of the same name, or to be versed in the Book of Revelation to comprehend and enjoy Rice's simple and clearly stated reference to the Four Horsemen. Thus, Rice the sportswriter was, in this story and in so much of his work, all things to all people. There were readers who cared not a whit for sports who enjoyed his writing; there were others who didn't know a poem from a point-after-touchdown who were equally avid.

Critics of Rice's prose accuse him of inflating the importance of athletic contests beyond reason. According to Fountain the cynics dismissed the Four Horsemen story as "the overwrought, overblown, meretricious utterings of an out-of-touch sentimentalist." One of the most acerbic of these critics, Robert Lipsyte, has written in *Sports World: An American Dreamland* (1975): "Painting the lily is not only presumptuous, but ultimately destructive. The flower dies. By layering sports with pseudomyth and fakelore, by assigning brutish or supernatural identities to athletes, the Rice-ites dehumanized the contests and made objects of the athletes." Lipsyte goes on to say:

> The writer who criticizes a ballplayer for muffing a grounder, no matter how nasty he gets about it, is still dealing with the ballplayer within his context. He is judging the athlete as a working professional. But the writer who likens a ballplayer to Hercules or Grendel's mother is displaying the ultimate contempt – the ballplayer no longer exists as a person or a performer, but as an object, a piece of matter to be used, in this case, for the furtherance of the sportswriter's career by pandering to the emotional titillation of the reader/fan. Rice populated the press boxes with lesser talents who insisted, like the old master, that they were just sunny fellows who loved kids' games and the jolly apes who played them.

Several years before the Four Horsemen rode their way into immortality, Rice had witnessed another force of nature when Jack Dempsey, outweighed by nearly sixty pounds, destroyed Jess Willard in a ring in Toledo, Ohio. From the time that he won the heavyweight title from Willard in 1919 until he lost it to Gene Tunney in 1926, Dempsey was probably the world's dominant sports figure. Rice rated him the greatest offensive athlete he ever saw. "I don't think there was ever a moment in any fight that Dempsey thought in terms of defense," Rice says in his autobiography. "He had a method of weaving and moving about that was partly defense but it always led to attack or headlong assault." Although Rice's criticism of Dempsey for failing to serve in World War I created some initial friction between the two men, they later became close friends. From the perspective of the early 1950s, Rice observes:

> [O]utside the ring, Jack is one of the gentlest men I know. . . . I've seen him in his restaurant at times when some customer, with more enthusiasm than good sense, would grab his vest or part of his shirt – strictly for a souvenir – with no kickback from Jack. I've known the man closely for more than 30 years and I've never seen him in a rough argument or as anything except courteous and considerate.

Rice was also fond of the other legendary heavyweight champion of his time – Joe Louis. However, the racial caste system that was still firmly in place in America meant that his relationship with Louis was different from his friendship with Dempsey. They always addressed each other as "Joe" and "Mr. Rice." "No one in the sporting world has helped his people advance with such great speed as Joe Louis," Rice writes. "In the first place, Louis was a great heavyweight. Aside from defending his title 25 times, more than all modern heavyweights combined, Joe was a clean, fair sportsman – above most white men in his own game. He was completely honest in a game where honesty is not the watchword. Louis was a beacon, a glowing example also for the white race to follow." The chapter on "The Negro Race" in Rice's autobiography is remarkably enlightened and fair-minded.

*The backfield of the Notre Dame football team, dubbed "The Four Horsemen" by Rice in 1924 – (left to right) Don Miller, Elmer Layden, Jim Crowley, and Harry Stuhldreher*

However, the language and stereotypes that one finds in some of Rice's earlier columns and stories seem anachronistic when measured by contemporary standards of political correctness. Although Rice and Louis discussed their respective golf games while sitting together at championship fights, Fountain notes that Rice never invited Louis out for a round of golf.

Like most Americans, Rice thrilled to the accomplishments of Jesse Owens at the 1936 Olympics in Berlin. Recalling those games nearly twenty years later, he writes: "I'll not forget the sight of those German storm troopers, in their severely cut black uniforms . . . looking every inch the super race. You would see them in the streets, out at the jam-packed Reich Sportsfield, at the Hofbraus. They didn't stroll; they marched . . . and gutturalized with quiet, confident bearing that betokened their Cheshire cat scorn of 'less endowed' mortals." When the Nazi propaganda machine tried to minimize the success of Owens and other black athletes with references to their superior ability to relax and their innate sense of rhythm, Rice shot back: "The ease and smoothness of championship is not the property of any race. It is the property of a champion." If one detects what seems condescension in Rice's rating Jesse Owens as the best of the "Negro athletes," it should be remembered that Rice considered an American Indian – Jim Thorpe – to be the best athlete of any race.

While Rice had relatively few opportunities to observe black athletes during his half century in sportswriting, he came in contact with even fewer women competitors. But the greatest female athlete of his time was a particular favorite of his. In 1932 Babe Didrikson was an obscure eighteen-year-old clerk-typist with Employers Casualty of Dallas. That year she took the Women's National AAU championships in Chicago by storm, winning the javelin, hurdles, shot put, baseball throw, and broad jump; tying for first in the high jump; and taking fourth in the discus. At the Olympics in Los Angeles she won gold medals and set world records in both the javelin toss and the eighty-meter hurdles. She also won a silver medal in the high jump, just missing the gold when her best jump was disallowed on a controversial ruling. Very much taken with her as both an athlete and a woman, Rice invited Didrikson to join him on the golf course. Although this may well have been her first golf game, it was immediately apparent to Rice that Didrikson had immense potential on the links. She later went on to be a champion professional golfer, even battling back from cancer in the early 1950s. Rice paid Didrikson the highest possible compliment when he wrote in *The Tumult and the Shouting:* "It is an odd turn that perhaps the two flashiest figures in sport – two of the immortals – were called 'Babe.' . . . They stand above the mobs and the multitudes. They will still be alone and above the others when you and I are dust."

In addition to helping shape the way the American people viewed important figures and contests in baseball, football, and boxing, Rice did more than any other individual to bring the game of golf to the nation's attention. He took up golf as recreation when he was working a grueling schedule on the *Nashville Tennessean* during the first decade of the century. At that time golf might be mentioned in the society or business sections of the daily newspaper but rarely on the sports page. Rice was a serious competitive golfer during his early years in New York. He won the advertising-industry tournament in Pinehurst, North Carolina, in 1916 and was runner-up in 1917. Rice and Rex Beach pretty much dominated play at the Writers and Artists Golf Association tournament in Palm Beach from its inception in the mid 1920s until some time in the late 1930s. Even after his days of tournament play were over, Rice rarely traveled anywhere without his golf clubs.

As Fountain notes, "Whatever the sport of golf gave to Rice, he more than repaid the debt. Its emergence as a popular sport – both as a participant and a spectator sport – was given shape and definition in Rice's writing in the years between Francis Ouimet's win in Brookline in 1913 and Bobby Jones's grand slam in 1930." In addition to covering the sport in his newspaper column, Rice collaborated on two books dealing with golf. He and Jerome D. Travers published *The Winning Shot* in 1915, the year Travers won the U.S. Open. Then in 1926 Rice provided the captions and text for cartoonist Clare Briggs's drawings in *The Duffer's Handbook of Golf.* He also edited *The American Golfer,* a highly literate magazine published by Condé Nast, from the mid 1920s until its demise in the Great Depression a decade later. Perhaps more important, Rice helped nurture the career of Bobby Jones, a sports hero whose fame and stature rivaled those of Ruth, Dempsey, and Red Grange.

Because he was a longtime friend of Jones's father, Bob Jones Sr., Rice had known young Jones from the time the boy was three. When he made his spectacular debut in the U.S. Amateur Tournament at age fourteen, it was apparent that Bobby possessed the skills to be a great golfer. It was just as apparent, however, that his temper stood between him and a successful career. Like a Dutch uncle, Rice admonished Jones to grow up and become the champion he was capable of being. The following year, when Jones and his friend Perry Adair were in New York to play war relief charity matches, they stayed with the Rices at 450 Riverside Drive. "Our living room reminded me of a set from *You Can't Take It With You,*" Rice recalls.

> We enjoyed having two partly grown-up "sons" about the place. How Kit, Floncy and I relished their stay. One evening I took the entire brood to Coney Island – a great trip. We didn't miss a ride! It was during their stay that Bob and I became acquainted in a way few persons with a gap of 20-odd years between them ever do.

After seven years of maturation, Jones finally won a major tournament – the U.S. Open.

From 1923 to 1930 Jones won nine major championships. Of the twenty major tournaments he entered during that period, he finished worse than second only four times. What is even more remarkable is the fact that he did this while devoting the lion's share of his time to the study and practice of law. He dominated the game of golf while playing it on an essentially part-time basis. In 1930 he won the four major tournaments of that era – the U.S. Open and Amateur and the British Open and Amateur. No other golfer has matched that feat before or since. That fall, in a letter to the U.S. Golf Association, Jones retired from the sport at the age of twenty-eight. (Like Alexander the Great, he had no more worlds to conquer.) Jones later conceived and designed the Augusta National Golf Course, which opened in 1932. Two years later he hosted an invitational tournament that later came to be called the Masters. Rice had suggested the dates of 22–25 March because Augusta was at its best in the early spring and because the baseball writers would be making their way north from spring training for the start of the regular season. From its inception until his death twenty years later, Rice was an abiding presence at the Masters.

Rice's favorite leisure activity, other than golf, was horse racing. Never a compulsive gambler who bet large sums of money, he relished an afternoon at the racetrack for the excitement and sociability it offered. Rice would have understood the sentiments of Winston Churchill, who once said that the thing he enjoyed most in life was winning money at gambling and the thing he enjoyed second most was losing money at gambling. Once a young native of California asked him whether his favorite racetrack was Santa Anita, Hollywood, or Del Mar. Rice responded that it was none of these but Sunshine Park in Oldsmar, Florida, which is about at the middle of a triangle made by Tampa, Saint Petersburg, and

Clearwater. In describing this park Rice's friend and fellow sportswriter Red Smith, in *To Absent Friends* (1982), reveals much about the simple tastes of the man nearly everyone called "Granny":

> The Sunshine Park that Grantland Rice knew was a Shoeless Joe among racetracks, a sort of slum-clearance project in a rattlesnake colony. The barns were sagging shanties of scrap lumber, the clubhouse and grandstand featured peeling paint, and there was a seedy, carpet-slippers informality that captivated Granny. The clubhouse dining room was bare, plain and tiny, but it served the superlative Stevens victuals, and immediately outside were the remaining necessities of life – an open-air bar and a mutuel battery maybe fifteen feet apart and not more than six paces removed from a view of the running strip.

Besides writing his newspaper column ("The Sportlight"), which moved to the Bell Syndicate in 1930 and later to the North American Newspaper Alliance, Rice was active in the newer media of radio and motion pictures. He was a fixture on the early network radio series *Cities Service Concert*. Also, in 1922 he provided the play-by-play descriptions for the first World Series game carried live on the radio. Two years earlier he had joined forces with moviemaker Jack Eaton to form Sportlight films. The company produced a series of ten-minute shorts, which eventually depicted 123 different activities. These included many novelty sports and outdoor scenes featuring wild game in various locations around the world. In addition, Rice was able to call upon his friendships with luminaries from the more conventional sports. His first talking picture, produced in 1929, was a film called *Four Aces,* in which the legendary promoter Tex Rickard introduced Dempsey, Jones, the tennis star Bill Tilden, and polo great Tommy Hitchcock. Over the years these films garnered six Academy Award nominations and won twice in the category of Best Single Reel.

By the end of the 1940s Rice had become a national institution. He was featured on Ralph Edwards's radio program *This Is Your Life* in 1948. Scholarships were established in his name at Vanderbilt University and the Columbia School of Journalism. Testimonials poured in from all directions. "It all amounted to a continuing retirement party for a man who had no intention of retiring," writes Fountain.

> It all must have been a little overwhelming, too; but Rice let it all pass by like a pleasing summer's zephyr. He continued to do his job, carrying his own typewriter up into the press box, generally plopping it down in an empty space next to some wide-eyed kid writer and giving the kid a lifetime's memory by ingenuously sticking out his hand and saying: "Hi. My name is Grantland Rice."

Rice's stature remained undiminished even as his influence was slipping. Newspapers felt it important to subscribe to his column whether they ran it often or not. In an age when sportswriters were becoming more incisive and adversarial, Rice's sentimental verse and philosophical reflections were beginning to seem stale and old-fashioned. Nevertheless, he provided his readers with an invaluable link to the past. He had known virtually every member of the baseball Hall of Fame and every heavyweight champion from John L. Sullivan to Rocky Marciano. He might not have been as topical as his younger colleagues, but he possessed infinitely more historical perspective. He was also a man of legendary courtesy and humility. Once when his press ticket to the Army–Notre Dame game was lost in the mail, he bought a ticket from a scalper on Broadway and watched the game from the stands with his typewriter between his knees. As the incredulous Red Smith observes in *To Absent Friends,* "this was Grantland Rice, who did as much for American football as any other man who ever lived; who practically invented the Army–Notre Dame game, who made it a part of American literature with his 'Four Horseman' story of 1924." The idea of throwing his weight around at the press gate simply would not have occurred to this southern gentleman.

While writing his column on 13 July 1954, Rice began to experience difficulty typing. Four hundred words into the piece, he left the paper in the typewriter, in midsentence, and had his chauffeur drive him to his office. Seeing how disoriented he was, Rice's secretary called his doctor, who summoned an ambulance. At the hospital he slipped into a coma and died shortly after six o'clock in the evening. For the next few months there was an outpouring of genuine grief from Rice's many friends and even larger circle of admirers. On Halloween night nearly every prominent athlete alive gathered for a party at Toots Shor's restaurant. It was the eve of what would have been Rice's seventy-fourth birthday and also the eve of the publication of his autobiography. Dempsey, Tunney, Gene Sarazen, Johnny Weissmuller, Lou Little, Tommy Henrich, Yogi Berra, Eddie Arcaro, Hank Greenberg, and the Four Horsemen were all there. So was James A. Farley from the world of politics, along with Ed Sullivan, Jackie Gleason, and Douglas Fairbanks Jr.

from show business. The ranks of Granny's fellow writers included everyone from Rube Goldberg and Bruce Barton to Bob Considine and Smith. "I've been offering eight-to-five all night that Granny will show up before the party's over," one guest, according to Fountain's biography, was heard to say, "and I ain't getting' no takers. Everybody's afraid to bet against it."

In the obituary he published for Rice in *The New York Times,* Arthur Daley recalled that "there was an era in the Twenties when every young writer tried to emulate Grantland Rice." By 1954 that era had long since passed. Fewer journalists were prone to put athletes on a pedestal and to invest sports with cosmic significance. If Rice embodied what has been called the "Gee Whiz" approach to sportswriting, the "Aw, Nuts" school epitomized by his friend Ring Lardner was becoming the preferred model for younger writers. It is perhaps significant that a year before Rice's death a young journalist named Howard Cosell began his thirty-nine-year career as a sports commentator for ABC radio and television.

If Rice chose to emphasize the ennobling aspects of sports, it was not because he was simply naive or sentimental. "He knew athletes and knew their failings," Fountain writes. "He knew that Babe Ruth was a sot and that Ty Cobb was a boorish churl. But he chose instead to celebrate their enormous athletic gifts." In describing the cozy relationship that existed between the media and public figures prior to the 1960s, Tom Wolfe has characterized the old-time press as "a proper Victorian gent." That is a description that fits Grantland Rice and one he would have taken as a compliment.

**Biography:**

Charles Fountain, *Sportswriter: The Life and Times of Grantland Rice* (New York: Oxford University Press, 1993).

**Reference:**

Mark Inabinett, *Grantland Rice and His Heroes: The Sportswriter as Mythmaker in the 1920s* (Knoxville: University of Tennessee Press, 1994).

# Damon Runyon

*(8 October 1880 – 10 December 1946)*

Paul J. Sandin

See also the Runyon entries in *DLB 11: American Humorists, 1800–1950* and *DLB 86: American Short Story Writers, 1910–1945, First Series.*

MAJOR POSITIONS HELD: Sports reporter and columnist, *New York American* (1911–1946); columnist and feature writer, King Features and International News Service (1936–1946); writer/producer, RKO Pictures (1941), 20th Century–Fox (1941–1943).

Damon Runyon

BOOKS: *The Tents of Trouble: Ballads of the Wanderbund and Other Verse* (New York: Fitzgerald, 1911);

*Rhymes of the Firing Line* (New York: Fitzgerald, 1912);

*Guys and Dolls* (New York: Stokes, 1931; London: Jarrod, 1932);

*Blue Plate Special* (New York: Stokes, 1934);

*Money from Home* (New York: Stokes, 1935);

*More Than Somewhat,* selected by E. C. Bentley (London: Constable, 1937);

*Take It Easy* (New York: Stokes, 1938; London: Constable, 1938);

*Furthermore: A Companion Book of Stories to "More than Somewhat,"* selected by Bentley (London: Constable, 1938);

*The Best of Damon Runyon,* edited by Bentley (New York: Stokes, 1938);

*My Old Man* (New York: Stackpole Sons, 1939; London: Constable, 1940);

*My Wife Ethel* (London: Constable, 1939); with different material (Philadelphia: McKay, 1940); republished as *The Turps* (London: Constable, 1951);

*A Slight Case of Murder* (New York: Dramatists Play Service, 1940);

*Damon Runyon Favorites* (New York: Pocket Books, 1942);

*Runyon à la Carte* (Philadelphia: Lippincott, 1944; London: Constable, 1946);

*Ten Stories* (New York: Avon, 1945);

*Damon Runyon Favorites* (Cleveland: World, 1946);

*The Three Wise Guys and Other Stories* (New York: Avon, 1946);

*In Our Town* (New York: Creative Age Press, 1946);

*Short Takes: Reader's Choice of the Best Columns of America's Favorite Newspaperman* (New York & London: Whittlesey House/McGraw-Hill, 1946; London: Constable, 1948);

*Poems for Men* (New York: Duell, Sloan & Pearce, 1947);

*Trials and Other Tribulations* (Philadelphia: Lippincott, 1947);

*Runyon First and Last* (Philadelphia: Lippincott, 1949); republished as *All This and That* (London: Constable, 1950);

*Runyon from First to Last* (London: Constable, 1954).

**Collections:** *The Damon Runyon Omnibus* (New York: Blue Ribbon Books, 1939);

*More Guys and Dolls* (Garden City, N.Y.: Garden City Books, 1951).

PLAY PRODUCTION: *A Slight Case of Murder,* by Runyon and Howard Lindsay, New York, 48th Street Theatre, 11 September 1935;

SELECTED PERIODICAL PUBLICATIONS: "Defense of Strikerville," *McClure's* (February 1907);

"Your Neighbor, the Gambler," *Cosmopolitan* (November 1921);

"Hold 'Em Yale!" *Collier's,* 88 (14 November 1931): 7–9;

"Little Miss Marker," *Collier's,* 89 (26 March 1932): 7–9;

"The Lemon Drop Kid," *Collier's,* 93 (3 February 1934): 7–9;

"Tight Shoes," *Collier's,* 97 (18 April 1936): 7–9;

"Baseball Hattie," *Cosmopolitan* (May 1936);

"All Horse Players Die Broke," *Collier's,* 100 (11 September 1937): 9–11;

"Little Pinks," *Collier's,* 105 (27 January 1940): 9–10.

The iconoclastic Damon Runyon was a highly respected sportswriter, but he was also more than a sportswriter. His style reflected the America of the turn of the century: brash, bold, and swaggering. Known as "The Chronicler of Broadway" because of books such as *Guys and Dolls* (1931), Runyon introduced to the nation the lingo and characters of the New York sports world and underworld. Baseball Hattie; Madame La Gimp; Sky Masterson (named after Runyon's friend Wild West legend Bat Masterson); Little Miss Marker; Dark Delores; Dave the Dude; Nathan Detroit; Regret, the Horse Player; Hot Horse Herbie; Nicely-Nicely – all these are products of Runyon's fertile imagination. His sports beat spanned the Polo Grounds, Yankee Stadium, Ebbets Field, and Madison Square Garden. He knew the gyms and fight rings where the great, not so great, and "wanna-be" pugilists trained. He was a regular at the racetracks of Saratoga, Pimlico, and Hialeah.

Runyon was a close friend of the gentlemanly Christy "Big Six" Mathewson as well as of the brutal fighter Jack Dempsey. He covered John McGraw's Giants and hunted with Babe Ruth. Jimmy Breslin reports in his biography *Damon Runyon* (1991) that Runyon lit his cigarettes from a gold lighter bearing the inscription "Damon Runyon. Amigo Mio. Salud. . . . Pancho Villa." He was a close friend of New York mayor Jimmy Walker and hobnobbed with Al Capone. In a 1940s article "Postscript on Sports Writing" Runyon writes:

> To be a great sports writer a man must hold himself pretty much aloof from the characters of the games with which he deals before his sympathy for them commences to distort his own viewpoint. There is nothing more engaging than an engaging rogue and there are many engaging rogues in professional sports. I fear I knew most of them, and that is not good for a sports writer.

Runyon was born Alfred Damon Runyan (a corruption of his family's French name, *Renoyan*) in Manhattan, Kansas, on 8 October 1880. His father, Alfred Lee Runyan, a former Indian fighter under George Armstrong Custer, was a drunkard and gambler who dabbled in small newspaper publishing. Alfred Lee married Elizabeth Damon, a direct descendant of John White of the Massachusetts Bay Colony. From the two parents Alfred Damon Runyan received his full name. (He later changed the spelling of his last name to Runyon.) Elizabeth Runyon died in Pueblo, Colorado, of tuberculosis when Damon was seven years old. His three sisters were sent back to Kansas to be raised by other family members, while young Runyon was left to run the streets of Pueblo. When he was not running errands for local gamblers or bar patrons, he accompanied his father to the Arkansas Saloon, where the elder Runyan regaled the customers with stories of Custer, Billy the Kid, and "Buffalo Bill" Cody. Since Runyon was expelled from school in the sixth grade (for "excessive horseplay"), the education he received at his father's elbow was his main source of learning. Young Runyon (still known as Al) began his newspaper career as a printer's devil at the *Pueblo Evening News.* By the time he was fifteen he was a reporter. He was also an accomplished drinker. He saw his first byline at the age of seventeen at the *Pueblo Evening Post.* Early in his career a printer's error caused the byline to read Alfred Damon Runyon. He liked the change and kept the *o* in his three-name byline until another paper would drop his first name.

Runyon was six months shy of his eighteenth birthday when the battleship *Maine* blew up in Havana harbor. When the United States declared war on Spain, he tried to enlist along with a contingent from Pueblo but was rejected because of his age. Undaunted by the refusal of the recruiting officer, Runyon went to Minnesota, exaggerated his age, and joined the Thirteenth Minnesota Volunteers. The outfit was sent to the Philippines to help put

*Runyon in the press box at the 1924 World Series*

down the Moro insurrection. Although the action brought forth no opportunity for heroics, medals, or acclaim for Runyon, it did give him plenty of ideas for writing. His first two books, *The Tents of Trouble: Ballads of the Wanderbund and Other Verse* (1911) and *Rhymes of the Firing Line* (1912), are poems inspired by his Philippines experience.

The years 1900–1906 are best known as Runyon's "hobo years" – years of unpaid railroad journeys. During this period of "riding the rods," Runyon traveled with what he considered to be quasi-legendary characters who rivaled those who dominated his father's saloon stories years before. According to biographer Tom Clark in *The World of Damon Runyon* (1978), Runyon once wrote that he "picked up much of the lore of the road through contacts with the real rovers and . . . accumulated considerable of their patter of that long-gone time." Much of that patter would make it into Runyon's writings in the years that followed. The trains took the twenty-year-old Runyon far and wide after he was mustered out of the army in San Francisco.

He returned to Pueblo in 1900 and began a short stint as a reporter for the *Chieftain*. In 1901 he went to Colorado Springs, where he worked for the *Gazette*, eventually losing the job because of his heavy drinking. In 1902 he moved to the *St. Joseph (Missouri) Gazette*. In 1903 he returned to the *Pueblo Chieftain*, then took a job at the *Trinidad (Colorado) Advertiser* in 1904, moving to the *Denver Republican* and the *Denver Post* in 1905. Fired later that year for being drunk on the job, he caught a train to San Francisco, where he worked at the *San Francisco Post* for a few months in 1906 (during the most violent earthquake the country had ever witnessed) before returning to Denver to write for the *Rocky Mountain News*. He remained at that Denver paper for four years.

While at the *Rocky Mountain News* Runyon met Ellen Egan, the paper's society reporter. They married in 1911 as Runyon was on his way to New York City to join the sports staff of the *New York American*, of the Hearst chain. Two children were born from this marriage: Mary in 1914 and Damon Jr. in 1918. During the period from 1911 to 1913 Runyon, deeply afflicted by alcoholism, swore off drinking, preferring instead to "get the money" – his phrase for staying sober enough to earn a living. Whether his conversion was inspired by his bride, his drive for success and money, or a determination not to end up like his father, he remained sober for the rest of his life. Apparently his dependency on alcohol was replaced by a dependency on caffeine: Runyon would drink up to forty cups of coffee per day after he quit drinking alcohol. Ellen Runyon became estranged from her husband and died on 9

November 1931 in Bronxville, New York. Damon Runyon Jr., in *Father's Footsteps* (1954), described his mother in her last years:

> My mother was always "sick" now, spending most of her time in bed at home. . . . She was in bad shape. Liquor hits a woman hard and fast . . . Such soft terms as "sick" were used to deceive. . . . Mother didn't go out much any more – she just took to bed and bottle.

On 8 July 1932 the fifty-one-year-old Runyon married Patrice Amiti del Grande, a twenty-five-year-old Spanish dancer. Whether because of their age difference, Runyon's excessive gambling, or reasons known only to her, Patrice obtained a divorce from her husband in June 1946 after she learned from his doctor that Runyon was dying of throat cancer. Patrice had fallen in love with Bill Coffin, a Massachusetts man forty years Runyon's junior.

Runyon's first sports assignment with the *New York American* was to cover the New York Giants, who were in spring training in Dallas, Texas. He filed his first story, signing it as usual "Alfred Damon Runyon." The sports editor, Harry Cashman, struck the "Alfred," saying that three names were no longer used for reporters. From now on, he explained, Alfred Damon Runyon would be known only as Damon Runyon. As Ed Weiner related in *The Damon Runyon Story* (1948), Al responded: "Yeah, Damon Runyon. It sounds kind of nice, and musical. Well, I guess that's the end of Al Runyon."

Runyon's first byline appeared over an interview he did with the famous heavyweight boxer John L. Sullivan, who by then had fallen on hard times. That Runyon obtained a byline at all was a coup. The New York papers at that time did not give sportswriters bylines. But the Sullivan interview guaranteed from that point that Runyon's name appeared over every story he wrote.

Meanwhile, Runyon had a colorful and talented Giants team to cover. "Big Six" Mathewson was the star pitcher for manager John McGraw. But it was a pitcher who chugged containers of cold beer in the Giants bullpen who attracted Runyon's attention. Bugs Raymond tossed baseballs into the stands in exchange for fans' beer – and became the first "Runyonesque" character. As the Giants toured the South, Runyon regaled his readers with the antics of Raymond, writing about the game in which Raymond crawled from the bullpen to the pitcher's mound on his hands and knees. Runyon delighted in telling the world that Giants manager McGraw cut off Raymond's salary because of the pitcher's tendency to spend his money on beer. McGraw directed that Raymond's money be sent directly to his wife, causing Raymond to say (according to Clark, in *The World of Damon Runyon*): "If she gets the money, let *her* pitch."

As Runyon's reports continued to come in to the *New York American,* so did phone messages and mail from the readership, praising Runyon's tales of Bugs Raymond. The reports also caught the attention of publisher William Randolph Hearst, who summoned Runyon to his home at Eighty-sixth Street and Riverside Drive, congratulated him on his work, and tripled his salary. According to Breslin's *Damon Runyon,* Runyon responded to this windfall by claiming, "That was fair of the Chief. He put his finger on exactly what I'm worth. Three times the usual grunt. I better get out and earn it."

Runyon's assignment at the *New York American* was to cover all the New York baseball teams, and he did so with humor, sarcasm, and sharp wit. His lead for the 16 July 1911 *New York American* was:

> Mathewson pitched against Cincinnati yesterday.
> Another way of putting it is that Cincinnati lost a game of baseball.
> The first statement means the same as the second.

Runyon then explained to his readers that the Cincinnati Reds had not beaten Mathewson since 1908.

In the fall of 1911 the *New York American* began to test Runyon's skills in sports other than baseball. He was sent to cover a horse show at Madison Square Garden, which originally had been constructed with the intention of housing horse shows. Instead of describing the feats of the horses, Runyon paid more attention to the pageantry and people in attendance. According to Edwin P. Hoyt in *A Gentleman of Broadway* (1964), instead of filing the expected reports from the show, Runyon penned a poem, "Mr. Deeters of Cheyenne," in which he offered his observations of the low-cut dresses of the wealthy matrons in attendance. In witty and whimsical style, Runyon concluded the lengthy poem with this comparison of the ladies' dresses to the thoroughbreds' blankets: "And [the horses would] freeze in the range from the sudden change / if they cut them blankets so low."

During the 1911 sports season, the testing ground for Runyon's unique style, he was a keen observer of his peers in the press box. He admired the work and jargon of Tad Dorgan, a cartoonist and boxing writer for the *New York American.* He borrowed from Ring Lardner, of the *Chicago Tribune,* a humorous, freewheeling approach to writing about baseball. In fact, Hoyt notes in *A Gentleman of Broadway* that Runyon got the reputation in later years of being a "second-rate Ring Lardner." Runyon also met the legendary "Wild Bill" Phelon (of the *Cincinnati Inquirer*) when he came to New York with the Reds. He was enthralled by the insane antics of Phelon but was more impressed by the way Wild Bill could carry on a conversation and write a story at the same time. He learned much from his associations with Grantland Rice and Heywood Broun, and he considered Irvin S. Cobb to be the greatest newspaperman in New York. The two became close friends. Runyon admired, too, the stories being told by the sports editor of the *New York Morning Telegraph,* Bat Masterson, the former sheriff of Dodge City and already a fixture in Wild West lore. Masterson's off time was spent at the Metropole Hotel bar, where Runyon soon became a nondrinking regular. It was at the Metropole that Runyon met a steady stream of bookies, breeders, jockeys, trainers, and promoters of New York horse racing.

By 1912 Runyon had the freedom to pick his own baseball and football assignments. He was also free to cover any event he wished. On 16 July 1912 a gambling-house proprietor named Herman Rosenthal was gunned down outside the Metropole. Runyon covered the trial and in the process began his exploration into, and fascination with, the New York underworld. But in 1913 he was back on his New York Giants beat, in a season that culminated in the Giants' World Series defeat by Connie Mack's Philadelphia Athletics. Runyon also covered the important college football games that fall, but it was a boxing story that gave him the opportunity to do some of his greatest writing.

The 1915 fight between Jack Johnson and Jess Willard ended in a controversy that has never been settled. In the twenty-sixth round Johnson fell to the floor of the ring and raised his hand as if to shield his eyes from the sun while he was being counted out. The fall was called a knockout, and Willard was declared the new heavyweight champion. From ringside in Havana and throughout the world came the cry of "fix." Johnson fanned the flames of controversy by alternately confirming and denying that he had thrown the fight. It was this match that inspired one of the longest columns of Runyon's career. He had concluded that the fight ended in an honest knockout. Many disagreed with Runyon's assessment, but Runyon was unrelenting, and his regular column defended his opinion for a week following the match. Runyon also believed that his story of the Johnson fight was the best individual piece of sportswriting of his career:

> In the twenty-sixth, Johnson took the full count shielding his eyes from the sun with his gloved hands and for 29 years he has pointed to this fact as implying that he knew what he was doing all the time, which no one who ever saw the fight ever doubted. He simply couldn't go farther. . . .
>
> The excuse Johnson is alleged to have given . . . for chucking the title is his mother was ill and some mysterious "boys" had promised to have a deportation ban on him lifted so he could return to the United States and visit his stricken parent but that the "boys" double-crossed him on that promise. I say nuts to that too . . .
>
> Well, about 95 per cent of all sports tradition is pure fiction. Lies if you like. But harmless. Who the hell cares if long after a sports event the facts get a little twisted?

On 16 March 1916 the revolutionist Mexican general Francisco "Pancho" Villa led four hundred soldiers into New Mexico, where they burned the town of Columbus and killed nine civilians and eight American soldiers. President Woodrow Wilson ordered Gen. John Pershing to take six thousand men into Mexico and capture Villa. Runyon was one of only four accredited journalists permitted to accompany Pershing into Mexico. The incursion became bloody, and a virtual standoff took place. Runyon, after writing in praise of the leadership qualities of Villa and presenting him the gift of a baseball autographed by the New York Giants, became acquainted with the general. They later became friends and attended horse races together.

Throughout this period of covering boxing matches, horse shows, football games, and international conflict, Runyon's interest in baseball never waned. He was increasingly impressed by the Giants' feisty manager McGraw. He admired McGraw more for his fire and combativeness than for his ability to lead his ballplayers. In baseball circles the five-foot-six-inch McGraw was known as "Little Napoleon," and his exploits on and off the field provided Runyon with ample material for his stories and columns. McGraw was also an active member of the Lambs Club, an actors' society, and socialized with the Broadway theater set. He and Runyon became good friends, and his fondness for playing the horses endeared him even more to Runyon. Meanwhile, Runyon noted that McGraw lost as many fights as he won. In a column titled "Café Gladiators," Runyon, after discussing McGraw's choice of opponents, marvels:

> Even when he was in the middle years and had a paunch, he was slugging it out with the ball players,

*Runyon at work in his apartment*

umpires, spectators and innocent bystanders. The New York newspapers made much of McGraw's fights, as they do of the battles of café society today, and which went something like this:

> Lost to Honest John Kelly.
> Lost to Ad Brennan.
> Lost to P. Sentell.
> And so on and so on.

The 1919 World Series match between the Chicago White Sox and the Cincinnati Reds profoundly affected Runyon's love affair with baseball, as it did with many of the sportswriters of his day. Eight White Sox players, including the great "Shoeless" Joe Jackson, conspired with gamblers (among them Arnold Rothstein, a friend of Runyon's) to throw the World Series to the Cincinnati Reds. Runyon and other sportswriters covering the series had written articles that described the curious play of some of the White Sox players, although Runyon himself primarily focused on the Cincinnati victory. In 1920 when the full story broke, writers had a field day with the accounts of what became known as the "Black Sox" scandal. In his 27 September 1920 editorial titled "Read This Story," Runyon scolded baseball officials for keeping quiet and doing nothing "to protect the great national game," but he believed that baseball "will be all the better for the cleansing it is now undergoing." Reflecting on the scandal twenty-one years later, Runyon, in a piece he called "Napkin Technique," wrote:

> The guilty ball players were given the hoovus-groovus from baseball for the rest of their lives. Among them were some of the greatest players that ever laced on cleated shoes – Buck Weaver, a third baseman, for one, and "Shoeless Joe" Jackson for another.
>
> Against the lively ball of today, "Shoeless Joe" would be hitting 'em across townships. He could hit .350 when he was just clowning and better than .400 when he was serious. You talk about hitters – why that old "Shoeless Joe" could drive the ball farther with the breeze on his missed strikes than most ball players can send it off their bats. Ask anybody that ever saw him in action, or look up the book on him.
>
> We felt sorrier for Joe than any of the others. He was not a smart fellow except up at the plate. There he was the wisest man in the world.

Of the myriad columns and stories about baseball written by Runyon, two are among his most famous. One, written for the *New York American* on 10 October 1923, is a piece about Casey Stengel's home run, which won the first game of that year's World Series. That game, in the first series ever played at Yankee Stadium, featured baseball at its finest. Runyon described Stengel's inside-the-park home run in a crescendo flourish designed to make the reader feel as though he was in an aisle seat with an unobstructed view of the action:

> This is the way old Casey Stengel ran yesterday afternoon running his home run home.
>
> This is the way old Casey Stengel ran running his home run home to a Giant victory by a score of 5 to 4 in the first game of the World Series of 1923.
>
> This is the way old Casey Stengel ran running his home run home when two were out in the ninth inning and the score was tied, and the ball still bounding inside the Yankee yard.
>
> This is the way –
>
> His mouth wide open.
>
> His warped old legs bending beneath him at every stride.
>
> His arms flying back and forth like those of a man swimming with a crawl stroke.
>
> His flanks heaving, his breath whistling, his head far back. Yankee infielders, passed by old Casey as he was running his home run home, say Casey was muttering to himself, adjuring himself to greater speed as a jockey mutters to his horse in a race, saying "go on, Casey, go on."
>
> The warped old legs, twisted and bent by many a year of baseball campaigning, just barely held out under Casey until he reached the plate, running his home run home.
>
> Then they collapsed.

Clark, in *The World of Damon Runyon,* observed that:

> Damon's account of the game, which runs to several thousand words, is representative in its thoroughness and lively style of the seventy or eighty such pieces he had composed during his dozen press box autumns, but the short "prose poetry" passage on Casey's homer is a singular gem that several anthologists have chosen to typify the Runyon mode of sportswriting.

Runyon's other defining baseball writing is found in the short story "Baseball Hattie." Published in *Cosmopolitan* in May 1936, "Baseball Hattie" tells the tale of a gifted but not too bright left-handed pitcher named Haystack Duggeler, a game-throwing player in cahoots with a shady gambler. A hybrid of Bugs Raymond, Joe Jackson, "Shufflin'" Phil Douglas and other real-life players, Haystack is conned by gambler Armand Fibleman into fixing a game. Foiling the scheme is Baseball Hattie, a baseball "groupie." Hattie, married to Duggeler and pregnant with his child, overhears the talk of the conspiracy between the southpaw and the gambler. Hattie, ever loyal to the game of baseball and the New York Giants, takes matters into her own hands by shooting her beloved Haystack. She slams a .38 bullet into Haystack's left arm, necessitating a medical amputation at the shoulder. Runyon muses that Haystack "will never pitch a baseball again, unless he learns to pitch right-handed." Fibleman, in an effort to avoid a similar fate, exits the premises without bothering to open the door. "Baseball Hattie" is perhaps the best example of the typical Runyon story. As Weiner notes in *The Damon Runyon Story:* "The stories usually have a happy ending, but always with cynical qualifications." Runyon's story of Baseball Hattie continues to be a baseball tale that is often anthologized.

Of Runyon's football journalism, one outstanding piece was written on 31 October 1925 when Runyon covered a University of Pennsylvania football game. The opponent was the University of Illinois, whose team featured the "Galloping Ghost," Red Grange. Sixty-five thousand spectators turned out to see the legendary Grange play, many doubting that all the things written about the Illinois team's famous number 77 could be accurate. They came away convinced, as did the cynical Runyon. Not a writer who often laid garlands at the feet of athletes, Runyon gushed about the heroics of the Illinois running back in his 31 October 1925 column: "What a football player; this man Red Grange! Say it again!" Runyon not only helped bring the exploits of the Galloping Ghost to fans, but he also foretold Grange's impact on the fledgling sport of professional football. In later years Grange would be widely credited for saving pro football with his speed, athleticism, and fan appeal.

Runyon not only covered boxing for the *New York American,* he also bet heavily on the sport. He wagered on underdogs with glass jaws and contenders, especially heavyweights, who had little chance to claim a crown. Runyon almost compulsively bought "pieces" of fighters as investments. It was his dream one day to own the heavyweight champion of the world. But instead of investing in a Joe Louis or a Primo Carnera, Damon's fighters were named Wyoming Warner, Sailor Hoffman, and Napoleon Dorval. None of those Runyon chose was successful. Not satisfied with betting on fights, reporting fights, or owning a few fighters, Runyon endeavored to become a boxing promoter. He was slightly more successful in this venture, being largely responsible for the negotiations that led to the Joe Louis–Max Baer bout in September 1935.

Runyon's first love, however, was horse racing. He was a regular at the Kentucky Derby, the Preakness, Saratoga in the summer, and Hialeah in winter. Encouraged by his second wife, Patrice, Runyon acquired a collection of thoroughbreds, the most notable of which was Tight Shoes. Although

*Runyon, Walter Winchell, and Sherman Billingsley at the Stork Club in Manhattan*

Runyon's horse was one of the favorites for the 1940 Kentucky Derby, the horse became ill prior to the race and was never a serious competitor again. Once, during a conversation at Lindy's, Runyon's favorite New York watering hole, a friend asked Runyon whom he would choose to own, if given the opportunity: the great thoroughbred Seabiscuit or the mighty pugilist Joe Louis? Runyon opted for Seabiscuit, claiming that he would not have to split the purse with a horse. Yet Runyon eventually lost more money in his ventures into horse racing than he did as an owner of prizefighters.

Runyon could afford his losses. Between 1932 and 1940 he sold to *Collier's* magazine thirty-two short stories for $5,000 apiece. While he wagered and lost more and more money each year, Runyon's losses at the track inspired some of his best writing. In what is perhaps his best-known poem (arguably the best-known piece in all his writing), Runyon describes the thrill of the race as he celebrated jockey Earle Sande. The poem, called "That Handy Guy Named Sande," was written on the occasion of Gallant Fox's winning the 17 May 1930 Kentucky Derby. Runyon revised the poem each time he wrote a piece about Sande in the following eight seasons. In the poem Runyon describes his joy in watching one of his personal sports icons ". . . Bootin' a winner in!" He concluded this serial poem with the lamentation that never again would the sporting public see a jockey with the skills to match the great Sande. "That Handy Guy Named Sande" remains one of the definitive examples of American sports poetry. In *Damon Runyon* Breslin describes the phenomenon of this Runyon poem that "lasted a lifetime. Sande believed it all so much he regarded himself as invincible, even against whisky, which he was not."

Runyon was able to write about the costly side of playing the horses as well. In his poem and short story "All Horse Players Die Broke" (*Collier's,* 11 September 1937), Runyon set out for future gamblers the lessons he learned from being a railbird. The element that makes Runyon's horse racing poetry and prose so powerful and enduring is that he was an active participant in the racing life. Runyon wrote baseball, football, and boxing stories from the perspective of a reporter-fan. He never played the games himself. He owned prizefighters, but he himself was not a fighter. "All Horse Players Die Broke" was written by Runyon the gambler, not Runyon the observer. To illustrate this point, one has but to read the conclusion of the first stanza of the poem:

.... Well, son, I've bet and won
And I've bet and lost and when all is done
I'm sure of one thing and only one –
All horse players die broke!

Only someone familiar with the anguish of losing a large amount of money at the track could have written "Monologue of a Horse Player," inspired when Runyon lost heavily at the 1935 Kentucky Derby, pockets cleaned of everything "but his keys and handkerchief." In this piece Runyon writes from the same perspective as would a long-suffering fan of a perennially inept sports franchise. Runyon's lament offers the words of a man who cannot understand why he has been visited by such great misfortune. Runyon's recurring theme is "What did I ever do to anybody?" The poem voices Runyon's frustration in losing so heavily on the horses. But it also sheds light on the reason Runyon was so popular with his readers. *His* words were *their* words. He wrote in a familiar style that made his readers nod their heads and mutter, "Man, do I know how this guy feels."

In the 1930s, when he was a successful and famous short-story writer, Runyon continued to write occasional baseball pieces. The year 1934 was particularly fertile and provocative for him. He had always disliked the way baseball had changed with the introduction of the lively ball, noting that "Any bush league outfielder could stick in his hip pocket most of the Polo Grounds hits that are counted as home runs" (*New York American,* 19 April 1934). The All-Star Game, first played in 1934 for a worthy cause ("relief of sick and distressed veterans of baseball"), also failed to interest him, because it "mean[t] nothing to competitive sport." He hoped only that the game might spark apathetic ballplayers to start thinking about their futures, since owners "never have done much to help superannuated players" (*New York American,* 10 July 1934).

The 1934 World Series, on the other hand, between the Saint Louis Cardinals and the Detroit Tigers, excited him (and stirred his alliterative talents) as no other series had in a long while. Finally, he wrote, this was baseball "as it ought to be . . . after a long stretch of pallid, perfunctory presentations of baseball's biggest event" (*New York American,* 3 October 1934). The series, which lasted seven games, was one of the finest in baseball history. Runyon's greatest thrill was watching Detroit's Schoolboy Rowe's gutsy twelve-inning victory over the Cardinals' hard-luck "Wild Bill" Hallahan, who had actually pitched better than the "long lean lad from the little town of El Dorado, Arkansas, with fire and magic in his pitching arm." Runyon saw in the game a bittersweet moral, in that "the world does not ask you how you won – but did you win? It cannot pause to listen to the sad, sad tale of a loser" (*New York American,* 5 October 1934).

In November 1934 he published "Undertaker Song," one of his best short stories about sports, in *Collier's.* The story is ostensibly a tale about a Harvard-Yale football match ("much pulling and hauling this way and that, to and fro") the day after an uneventful boxing match (in Runyon's jargon, "a contest of skill and science"). The piece is filled with Runyon's usual gambling lore and is built around a "wrong gee" named Joey Perhaps, who has come to Cambridge, Massachusetts, to blackmail, or perhaps reclaim, "quite a chump doll" named Doria Logan. Logan is engaged to a Harvard football star who knows nothing of her regretful, youthful past. As it turns out, her fiancé is the son of the wealthy Mr. Phillips Randolph (the true object of the blackmail), who was rudely brushed aside on the train by Joey Perhaps the day before.

In typical Runyon fashion the story climaxes with a series of ironic twists, with Joey, although no one seems to care, sitting upright, dead in the stands. During the prevailing confusion when the Yale team makes its tallies, he is knifed by the brother of the man he had recently double-crossed and sent to the electric chair, whom one of the two narrators in this story has sent for after Doria sought his help ("figuring he may love to see a nice football game"). The blood across Joey's throat, Mr. Randolph assumes, is "a big, broad crimson ribbon where he once wore his white silk muffler." The title refers to the howling "Undertaker Song" the Yales "sing when they have the other guy licked" – "a very lovely song, to be sure, and quite appropriate all the way around." Doria is thus saved, and the two hoodlums who save her also win back more than the amount of the money they lost on the boxing match by betting heavily against a couple of Harvards.

Runyon was diagnosed with throat cancer in November 1944 at New York Memorial Hospital. His larynx was removed, and as a result he could communicate only by writing notes. Patrice divorced him in June 1946, and on 10 December 1946 Runyon died at the age of sixty-six. At his bedside were his son, Damon Jr., and Runyon's agent, Paul Small. Eight days later, in compliance with his will, Runyon's ashes were scattered over Broadway from an airplane. On board the plane were Mike Todd, a producer, and Damon Runyon Jr. Flying the plane was the famed Eddie Rickenbacker.

Runyon's death provides one other note of interest. Runyon is thought to be the first person in the history of the United States whose official cause of death was listed as "cancer." Prior to that time, people with cancer were listed as having died from "complications after a long illness." In 1954 Damon Runyon Jr. wrote in his book *Father's Footsteps:*

> As far as I know, my father was the first person of note whose death was attributed publicly and bluntly to cancer. Even as a newspaperman I was astonished at the space given throughout the nation to his passing – many papers ran several columns of type on him – so the cause of his demise was widely circulated. Now it is common usage to mention cancer in obituaries.

Shortly after Runyon's death Walter Winchell spearheaded a drive to establish the Damon Runyon Memorial Fund for Cancer Research. Eighteen months after Runyon's death, contributions to the fund topped the $2 million mark. Twenty-five years later the Runyon Fund had exceeded $30 million.

There is another interesting postscript to Runyon's death. A gambler to the end, Runyon bet that he would return after his death. A Broadway friend laid odds that Runyon was wrong. In one of his last columns, titled "Shall I Return?," Runyon tells the story of that wager. In true gambler's style Runyon wagered one dollar against his friend's ten dollars that he (Runyon) would be resurrected from death and return to New York. Runyon underscores the theme of this column with the words: "Most persons will no doubt say that I have a bad bet. I think I have a good one. In any event, could I resist the price?"

If there is a definitive Runyon piece it would be the story relating a gambler's deathbed advice to his son. Runyon claimed that his own father told this story to him when he graduated from the third grade. It was Runyon's favorite story.

> "Son," the old man said, "as you go around and about this world, some day you will come upon a man who will lay down in front of you a new deck of cards with the seal unbroken and offer to bet he can make the jack of spades jump out of the deck and squirt cider in your ear."
>
> "Son," the old man said, "do not bet that man, because just as sure as you do you're going to get an earful of cider."

It may be argued that Runyon is better known for his stories about the life and characters of Broadway than for his acumen as a sportswriter. To be sure, Runyon made the avenues of the big city come alive with his vivid prose and colorful jargon. Yet to this day he is recognized as one of America's greatest chroniclers of sporting life. Runyon shunned the box scores, round-by-round accounts, and the win-place-shows in favor of more colorful descriptions of the people involved in the events. He noticed the small details, the behind-the-scenes minutiae. And readers could not wait to read his column. For many sports fans Runyon was the "man on the scene." The scores and mundane details of a sports event could be read in other places. Runyon gave his readers the opportunity to see sports from an angle the common fan might never be able to see. He was a "color commentator" decades before the term was even invented.

In *The World of Damon Runyon* Tom Clark writes: "To a later generation of sportswriters, Ring (Lardner) and Damon loomed as twin masters of the sharp, cunning, tongue-in-cheek style." "Damon was a cynical humorist," Jimmy Cannon told Jerome Holtzmann in *No Cheering in the Press Box* (1992). "He was like Ring Lardner. I don't think [Runyon] thought very much of the human race. The same was frequently said of Ring."

In his 12 December 1946 obituary in the *New York Herald Tribune* Red Smith wrote of Runyon:

> Above all, he was a reporter. Perhaps the world at large will remember him longest for his short stories, but the men in his own business will remember the job he could do on a news assignment, whether it was a heavyweight championship or a political convention.

A prolific versifier, Runyon is not taken seriously as an American poet, but his poetry is still widely read. Anthologist Clark Kinnard writes in the "Editor's Note" for Runyon's *Poems for Men* (1947):

> [Runyon's poems] were recited in Army forts, in wayfarers' "jungles," on seamen's "beaches." They appeared in prison papers anonymously. Two were quoted regularly in columns of rival sports writers, for they had passed into the national folk-lore like Thayer's "Casey at the Bat." Lines from others became vernacular phrases in everyday speech. No poet since Kipling was taken to heart by so many ordinary men.

Kinnard's comparison of Runyon with Kipling is perhaps overstated; but the fact remains that the poetry of Runyon is rugged, durable, and still appealing to, as Kinnard states, "the common man."

Runyon found his short stories and tales of Broadway to be more profitable than either his sports articles or his poetry. Indeed, several of his stories were snapped up by Hollywood. The public, it might be argued, was drawn to the work of Runyon because he spoke its language. He created vivid and unforgettable characters.

Runyon continued to write about sports and remained a member of the Baseball Writers Association of America throughout his career. He is enshrined in the baseball writers' wing of the Baseball Hall of Fame in Cooperstown, New York. Runyon always spurned friends' suggestions that he write a novel. He believed that his talents lay in the realm of newsprint, and that was the medium to which he committed most of his creative energy. He was first and foremost a sportswriter. Stanley Frank, writing in *Sports Extra* (1944), made this observation of the influence of Runyon on sportswriting: "Novices studied his simple, straightforward style with profit, but those who attempted to copy his subtle tricks fell flat on their foolish faces."

Still, Runyon critiqued his work more accurately than anyone else. In a self-analysis he once said that he used humor and jocularity as the primary vehicle to convey his thoughts. He considered himself a rebel among the press herds and toward the end of his life expressed regret that he compromised that rebel spirit for the sake of staying employed and well paid. He wrote in his self-critique:

> He is a disguised defeatist in that he is always saying with a sort of smirk as if he is half kidding that the world will not change as long as we have human greed and man's inhumanity to man, that dames will continue to doublecross their husbands and sweethearts and vice versa, that politicians will continue to rob the public and that the public will remain as dumb as ever.

Runyon often expressed his humor in "The Brighter Side," his column written for King Features. In 1936 he became the mainstay of the powerful and influential Hearst organization, replacing Arthur Brisbane. Brisbane, one of the most famous American journalists, regarded Runyon as the best reporter in the world.

As Fred Lieb wrote in *The Sporting News* (18 December 1946), reflecting on Runyon's death: "[Runyon] found people in all walks of life perpetually good copy, whether they were a Giants pitcher, a Broadway tout, or the little grocer of the Colorado town in which he grew up."

**Biographies:**

Ed Weiner, *The Damon Runyon Story* (New York, London & Toronto: Longmans, Green, 1948);

Tom Clark, *The World of Damon Runyon* (New York: Harper & Row, 1978);

Jimmy Breslin, *Damon Runyon* (New York: Ticknor & Fields, 1991).

**References:**

Stanley Frank, *Sports Extra* (New York: Barnes, 1944);

Edwin P. Hoyt, *A Gentleman of Broadway* (Boston & Toronto: Little, Brown, 1964);

Damon Runyon Jr., *Father's Footsteps* (New York: Random House, 1954);

Red Smith, *To Absent Friends* (New York: Atheneum, 1982);

Jean Wagner, *Runyonese: The Mind and Craft of Damon Runyon* (New York, Paris & London: Stechert-Hafner, 1965).

# Chet Smith

*(1899 – 28 January 1973)*

Timothy E. Walsh
*Butler County Community College*

MAJOR POSITIONS HELD: Writer, *Pittsburgh Dispatch* (1920–1923); sportswriter, *Pittsburgh Gazette-Times* (1923–1927); sportswriter, *Cleveland Press* (1927–1931); sports editor, *Pittsburgh Press* (1931–1966).

BOOK: *Pittsburgh and Western Pennsylvania Sports Hall of Fame: The Story about the Men and Women Who Made Us a Great Sports Area,* edited by Marty Wolfson (Pittsburgh: Wolfson, 1969).

OTHER: "Uniforms Shouldn't Be – Uniform," in *Baseball: A Treasury of Art and Literature,* edited by Michael Ruscoe (New York: Levin, 1993), pp. 298–299.

As sports editor of the *Pittsburgh Press* from 1931 to 1966, Chet Smith hammered out a living on the typewriter that earned him a niche in the annals of sports journalism. Few sportswriters were more versatile, as his talents ran from heading a top sports department to writing his nationally acclaimed column, "The Village Smithy." He wrote with a wit that many readers relished, his peers admired, and his few adversaries envied. He was a wry humorist, razor-sharp toastmaster, and originator of everything from the syndicated sports cartoon "Time Out," which appeared in more than one hundred papers, to an assortment of organizations that brought new meaning to the social milieu of Pittsburgh.

Smith remains one of the greatest and best-known sportswriters. Stanley Woodward identified Smith as one of a handful of "excellent sports columnists" who possess "gifts above and beyond those of the average newspaper writer." He was one of only a few to have been elected president of both the Baseball Writers Association of America (1950) and the National Football Writers Association (1956). He was also a member of the National Football Foundation and Hall of Fame, and he was the first sportswriter to receive from that foundation a citation for distinguished service to amateur football. On more than one occasion Smith turned down offers from New York newspapers to write a syndicated column, for his life and interests were too deeply rooted in Pittsburgh.

Smith was born in Springdale, Pennsylvania, just outside Pittsburgh, in 1899, but Pat Livingston, Smith's longtime friend and colleague at the *Pittsburgh Press,* recalled how Smith would say that he was "brung up" in Salem, Ohio. Smith's father, Dr. Charles L. Smith, was a Methodist minister as well as a newspaper editor. After attending Dartmouth for two years, Chet Smith enlisted in the navy, where he served as a first-class quartermaster in World War I. Following the war he broke into journalism with the *Pittsburgh Dispatch* in 1920. "My dad got tired of seeing me loaf," said Smith, and the young man first took a job at a local lumberyard, even though his father had lined up a position for him at the *Dispatch.* After only one day at the lumberyard Smith was convinced that he had made the wrong choice. He contacted Dave Davies, a city editor at the *Dispatch,* took a reporter's job for eighteen dollars per week, and never even collected his day's pay from the lumberyard.

Smith's first assignment at the *Dispatch* was a column that allowed him to write about the rich history of Pittsburgh's rivers. He was interested in history and enjoyed this opportunity, even though Livingston said that Smith "just wanted to write anything for a paper." After his next assignment of reporting on fires and murders, he was drafted into duty in a depleted sports department. Yet he was more than glad to fill in, and he soon blossomed into a great sportswriter.

His first sportswriting assignment was to cover college sports and golf. In 1923 he joined the *Gazette-Times* and continued to cover his beat but was allowed to focus more on college football, which was rapidly becoming his passion. Smith married during these years with the *Gazette-Times,* and while he was covering college football, Glenn "Pop" Warner, coach at the University of Pittsburgh, gave the young sportswriter his first big lift by inviting him to his apartment, where Smith met many of the football greats who gathered around Warner's coal-burning stove.

In 1927 Smith took a position with the *Cleveland Press,* where he continued to cover college football. Soon Smith met coach Knute Rockne, who took the young sportswriter under his wing. As Roy McHugh reported on 27 February 1966, following Smith's first visit to Notre Dame University, Rockne drove him to catch the train back to Cleveland and reportedly said, "anytime you come to South Bend just throw your bags in the vestibule and stay with us – the door's always unlocked." Smith and Rockne were to spend many fall weekends together after that.

Late in 1931 Smith was offered a job as sports editor of the *Pittsburgh Press,* where he was to replace W. W. Forster. Being back in his hometown, Smith flourished in this position. He began to develop a friendship with Jock Sutherland, who was to be the football coach of the University of Pittsburgh. Their relationship had reportedly begun at a speakeasy on Fifth Avenue in Pittsburgh during Prohibition. Smith and Sutherland, whom Smith aptly nicknamed "the Great Stone Face," became close friends, and many consider this friendship with Sutherland to have provided Smith's education in football. They spent many nights at the "Pull-Em-Up" speakeasy across the street from the *Dispatch,* where, an anonymous source reported to McHugh, they "plunked down two bucks in a box and pulled up [their] own beer from the cellar with a rope." There they spent evenings discussing football, and Smith often received exhaustive briefings on weekly game plans.

This knowledge enabled Smith to write analytically as well as descriptively in the Sunday paper following the weekly game. In fact, McHugh wrote in 1973 that, with the schooling Smith had received from Sutherland, Smith felt that his own interpretation of the game was "more edifying to the reader than anything the players or coaches might have to say" – although Smith rarely second-guessed them in print. In fact, added McHugh, "An intended compliment [Smith] heard more than once was the statement that in his 35 years as sports editor of [the *Pittsburgh Press*] he 'never said a bad thing about an athlete or a coach if he could say a good thing.' The key phrase was 'if he could say a good thing.' "

McHugh and Livingston were good friends as well as writers under Smith during his tenure as sports editor. McHugh fondly remembered in 1973 that "Chet Smith knew and wrote about the people in sports who fascinated a public less jaded and cynical than today's. He wrote to entertain, not pillory, condemn or set the world straight, and most of the time he succeeded."

Smith may not have tried to set the world straight, but Smith's stories written during World War II are among the columns that became classics. In one that appeared on 6 December 1942 Smith lamented:

> There is a memo on the desk from the Treasury Department suggesting it would be nice if we neglected the touchdown and plug industries for a day to put in a good word for the bond drive.
>
> Why not? As we get it from our banker friends, a war can't be run off the cuff. And we're in this one up to our ears. Furthermore, it should be fairly obvious that we have to win it. There are, you know, no ties or playoffs in war. Either you win or you lose, and who wants Tokyo in the National League? Heaven knows, the Phils are bad enough.

Smith often concealed whatever tensions he may have felt by relying on his casual, flippant manner. In another such column Smith embarked whimsically on a "long and fruitless campaign to dress ball players like individuals and with some degree of originality." He could remember the days "when the old New York Highlanders and also the [Chicago] White Sox were natty in blue with white piping." During the middle of the twentieth century, however, home teams always wore white and the visiting teams wore gray, and baseball had grown steeped in uniformity before the days of the variegated and "exclusive suiting[s]" of the Astros, Athletics, and White Sox that Smith hoped for were to reappear. "A baseball team," Smith wrote in 1948, "expresses as much personality in its uniforms as a crate of oranges."

His sense of humor was also evident when, in the 1930s, Smith and Sutherland joined Bill McClintock – a Harrisburg, Pennsylvania, paper-box manufacturer and "godfather" of University of Pittsburgh teams – to form a mythical "College of the North," a school composed of leading eastern sportswriters and close friends. Their college had its own stationery, with Sutherland listed as "head Coach and dean of Women," and a motto: "Not One Cent for the Faculty – Everything for Football."

On Mondays, Smith – through the eyes of Joe Punt, a fictional character who always agreed to talk to him in exchange for a cup of coffee and two tickets to the next week's games – would take retrospective, humorous looks at football games of the past weekend. Speaking through Punt, Smith offered mordacious assessments to his readers. For example, in talking to Smith one late autumn about the end of the football season and the approaching holidays, Punt declared:

> It's a funny way we Americans have. I like football as well as the next fellow, but I get awfully tired of what comes between the last of the big games and bowl battles. We seem to think we have to eat our way back to normalcy, and also listen while everybody tells us who belongs on the All-America teams and what the upsets have been, and the good and bad plays that were made. I have been invited to three banquets this week. The same speakers will no doubt be at them all, telling the same stale stories they told at the same dinners last fall. By Christmas I will be able to balance a dozen or more peas on my knife and carry on a conversation to my right and left without dropping a single pea.

Smith's readers both revered and detested Joe Punt. Some sports fans enjoyed Joe's on-the-street insights about the college game as well as his sarcastic comments on professional football and social topics, but the more intense sports fans and those dedicated to professional football demanded no-nonsense analyses of the Pittsburgh Steelers in Smith's Monday columns. The mixed mail that Smith received prompted him to ask his readers to vote, by postcard, on whether or not he should continue reporting his clandestine meetings with Punt. Having counted the ballots himself, Smith announced in "The Village Smithy" that Punt would stay.

This mixed reaction that Punt had aroused among readers was occasioned by Smith's attitude toward professional football. He had nothing against it except that it conflicted with college football. Nevertheless, Smith eventually grew to love the Pittsburgh Steelers, mostly because his good friend Sutherland later became coach of that pro team.

What Smith truly loved were the fall weekends, the color and excitement of college football, the Friday nights with coaches, the hours in the press box overlooking the playing fields on Saturdays and, eventually, Sundays. His love for the game was the basis for the Curbstone Coaches, a loose organization that he formed in the 1930s and presided over well after his retirement from the newspaper business in 1966. This self-proclaimed group of football nuts met every Monday for a downtown luncheon during the football season. Smith, an excellent public speaker who was fast with ad-libs, always served as the master of ceremonies, and, as McHugh recalled in 1973, reporters covering the luncheons quoted him "almost as extensively as the [football] notables he was there to introduce."

Although Smith loved football, he felt otherwise about basketball and made no effort to hide that fact. Livingston recalled how Smith often stated that if the championship game of the National Invitational Tournament were being played in Smith's front yard, he would "pull the window shade down!" Livingston speculated that Smith's attitude toward basketball stemmed not from his dislike of the game but from the fact that it was played at night – and indoors. Livingston said, "Chet felt that real sports were played outdoors during the day, under the sun. Not indoors and at night like basketball." Yet Smith's columns dutifully and regularly commented – usually through another fictitious character, Basketball Operative 00 – on the game that Smith claimed he could not watch.

Smith covered baseball with more enthusiasm than he had for basketball. He was a fan of baseball, but he believed that, as a real sport, it was a game to be played outdoors on drowsy summer afternoons. Although he lived for fall football, Smith looked forward to each spring training and the grapefruit leagues, where major league baseball players worked off the extra pounds they had accumulated during the winter. He enjoyed the relaxed atmosphere that characterized baseball in the 1930s, postwar 1940s, and 1950s, but as air travel, arc lights, and corporate technology and marketing methods crept into the game, McHugh noted in 1973 that Smith's feeling for baseball diminished from what it had been when baseball was played in those earlier years.

In 1969 Smith published *Pittsburgh and Western Pennsylvania Sports Hall of Fame: The Story about the Men and Women Who Made Us a Great Sports Area,* edited by Marty Wolfson. Covering the century that began when a baseball team called the Pittsburgh Olympics boated down the river to Cincinnati and lost to the 1869 Red Stockings, 54–2, this is a history of sports in western Pennsylvania. Smith recounts his stories in a light way, with chapters on players and sports figures such as Honus Wagner and Pie Traynor, Sutherland, Roberto Clemente,

Arnold Palmer, and Johnny Unitas, and he devotes whole sections to particular periods of sports history and specific sports, including basketball. The book is filled with sketches by Wolfson and with many photographs of sports heroes. Smith skillfully avoids incorporating stories that are too familiar, and he does so without omitting folklore worthy of preservation.

Smith wrote rapidly and with exceptional grace. McHugh recalled in 1973 that the "act of composition never gave [Smith] much trouble." Livingston added that Smith invariably finished and filed his postgame stories while other reporters were still working on theirs: in forty-five minutes or less Smith would have his daily story finished.

When Smith retired from the *Pittsburgh Press* in 1966, he remained active. Within three years of his retirement he had written for a chain of suburban weeklies, taught sports journalism at Point Park College in Pittsburgh, become the author of a weekly football newsletter for the University of Pittsburgh, and continued serving as master of ceremonies for the Curbstone Coaches' luncheons. Relishing every fall weekend until his death, Smith stayed as active as possible throughout his later years.

**References:**

Roy McHugh, "Coaches to Honor Chester L. Smith, March 18," *Pittsburgh Press,* 27 February 1966, section 4, p. 2;

McHugh, "In a Less Uptight Sports Era, Chet Smith Took Time to Savor It," *Pittsburgh Press,* 30 January 1973, p. 25;

Stanley Woodward, *Sports Page* (New York: Simon & Schuster, 1949).

# Red Smith

*(25 September 1905 – 15 January 1982)*

Lawrence Baldassaro
*University of Wisconsin – Milwaukee*

See also the Smith entry in *DLB 29: American Newspaper Journalists, 1926–1950.*

MAJOR POSITIONS HELD: Reporter, *Milwaukee Sentinel* (1927–1928); copy editor, sportswriter, *St. Louis Star* (1928–1933); rewrite man, *St. Louis Star-Times* (1933–1936); sports reporter and columnist, *Philadelphia Record* (1936–1945); sports columnist, *New York Herald Tribune,* merged to form *World Journal Tribune* (1945–1967), Publishers-Hall Syndicate (1967–1971), *New York Times* (1971–1982).

BOOKS: *Terry and Bunky Play Football,* by Smith and Dick Fishel (New York: Putnam, 1945);
*Out of the Red* (New York: Knopf, 1950);
*Views of Sport* (New York: Knopf, 1954);
*Red Smith's Sports Annual* (New York: Crown, 1961);
*The Best of Red Smith,* edited by Verna Reamer (New York: Watts, 1963);
*Red Smith on Fishing around the World* (Garden City, N.Y.: Doubleday, 1963);
*Strawberries in the Wintertime: The Sporting World of Red Smith* (New York: Quadrangle, 1974);
*Press Box: Red Smith's Favorite Sports Stories* (New York: Norton, 1976);
*To Absent Friends from Red Smith* (New York: Atheneum, 1982);
*The Red Smith Reader,* edited by Dave Anderson (New York: Random House, 1982).

OTHER: *Sports Stories,* edited, with introductions, by Smith (New York: Barnes, 1949);
*This Was Racing,* edited by Smith and Joseph Hill Palmer (New York: Barnes, 1953).

Red Smith

SELECTED PERIODICAL PUBLICATIONS – UNCOLLECTED: "Doghouse to Let: Apply Newhouser and Trout," *Saturday Evening Post,* 217 (31 March 1945): 22–23;
"Little Man with Big Horses," *Saturday Evening Post,* 218 (25 May 1946): 20;
"What Broke Up the Yankees?," *Saturday Evening Post,* 219 (29 March 1947): 23;
"Has Baseball Forgotten the Fan?," *Saturday Evening Post,* 220 (4 October 1947): 25;
"Reluctant Yankee," *Collier's,* 126 (8 July 1950): 30–31;
"Anyone for Basketball?," *Collier's,* 127 (3 March 1951): 24–25;
"Paradise for Horseplayers," *Saturday Evening Post,* 229 (29 June 1957): 26–27;
"Good Clean Fun in Finland," *Reader's Digest,* 92 (June 1968): 133–135;
"Tequila," *Holiday,* 44 (July 1968): 77–78;
"Red Smith on Politics," *Harper's Bazaar,* 105 (January 1972): 94;

"Super Sunday," *New York Times Magazine,* 12 January 1975, p. 79;

"The House that Ruth Built-Rebuilt," *New York Times Magazine,* 11 April 1976, pp. 60–61;

"Fan as Designated Sitter," *Saturday Review,* 3 (26 June 1976): 26–27;

"Olympic Games," *Horizon,* 18 (Summer 1976): 109–112;

"Four!," *American Heritage,* 31 (August/September 1980): 76–80.

In a column written after the death of Frank Graham in 1965 and reprinted in *To Absent Friends from Red Smith* (1982), Red Smith eulogized his friend and fellow sportswriter: "His incredibly accurate ear and implausibly retentive memory (he almost never took notes), his faultless taste . . . and the purity of his prose combined to produce – it says here – the finest sports column of all time." What he wrote in generous tribute to Graham is, in fact, an accurate assessment of Smith's own talent. Gifted with a keen ear and a remarkable memory, Smith was able to quote verbatim from interviews without using notes; more important, he was considered not only the best sportswriter but one of the finest prose stylists of his time. Seven years before he wrote his eulogy of Graham, Smith's picture had appeared on the cover of *Newsweek* magazine, which identified him as "the world's most widely read sportswriter." By the time of his death in 1982 his column was syndicated in 275 newspapers in the United States and in another 225 abroad.

Smith was witness to, and chronicler of, much of the sports history of America in the twentieth century. He grew up in Green Bay, Wisconsin, when the Packers were formed; he attended Notre Dame during the years of Knute Rockne's Four Horsemen; and, as a sportswriter, he covered most of the major events from 1929 to 1982. He also wrote about the Tour de France, the Westminster Dog Show, rodeos, and even cockfighting. He was at the 1932 World Series when Babe Ruth supposedly called his home-run shot – Smith was not sure whether he did – and he was at the 1977 World Series when Reggie Jackson hit three home runs in one game. Smith covered sports from the era of sixteen-team major leagues – he was the beat writer for the Saint Louis Browns and Cardinals when they were the westernmost teams in baseball – to the age of expansion, divisional play, and the designated hitter. His subjects ranged from Connie Mack and George Halas to Don King and George Steinbrenner.

Walter Wellesley Smith was born in Green Bay on 25 September 1905. His parents had met when his mother, the former Ida Richardson, a native of New York, was on vacation in Green Bay. His father, Walter P. Smith, ran the wholesale produce and retail grocery business that had been started by Smith's great-grandfather. His mother claimed to be a descendant of Arthur Wellesley, the Duke of Wellington, who defeated Napoleon at Waterloo, thus accounting for Smith's middle name and the first name of his older brother, Arthur. (His only other sibling – a younger sister, Catherine – would die of tuberculosis when Smith was in college.) Smith was uncomfortable with the ring of formality of his names and preferred the nickname "Brick," given him by his father (The more enduring nickname of "Red" came after his college years). As a boy he played all sports, though he was far from gifted athletically, and spent much of his time reading.

Smith's decision to attend Notre Dame and study journalism stemmed from a desire to emulate his hero, Vince Engels, an older hometown friend. It was Engels who introduced Smith to fishing, the one sport in which he would participate throughout his life. He graduated from high school in 1922 and was accepted by Notre Dame; but his father's business was not doing well, and Smith had to delay the beginning of his college career for a year while he worked as a clerk in a Green Bay hardware store.

When Smith entered Notre Dame in the fall of 1923 it had fifteen hundred students, all male; a journalism program that consisted of one professor, John Michael Cooney; and a famous football coach, Rockne. Hired in 1918, Rockne had turned the small midwestern school into a football powerhouse. The team had a record of 35–4–1 during the four years Smith was there, winning the Rose Bowl and the national championship in 1924 with the legendary Four Horsemen backfield. Rockne also served as the track coach. Smith, anxious to escape calisthenics class, signed up for the track team in his freshman year but quit after the first race when, lagging far behind the other runners and two laps short of finishing the mile, he was told by Rockne to get off the track so the next race could begin.

Smith graduated cum laude in 1927 and was hired by the *Milwaukee Sentinel,* a Hearst newspaper, as a general-assignment reporter for twenty-four dollars a week. In 1928, lured by an offer of forty dollars a week, he moved to the *St. Louis Star* as a copyreader. It was there that Smith first became a sportswriter, but not by choice. The managing editor had fired half of his six-man sports staff for ac-

*Smith on vacation in Sturgeon Bay, Wisconsin, in 1947 with his first wife, Catherine (Kay) Cody Smith, and their children, Catherine (Kit) and Terence (Terry) (photograph courtesy of Phyllis Smith)*

cepting payoffs from a wrestling promoter, and Smith was the most dispensable man on the copy desk. After covering basketball, boxing, and track through the winter, Smith got his first major assignment in the spring of 1929: covering the Browns. It was just before the beginning of spring training that he got his first byline, as "W. W. Smith."

In 1930 Smith was assigned to cover the Cardinals – a promotion, since the team was the more successful and popular of the two Saint Louis franchises. The following year he turned down an offer of a job at *The Sporting News.* He covered the Cardinals for three years, and although he enjoyed his baseball work, he was bored by the winter sports he had to cover. In 1933 Smith left sportswriting to become a rewrite man for the paper, which had recently merged to become the *Star-Times.* On 11 February of that year he married Catherine (Kay) Cody, and in November they became the parents of their first child, Catherine (Kit) Wellesley; a son, Terence (Terry), would be born in 1938.

In 1936 Smith moved to the *Philadelphia Record* and back to the sports desk, covering baseball, football, and various winter sports. When Smith handed in his first story without a byline, a slot man inserted the byline "Walt Smith." Smith, who was not fond of his first name, asked that it not be used but gave no other preference. The slot man then came up with "Red," and Smith's byline never changed again. In September 1939 he got his first full-time column, which appeared seven days a week. He established a solid reputation while writing for the *Record,* and his piece on the Army–Notre Dame game was included in the collection *Best Sports Stories 1944.* In 1945 Stanley Woodward, sports editor of the *New York Herald Tribune,* asked Smith to join his staff. Even though he was being hired as a sportswriter and not as a columnist, Smith readily accepted the offer since it had been his goal to work for a New York newspaper ever since he had gone to Saint Louis.

Within ten weeks of joining the *Herald Tribune* Smith began writing a column. In 1946 he won the National Headliners Club Award for excellence in newspaper writing and the following year the award for the best feature story in the *Best Sports Stories* anthology for a story about the Harvard-Yale football game titled "Holy Sight." In 1950 a collection of Smith's *Herald Tribune* columns was published as *Out of the Red.* A second collection, *Views of Sport,* appeared in 1954. When Grantland Rice died on 13 July 1954, Smith, whose work was then carried by some seventy papers, succeeded his friend as the most widely syndicated sports columnist in the country.

The *Herald Tribune* became the *World Journal Tribune* in 1966, but the paper collapsed in May 1967, only eight months after its creation. The Publishers-Hall syndicate bought Smith's contract from the *World Journal Tribune,* but for the first time in twenty-two years his column was not appearing in a New York daily. His professional misfortune was compounded by personal tragedy. On 15 February 1967, four days after their thirty-fourth anniversary, Kay Smith died.

In 1968 Smith returned to New York journalism when the publisher of *Women's Wear Daily* subscribed to his syndicated column. On 8 November of that year he married Phyllis Warner Weiss, an artist and widowed mother of five children. In 1971 Smith was hired by *The New York Times,* the paper for which he would write until his death. Smith says in *No Cheering in the Press Box* (1974), a collection of interviews with sportswriters edited by Jerome Holtzman,

> I wasn't one of those who dreamed of being a sportswriter and going around the country traveling with ball players and getting into the games free.... I'm not pretending I haven't enjoyed this hugely. I have. I've loved it. But I never had any soaring ambition to be a sportswriter, per se. I wanted to be a newspaperman, and came to realize I didn't really care which side of the paper I worked on.... The guy I admire most in the world is a good reporter. I respect a good reporter, and I'd like to be called that.... I like to report on the scene around me, on the little piece of the world as I see it, as it is in my time. And I like to do it in a way that gives the reader a little pleasure, a little entertainment.

Smith kept the games and athletes he wrote about in perspective by remembering, as he told Holtzman, that "these are just little games that little boys can play, and it really isn't important to the future of civilization whether the Athletics or the Browns win. If you can accept it as entertainment, and write it as entertainment, then I think that's what spectator sports are meant to be." But while he could refer to the sports department as the "fun and games department" of a newspaper, and to himself as "the tenant in this literary flophouse," he also acknowledged the cultural significance of sports and sports journalism: "I think sports constitute a valid part of our culture, our civilization, and keeping the public informed and, if possible, a little entertained about sports is not an entirely useless thing."

*Smith (second from left) at spring training in Saint Petersburg, Florida, in 1949 with sportswriter Grantland Rice, Saint Louis Cardinals manager Eddie Dyer, and New York Yankees manager Casey Stengel (photograph: H. G. Tucker/Houston Press)*

While he covered virtually the entire spectrum of sporting events in his long career, he enjoyed some more than others. A 1946 column, he confessed, was "written by one who would rather drink a Bronx cocktail than speak well of basketball." As for track, "if God had intended man to run He would have given him four legs." Among his favorite sports were baseball, boxing, and horse racing. To the question of whether covering baseball every day got dull, Smith replied, "It becomes dull only to dull minds." This comment to Holtzman, though specifically referring to baseball, summarizes the approach to sportswriting that sustained Smith through his long career: "If you have the perception and the interest to see it, and the wit to express it, your story is always different from yesterday's story."

Following the "no mas" fight in 1980, when Roberto Duran refused to answer the bell, in effect surrendering to Sugar Ray Leonard, Smith wrote: "The Sweet Science is a harsh mistress, and under her cruel rules the deadliest sin is to give up under punishment." Smith wrote about boxing in those terms: he recognized its cruelty and never softened its hard edges with sentimentality, but he also admired the sport as a test of human will and endurance. His descriptions of classic bouts such as Tony Zale–Rocky Graziano, Sugar Ray Robinson–Randy Turpin, and Muhammad Ali–Joe Frazier portray the brutal reality of the sport as seen from ringside, where the hype and glamour are replaced by raw force and violence. Of the 1947 Zale-Graziano rematch, he wrote: "This time it was Graziano who took the merciless beating. Zale punched him stupid, clubbing the body, cutting up the face. When an eye swelled shut, Rocky's trainer broke the swelling with a 25-cent piece." But, as with every sport he covered, it was the human dimension of boxing that most interested Smith. When Graziano fought Sugar Ray Robinson for the title, Smith described the scene as Zale, then retired, entered the ring to wish both fighters luck: Zale "strolled over to Rocky's corner with a smile that lit up his face. When two men have had the sort of fights Tony and Rocky had, there is a bond between them that nobody else in the world can share. The pain and triumph were theirs and theirs alone."

Horse racing evoked a reverence in Smith that may have been only partially tongue in cheek. In a 1957 column about racing at Saratoga, Smith wrote: "It has often seemed here that there is a happy affinity between horse playing and piety, and that it is an established fact of theology that men who live on the race track live long and do good deeds." Some of Smith's favorite characters were those he met at the track – owners, trainers, jockeys, and grooms. In his last column, noting that two questions he was invariably asked were who was the best athlete and which one did you like the best, Smith wrote: "Both questions are unanswerable, but on either count Bill Shoemaker, the jockey, would have to stand high."

The only sport Smith participated in as an adult was fishing, a subject he wrote about with affection and humor. In describing a 1953 fishing trip to Chile, he compared his prey to a couple of oversize heavyweight champions: "This typewriter is being beaten with fingers whose knuckles are bleeding and nails broken after a hand-to-fin struggle with trout exactly the size, shape, and disposition of Tony Galento. . . . Between backlashes, a distant relative of Primo Carnera sprang up for a look at the latest New York styles in outdoor apparel, made a face of hideous disapproval, and spat out the spoon." In a 1978 column about an outing in New Brunswick with Ted Williams, he described in the third person his feelings on catching a salmon on a day when even Williams, one of the world's top sport fishermen, did not: "His glee was unworthy, and secret."

Smith's stories are enlivened by anecdotes about both major and minor events and characters, and his eye for telling detail and his ear for dialogue give him the ability to make his readers feel as if they are eavesdropping behind the scenes. In his 1986 biography, *Red,* Ira Berkow quotes Smith's son, Terry, on his father's concern for a "sense of presence" in writing: "He'd say, 'The one thing you can do is try to get into every piece, whether it's politics or sports, the sense that you experienced whatever it is you're describing, even by some small detail of description that will bring it off and give it color. It puts you into the scene without artificially becoming a player.' "

The telling detail is evident in Smith's 8 January 1971 description of the scene at Sonny Liston's press conference in Chicago after he won the heavyweight title: "A visiting author came weaving into the room, took the floor, and held it, making an ass of himself. Working stiffs tried to shout him down, but the new champion lifted a restraining paw. 'Leave the bum speak,' he commanded, benevolence itself." The eye for detail is also evident in his 1955 recollection of another large pair of hands: "If you were a Johnny-come-recently who never saw Honus Wagner until his playing days were long past and he was a coach traveling with the Pirates, it was always those hands of his that you thought about when his name was mentioned. Those tremendous mitts and the twinkling good humor in his ruddy old face."

Smith came to know many of his subjects as more than sports figures witnessed from the sanctuary of the press box, and he wrote with affection about those he admired and respected as human beings. When James E. Fitzsimmons, the legendary horse trainer known as "Mr. Fitz," retired in 1963 at ninety, Smith asked: "What's he going to do with all that bustling energy, all that sparkling enthusiasm for living, all that affection for people?" In 1956 he called Rockne "a great man who happened to choose football as a career." Mack was described in a 10 February 1956 column as "tough and warm and wonderful, kind and stubborn and courtly and unreasonable and generous." On 16 November 1965 he wrote, "To say that Branch Rickey has the finest mind ever brought to the game of baseball is to damn with the faintest of praise, like describing Isaac Stern as a fiddler." Even a horse could be a personality worthy of his admiration. In an 8 April 1953 column he said Whirlaway "was Babe Ruth, Jack Dempsey, Bill Tilden, Bobby Jones – not just a

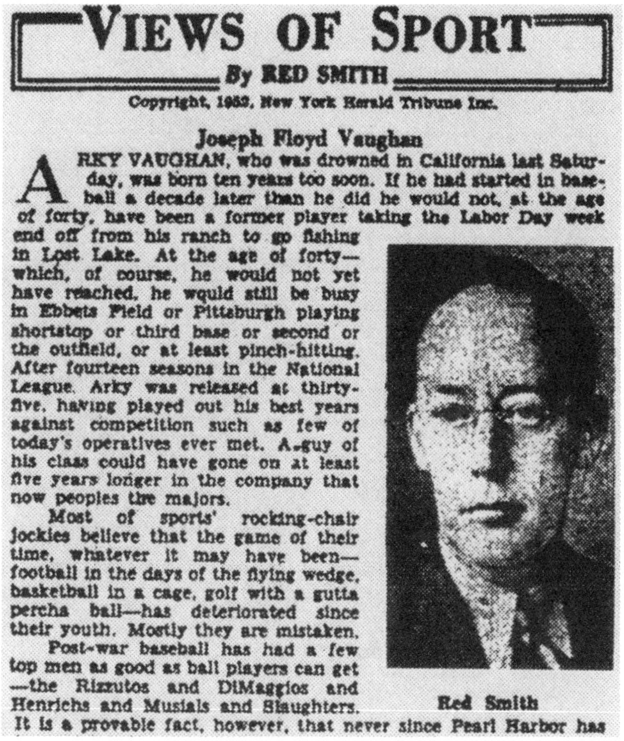

VIEWS OF SPORT

By RED SMITH

Copyright, 1952, New York Herald Tribune Inc.

Joseph Floyd Vaughan

ARKY VAUGHAN, who was drowned in California last Saturday, was born ten years too soon. If he had started in baseball a decade later than he did he would not, at the age of forty, have been a former player taking the Labor Day week end off from his ranch to go fishing in Lost Lake. At the age of forty—which, of course, he would not yet have reached, he would still be busy in Ebbets Field or Pittsburgh playing shortstop or third base or second or the outfield, or at least pinch-hitting. After fourteen seasons in the National League, Arky was released at thirty-five, having played out his best years against competition such as few of today's operatives ever met. A guy of his class could have gone on at least five years longer in the company that now peoples the majors.

Most of sports' rocking-chair jockies believe that the game of their time, whatever it may have been—football in the days of the flying wedge, basketball in a cage, golf with a gutta percha ball—has deteriorated since their youth. Mostly they are mistaken.

Post-war baseball has had a few top men as good as ball players can get—the Rizzutos and DiMaggios and Henrichs and Musials and Slaughters.

Red Smith

It is a provable fact, however, that never since Pearl Harbor has

*Opening of the 2 September 1952 Smith column*

champion but a champion who was the most colorful figure in his game."

Smith, who often cited the admonition of his editor and close friend Woodward against "godding up" the athletes, generally avoided hyperbole in assessing their skills or their character. He was, however, less cautious in his praise of longtime friends and associates: Rice was "the greatest man I have known, the greatest talent, the greatest gentleman"; Woodward was the "best and kindest man I ever knew."

His praise of those he admired was matched by his acerbic jibes at those he did not. Of Happy Chandler's appointment as baseball commissioner in 1945 Smith wrote that the owners "had bought a turkey" and that they hired Muddy Ruel, an attorney and former ballplayer, to show Chandler "where first base was and try to keep Happy's big foot out of Happy's big fat mouth." When Chandler accused Smith and other sportswriters of conspiring against him for money, Smith wrote: "If I can get paid for thinking Happy Chandler has performed like a clown and a mountebank, then I want all of that kind of money I can get. Ordinarily I have to work for mine."

"*De mortuis nil nisi bonum,*" Smith wrote more than once, but he could not hold to that pledge in the case of Avery Brundage, the longtime president of the International Olympic Committee. "One pictures him still haughtily erect at the end, shoulders squared, staring coldly down his nose at an apologetic angel of death," Smith wrote on 12 May 1975 after Brundage died, concluding: "Although Avery was frequently wrong-headed, he could also be arrogant and condescending." When the legendary Woody Hayes was fired for punching a Clemson player during the 1978 Gator Bowl, Smith referred to the incident as "only the last degrading incident in a pattern of behavior that had long distinguished the Ohio State coach. For years, Hayes had been throwing tantrums, screaming abuse and striking out at anyone within reach when his team was losing."

Notably absent in Smith's writing is any sense of self-importance. His inclination to self-effacement may account for his willingness to expose what he perceived to be pretentious or self-righteous behavior in others. But even when deflating pomposity and hypocrisy, he preferred subtle sarcasm over righteous indignation.

One of his favorite targets was what he saw as the hypocrisy of amateur sports. In 1960 he ridiculed the sham of "amateur" tennis promulgated by those who "truly believe it is nobler for an amateur to take money on the sly and lie about it than to accept pay openly and be branded a professional." In a 1962 column on a dispute between the Amateur Athletic Union and the National Collegiate Athletic Association, his disdain for both governing bodies is obvious. He characterizes both groups as "infantile in their approach to the world," adding: "If the wowsers of the N.C.A.A. think they can do a better job than the waxworks effigies of the A.A.U., then let them try."

His greatest indignation with regard to amateur sports was reserved for the International Olympic Committee, which he characterized on various occasions as "the groundhogs," "that clutch of overripe playground directors," and "the waxworks," and in particular for Brundage, whom he described as a "rich and righteous anachronism . . . a vestigial remnant of an economy that could compete in athletics for fun alone." In 1972 he sneered at the decision by the IOC – "the self-appointed, self-perpetuating kangaroo court" – to banish Austrian skier Karl Schranz for endorsing ski equipment: "The simple truth is that the whole notion of amateurism is archaic." In 1976, fed up with the hypocrisy of "amateurism" as well as the nationalistic fervor of the Olympics, Smith advocated the abolition of all team sports and of national anthems and colors: "Olympic athletes were pros 2,500 years ago and they are pros today. Knock off the hypocrisy, stop telling kids what an honor it is to represent their country and give them a chance to play games for the fun of it."

Smith, however, was not a crusading journalist; for most of his career he remained faithful to his mission to entertain his readers. He wanted to get the facts straight, but how he wrote was as important to him as what he wrote. He constantly worked to polish his style, to find the right word. It was not a painless struggle. Smith admitted that the preparation for writing – watching the event and talking to the athletes – was more enjoyable than the act of writing. He referred to his office as "The Torture Chamber," admitting to Holtzman: "I don't enjoy the actual labor of writing. I love my job, but I find one of the disadvantages is the several hours at the typewriter each day. That's how I pay for this nice job. And I pay pretty dearly. I sweat. I bleed." Yet, for all the pressure of meeting a deadline six or seven times a week, Smith was convinced that sportswriting was "the most pleasant way of making a living which man has yet devised."

As generous in his praise of others as he was modest in the assessment of his own work, Smith readily acknowledged his gratitude to those who

*Smith, around 1960, enjoying the only sport in which he participated throughout his life (photograph by Joan Sydlow)*

paved the way in his profession and who influenced his own style. He revered Rice, referred to Woodward as "my idol," and credited Graham with revolutionizing the sports column by changing it from editorial opinion to conversation and mood pieces. He cited Ring Lardner, Damon Runyon, Westbrook Pegler, and Joe Williams as writers who influenced his style when he was young. As he matured he strove for a clean, spare style that reflected the clarity and precision of writers such as Ernest Hemingway and E. B. White. "In my later years," he told Holtzman, "I have sought to become simpler, straighter, and purer in my handling of the language."

Smith used words sparingly and precisely, crafting a style marked by graceful simplicity and a straightforward syntax. His aversion to overblown rhetoric was as strong as his distaste for pomposity. While he refused to exaggerate the significance of sports, he took the language in which he wrote about sports seriously. In January 1960 he proposed as a New Year's resolution that "all sports reporters, including those who write for the papers and especially those who broadcast by radio and television, undertake the revolutionary experiment of delivering their reports in English. . . . Murder of the mother tongue is a form of matricide committed with premeditation by football coaches and encouraged by writers and broadcasters as accessories after the fact."

Smith had the talent to play with words, to invent the unexpected turn of phrase, while avoiding the trap of producing prose that was too clever or overwrought. He enlivened his writing with metaphors without churning out the cliché-riddled figures of speech that can easily infect sportswriting. His description of the Olympics as a "quadrennial muscle dance" or a "sweaty love-in" was not whimsical wordplay for its own sake but a colorful expression of his disenchantment with the event. He described the participants in the twenty-thousand-meter walk at the 1960 Olympics as "Thirty-seven soft-shoe shufflers in their underwear" who "went wriggling and squirming away like night crawlers after a rain."

Smith was one of the few newspaper writers of his generation with a college degree, and although he worked to keep his style free from adornment he did not hesitate to employ allusions to literature, ancient history, or classical mythology to enliven his prose or to make a point. On the opening of Aqueduct racetrack in 1959, he wrote: "Most of the guests wore lapel pins on which appeared a large "A," similar to that worn by Hester Prynne." He referred to the New York Giants' defense as "the noblest band of heroes since Leonidas at Thermopylae."

Smith delighted in approaching his subject matter from an oblique angle, looking for the unusual, usually humorous, twist that would surprise and entertain his readers. Covering the 1948 Westminster Kennel Club show at Madison Square Garden, he wrote that one of the judges "had splendid conformation – broad shoulders, white hair and an erect carriage – and was beautifully turned out in an ensemble of rich brown. One was inclined to hope that he would, in the end, award first prize to himself." He juxtaposed the fashion statements made by Princess Grace of Monaco and the racehorse swaps at the Atlantic City track in September 1956: "The Princess was fetching in a gray suit and bright red cossack turban. The horse wore Willie Shoemaker in red and black silks." When Ted Williams was fined $5,000 for spitting during a game in 1956, Smith opened his column: "By now some modern Dickens, probably in Boston, must surely have brought out a best seller entitled *Great Expectorations*. It was a $4,998 mistake when Ted Williams chose puritanical and antiseptic New England for his celebrated exhibition of spitting for height and distance. In easygoing New York's insanitary subway the price is only $2."

Smith was particularly adept at setting the tone of his column in its opening lines, capturing the essence of the moment or of a personality. One of the most dramatic events he covered was the third and final game of the 1951 playoffs between the Giants and Dodgers, which ended with Bobby Thomson's home run off Ralph Branca, "the shot heard 'round the world." In the classic opening of his column, Smith wrote what every writer covering the event must have been thinking: "Now it is done. Now the story ends. And there is no way to tell it. The art of fiction is dead. Reality has strangled invention. Only the utterly impossible, the inexpressibly fantastic, can ever be plausible again."

A master of the anecdote, Smith excelled at opening and closing his columns with brief narratives that bring his subject to life for the reader. In the opening of a 14 November 1968 column on Charlie Goldman, Smith not only paints a vivid portrait of the former fighter but also expresses his admiration for the man behind the scars: "Charlie Goldman weighed about as much as thirty-five cents' worth of liver. He had a flat nose, a pair of outsize walnuts for ears, a stripe of scar tissue over each eyebrow, and a rather battered look around the mouth. He had earned those badges honorably in something like four hundred fights." On the twenty-fifth anniversary of the death of Babe Ruth, Smith ended his column with a story: "He was buried on a sweltering day in August 1948. In the pallbearer's pew, Waite Hoyt sat beside Joe Dugan, the third baseman. 'I'd give a hundred dollars for a cold beer,' Dugan whispered. 'So would the Babe,' Hoyt said."

It was not only his writing style that changed as Smith matured. "Unlike the normal pattern, I know I have grown more liberal as I've grown older," he told Holtzman in 1973. "I have become more concerned that there is room for improvement in the world." This is not to suggest that Smith –

*Smith at work in the pressroom at Yankee Stadium during the 1981 World Series (photograph by Sara Krulwich/NYT Pictures)*

who always championed the underdog – ignored issues in his early years. He criticized baseball owners for raising the price of admission during the Depression, and he was consistently critical of the International Olympic Committee for what he considered its hypocritical attitude to amateurism in sports. But Smith was slow in responding to the most important social issue in sports in his lifetime: the integration of baseball. While he believed that Jackie Robinson deserved to be given a fair chance and to be judged according to his ability, he did not condemn the racism of those who openly opposed Robinson's entrance into the big leagues in 1947. In 1956, however, Smith put the issue in the proper perspective, writing that Robinson's "arrival in Brooklyn was a turning point in the history and the character of the game; it may not be stretching things to say it was a turning point in the history of this country." Two years earlier, during spring training, he had urged the Yankees, whose only black player was Elston Howard, to defy a Birmingham, Alabama, segregation policy forbidding sporting events involving whites and blacks.

Smith's concern with social issues became more evident, and consistent, in the 1960s. He attributed his fresh outlook, in part, to his relationship with his second wife, who was fifteen years younger than he, and her five children. His new attitude was evident in his coverage of the 1968 Summer Olympics. Following the two-hundred-meter race, Tommie Smith and John Carlos stood on the victory platform and raised a black-gloved protest salute while refusing to look at the American flag. Smith's response to their gesture was in marked contrast to his initial reaction to Robinson: "Dignity and justice and angry courage stood there in the dusk." When Smith and Carlos were thrown off the United States team for their actions, Smith commented: "The simple little demonstration by Smith and Carlos had been a protest of the sort every black man in the United States had a right to make." He went on to scold Olympic officials as "honorable men who consider children's games more sacred than human decency."

Smith was outraged at the IOC's decision to let the 1972 Munich Games continue after Arab terrorists took Israeli athletes hostage and killed two of them (in all, eleven Israelis, five terrorists, and a policeman would die in Munich). He wrote: "Walled off in their dream world, appallingly unaware of the reality of life and death, the aging playground directors who conduct this quadrennial muscle dance ruled that a little bloodshed must not be permitted to interrupt play. . . . The men who run the Olympics are not evil men. Their shocking lack of awareness can't be due to callousness. It has to be stupidity."

At a time when most sportswriters were uninterested in labor issues, Smith was one of the first to speak out in favor of ending baseball's reserve clause. In 1957 he referred to it as "the slave trade," and in the 1970s he devoted several columns to the reserve clause and free agency. He was also the first sports columnist to advocate a U.S. boycott of the 1980 Moscow Olympics in protest of Soviet aggression in Afghanistan. Smith's column did not appear in *The New York Times,* whose sports editor found it inappropriate, but it was run by the 350 publications that subscribed to the *Times* news service. The outcry over the decision by the *Times* not to publish the column focused national attention on the issue of the proposed boycott.

Unlike many sportswriters who achieve fame, Smith was never tempted to move into other areas of journalism. On rare occasions he was persuaded to do some political reporting, such as covering the Democratic and Republican conventions in 1956 and 1968, but he treated those events as if he were a sportswriter watching a spectator sport.

In 1976 Smith was awarded the Pulitzer Prize for distinguished commentary. The citation read: "In an area heavy with tradition and routine, Mr. Smith is unique in the erudition, the literary quality, the vitality and freshness of viewpoint he brings to his work and in the sustained quality of his columns." Smith never lost his love for his work. "My enthusiasm is self-generating, self-renewing," he told Holtzman in 1973, when he was sixty-eight years old. In spite of failing health, he had long resisted the pleas of his family to reduce his workload. In his last column, titled "Writing Less – and Better?," Smith informed his readers that he was reluctantly cutting back from four to three columns a week. He recalls that the first time he met John S. Knight, his publisher, Knight offered to reduce his workload from six to three columns a week, reasoning that "nobody can write six good columns a week." Smith quotes himself as replying: "Suppose I wrote three stinkers? I wouldn't have the rest of the week to recover." He adds: "One of the beauties of this job is that there's always tomorrow. Tomorrow things will be better." The column appeared on 11 January 1982; Smith died four days later.

**References:**


Pete Axthelm, "Masters Touch," *Newsweek,* 87 (17 May 1976): 75;

Ira Berkow, *Red: A Biography of Red Smith* (New York: Times Books, 1986);

Charles Einstein, "The Case for the Red Smith Irregulars," *Harper's,* 210 (March 1955): 82–86;

Jerome Holtzman, ed., *No Cheering in the Press Box* (New York: Holt, Rinehart & Winston, 1974);

Dennis Jackson, "Sportswriter Red Smith's 'Jousts with the Mother Tongue,' " *Style,* 16 (Fall 1982): 414–435;

Roger Kahn, "Red Smith of the Press Box," *Newsweek,* 51 (21 April 1958): 77–80;

Kahn, "Smith the Bleeder," *Newsweek,* 34 (10 October 1949): 62;

"One of a Kind," *New York Times Magazine,* 16 September 1973, pp. 42–44;

"One Was the Brightest," *Newsweek,* 48 (27 August 1956): 70;

R. Poe, "Writing of Sports," *Esquire,* 82 (October 1974): 176;

"Red from Green Bay, *Time,* 55 (15 May 1950): 82–83;

J. D. Scott, "Big Money Writing," *Cosmopolitan,* 147 (August 1959): 38;

Wilfrid Sheed, "Reds," *New York Review of Books* (23 September 1982): 45–48.


**Papers:**

Smith's papers have been given to the University of Notre Dame.

# Wendell Smith

*(27 June 1914 - 26 November 1972)*

Mark W. Schraf
Spitball *Magazine*

MAJOR POSITIONS HELD: Sportswriter, assistant sports editor, city editor, columnist, and sports editor, *Pittsburgh Courier* (1937–1947); sportswriter, *Chicago American* (1947–1963); sports editor, WBBM television (1963); reporter, WGN television (1964–1972); sports columnist, *Chicago Sun Times* (1964–1972).

BOOK: *Jackie Robinson: My Own Story,* ghostwritten by Smith (New York: Greenwood Press, 1948).

Most often remembered for his role in the integration of major league baseball, Wendell Smith was one of the most influential and talented sports journalists of his time. While perhaps not as widely read as many of his white counterparts, Smith's work as sports editor and columnist for the *Pittsburgh Courier,* at the time the nation's largest black newspaper, remains important as a historical account of the Negro league and the dissolution of the color barrier in baseball as well as an example of sportswriting that Jules Tygiel characterizes as "among the best of [the] generation." The defining thrust of his journalistic focus was such that accounts of Smith's career are inextricably linked to the events leading up to and including the entry of the black ballplayer into the major leagues. Jim Reisler acknowledges that Smith's career has been characterized by his contemporary Sam Lacy of the *Baltimore Afro-American* as the greatest of all the black sportswriters.

*Wendell Smith (courtesy of* The Chicago Defender*)*

Wendell Smith was born in Detroit on 27 June 1914 to an American mother, Gertrude Thompson Smith, and a Canadian father, John Henry Smith, who worked as a cook for famed automobile magnate Henry Ford. Despite growing up in an all-white neighborhood and being the only Negro in Detroit's Southeastern High School, Wendell Smith did not experience his first encounter with prejudice until 1933, during an American Legion playoff game. After pitching his team to a 1–0 triumph, Smith's battery mate (and future major leaguer), Mike Tresh, as well as the losing pitcher – both white – were signed to professional contracts by Detroit scout Wish Egan. Jerome Holtzman writes that Egan told Smith, "I wish I could sign you too, kid. But I can't," and Smith vowed then that he would dedicate himself to trying to stop the discrimination against black ballplayers. This incident was crucial in his decision to become a sportswriter.

A high school star in both baseball and basketball, Smith attended West Virginia State College on an athletic scholarship. There he was captain of both the baseball and basketball teams, and he has been credited as the first practitioner of the one-handed jump shot. His focus began shifting to journalism in his junior year, when he became sports editor of the school newspaper as well as the campus correspondent for the Charleston, West Virginia, newspapers, where his first byline appeared. After graduating in 1937 with a bachelor's degree in physical education, Smith was offered a position at the *Pittsburgh Courier* for about seventeen dollars a week, the average starting salary. His first year comprised a basic journalism education: learning how to write fillers and captions and to make up pages, in addition to covering high school sports. In 1938 Smith was promoted to assistant sports editor as well as city editor and columnist.

While black newspapers were often run on a shoestring budget and quick promotions such as Smith's resulted less often from merit than from necessity, he soon proved more than equal to the task. The *Courier* had routinely championed various civil rights causes, and, recalling his own experiences, Smith suggested undertaking a campaign to include Negro ballplayers in the major leagues. Chief editor Robert L. Vann was enthusiastic about the idea, and Smith began his assault on the color barrier. Under the byline "Smitty's Sport Spurts," his first column to deal with the subject on 14 May 1938 was aimed not at the white major league owners but at black patrons of the white major leagues, and in the piece "A Strange Tribe" Smith's aggressive, direct style was evident:

> We have been fighting for years in an effort to make owners of major league baseball teams admit Negro players. But they won't do it, probably never will. We keep on crawling, begging and pleading for recognition just the same. We know that they don't want us, but we still keep giving them our money. Keep on going to their ball games and shouting till we are blue in the face. Oh, we're an optimistic, faithful, prideless lot – we pitiful Black folk.

Smith clearly had grasped the economic dimensions of this integration issue and was challenging black baseball fans to consider the implications of their actions from a racially responsible viewpoint. These two themes were recurrent throughout his campaign to integrate baseball.

Smith's next battle in this fight was to debunk the myths that no black ballplayers were qualified to pitch in organized baseball and that no white players would agree to play with blacks. He decided to interview white ballplayers, and Holtzman recalls that Smith usually started these sessions by asking, "Have you seen any Negro ballplayers who you think could play in the major leagues?" Since Pittsburgh was a National League city, Smith was able to interview all eight National League managers and about forty players, including stars such as Dizzy Dean and his brother Paul, Gabby Hartnett, and Carl Hubbell. As a black man, Smith could not obtain a baseball writers' card and was therefore unable to conduct the interviews at Pittsburgh's Forbes Field. Undaunted, he pursued the players at the Schenley Hotel, where the visiting teams stayed, and used its lobby as his interview room.

His findings were favorable. About 75 percent of the players said that they had no objection to playing with blacks, while 20 percent were opposed and 5 percent had no reaction. The story was picked up by *The Sporting News* and many other daily newspapers. Smith's managing editor was elated by both the results of the poll and the subsequent publicity for the *Courier*. Given a raise, named sports editor (in 1940), and encouraged by his success, Smith believed that these stories established him as a sportswriter and provided momentum for him to continue to fight for the integration of baseball.

Smith did not devote his sportswriting talents exclusively to the question of race equality. He covered Negro league baseball extensively and reported most often on the Pittsburgh-based teams, the Homestead Grays and Pittsburgh Crawfords. Considered one of the country's top boxing experts, Smith also covered the bouts of Joe Louis, Ezzard Charles, and Rocky Marciano. However, his passion remained with the integration issue, and David K. Wiggins writes that Smith was "spirited and tenacious" in his efforts.

Once Smith had sufficiently addressed the issues of the abilities of black ballplayers and the willingness of white ballplayers to play alongside black men, he turned his attention to other topics calculated to bolster Negro American pride and unity. He perceived correctly that in order for black Americans to make concerted efforts to break the color barrier, they needed first to develop self-reliance, racial self-esteem, and the confidence that would result from strong group unity. At the beginning of 1939 Smith therefore began proposing the creation of an organization, like the National Association for the Advancement of Colored People (NAACP), to help the cause of the black ballplayer, for he felt that it was time for black people nationwide to unify and support the integration of the major leagues.

He also realized that an opportunity existed to mobilize liberal white Americans to aid in this fight, for Smith saw a distinct similarity between the treatment of certain peoples by Nazi Germany and that of blacks in the United States. He wrote in February 1939 that, when compared with the white leadership of baseball, Hitler's principles really were no different – his methodology was only cruder. Again, Smith displayed his ability to discern the essential crux of an argument and then hammer away until his desired message was recognized.

Positive results of his efforts followed, as many white sportswriters – such as Jimmy Powers of the *New York Daily News,* Shirley Povich of the *Washington Post,* Dan Parker of the *New York Daily Mirror,* and Dave Egan of the *Boston Record* – furthered the campaign with columns of their own. When the United States entered World War II, the *Courier* was able to intensify its attack against the racial policies of the major leagues, and prior to the revelation of wartime genocide Smith wrote on 25 July 1942 that "big league baseball is perpetuating the very things thousands of Americans are fighting overseas to end, namely, racial discrimination and segregation."

Despite the successes Smith's campaign had enjoyed and the progress it had made, several setbacks occurred. In August 1942 Pittsburgh Pirates president William E. Benswanger approached Smith and asked for the names of four black ballplayers worthy of tryouts. Although Smith first saw this interest by a member of organized baseball as an unabashed triumph and called Benswanger "the greatest liberal in baseball history," it later became apparent that Benswanger's overture was merely an attempt to placate the Communist Party's *Daily Worker,* which had been pressuring him to sign black ballplayers. The tryouts never took place, and Smith later responded bitterly by calling the Pirates' president "an unmitigated storyteller . . . baseball's #1 phony."

In 1943 President Franklin Delano Roosevelt also failed to respond to Smith's appeal to enforce the "Fair Employment Practice Policy" in big league baseball as Roosevelt had done in war and government agencies. Roosevelt's insistence that the quality of wartime baseball should be maintained was also particularly offensive, because an entire league of able-bodied ballplayers was available for the asking. He pushed throughout 1943 for the moguls of Negro league baseball to make an official statement about the possibility that blacks might be able to play in the major leagues, but no statement was forthcoming for nearly the entire year. Smith interpreted this as a result of the selfish interests of owners, their views motivated not by racial pride but by greed. Clearly the owners of Negro league baseball did not share Smith's belief that integrating the majors would benefit black baseball.

Finally, in December 1943 Smith asked Baseball Commissioner Kenesaw Mountain Landis to meet with the black Newspaper Publishers' Association at the winter baseball meetings. Landis agreed, and Smith was encouraged by the opportunity for influential black leaders such as *Courier* president Ira Lewis and activist black athlete-singer Paul Robeson to debate the discrimination issue face-to-face with the leaders of organized baseball. However, Landis used the opportunity to deny officially the existence of any barriers to black ballplayers in the major leagues, and despite a concluding statement from the major league owners that "each club is entirely free to employ Negro players to any and all extent it pleases . . . without restriction," the black delegation knew that none of the owners had any intention of signing black ballplayers.

The following year also proved frustrating for Smith in his integration efforts, yet he vigorously persisted. When Landis died late that year, Smith candidly voiced the attitudes of black Americans toward the late commissioner by pointing out that through Landis's self-aggrandizing style he had promoted himself as the symbol of honesty and integrity. Yet, Smith wrote on 2 December 1944, when Landis had been confronted with the issue of racial discrimination in baseball, he had repeatedly ignored it and displayed none of the "courage and exactness [with which] he faced other" problems during his tenure.

However difficult this period was for Smith, he still managed to write some remarkable prose. "Introducing 'El Diablo' Wells of Mexico," another column in "Smitty's Sport Spurts" that appeared 6 May 1944, presents an interview that Smith conducted with Willie "The Devil" Wells, who had jumped his Newark team in the Negro league to play for Vera Cruz in the Mexican league:

> "One of the main reasons I came back to Mexico," [Wells] said, "is because I've found freedom and democracy here, something I never found in the United States. I was branded a Negro in the States, and had to act accordingly. Everything I did, including playing ball, was regulated by my color. They wouldn't even give me a chance in the big leagues because I was a Negro, yet they accepted every other nationality under the sun. Well," the famed shortshop said, "here in Mexico, I am a man. I can go as far in baseball as I am capable of going. I can live where I please and will encounter no

> restrictions of any kind because of my race. So, you see, that also has a lot to do with my decision to return here."

That an American believed he had to leave the United States to find democracy illustrates how difficult life was for blacks in America, and Smith's simple but elegant style reveals Wells's happiness as well as his disappointment.

In 1945 patience was wearing thin in the black community, and in April, Smith contacted Isadore Muchnick, a white city councilman from Boston. Running for reelection in a predominantly black area, Muchnick was having trouble garnering support, and Holtzman has recounted how Smith suggested that the councilman might galvanize the Negro vote if he threatened to protest Sunday baseball in Boston unless the Braves and Red Sox agreed to hold tryouts for black ballplayers. Muchnick did so, and with the possibility of no Sunday baseball looming in Boston, the two major league teams agreed to hold the tryouts. Smith chose three young Negro league stars – Sam Jethroe, Marvin Williams, and Jackie Robinson – and traveled to Boston expecting to complete the tryouts the next day and return home.

However, the Braves did nothing, and the Red Sox, obviously hoping the black contingent would simply give up, delayed the tryouts without giving any explanation for more than a week and a half. Smith responded by writing an imaginative column comparing the young ballplayers to the American rebels of the Revolutionary War and the Red Sox and the Braves to the British:

> We came here to Boston – the cradle of democracy – to see if perchance just a spark of the "Spirit of 1776" still flickers in the hearts and minds of the owners of the Boston Red Sox and Boston Braves.... Neither ... have displayed so much as a semblance of that indomitable spirit we had anticipated here in the shadow of Bunker Hill. They stand before us as uncompromising despots. They loom as modern Red Coats, resolved to deny us opportunities which are rightfully ours by virtue of heritage.
>
> But we are not giving up! We are Americans, the color of our skins to the contrary ... and we're going to stick to our guns! We are looked upon as "confounded rebels" who would challenge the divine right of baseball's kings.
>
> Nevertheless, we stay on, playing the fife of Americanism, beating the drum of democracy, and waving our tattered flag of citizenship.

Many typical Smith techniques are evident in this column. The appeal to the men responsible for the delay to act with courage and conviction tacitly reproaches them for their cowardice in refusing to grant the tryouts. By cloaking the cause of the black players in the robes of the colonists, Smith's analogy exposes those Boston officials as embodying the forces of tyranny and standing opposed to democracy. His jingoistic rallying also serves to inspire pride and admiration for the oppressed combatants in the fight as well as for their cause. In fact the *we* in the column includes not only the men attempting to play baseball but the entire black race – and all Americans in general. Using language brilliantly and effectively, it illustrates Smith's particular skill: reporting a story and simultaneously delivering a persuasive argument.

The tryout was finally held on 16 April 1945 before Red Sox manager Joe Cronin and coach Hugh Duffy. As Robert W. Peterson writes, Jackie Robinson recalled that the players performed magnificently, and even Red Sox officials admitted that the players looked good. However, no contracts were offered, and the death of President Roosevelt the following day buried any publicity that the tryouts might have been given.

Obviously angry at the turn of events, Smith knew that his pique would not be assuaged as he traveled to Brooklyn to attend the press conference called by Dodgers general manager Branch Rickey to announce his plans for a new United States League (USL). Comprising six black baseball teams and including Rickey's own Brown Dodgers, the USL was actually a subterfuge designed to conceal his goal of scouting black ballplayers for the major leagues. However, none of the sportswriters in attendance knew of Rickey's actual aims, and Smith and the rest of the black press were understandably outraged that, despite his smooth assurances that the league would eventually be incorporated into organized baseball, Rickey would apparently be seeking to exploit black ballplayers by establishing a separate black league for them.

When Rickey asked to speak to Smith in private, Smith expressed his skepticism about the new league. Rickey asked Smith – whom he recognized as the leader of the campaign for integration, the chief organizer of the recent Boston tryouts, and the man perhaps most knowledgeable about black ballplayers – if he knew of any black athletes capable of playing in the major leagues. Murray Polner writes that Smith, both exasperated and hopeful, responded, "If you aren't serious about this, Mr. Rickey, I'd rather not waste our time discussing it, but if you are serious, I do know of a player who could make it. His name is Jackie Robinson."

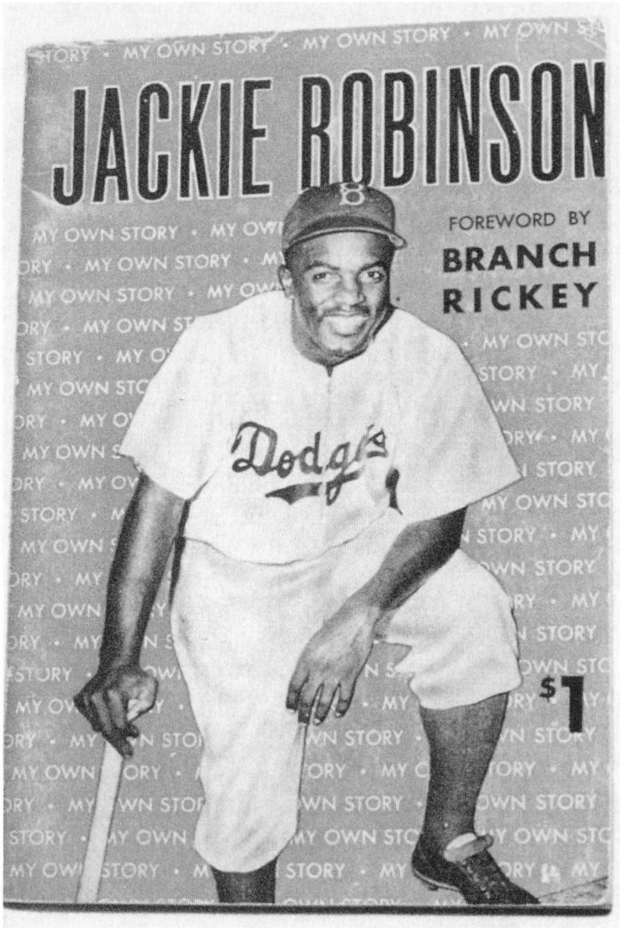


*Dust jacket for Wendell Smith's only book, a heavily illustrated biography of Jackie Robinson written to ease integration in major league baseball*

Rickey had heard of Robinson before, and after Smith's recommendation he sent scout Clyde Sukeforth to scout the former UCLA athlete extensively. Without completely revealing his plans about Robinson, Rickey remained in contact with Smith concerning Robinson and other black players through summer 1945. To a journalist who had been crusading for integration, the story was a bombshell waiting to be dropped. Yet, in an unselfish effort to avoid provoking controversy, Smith never fully reported what he knew and only hinted at the role that Branch Rickey was to play in the coming months. On 28 April 1945 Smith commented, "From this perch, it appears to me that Branch Rickey, one of the wisest and shrewdest men in baseball, looms as a valuable friend, both for organized Negro baseball and the cause of the Negro player in the majors." Within months, Jackie Robinson was signed to play for the Montreal Royals, the Brooklyn Dodgers' top minor league team.

While the response of the *Courier* was as wildly enthusiastic as that of other black newspapers, Smith had further responsibilities on his mind. Rickey had asked Smith to work as a scout for Brooklyn, although Smith's real job would be to accompany Robinson as a roommate and road companion at the Dodgers' spring training site in Sanford, Florida. Although he lacked experience as a baseball scout, he recommended several black players to Rickey – including John Wright, Robinson's Montreal teammate, and future stars Larry Doby and Monte Irvin. Smith knew that he would not be paid as a scout, but he accepted Rickey's offer, adding to his newspaper salary and witnessing firsthand the historically significant initial steps toward the integration of major league baseball.

Writing daily stories for his newspaper back in Pittsburgh, he continued to report, and many other writers, realizing that he was Robinson's roommate and trying to gather as many details as they could about the momentous events, would ask questions of Smith. According to Holtzman, Smith soon realized that he was rapidly becoming Robinson's James Boswell. Yet, in an effort to enhance Robinson's and Wright's chances of succeeding and to avoid encouraging further offenses from imitators, Smith did not detail much of the racial hatred and discrimination that he saw the two black players encounter.

Soon other black ballplayers were signed by both the Dodgers and other major league clubs, and "the great experiment" seemed to be proceeding well. However, Smith and the other architects of the integration movement remained cautious, for they realized that any hint of difficulty involving a black player could be disastrous. When Roy Partlow, a black Dodger minor league player, objected to being demoted from Montreal to Trois-Rivières of the Class C Canadian American League, Tygiel notes that Smith devoted a 20 July 1946 *Courier* column to "an attack on the former Negro League star and a sermon on the responsibility of racial pioneering."

> It looks as through Partlow has turned out to be an eccentric "prima donna" and a problem child of no small means. . . . Apparently he fails to realize that whatever he does, good or bad, is a reflection on Negro ballplayers. . . . Thousands of people will welcome Partlow's defection with glee and say, "See, I told you Negro players can't take it in organized baseball." [He] . . . needs to sit down and think about the significance of his position . . . needs to think about those 14 million Negroes from coast to coast who are pulling for him to

> make good in white organized baseball . . . and less about himself.

Smith may have overreacted. Tygiel points out that Montreal newspapers paid little attention to the incident, and, after meeting with Rickey, Partlow did report to Trois-Rivières. But the nearly religious fervor that Smith displayed toward this event in his column underscored his commitment to integration and the steps he was willing to take to ensure that it succeeded.

After Robinson completed his successful season in Montreal in 1946, it became obvious that he would have a legitimate shot to make the Brooklyn team in 1947. Rickey again asked Smith to accompany and room with Robinson during spring training. When Robinson was transferred from the Montreal Royals to the Dodgers near the end of spring training and was in the lineup for an exhibition game against the Yankees in New York, Smith was at Yankee Stadium to report on the historic game.

In his 19 April 1947 article Smith combined details about the unusual event of the first black man to play major league baseball with details about the familiar event of a rookie's first game. He wrote of the smile on the face of "a powerfully built fellow" as Robinson ran off the field "with the hordes of Brooklyn idols." He quoted Robinson's response to a reporter's question about how it felt to be a big leaguer: "Gee, fellows, it's great!" Then, Smith wrote, "The writers smiled knowingly and moved in a bit closer. Robinson's answer was an old one. He had just said what all rookies say their first day against major league competition. And they all smile, and they all seem just a bit awed at the thought of it."

Smith recognized the reporters' acceptance of Robinson by noting their movement closer to him, and by placing his response in the context of all other rookie ballplayers, Smith conveyed the idea that a black ballplayer is really no different from a white one. Building on this theme, Smith added that "Robinson . . . started to take off the heavy, woolen stockings that all big leaguers wear." Smith recounted the scene later when a happy crowd rushed to Robinson as he left the ballpark, and he quoted the gratified Robinson's statement, "I love 'em. Every darn one of 'em!" Smith carefully depicted this first black major league ballplayer as a warm, friendly, and gracious man who genuinely appreciated the opportunity to play, and his story expertly imparted how players, writers, and fans alike accepted Robinson.

In 1947 Smith began to commute between Pittsburgh and Chicago when he was hired to cover professional boxing and Big Ten football for *Chicago Today,* later called the *Chicago American.* He continued to commute until 1949, when he married *Courier* secretary Wyonella Hicks and settled in Chicago, where, as a member of the *American* sports staff, Smith was to begin breaking some of the same ground in journalism that Robinson had broken in baseball. Having made repeated attempts for many years to join the Baseball Writers Association of America, Smith finally became the first African American to be granted membership in the organization after he began to cover the Chicago White Sox full-time for the *American* in 1948. That year was especially momentous for Smith, for by traveling to London for the Summer Games of that year he became one of the first black reporters to cover the Olympics.

Smith completed *Jackie Robinson: My Own Story,* the "as-told-to" biography of Robinson, that year. With a foreword by Rickey, this first biography of Robinson recounts his life from his birth in Cairo, Georgia, through the 1947 World Series, although only the barest details of his life before his involvement in the integration of baseball in 1945 are provided. Primarily an account of events leading up to and including Robinson's entry into organized baseball, the book includes forty-eight photographs.

Written as an as-told-to narrative, the book reveals the clearly discernible voice and perspective of Wendell Smith throughout. Robinson never fails to be humble, gracious, and courageous, although Smith carefully portrays him as a man with typical struggles and transgressions. Smith and Robinson intended to produce a book that would educate as well as entertain and would be accessible to baseball fans of nearly all ages. Much of the book therefore consists of photographs, and an entire chapter is devoted mostly to white sportswriters who supported the integration movement in baseball as far back as 1888. Smith felt a responsibility to thank publicly those journalists who supported his campaign and to inform readers that many white Americans had always supported removal of the color bar from baseball.

While much game action is described in detail, most of the book focuses on off-the-field events and on his state of mind during his struggle to reach the major leagues. However, Smith's narrative cautiously avoids revealing many of the harrowing details of the racial discrimination Robinson faced. Despite Robinson's remarkable rookie seasons in both Montreal and Brooklyn, Smith's prose remains characterized by restraint and aims to report from a positive angle. As a consequence, Robinson often appears to be somewhat naive, especially in light of details that were revealed later in other, more scholarly accounts. For instance, Smith's presentation of the first meeting between Robinson and Clay Hopper, the southern-born Montreal manager, portrays Robinson as uncertain about whether or not Hopper resents his presence in the ball club. Smith gives no further attention to this question. Yet later books, particularly Carl Rowan's *Wait Till Next Year: The Life Story of Jackie Robinson* (1960), reveal that Hopper had not supported Robinson's presence on the team at all and at first had not even considered blacks to be human.

Although *Jackie Robinson: My Own Story* reveals some disturbing incidents of bigotry and racial discrimination, the light focus and positive aim of the book preclude it as a serious record of how baseball was integrated, and it is not mentioned in Robinson's other biographies. While it is notable as his first biography and as part of Wendell Smith's extensive sportswriting career, *Jackie Robinson* remains interesting mostly to baseball-book collectors. In a 27 November 1972 obituary *The New York Times* reported that Smith had also ghosted books by Ernie Banks, Joe Louis, and Roy Campanella, but no such books are listed in standard bibliographies, nor do specialized baseball bibliographies credit Smith with such works. Perhaps the *Times* may have been referring to newspaper stories that Smith had ghosted and that had appeared in the *Courier* as columns "written" by the three athletes. In any case, Smith's widow indicated that he had written only one book.

Smith continued to cover the major leagues for the next fifteen years, and although he no longer had the issue of integration to fire his creative talents, he continued to discover issues needing his support. On 14 July 1948, for instance, Tom Spink's editorial in *The Sporting News* attacked Bill Veeck, owner of the Cleveland Indians, for having signed aging black pitcher Satchel Paige. Spink charged that Veeck's action was a publicity stunt that demeaned the quality and dignity of major league baseball, and Smith responded on 24 July in the *Courier* with an eloquent open letter that likened the signing of Paige to that of forty-one-year-old Ty Cobb and forty-four-year-old Cy Young. Smith acknowledged both of them as "immortals" and stated that they had been asked to play at their advanced ages "because a venerable halo adorned their persons as, heroically, they approached the golden sunsets of glorious voyages."

Yet when Paige later began to miss curfews and report late for games, Smith's criticism was typically sharp. "If you were Satchel Paige" and were in a position to jeopardize the future of black ballplayers, Smith rhetorically asked readers, "would you represent your people admirably or would you remain Satchel Paige?" Fearing that Paige's behavior "could have major repercussions and fatal effects on the future of the Negro player in the majors," Smith admonished him in a 20 November 1948 article in the *Courier*. Again Smith was assuming his role as watchdog of the integration movement: fiercely protective of black ballplayers overcoming the color line, he was equally fierce in criticizing black ballplayers who did not, in his eyes, uphold the sacred trust that their race had placed in them.

In 1961 and 1962 Smith began another campaign, one decrying the Jim Crow practices of major league baseball teams during spring training in Florida. In his usual low-key, nonconfrontational yet relentless style Smith exposed the segregationist policies of many teams, which he criticized for giving in to the exclusionary customs of the southern towns that provided facilities for them. His efforts were doubly rewarded: by 1962 every spring training camp was completely integrated, and Smith was nominated for a Pulitzer Prize in journalism.

In 1963 Smith moved to television as sports editor for WBBM, and the following year he became a television news reporter for the WGN *People to People* Sunday program in Chicago. Despite the change in journalistic venue, Smith still managed to continue writing a weekly sports column for the *Chicago Sun-Times* and freelance articles for periodicals and books such as *Baseball Review, 1962; Baseball Monthly; Baseball Digest;* and *The Encyclopedia of Sports.*

Smith earned many other accolades following his postwar efforts to integrate major league baseball. These honors included the Alumnus of the Year Award from West Virginia State College in 1951, three top sportswriting awards from the William Randolph Hearst organization, and the Best Sportswriter Award for 1959 from the Chicago Press Club, an organization for which Smith was serving as president at the time of his death in 1972. Smith died of cancer at age fifty-eight, just one month and two days after the death of Jackie Robinson, whose professional career had been so closely linked with Smith's interests. Shortly thereafter an elementary school in Chicago was named in Smith's honor, and the University of Notre Dame as well as the Chicago chapter of the Baseball Writers Association of America continue to present annual awards in Smith's memory.

His reputation and legacy were recognized posthumously when he was inducted into the Chicago Journalism Hall of Fame in 1982, and another chief honor accorded Smith also came posthumously when he was named recipient of the 1994 J. G. Taylor Spink Award for meritorious contributions to baseball writing – yet another groundbreaking recognition, for Smith was the first black sportswriter to be so honored. In her acceptance speech on his behalf, Wyonella Smith recounted that shortly before his death he had been pleased that many of the Negro league stars of the past were finally achieving recognition by being inducted into the Baseball Hall of Fame, an honor that the 1994 J. G. Taylor Spink Award also brought him.

Smith's vision of American society, expressed succinctly and beautifully in the *Chicago Defender* after Smith's death, was one in which "ability, skill and character are the sole measures of a man and not the color of his skin," a vision that Smith "pursued . . . not with the militancy of the new breed of Black spokesman, rather with the calm and patient logic of a wise man whose vision was sharp enough to see the light at the end of the tunnel. . . . History will not pass him by." Wiggins hails Smith as "the catalyst" behind the success of the *Courier* in integrating major league baseball, and he believes that Smith was especially effective at correlating the discrimination of blacks in organized baseball with the deplorable conditions that black Americans were forced to endure. Wiggins praises him for recognizing that the integration of major league baseball, initially only a symbolic pathway for black participation in American society, would foster black dignity, pride, and self-esteem, the components needed to destroy segregation and to elevate blacks to full citizenship in this country.

Tygiel observes in *Baseball's Great Experiment* that Smith "could be bitterly sarcastic and vitriolic in his rage against Jim Crow, yet lyrical in his descriptive prose" and calls him "the most talented and influential of the black journalists" of the era. For a man who dedicated his life to the cause of the breaking of major league baseball's color barrier, perhaps no greater tribute could be voiced than that of Jackie Robinson, who believed that he never would have made it to the Brooklyn Dodgers without the help of Wendell Smith. In presenting Smith with a copy of his book, *I Never Had It Made* (1972), Robinson wrote, "To Wendell who did so much to break the barrier in baseball and who I have always appreciated as a friend."

**References:**

Jerome Holtzman, "Wendell Smith – A Pioneer for Black Athletes," *Sporting News,* 22 June 1974, pp. 15–16;

Robert W. Peterson, *Only the Ball Was White* (Englewood Cliffs, N.J.: Prentice-Hall, 1970);

Murray Polner, *Branch Rickey: A Biography* (New York: Atheneum, 1982);

Jim Reisler, *Black Writers / Black Baseball: An Anthology of Articles from Black Sportswriters Who Covered the Negro Leagues* (Jefferson, N.C.: McFarland, 1994);

Jackie Robinson, *Baseball Has Done It,* edited by Charles Dexter (New York: Lippincott, 1964);

Robinson and Alfred Duckett, *I Never Had It Made* (New York: Putnam, 1972);

Donn Rogosin, *Invisible Men: Life in Baseball's Negro Leagues* (New York: Atheneum, 1983);

Carl Rowan and Jackie Robinson, *Wait Till Next Year: The Life Story of Jackie Robinson* (New York: Random House, 1960);

Rob Ruck, *Sandlot Seasons: Sport in Black Pittsburgh* (Chicago: University of Illinois Press, 1987);

Jules Tygiel, *Baseball's Great Experiment: Jackie Robinson and His Legacy* (New York: Oxford University Press, 1983);

David K. Wiggins, "Wendell Smith, the *Pittsburgh Courier-Journal* and the Campaign to include Blacks in Organized Baseball, 1933–1945," *Journal of Sport History,* 10 (Summer 1983): 5–29;

A. S. "Doc" Young, *Great Negro Baseball Stars: And How They Made the Major Leagues* (New York: Barnes, 1953).

# John R. Tunis

*(7 December 1889 - 4 February 1975)*

Leverett T. Smith Jr.
*North Carolina Wesleyan College*

See also the Tunis entry in *DLB 22: American Writers for Children, 1900-1960.*

MAJOR POSITIONS HELD: Freelance writer (1925-1975).

BOOKS: *\$port\$: Heroics and Hysterics,* illustrated by Johan Bull, introduction by Grantland Rice (New York: John Day, 1928);
*American Girl* (New York: Brewer & Warren, 1930);
*Was College Worth While?* (New York: Harcourt, Brace, 1936);
*Iron Duke,* illustrated by Bull (New York: Harcourt, Brace, 1938);
*The Duke Decides,* illustrated by James MacDonald (New York: Harcourt, Brace, 1939);
*Champion's Choice,* illustrated by Jay Hyde Barnum (New York: Harcourt, Brace, 1940);
*Choosing a College* (New York: Harcourt, Brace, 1940);
*The Kid From Tompkinsville,* illustrated by Barnum (New York: Harcourt, Brace, 1940);
*Sport for the Fun of It,* illustrated by Bull (New York: Barnes, 1940; revised, 1950);
*This Writing Game* (New York: Barnes, 1941);
*World Series,* illustrated by Barnum (New York: Harcourt, Brace, 1941);
*Democracy and Sport,* introduction by Raymond Graham Swing (New York: Barnes, 1941);
*All-American,* illustrated by Hans Waleen (New York: Harcourt, Brace, 1942);
*Million-Miler: The Story of an Air Pilot* (New York: J. Messner, 1942);
*Keystone Kids* (New York: Harcourt, Brace, 1943);
*Lawn Games* (New York: Barnes, 1943);
*Rookie of the Year* (New York: Harcourt, Brace, 1944);
*Yea! Wildcats!* (New York: Harcourt, Brace, 1944);
*A City for Lincoln* (New York: Harcourt, Brace, 1945);
*The Kid Comes Back* (New York: Morrow, 1946);
*Highpockets* (New York: Morrow, 1948);

*John R. Tunis*

*Son of the Valley* (New York: Morrow, 1948);
*Young Razzle* (New York: Morrow, 1949);
*The Other Side of the Fence* (New York: Morrow, 1953);
*Go, Team, Go* (New York: Morrow, 1954);
*Buddy and the Old Pro,* illustrated by Barnum (New York: Morrow, 1955);
*The American Way in Sport* (New York: Duell, Sloan & Pearce, 1958);
*Schoolboy Johnson* (New York: Morrow, 1958);
*Silence Over Dunkerque* (New York: Morrow, 1962);
*A Measure of Independence* (New York: Atheneum, 1964);
*His Enemy, His Friend* (New York: Morrow, 1967);
*Grand National* (New York: Morrow, 1973).

John R. Tunis was an unusual sportswriter for his or any other time. During the years between the two world wars, when he did most of his sportswriting, he made his living by questioning the values underlying sport in America in article after article. He continued such questioning in the children's books, most of which concerned sports, that he began writing in the late 1930s. Tunis saw the position of sports in American life as central, describing it in his autobiography, *A Measure of Independence* (1964), as "a kind of substitute for the theater.... Our great sports events mean what the comedies and tragedies of the Greek playwrights meant to the citizens of Athens as they packed the theater of Dionysus on the slopes of the Acropolis after the passing of the winter solstice." A close reader of the central works of the cultural historians Vernon Parrington and Richard H. Tawney, Tunis had a keen sense of the relation between sport and economics. All his life he argued for amateurism.

Tunis went his own way in all his endeavors. Though his family and college classmates were conservative Republicans, Tunis became a liberal Democrat. He embraced the New Deal and Franklin Delano Roosevelt, enthusiastically involved himself in local politics, and at the age of seventy-eight produced an antiwar novel, *His Enemy, His Friend* (1967), which he dedicated to "Will Lawless, Tommy Rodd, and the many other young Americans whose conscience has led them to prison in these years."

John Roberts Tunis was born in Boston on 7 December 1889, "the product of a parson and a teacher," as he put it in his autobiography. His father, John Arthur Tunis, died when Tunis was still a boy, but his mother, Caroline Greene Roberts Tunis, lived into her nineties and had a profound influence on her son's life. When she was ninety, as Mel R. Allen tells it in his 1989 essay "The Kid From Rowayton," Tunis treated her to lunch at the Ritz-Carlton. Outraged that the restaurant charged them fifteen cents for bread and butter whether they wanted it or not, she took on the entire hotel bureaucracy, one by one, until the charge was removed. When she left, she did so to the applause of the other diners. Her passion for education got Tunis to Harvard, and he graduated with an A.B. in 1911.

But neither his mother's passion nor Harvard inspired Tunis to learn while he was at the college. "What did interest me in college? " he asks in his autobiography: "Two things, athletics and the theater." He remembered "working out with the cross-country team in the autumn and the tennis team in the spring" at Harvard. He remained an enthusiastic and talented tennis player throughout his adult life.

After serving in Europe during World War I – his daily letters to his mother helped to convince him he could become a writer – he married Lucy Rogers in Cambridge, Massachusetts, on 19 February 1918. Though the couple was childless, Tunis gave considerable aid to the children of one of his wife's brothers after the brother's sudden death. In the early 1920s Tunis began the slow process of learning to support himself as a freelance writer. His autobiography provides a list of magazines that testifies to the extent of his labors:

> Strange and peculiar were the magazines I pestered and sold copy to those days: *Good Hardware* (a publication for which I must have written, though just what is hard now to imagine), *Motor Boating, The Younger Set, American Lawn Tennis, Arts and Decorations, The Railway Man's Monthly, The Open Road, Charm, Golf Illustrated.*

It was not until the middle of the decade that he established himself, contributing regularly and mostly on sports subjects to *The New Yorker*, the *New York Evening Post, American Lawn Tennis,* and the *American Legion Monthly*. He also began to be published in *The Saturday Evening Post, Collier's,* and *Harper's*.

Tunis was a well-known journalist in the 1930s. In addition to his freelance writing, he became a pioneer broadcaster as he covered tennis matches from France and England for NBC beginning in 1934. In the late 1930s he began to focus his writing more on educational subjects; he published articles in *Redbook, This Week,* and the *Atlantic*. Increasingly, though, Tunis's income came from the sales of his children's books, and he devoted more and more of his time to them. His concern for the meanings of sport in American society remained central.

Tunis's ideas about sport, developed in his magazine articles between the wars, were organized in a series of books published during a thirty-year period, beginning with *$port$: Heroics and Hysterics* in 1928 and ending with *The American Way in Sport,* published in 1958. These same ideas inform what he called in his autobiography his "sports novels for high school groups," which he published between 1938 and 1973.

In his introduction to Tunis's first book, *$port$: Heroics and Hysterics,* Grantland Rice expresses some apprehension about the book's thesis, asserting that Tunis "has tackled one of the great problems of our times and whether or not one can agree with all his conclusions he has handled the situation with courage and skill." The great problem, according to Rice, was "the entire amateur situation" in sports. The critic for the 13 October 1928 *Saturday Review of Literature* complained not about Tunis's conclusions but about the tone of the book: "The horrors of the situation are flung out in a merciless bombardment; and the resulting effect, although diverting always, is frequently confusing." There could be little confusion, however, about Tunis's thrust. The bombardment of horrors amounted to a single overarching horror: the relation of sports and money.

Tunis asserts the problematic relationship of sports and money right from the title page, where both esses in "Sports" are dollar signs. Tunis makes a well-organized four-part argument to indict the culture of American sports and suggest its reformation. First, he argues that America's perception of sport is controlled by what he calls "the Great Sports Myth." Second, he describes in detail the actualities of contemporary sport culture. Third, he articulates his own understanding of how sport ought to operate in human culture. Finally, he has some suggestions about how that ideal might be achieved. In *$port$: Heroics and Hysterics* Tunis sounds many of the themes he will be concerned with for the rest of his career.

Tunis spends the second chapter of the book defining "The Great Sports Myth." He admits some of the many benefits of sport before addressing its corruption:

> Let us grant that sport between individuals is a working laboratory for the building of character. Let us admit freely that the health of nations is being improved by friendly outdoor games. . . . But let us not confound the precious informality of individual sport with the huge, widely advertised sporting contests with which we are being inundated from year's end to year's end.

The myth is that these "widely advertised sporting contests" build healthy bodies and good character and that they promote goodwill among people and peoples. Tunis argues that such events do neither. He points out that the pressure championship athletes are under often causes strange and childish behavior and that these athletes often have limited outlooks and narrow horizons outside their specialties. Tunis observed such deficiencies in the Olympic athletes he covered in 1928 and further notes that the games produced bickering among nations rather than the advertised goodwill.

In a chapter on intercollegiate football that epitomizes his understanding of the realities of sport in America, Tunis concludes that college football is "a huge, commercial magnificent spectacle" that has little to do with either amateurism or sport as he understands it. Tunis speaks of American sports as "cursed with the materialism of America" in a chapter on international attitudes in sport, but it is his fourth chapter, titled "The Lawn Tennis Industry," in which he best sets forth his view that sport in America has been destroyed by "the spirit of commercialism." Professional athletes, "the Ruths, Cohens, Cobbs, and Hornsbys are primarily business men, not sportsmen," he argues, and tennis as practiced in America is an industry, not a sport. He asserts that everyone connected with tennis – players, officials, promoters – are "in it for what they can get out of it" and "finds it difficult to condemn amateur stars for taking what is offered them," since "the overlords of the game . . . [make] profits out of tennis far beyond the reach of the average player."

In tennis Tunis can find "hardly any difference or distinction between the amateur and the professional." He asserts that the class distinction "between a man who earns his living in sport and a man who does not is out of date" and should be abolished. Tunis offers Bill Tilden as an example of the amateur as professional, saying that "Mr. Tilden professes tennis as a doctor professes medicine, or a lawyer professes law. I mean that life to Mr. Tilden is tennis; that he eats, breathes, sleeps and thinks tennis; that tennis is his be-all and end-all." The real sportsman, for Tunis, "knows the thrill of real sport, of playing, not for championships, for titles, for cash, for publicity, for medals, for applause, but simply for the love of playing."

While listing sports he regards as amateur – handball, squash, badminton, archery, and croquet, among others – Tunis identifies "the danger point" between sport and commercial spectacle as being when sports "develop associations, officials, national championships, ranking lists, columns of publicity in the Metropolitan dailies, and all the rest of the sporting flapdoodle so dear to the believers in the Great Sports Myth." For those sports that have passed that danger point – such as amateur tennis – Tunis suggests eliminating not only the publicity surrounding the stars and the championships but also the practice of paying a player's expenses.

Although he did not expect immediate action on his ideas for reform, Tunis saw reason to hope for a change in attitude toward sport among young people. Veterans of World War I, he believed, had seen too much to be taken in by the hoopla of commercialized sport. He also had hopes for undergraduates. The development of large universities, he felt, militated against strong interest in intercollegiate athletics on the part of the student body. He sensed, too, that this generation of college students was "sounder" about commercialized sports than their fathers. The future, however, would prove the younger generations even more enamored of commercialized sports than their elders.

Two years later Tunis again addressed the problem of commercialization in his only book-length adult fiction, *American Girl* (1930). As William Soskin observed in his 6 August 1930 review in the *New York Post, American Girl* integrated two closely related purposes: to depict the economic structure of a commercialized amateur sport (Tunis was later to do this successfully for high school football and basketball fiction for young adults) and to tell the story of the spiritual cost of an American girl's becoming a tennis champion. Unlike Ring Lardner's Midge Kelly, the one-dimensional protagonist of his much better known story "Champion," Tunis's "Happy Florence Farley, the Girl With the Laughing Voice" is a multidimensional human being from first to last. The reader is both outraged by the effect the system of organized amateur tennis has upon her and moved by the signs of humanity remaining in the hardened champion at the end of the book. Florence can still help a fellow tennis player in distress, fly to the bedside of a sick relative, and acknowledge that the love of her life has passed her by, even while she ignores former friends in favor of more financially rewarding contacts and takes money for ghostwritten articles and stolen dress designs. The reader believes the description of "her tortured face" near the end of the book.

On the final page of the novel, Florence finally admits, "I'm not in this game for fun." Tunis's experience covering tennis both in the United States and abroad stands him in good stead, for certainly one of the novel's strengths is his detailed account of the way the business of championship amateur tennis works. Beginning in Florence's hometown of Millville, Tunis shows how her ability is exploited economically by herself, her family, and others. When she is still quite young, the family buys a new car that her father knows it cannot afford but that her mother senses is a key to her developing tennis career. The local country club is eager to send Florence to a national tournament because of the good publicity the town will then receive.

When Florence enters the world of championship tennis, "the politics of the sport" take over. As a successful tournament player, she finds free tennis racquets available to her. As she becomes more successful, newspapers vie to pay for the use of her name on articles that will be ghostwritten. Later she finds that hotels, counting on her celebrity to entice other customers, will put her up for free. She even receives a new family home. She begins to make money from her modest singing ability and discovers "an astonishing talent for making money." Florence evenually needs a secretary to keep track of all her ventures and becomes skilled at balancing on "the narrow ridge between amateurism and professionalism," assuring one of her suitors that "lawn tennis is an attempt to bring sport and business together in the holy bonds of matrimony. It succeeds."

The novel excited a good deal of controversy. A critic in the 24 August 1930 *New York Times* called it "the 'Uncle Tom's Cabin' of American amateur tennis" and commented that "Mr. Tunis has provoked a civil war of sorts." The reviewer for the 11 August 1930 *Time* magazine thought that sportswriters were generally opposed to the commercialization of sport, saying, "like most successful sportswriters, harping on and exposing professionalism is his forte." But a negative attitude toward professionalism was far from universal in sports departments. W. O. McGeehan wrote disparagingly of Tunis's crusade in the *New York Herald Tribune* of 7 August 1930:

> in his book "Sports," spelled with the dollar sign, Mr. Tunis indicated a passion for kicking all of that quality which the boys call sordid commercialism out of amateur athletics of all sorts. I do not think that he struck a particularly sympathetic note. Nobody seemed to care about the loose change picked up in devious ways by the professional amateurs.

Bill Cunningham portrayed Tunis in the 8 August 1930 *Boston Post* as an embattled outsider:

> He's been smiting his bloomin' lyre for some 15 years in celebration of tennis, polo and those far politer pastimes, and for a long time he was a voice singing in the wilderness, although singing isn't exactly the word either – he was a voice lifting the razzberry in the wilderness whence all else were mocking birds or dulcent-toned thrush.

Later in the article Cunningham characterized Tunis as "just a little goofy." Boyleston Adams vigorously defended commercialization in the 28 August 1930 *Boston Herald* on the grounds that it produced excellent tennis.

Tunis was not daunted by the criticism and seems to have enjoyed his status as a prophet and outcast. In *This Writing Game* (1941) he rather proudly lists some of the names he had been called:

"I have been described in print as an 'agitator,' a 'bellyacher,' a 'grouch,' a 'liar,' and a 'serpent,' besides being characterized as 'wrong-headed,' 'tiresome,' 'petulant,' 'insincere,' 'jaundiced,' 'violent,' 'venomous,' and 'cowardly.'"

Tunis continued to report on tennis for several years after the publication of *American Girl,* but in the 1930s college football and the larger question of the value of education in American life became his dominant concerns. Tunis published a series of six essays between November 1929 and June 1937 in *Harper's* magazine about an institution he calls "Mammoth University." Taken together, the satiric, often funny articles form a comprehensive critique of American higher education. The first article published, "Maguire: Builder of Men," is on the college football coach. The second, "Portrait of an Intellectual," deals with an education professor who spends most of his time and energy on remunerative consulting. The third, "Eddie Stands for Good Clean Sport," draws a portrait of the director of athletics. The fourth, "Fellow Alumni of Mammoth," concerns the college's alumni secretary. The fifth focuses on the "College President," and the sixth, "Honoris Causa," treats the awarding of honorary degrees. Through these six perspectives, Tunis paints a devastating portrait of American college life and laments the dominant position of intercollegiate athletics.

It is clear that Tunis liked none of what he saw in the modern university, which he argued had undergone a process of bureaucratization, commercialization, and rationalization comparable to championship amateur athletics. The thirty-year-old Mammoth University he creates is mammoth, the largest state university in the country. Football coach Maguire's office, for example, is like the office of any big corporation executive; he prepares athletes to succeed in the industrial world. Tunis marvels at the skills demanded of the athletic director, a new college position in the 1930s. He calls Mammoth's Eddie Morgan "Secretary of Adolescents" and points out that he has to be a diplomat, financial genius, real estate operator, bureaucrat, and labor employer.

Throughout the essays Tunis contrasts Mammoth University with a premodern college as an ideal. Coach Doc Maguire constantly uses trickery to outscore his opponents on the football field, and as a consequence "Dear old Professor Jenkinson of the Greek department . . . asked timidly at a faculty meeting whether it was in this spirit that young minds . . . were to be prepared to join the fellowship of educated men?" Coach Maguire soon sets him straight. Jimmy Raymond – the modern intellectual – is contrasted with his uncle "old Professor Raymond of the English department at Cambridge," who helped out with athletics at the college too:

> Thoroughly unbusinesslike was old Professor Raymond, yet a character withal. Those were the days before it was considered necessary to hire a Graduate Manager of Athletics at fifteen thousand dollars a year plus twice that sum for assistants, clerks, and various other supernumeraries. The Professor, in the role of Faculty Advisor on Athletics, performed most of those duties himself; with a single helper he received moneys, arranged schedules, settled eligibility problems, and otherwise attempted to keep the sport of the University within the bounds of common sense. He was, however, always an idealist; he persisted in the belief that intercollegiate athletics were merely contests between amateur sportsmen played for the sake of the game alone. Consequently, it must be admitted that he was somewhat less than efficient and that toward the end of his regime things fell into a chaotic condition quite beyond his lovable but unpractical nature.

Clearly, old Professor Raymond's influence is personal rather than professional.

Tunis's critique of modern universities proceeds from his sense of their importance to modern society. He does not dislike just intercollegiate athletics; he also dislikes the modern curriculum that is dominated by the social sciences and professional programs. He presents the athletic director, Eddie Morgan, as a man whose message was that there was no "overemphasis on athletics" at Mammoth. One of Eddie's proofs was the fact that his salary was the same as that of a full professor. "He sees no reason why a man who teaches baseball or basketball should receive more than a man who teaches chemistry or bee-keeping or cafeteria management or some other course beneficial to the intellect." Tunis's criticism here echoes that of the 1929 report on American college athletics published by the Carnegie Foundation, which said about football, "It is doubtful if in any other department of the American College curriculum a single subject receives more thoroughgoing attention." Tunis imagines a new president who would "make Mammoth University an intellectual center, not an educational factory" and "loosen the hold of football and extracurricular activities" on the institution. Specifically he asks, "How can we make football a game?" as distinguished from the intercollegiate spectacle it has become.

In 1936 Tunis produced another denunciation of the modern university, *Was College Worth While?,* which he based on the autobiographies of his classmates at Harvard University who graduated in 1911. While there are few references to intercollegiate athletics in the book, those that are included are highly charged. Tunis reports ruefully that "the number of us who desire our sons to become halfbacks, as against the number who express a desire that their sons shall receive an education or standards of value in college, is in the proportion of fifty to one." His language crackles when he considers the preferences of alumni groups for speakers: "the group clamors for the football coach . . . and greets his banalities with cowlike approval. Like most college graduates we get more steamed up over football and the football coach than anything else." His next sentence changes the subject and makes the point: "We are not, on the whole, especially interested in or articulate about education, though occasionally someone refers to it."

In 1940 Tunis published a guidebook called *Choosing a College,* in which he observes the strengths and limitations of private colleges, two-year colleges, and state universities. The subject of intercollegiate athletics rarely comes up, even when Tunis describes a mythical institution, Pofflebury College, of the sort he wanted readers to avoid attending. But there is one moment in the chapter "Ten Reasons for Choosing a College" in which the combative Tunis emerges:

> Up to the last minute it seemed possible to write a book on college without any reference to athletics. Vain and illusory hope! . . . If you find yourself wavering between a college that wins the Rose Bowl every New Year's Day and one that has departed from the entertainment business and abandoned football, please don't hesitate. You will be fortunate enough if the latter institution looks with favor on your application.

Between *Was College Worth While?* in 1936 and *Choosing a College* in 1940, Tunis took a fictional look at college life in his first two juvenile novels, *Iron Duke* (1938) and *The Duke Decides* (1939). These books, in which intercollegiate sport is central, treat the adventures of Jim Wellington from Waterloo, Iowa, and deal with some of the same issues as the adult books did from the perspective of the college student. *Iron Duke* begins as Jim leaves Waterloo for his freshman year at Harvard University and follows him as he makes a place for himself at the school. Failing in his efforts to make the football team and struggling with his studies and with the social life at Harvard, he is lured onto the track team by a sympathetic coach named Slips Ellis. By the end of his junior year, he has made the dean's list and been elected captain of the track team. *The Duke Decides* covers Jim's senior year. The replacement of his track coach plays havoc with both Jim and the team, as he is introduced to big-time intercollegiate track and then, as he makes the 1936 Olympic team, to international amateur athletics. The novel ends after the 1936 Olympics in Berlin with Jim's decision not to continue his career in amateur track.

These books allowed Tunis both to show his love of sport and to dramatize his criticism of it. Of the two books, *The Duke Decides* is the more critical because in sending Jim to the 1936 Olympic Games, Tunis was able to contrast athletics in the United States with those in Nazi Germany. Tunis wrote the book before World War II and after the 1936 Olympics, which Adolf Hitler used to demonstrate – some historians believe successfully – that, far from being a lunatic fringe party, he and the Nazis were the skilled managers of a modern state. Tunis also compared as well as contrasted the cultures of the United States and Nazi Germany. He uses Jim's new track coach, Dutch Coffman, to suggest both the German-ness of American amateur athletics and its basis in the economic order. The villain is "the great octopus of organized athletics" that Jim escapes.

Characteristically Tunis concludes *The Duke Decides* with an ethical decision rather than an athletic triumph. He found the standard claims for athletics offensive. In *This Writing Game* (1941) he makes his attitude clear: "Whenever you hear ponderous statements about the moral values of football, when anyone descants upon self-control and chivalry through sport, when you hear it said that sport is useful for the building of character – laugh."

*This Writing Game,* subtitled "Selections from Twenty Years of Freelancing," marked the end of one period of Tunis's writing career and the beginning of another, for though Tunis would continue to do freelance work, his juvenile fiction would become his principal source of income. While *This Writing Game* contains considerable nonsports material, mainly the articles are about sports. It includes all six articles concerning Mammoth University; the germ of *American Girl,* "Mother of a Champion"; articles in which Tunis investigates the implication of sport on the national character and political organization, such as "Derby Day," "Crowds at Court," and "The Dictators Discover Sport"; and attacks on the competitive spirit and big-time athletics in America in "Play Hard" and "What They Call Sportsmanship."

Other articles in *The Writing Game* anticipate three other books on sports that Tunis would publish at about the same time. Parts of the foreword to *Sport for the Fun of It* (1940) were reprinted. In that book Tunis considers twenty sports, in alphabetical order from archery to volleyball, that could still be played for fun. Tunis describes each sport briefly and then prints its official rules, which often take up much more space, graphically suggesting a bureaucratization of the sport Tunis would prefer to deny. Tunis's description of bowling, for example, takes seven pages, while twenty pages are needed for the "Rules and Regulations of the American Bowling Congress." In 1943 Tunis's *Lawn Games,* organized much the same way as *Sport for the Fun of It* though aimed at younger readers, was published as a volume in the Barnes Sports Library.

Tunis borrowed a good deal from "The Dictators Discover Sport" and from the last essay in *This Writing Game,* "Who's Next," in writing a small book titled *Democracy and Sport* (1941). The rise of European dictatorships in the 1930s led Tunis to contemplate the relation of political democracy and sport, a theme that would engage him in his future juvenile fiction. In introducing the book Raymond Graham Swing praises Tunis's credibility:

> It is no minor contribution to American citizenship to have striven, as Mr. Tunis has done through his writing career, to keep sports within their natural and healthy bounds. He has persistently and annoyingly assailed the evils of idolatry and commercialism in sport. He has battled against special privilege, and the formation of vested interests. No one is better qualified to write of the relationship between democracy and sport, and no one is entitled to a more respectful reading.

In addition to his already established themes, Tunis promotes the idea that true sport is an essential ingredient for an embattled democratic society.

Tunis sets forth his purpose in the book's first paragraph, asserting that he would "explain why real sport is the essence of the democratic process; why much that the public today regards as sport in the United States is undemocratic in principle; how sport can be made a testing ground for the precepts of democracy; what some of its values are; and, especially, how the valid elements of sport are part of the heritage of a democratic nation." Tunis feels that "through sport we can teach democratic behavior in practice, we can show youth when their acts offend the democratic principle in their games." Then he lists six ways democracy could be taught through sport: eliminate discrimination by race or social class, respect "inept or awkward players," develop leadership qualities, encourage tolerance, respect individual personality differences, and foster individual rather than team sports because "these sports force people to think."

Fascist Europe opposed all these practices, but Tunis was more concerned with practices in the United States. He singles out college football as a bad influence, eventually focusing on "football's treatment of the Negro player," concluding that "colored athletes are really welcomed in none of our sports." This leads him to observe that in organized baseball all players are, in effect, enslaved, and "until there is a trade union to remedy the existing injustice, this situation will doubtless continue." For Tunis, organized sport was always undemocratic sport.

Another of Tunis's concerns in *Democracy and Sport* is sport in the nation's high schools. In 1941 he felt that high school sport was more democratic than college or professional (or championship amateur) sport. In high school, he argues, "the welfare of the many as against the expert coaching of the few, is being increasingly promoted." The Works Progress Administration's (WPA) building projects of the 1930s had encouraged this spirit by constructing various grounds on which the many could play.

Tunis saw several clear parallels in the nature of real sport and the nature of democracy. In both law (or rules) reigned; equality of treatment should prevail; and the best were supposed to rise to the top. He insisted that "democracy means a share in responsible government by the people. Sport means a share in playing for everyone." Tunis elaborates on this point throughout his book. He discusses team sports, which he thought had a great potential to develop leadership and promote loyalty to a cause greater than personal gain. He also saw the "British" traits of gameness and sportsmanship as important qualities. Thus he concludes, "*Never before in the history of the United States have we so needed the spirit of democracy in our sport and the spirit of sport in our democracy.*"

While speaking of the ideals of sports, Tunis recognized in the United States the facts of racial prejudice, economic inequality, and unrestrained individualism. The conflict of these facts and Tunis's ideal animated the best of his juvenile fictions of the next few years: *All-American* (1942), *Keystone Kids* (1943), *Yea! Wildcats!* (1944), and *A City for Lincoln* (1945). Most of Tunis's juvenile works, of which there are some twenty-one, beginning in 1938 with

*Iron Duke* and ending in 1973 with *Grand National,* are concerned with sports.

Nine of Tunis's juvenile books, beginning with *The Kid From Tomkinsville* (1940) and ending with *Schoolboy Johnson* (1958), are about baseball. One, *Buddy and the Old Pro* (1955), has a little league setting, and eight are set in the major leagues, seven of them focusing on the Brooklyn Dodgers. Tunis uses the Dodgers to celebrate the ideals he had proposed in *Democracy and Sport.* Particularly the first four books – *The Kid From Tomkinsville, World Series* (1941), *Keystone Kids,* and *Rookie of the Year* (1944) – form a single action in which first the nature of team membership and leadership are defined. The fight for the pennant becomes analogous to the struggle for democracy. The remaining five novels are less compelling as a group, though *The Kid Comes Back* (1946) is a particularly successful depiction of the difficulties of a returning veteran. In Roy Tucker, the protagonist of this and the first two novels and a character in all but one, Tunis created one of the most memorable of fictional baseball players.

Tunis more effectively explores the range of his themes in nonprofessional settings. Although the conflict of sport and commerce certainly appears in Tunis's professional baseball novels, it is not as central there as in his two college novels or as in his novels of high school sports. In addition, Tunis was able in his high school fiction to address the question of race in a way not possible in the setting of professional baseball before 1948.

Tunis's three novels of high school basketball in Indiana – *Yea! Wildcats!, A City for Lincoln,* and *Go, Team, Go* (1954) – are of primary interest. The first two together tell the story of democratic leadership. The focus is on Don Henderson, a high school basketball coach who, by the end of the second book, after both losing and winning in the finals of the Indiana high school basketball tournament, is elected mayor of the city. Tunis portrays racism, commercialism, and a fixation on winning as characteristics of Indiana high school basketball, but he also celebrates the team as a metaphor for the properly functioning community and shows the experience of defeat in sports as essential in gaining maturity. For *A City for Lincoln* particularly, Tunis was able to draw on his experience in grassroots political organizing for material. Written almost a decade later, *Go, Team, Go* is a revision of the first two novels, focusing on the same problems but from the perspective of a player rather than the coach.

In *All-American,* his only book about high school football, Tunis gathers most of his main themes about sport into a single fiction. Critic Richard Shereikis cites "racism and the over-emphasis on winning," "the issue of discrimination," "the hypocrisy and commercialism of high school sports," and "anti-Semitism." The title of the book suggests the national interest in elite sport, in all-Americans, but Tunis gives the phrase social and political meaning.

*All-American* tells the story of Ronald Perry's growing up. The book begins with a big game in which Ronald's academy defeats another high school. The great victory is marred by the deliberate injuring of an opposing player, Meyer Goldman. When Ronald, appalled by what he and his teammates have done to Goldman, leaves the academy and enters Goldman's high school, Tunis makes it clear that he is leaving a world that is both snobbish and racist and entering one where political democracy is the norm, equality the ideal. The message is emphasized in a speech by Ronald's father, the only passage in the book that includes the phrase "All Americans":

> Calling names like peasants and clunks and all that is no good. It's no good in the United States. There aren't any peasants in this nation. There are just citizens, one as good as another and no better than the others, you and Goldman and Keith and all the rest of you. All Americans. Americans, all of you.

Tunis believes in the melting-pot ideal and believes that the experience of team sports promotes it. But he subjects American high school reality to severe criticism for the rest of the book. Roland finally learns that his duty as a citizen is to denounce the moral attitudes that undermine the ideal. The last fifth of the book, where the big game conventionally occurs in a boy's sport story, is given over to Roland's successful attempt to prevent his high school team from playing an intersectional game without its star black player in a segregated city.

In *The American Way in Sport,* a book he terms "personal and subjective" in his foreword, Tunis again addresses an adult audience with his thoughts on the development and meaning of sport in American life. Sixty-nine years old in the year of the book's publication, Tunis was able to speak from personal experience about the development of sport in the twentieth century. As the approach of World War II had occasioned *Democracy and Sport,* so the perceived crisis in American education following the Russian success with Sputnik in 1957 seems to have been the occasion for *The American Way in Sport*. While Russian schools pay students to become "linguists, administrators, scientists, economists," says Tunis, American schools spend their money on "football halfbacks and basketball forwards." As a solution he suggests "the divorce of education and entertainment for the masses."

While Tunis emphasizes that *The American Way in Sport* is a book of personal impressions, he believes these impressions are rooted in an understanding of American history, asserting, "Sport is a great clue to national character." Tunis remarks that in his lifetime he had seen the growth of spectator sports as well as individual participatory sports and the exploitation of the commercial potential of sport. Though Tunis is often critical of such developments, he yet believes that sports are "pregnant with greatness" and dedicates the book to the many unknown physical educators who work with youth. He announces that he will confine his focus to "the growth of athletics in this century" and show "why we Americans are as we are in sport."

For Tunis, the answer to that question lies in Frederick Jackson Turner's thesis about the role of the frontier in American history. "Sport is the American adventure of the twentieth century, and the extension of the American dream." "In athletics the American is still a frontiersman, still, like his ancestors, pressing on. As time passed, and many frontiers closed down, the heroes of American life changed from the pioneer to the athlete." In addition "any boy without money, background, or pull may rise to the top in sport." Tunis presents Theodore Roosevelt as the public figure who, at the turn of the century, "made sport respectable." Roosevelt "publicized athletics before the era of the sports page, made exercise fashionable, and, . . . promoted and furthered outdoor life all over the nation."

Tunis begins his historical accounts by examining baseball and football. He recounts his own experience at the 1903 World Series in order to show that though in midcentury America baseball has become a business, "sixty years ago the game was not fixed and rigid; its future road was in doubt." Its road in the twentieth century, according to Tunis, has proven to be from national sport to mass entertainment. Football, he argues, followed a similar path. He describes its development as being "from a rather insignificant game played by a few eastern colleges to a systematized, organized, and highly complex industry." To back his contention that football has confused "the purposes of education in the minds of the general public," Tunis calculates the costs: "Three thousand dollars per player to ready a man for sport; $200 or less per student for library service. . . . Yet we wonder why the Russians are leaving us behind in the modern world." College football star Red Grange was Tunis's principal example of the heroic athlete of the 1920s. Tunis felt that this kind of hero worship was a disease and that "it tumbled along with other illusions when the stock market crashed in 1929 and 1930."

Tunis spends a chapter on high school basketball in America, making many of the same points he dramatizes in his fiction. From his perspective high school basketball was the sport of the American small town, but he sees its progress as similar to that of baseball and football. He is fully aware of the importance of high school basketball to the community: "The high school gymnasium today in the Midwest has the role of the playhouse of the Elizabethans, the bull ring of Seville, the theatre of the ancient Greeks. Indeed the Greeks would have been entirely at home watching a basketball game in Kokomo, Indiana." But he also sees the "professional approach" seeping down to even the high school level and concludes: "Call this sport if you wish. The fact is they are there to furnish entertainment for the town."

While Tunis also argues that "the cult of golf has changed to a sober business," he spends considerable time in the same chapter praising the accomplishments of the WPA in the 1930s: "It left a mark upon our landscape. Nowhere more than in our recreation. Three thousand and twenty-six athletic fields, 1,817 handball courts, 318 ski trails, 2,261 horseshoe courts, 805 municipal swimming pools, 10,000 tennis courts, and 254 golf courses with 2,797 holes covering 18,463 acres, were built by those derided leaf-rakers in seven years." Always the democrat and lover of sport, Tunis found true sport in the recreation of millions, not in the competition for championships.

The meaning of competition is one of three contemporary tendencies in American sport to which Tunis devotes a chapter in *The American Way in Sport,* the others being government interference in sport for its own purposes and the growth of sport bureaucracy. Though Tunis is clearly concerned with the propagandist uses of the Olympic Games for Cold War purposes, his personal experience leads him to draw his principal example from the growth of the Davis Cup matches into international sport spectacles during the course of the century; he laments "that first big business and then politics should fasten upon what was merely an entertaining interlude in life." Tunis contrasts his own experience at the Longwood tennis club in 1903 with the 1937 Davis Cup battle between Germany and the United States. Bureaucracy is another concern.

Tunis views competition as both a positive and negative force in American sport. The problem he sees is an "excessive devotion to the competitive spirit at the expense of the cooperative spirit." Tunis further aligns competitiveness with conformity, a "keeping up with the athletic Joneses," and asserts, "the curse of this pointless competitive living permeates American life." On the other hand, Tunis asserts that "competition is the life-blood of sport, whether man is competing against par, against time, against a mountain or against another human being." And, for Tunis, competition is an essential ingredient of the frontier spirit that defines America. But overdone, it is harmful. In contemplating the competition between Russia and the United States that was the Cold War, Tunis concludes that "this competitive spirit has brought each country power and riches. To date there is no evidence to suggest that it has brought either happiness."

Tunis believes that there needs to be more focus on education and less on athletics. He insists he is in favor of sport, only opposed to "the excesses we have created in our sports." Sport has its important lessons to teach the young: self-confidence and determination in the face of defeat. But "most of the lessons of sport cannot be obtained from the vast spectator entertainments in our schools and colleges where winning is important and victory the goal. American education today is not bad; it is terrible." Tunis concludes his book on this note of the necessity of educational reform in the context of the Cold War.

Although Tunis may have considered *The American Way in Sport* to be his last word on the subject, six years later in his autobiography he would return to some of the same themes. Tunis is reluctant to present himself as a sportswriter in his autobiography. He says that "sport must be the theme of this book because it was my great passion as a boy, and a means of earning a living as a young man." But he presents himself not as a sportswriter but as a freelance writer. "Much of my life," he writes, "has been spent as a small businessman, as a local politician occasionally, always as an interpreter of American life." *A Measure of Independence* presents his credentials for the last and largest of these occupations.

Tunis takes pains in the course of his autobiography to document his development as a student of American history. He writes of attending evening school at Columbia after World War I and talks of the influence Frederick Lewis Allen had on him: "Fred developed in me something of his concern for our country, a concern that deepened and strengthened through my study and reading in the thirties and forties." He followed the "Harvard Reading List in American History" and also took summer-school courses in American history at Columbia. The books he cites as his main influences are Vernon Parrington's *Main Currents in American Thought* (1927–1930) and Richard H. Tawney's *The Acquisitive Society* (1920). Tunis saw himself as an interpreter of American life who just happened to be looking at American sport.

In speaking to youth groups, Tunis was always forthright in his criticism of sports: "I had to rise and explain I was merely a reporter who cared about his country, described what I saw; that I only tried to write what I observed, felt and believed about organized athletics in schools and colleges, which is seldom good and rarely clean in its effect on those who watch and those who compete." But he also asserted that "despite its excesses, sport is still the open road to the heart of a boy." He felt that the most valuable lesson sports taught was that "to accept defeat is an element in the coming of maturity, that you cannot always or even often have your own way, that the manner in which we face up to defeat and disappointment is a test of growth and part of the development of character."

In his 1974 talk with interviewer Jerome Holtzman, Tunis proved both lively and testy at the age of eighty-five, less than a year before his death. Claiming never to have been interviewed before, except over the phone once by a schoolboy from Texas, Tunis freely gave his settled opinions about sports in America, that there is "too damn much organization" and too much emphasis on winning. As throughout his career, Tunis argued for the potential value of sport in building character. Speaking of his juvenile novels, he observed, "my heroes have been the losers. All my books have been in that vein. Every book I've ever written." But it is through losing that Tunis's characters grow.

**Interview:**

Jerome Holtzman, "John R. Tunis," in *No Cheering in the Press Box* (New York: Holt, Rinehart & Winston, 1974), pp. 260–272.

**References:**

Mel R. Allen, "The Kid From Rowayton," *Yankee Magazine* (December 1989): 77–81, 116–119;

Philip Bergen, "Roy Tucker, Not Roy Hobbs: The Books of John R. Tunis," *SABR Review of Books,* 1 (1986): 85–97;

Loretta Clark, "*His Enemy, His Friend,* A Novel of Global Conscience," *English Journal,* 62 (May 1973): 730–736;

Larry Engelmann, *The Goddess and the American Girl: The Story of Suzanne Lenglen and Helen Wills* (New York: Oxford University Press, 1988);

Joseph Epstein, "A Boy's Own Author," *Commentary,* 84 (December 1987): 50–56;

Adam Hammer, "Kidsport: The Works of John T. [*sic*] Tunis," *Journal of Popular Culture,* 17 (Winter 1983): 146–149;

William Jay Jacobs, "John R. Tunis: A Commitment to Values," *Horn Book,* 43 (February 1967): 48–54;

Geraline N. Nerney, "The John Tunis *I* Know," *Top of the News,* 20 (October 1963): 31–34;

Richard Schickel, "John R. Tunis," *New York Times Book Review,* 6 April 1975, p. 43;

Richard Shereikis, "How You Play the Game: The Novels of John R. Tunis," *Horn Book,* 53 (December 1977): 642–648;

Leverett T. Smith Jr., "John R. Tunis's American Epic; or, Bridging the Gap Between Juvenile and Adult Sports Fiction," in *The Achievement of American Sport Literature: A Critical Appraisal,* edited by Wiley Lee Umphlett (Rutherford, N.J.: Fairleigh Dickinson University Press, 1991), pp. 46–61;

Smith, "John R. Tunis's Indiana Basketball Novels," *Aethlon: The Journal of Sport Literature,* 10 (Fall 1992): 9–27;

Smith, "The Realism of Roy Tucker," *SABR Review of Books,* 5 (1990): 110–120;

Robert Unsworth, "Tunis Goes Down Swinging . . . Who'll Keep the Ball Rolling?," *School Library Journal,* 24 (May 1978): 38–39;

Jerome Weidman, "John R. Tunis," *Horn Book,* 64 (February 1968): 48–50.

**Papers:**

The John R. Tunis Collection, including manuscripts, correspondence, scrapbooks, visual materials, financial records, and memorabilia, is housed in Mugar Memorial Library, Boston University.

# Charles E. Van Loan

*(29 June 1876 - 2 March 1919)*

R. C. Brignano
*Pennsylvania State University, Beaver Campus*

MAJOR POSITIONS HELD: Sports reporter and editor, *Los Angeles Morning Herald* (1904–circa 1907); reporter, *Denver Post* (circa 1907–1909); sports reporter, *New York American* (1909–1911); freelance contributor (1911–1918), associate editor, *Saturday Evening Post* (1918–1919).

BOOKS: *The Big League* (Boston: Small, Maynard, 1911);

*The Ten-Thousand-Dollar Arm, and Other Tales of the Big League* (Boston: Small, Maynard, 1912);

*Inside the Ropes* (Boston: Small, Maynard, 1913);

*The Lucky Seventh: Tales of the Big League* (Boston: Small, Maynard, 1913);

*Buck Parvin and the Movies* (New York: Doran, 1915); republished as *Buck Parvin and the Movies: Stories of the Moving Picture Game,* with an introduction by George Horace Lorimer (New York: Doran, 1919);

*Taking the Count: Prize Ring Stories* (New York: Doran, 1915; republished, 1919);

*Old Man Curry* (New York: Doran, 1917); republished as *Old Man Curry: Race Track Stories,* with an introduction by L. B. Yates (New York: Doran, 1919);

*Fore!* (New York: Doran, 1918); republished as *Fore! Golf Stories,* with an introduction by Robert H. Davis (New York: Doran, 1919);

*Score by Innings* (New York: Doran, 1919); republished as *Score by Innings: Baseball Stories,* with an introduction by Grantland Rice (New York: Doran, 1919).

SELECTED PERIODICAL PUBLICATIONS – UNCOLLECTED: "Baseball as the Bleachers Like It," *Outing Magazine,* 54 (September 1909): 642–652;

" 'Kill the Umpire!,' " *Munsey's Magazine,* 42 (October 1909): 150–152;

"Comedians of the Diamond," *Outing Magazine,* 56 (May 1910): 215–220;

"Making Good in the Big League," *Outing Magazine,* 56 (June 1910): 314–321;

"The Greatest Trout-Fishing Town in the World," *Outing Magazine,* 55 (July 1910): 387–397;

"Big Leaguers in the Spangles and Out," *Munsey's Magazine,* 47 (July 1912): 527–535;

"The National Pastime – Indoors and Out," *Saturday Evening Post,* 186 (30 May 1914): 9–11, 37–38;

"Ghost Cities of the West: Virginia City," *Saturday Evening Post,* 188 (31 July 1915): 3–5, 36–38;

"Ghost Cities of the West: The Camp of Nine Graveyards – Eureka, Nevada," *Saturday Evening Post,* 188 (4 September 1915): 6–7, 30–31;

"Ghost Cities of the West: She Came Back: Aurora," *Saturday Evening Post,* 188 (18 September 1915): 8–9, 45–46, 49–50;

"Ghost Cities of the West: Bad, B-a-d Bodie," *Saturday Evening Post,* 188 (25 September 1915): 18–19, 55–58;

"A Rank Quitter," *Saturday Evening Post,* 188 (1 April 1916): 16–18, 73–74, 77–78;

"John Jones – Goat," *Saturday Evening Post,* 188 (3 June 1916): 5, 7–9, 84–85;

"How I Broke into the Magazines," *American Magazine,* 86 (December 1918): 39–40, 118–122;

"The Hasher," *Saturday Evening Post,* 191 (8 March 1919): 5–7, 68, 70.

Not yet forty-three when he died in 1919, Charles E. Van Loan had become, in only ten years, what an obituary in the *Philadelphia Public Ledger* called the figure "who, in the opinion of the critics, had the largest following of men readers of any magazine fiction writer." Moving his focus back and forth among the baseball diamond, the boxing ring, the racetrack, and the golf course, Van Loan wrote scores of short stories for national magazines, starting in 1909. During his last seven years most of his stories, not all of which were about sports, appeared in *The Saturday Evening Post,* where Van Loan was appointed associate editor a few months before his death. Many of his sports stories were collected in eight books between 1911 and 1919; a ninth book (1915) comprises stories about the booming Hollywood film industry.

Charles Emmett Van Loan was born in the then small town of San Jose, California, on 29 June 1876 to Richard and Emma J. Blodgett Van Loan. His first job was as a stenographer at the Standard Oil Company in Los Angeles; later, working as a secretary to the manager of a Los Angeles meat-packing house, he often accompanied the manager to afternoon minor league baseball games. While taking dictation from his boss during the games, Van Loan also jotted down notes about odd happenings or humorous events on the field or in the grandstand. In his free time he expanded the notes into newspaper articles and sent them to the *Los Angeles Examiner,* which published some of them. On 20 November 1902 he married Emma C. Lenz.

Van Loan describes his entry into full-time newspaper work in his only autobiographical piece, "How I Broke into the Magazines," in *The American Magazine* (December 1918). Driven by what he calls an "author's itch," he pestered newspaper editors for assignments until, in 1904, "one day a city editor offered me a job. I guess he did it to get rid of me, but I fooled him. I accepted it." The paper was the *Los Angeles Morning Herald.* The editor immediately sent Van Loan to interview the jockey who had ridden the English Derby winner. The next morning Van Loan, who admits that he knew "no more of the technical end of that job than the rawest cub that ever manhandled a typewriter," was named sports editor in place of one who had resigned two days earlier. Around 1907 he moved to the *Denver Post* as a reporter. Of his years with the *Herald* and the *Post* he said in "How I Broke into the Magazines":

> Six years, and oh, what bully years they were.... Six years of variety – for it wasn't always baseball games and horse races – six years of looking at Life through a kaleidoscope; six years of watching the Great Human Comedy from a front seat, and getting money for it! A job? No, it was a picnic, a three-ring circus!

Van Loan first met Damon Runyon in the editorial room of the *Post,* and they continued their conversation at the bar of the Denver Press Club. They became regular drinking partners at the club, where, a few weeks later, Runyon announced to Van Loan that he was going to be a poet. Van Loan retorted by proclaiming that he wanted to be a fiction writer.

By 1909, when Van Loan became a sports reporter for the *New York American,* he had achieved a nationwide reputation. Sent to Philadelphia to cover the middleweight championship boxing match between Stanley Ketchell and Jack O'Brien, Van Loan found himself seated at ringside next to a knowledgeable and intense fan who imitated many of Ketchell's feints, jabs, and punches. Just as Ketchell was landing the left hook that knocked O'Brien out, Van Loan was knocked off his chair by a blow to his ribs from his neighbor. Apologizing, the man offered Van Loan his card, and the man turned out to be Robert H. Davis, editor of *Munsey's Magazine,* to whom Van Loan had been supplied letters of introduction by friends in Denver. On the train back to New York, Van Loan and Davis agreed to meet the next morning at Davis's office. Davis recalled: "The next morning bright and fair, he presented himself and offered a manuscript. I don't think I shall ever forget the expression on his face. His eyes were very blue, the eyebrows so blond as to be almost invisible at a certain angle; a strong, protruding lower jaw, with a tremendous physical vitality marking his every movement. When he sat down at my desk he completely surrounded it."

With Davis's help, Van Loan sold his first piece of fiction on 15 June 1909 to *The All-Story Weekly.* A racing tale, "The Drugstore Derby," it appeared in January 1910. By that time *Munsey's Magazine* had already published Van Loan's first baseball story, "The Golden Ball of the Argonauts," in September 1909. Although Van Loan would not leave his newspaper job until 1911, he had already commenced what was probably the most concentrated period of sports fiction writing by any figure in American literature – an immensely productive stretch of ten years.

In San Francisco to cover a boxing match between James Jeffries and Jack Johnson, Van Loan encountered Runyon, who by then had also published some sports stories. Van Loan, who was unsure of his ability to sustain plots, offered Runyon a deal: they would collaborate on stories, with Runyon supplying the plots; until money actually appeared, Runyon could live with Van Loan and his wife. Runyon moved into the Van Loan household in Brooklyn in 1910, and Van Loan found him a position with the *New York American* reporting on New York Giants baseball games. Whether Van Loan and Runyon ever actually collaborated on stories is unclear, and by 1911, the year Van Loan published *The Big League,* a collection of baseball stories, Runyon had moved out and gotten married.

By this time Van Loan's magazine credits had grown beyond *Munsey's Magazine* and *The All-Story Weekly* to *Popular Magazine* and *Outing Magazine*. Van Loan's second collection of baseball tales, *The Ten-Thousand-Dollar Arm, and Other Tales of the Big League* appeared in 1912; a collection of boxing stories, *Inside the Ropes,* and a third volume of baseball stories, *The Lucky Seventh: Tales of the Big League,* were published the following year. A review of *Inside the Ropes* in *The New York Times* (27 July 1913) said that nearly all the boxing tales "are better done, told with more dash and go and with better art as well, than were the baseball stories"; of *The Lucky Seventh, The New York Times* (19 October 1913) said that the book would fail to interest only "the benighted few who care nothing about our national game."

Van Loan's tales are basically character oriented; the sports action itself is less important, although Van Loan could easily and accurately portray the drama and excitement of the buildup to a knockout punch or a close stretch drive on the racetrack. The most common feature in his stories is humor, but he could also deal effectively and poignantly with serious and sad topics. Many of his tales are morally instructive; especially in his later works a recurrent theme is that justice must prevail, even if the heroes must resort to tricks and schemes that rival the chicaneries of their antagonists. Some of his stories border on sentimentality, emphasizing kindness, compassion, forgiveness, and rebirth.

An early Van Loan baseball story, "The Crab," included in *The Big League,* combines poignant, moralistic, and sentimental elements. Henry Gilman, a ten-year veteran third baseman noted, and usually denigrated, for his lackluster, quiet, and distanced personality – in his off-season life he is a hardworking, frugal Kansas farmer and a responsible family man – must confront the waning of his career because his throwing arm is losing its strength and accuracy. Relegated to the bench as his team fights for the pennant, he is put onto the field in the late innings of the deciding game by a manager who has no other choice. Gilman makes the game-saving sensational catch, emerges briefly as a hero, and, in retirement, returns to his farm to serve as a school-board member and a deacon in his church.

This idea of the noble but quiet rebirth, which sometimes occurs off the field or behind the scenes, is inherent in many other Van Loan tales. In "The Comeback," in *The Ten-Thousand-Dollar Arm,* Solomon Lee is a hugely successful pitcher during his first two years in the big leagues but then disappears from baseball because of his high living and drinking. Years later he teaches a new pitching whiz, Bud Marsh, the pitch that had been Lee's trademark. Where Lee failed, the reader is led to believe, Marsh – the vicariously reborn Lee – will not.

What Van Loan stood for in his stories was perhaps best summarized by the sportswriter Grantland Rice in his introduction to Van Loan's *Score by Innings* (1919): "Van displayed the faculty of scenting from far away any story that carried a humorous angle or a human interest twist. He made it his business not only to know the game, but to study at close range the human emotions of the players as well. . . . He soon got to know all their habits, their eccentricities, their superstitions, their general attitude towards life. . . . He found the game rich in a lore that had hardly been tapped." Rice notes that athletes were happy to provide Van Loan with material, so that "in almost every story Van wrote you could picture an actual character or living actor from the field." Furthermore, he was respected rather than resented when he focused on such painful topics as "the tragedy of the town hero one week hooted off the field a few days later – the tragedy of the has-been who doesn't see that he has come to the end of the road."

In February 1913 *The Saturday Evening Post* printed a Van Loan story about a circus clown. George Horace Lorimer, the editor who was molding the *Post* into one of the country's most influential and taste-setting magazines, took a liking to the fun-loving, robust, six-foot-two-inch-tall Van Loan, whom he swept into his inner circle of such notables as Sen. Albert Beveridge and the writers Irvin S. Cobb, Mary Roberts Rinehart, and Kenneth Roberts. Lorimer and Van Loan shared a love for the Grand Canyon, and they made several trips to the South Rim. John Tebbel says in his *George Horace Lorimer and the Saturday Evening Post* (1948) that Van Loan "enjoyed everything that was to be done at the Canyon. . . . He and Lorimer were forever moving about from one place to another, as though they were afraid they might miss something of that spectacle." Lorimer did not, however, appreciate Van Loan's boyish game of pushing boulders, some as large as pianos, down the precipitous sides of the canyon, accompanying them with yells of encouragement.

By the end of 1913 Van Loan had moved back to Los Angeles, where he could indulge himself year-round in his passion for golf; but early in 1914 he encountered financial difficulties, and Lorimer assisted him by briefly employing him on the *Saturday Evening Post* staff in Philadelphia. There Van Loan called Lorimer's attention to a short story in letter form by a Chicago sportswriter, Ring Lardner. Lorimer published that piece and five more that were later worked into Lardner's comic epistolary novel, *You Know Me Al* (1916). Lorimer also dispatched Van Loan to Chicago to entice Lardner to write regularly for the *Post.* Van Loan returned to California in the latter part of 1914 to write, play golf, and enjoy the outdoors – but a tragedy would interfere.

Returning from a two-week outing and hunting trip to Bear Valley and Deep Creek in southern California with Clyde Bruckman of the *Los Angeles Times,* Van Loan was driving down the Skyland Grade above San Bernardino when his car went out of control and plunged down a thirty-foot embankment. Van Loan was thrown from, and then run over by, the vehicle. While Bruckman received only scratches and bruises, Van Loan suffered a fractured skull, broken ribs, and a compound fracture of the left wrist. His bones were set but did not heal properly and had to be broken again and reset. Van Loan never regained the use of his left arm and hand despite two agonizing bone-graft operations in 1917 and 1918. Amazingly, he never gave up golf, and he reportedly sometimes shot in the eighties playing with only his right hand.

Van Loan's brush with death may have caused him to become more concentrated in his work as a writer. The period between late 1914 and early 1918, when he underwent his second unsuccessful bone graft, was astonishingly productive. He became a fixture on the pages of *The Saturday Evening Post* and was often featured as the lead writer. He also wrote for *Cosmopolitan Magazine* and *Collier's.* His scores of stories dealt with baseball, horse racing, boxing, track and field, newspaper reporting, circus life, and the Hollywood film industry. His Hollywood stories were collected in *Buck Parvin and the Movies* (1915), boxing tales in *Taking the Count* (1915), and horse racing stories in *Old Man Curry* (1917). For the *Post* he wrote a four-part series of nonfiction articles about ghost towns of the West that he had visited. Lorimer enlisted Van Loan in his campaign to have the Grand Canyon designated a national park, and Van Loan's article opposing private development of the canyon appeared in the 9 June 1916 issue of the *Post.*

Nearly all of the horse racing tales in *Old Man Curry* end with the victory of right over might or deceitfulness. In "Levelling with Elisha" the evil characters are thwarted by a verbal trick during a race. In "Sanguinary Jeremiah" Old Man Curry outwits selfish interests and wins a large amount of money in the process. "Playing Even for Obadiah" ends with retribution for a horse destroyed because of an injury inflicted by white men to force Old Man Curry's African American jockey out of racing. *Old Man Curry* is unusual for its time in its depiction of the dignity and humanity of blacks and its lack of white condescension. Unfortunately, although Old Man Curry names his horses after Old Testament figures and quotes and is guided by the wisdom of Solomon, Van Loan's representations of Jewish characters are frequently unflattering.

In January 1918 the second bone-grafting operation left Van Loan's arm in a plaster cast for nearly eight months and in a leather case thereafter. The year was relatively unproductive; fewer than a half dozen of his stories reached magazine pages. In June 1918 he published a collection of golf stories, *Fore!* Most of the stories in the book feature a first-person narrator who is an observer of the events rather than a participant in them. Humor remains a trademark, and people, rather than play, dominate. For instance, in "Gentlemen, You Can't Go Through" a resourceful college student teaches a lesson in courtesy and consideration to a self-centered power group at a country club. An arrogant, cocky golfer in "Little Poison Ivy" must win a crucial match before club members will accept him. In "Major, D.O.S." a combination of trickery and deftly applied psychology lead to a comic ending that hinges on the deafness of the main character.

In early November 1918 Van Loan and his family moved to the Philadelphia suburb of Wyncote so that Van Loan could take a position as associate editor at *The Saturday Evening Post.* Among his duties was headhunting for writers. Early in the following year he suffered a retinal hemorrhage, and the specialist treating him discerned symptoms of larger problems. At the Philadelphia Orthopedic Hospital, Van Loan was diagnosed with high blood pressure and a kidney condition that may have stemmed from the automobile wreck and the ensuing operations. He was transferred to Abington Hospital on 18 February and went into a continuous decline, rallying only once. On Sunday morning, 2 March 1919, he died of nephritis. At his bedside were his wife and Lorimer. Also surviving were a teenage son, Richard, and daughter, Virginia. Later that day word of Van Loan's death reached his seventy-one-year-old father, who left the church service he was attending with his second wife, went home, collapsed with grief on his bed, and died. That day's *New York Times* included a favorable review of Van Loan's final collection of baseball stories, *Score by Innings*. The reviewer commended Van Loan's "expert knowledge of the great national sport" and lauded him for being "gifted with humor of the spontaneous kind, with very often a background of pathos that gives the most satisfactory results." The following week's issue of *The Saturday Evening Post* carried Van Loan's last story, "The Hasher," a romance about a counter woman in a Mojave Desert railroad station lunchroom. In a eulogy Davis wrote of Van Loan: "He possessed the peculiar gift of characterization developed to a high degree and could cover a baseball game, a horse-race, a prize-fight, or any sporting event with fine grace and distinction. In his hands the brawn of life, the animated, playful mob, the lusty-throated fans, the Vikings of the diamond, became personalities in literature." Later in 1919 the George H. Doran Company, capitalizing on Van Loan's popularity, reprinted five of his story collections with newly written introductions by his friends and colleagues.

**References:**

Jimmy Breslin, *Damon Runyon* (New York: Ticknor & Fields, 1991; London: Hodder & Stoughton, 1992);

Clifford M. Caruthers, ed., *Letters from Ring* (Flint, Mich.: Walden Press, 1979);

Robert E. Davis, "The Late Charles E. Van Loan," *Bookman,* 49 (May 1919): 280–285;

Patricia Ward D'Itri, *Damon Runyon* (Boston: Twayne, 1982);

Clark Kinnaird, Foreword to *Runyon First and Last,* by Damon Runyon (Philadelphia: Lippincott, 1949);

Grantland Rice, *The Tumult and the Shouting: My Life in Sport* (New York: Barnes, 1954; London: Cassell, 1956);

John Tebbel, *George Horace Lorimer and the Saturday Evening Post* (Garden City, N.Y.: Doubleday, 1948);

Jonathan Yardley, *Ring: A Biography of Ring Lardner* (New York: Random House, 1977).

# Herbert Warren Wind

*(11 August 1916 – )*

David Sanders
*Harvey Mudd College*

MAJOR POSITIONS HELD: Staff writer, *New Yorker* (1947–1954, 1962–1989), *Sports Illustrated* (1954–1960).

BOOKS: *The Story of American Golf: Its Champions and Championships* (New York: Farrar, Straus, 1948; revised edition, New York: Simon & Schuster, 1956; revised edition, New York: Knopf, 1975);

*Thirty Years of Championship Golf: The Life and Times of Gene Sarazen,* by Wind and Gene Sarazen (New York: Prentice-Hall, 1950; London: Black, 1990);

*Five Lessons: The Modern Fundamentals of Golf,* by Wind and Ben Hogan (New York: Barnes, 1957; revised edition, New York: Cornerstone Library, 1962);

*On the Tour with Harry Sprague* (New York: Simon & Schuster, 1958);

*The Gilded Age of Sport* (New York: Simon & Schuster, 1961);

Jack Nicklaus, *The Greatest Game of All: My Life in Golf,* ghostwritten by Wind (New York: Simon & Schuster, 1969);

*Herbert Warren Wind's Golf Book* (New York: Simon & Schuster, 1971); republished as *The Lure of Golf* (London: Heinemann, 1971);

*The World of P. G. Wodehouse* (New York: Praeger, 1972; London: Hutchinson, 1981);

*Game, Set, and Match: The Tennis Boom of the 1960s and 70s* (New York: Dutton, 1979);

*Herbert Warren Wind's Golf Quiz* (Norwalk, Conn.: Golf Digest/Tennis, 1980);

*Following Through* (New York: Ticknor & Fields, 1985).

OTHER: *The Complete Golfer,* edited by Wind (New York: Simon & Schuster, 1954);

*Sports Illustrated: Tips from the Top,* 2 volumes, compiled by Wind (Englewood Cliffs, N.J.: Prentice-Hall, 1955, 1956); abridged as *Golf Tips from the Professionals* (Greenwich, Conn.: Fawcett, 1958);

*Great Stories from the World of Sport,* edited by Wind and Peter Schwed (New York: Simon & Schuster, 1958);

*The Realm of Sport: A Classic Collection of the World's Greatest Sporting Events and Personalities as Recorded by the Most Distinguished Writers,* edited by Wind (New York: Simon & Schuster, 1966).

Herbert Warren Wind has been the preeminent American golf writer since he published *The Story of American Golf: Its Champions and Championships* (1948). His golf pieces, along with Roger Angell's on baseball, for "The Sporting Scene" column in *The New Yorker* are significant not only for their merits but because *The New Yorker* chose to give them space commensurate with that allotted to its other articles. The best of Wind's work for *The New Yorker* takes the form of personal essays after the example of Bernard Darwin, the English golf writer who was Wind's acknowledged master. Like Darwin, Wind excels at re-creating the "feel" of a match, hole by hole and stroke by stroke, with history and tradition entering the account as naturally as the writer's meticulous notice of the terrain underfoot and the weather overhead. Wind has written with distinction about other sports, especially tennis, in "The Sporting Scene," but it is golf, which he calls "perhaps the most difficult of all the major games to play well," to which he is devoted.

Wind was born in Brockton, Massachusetts, on 11 August 1916, one of five children of Max and Dora Wind. The old shoemaking town had four golf courses during Wind's boyhood, and he had been playing on them for a few years by the time he saw Walter Hagen and Horton Smith in an exhibition match at the local Thorny Lea Club in 1929. Almost sixty years later he wrote in *The New Yorker* (18 May 1987) of Hagen that "both his ability and his presence surpassed anything I had expected." After graduating from Brockton's public schools

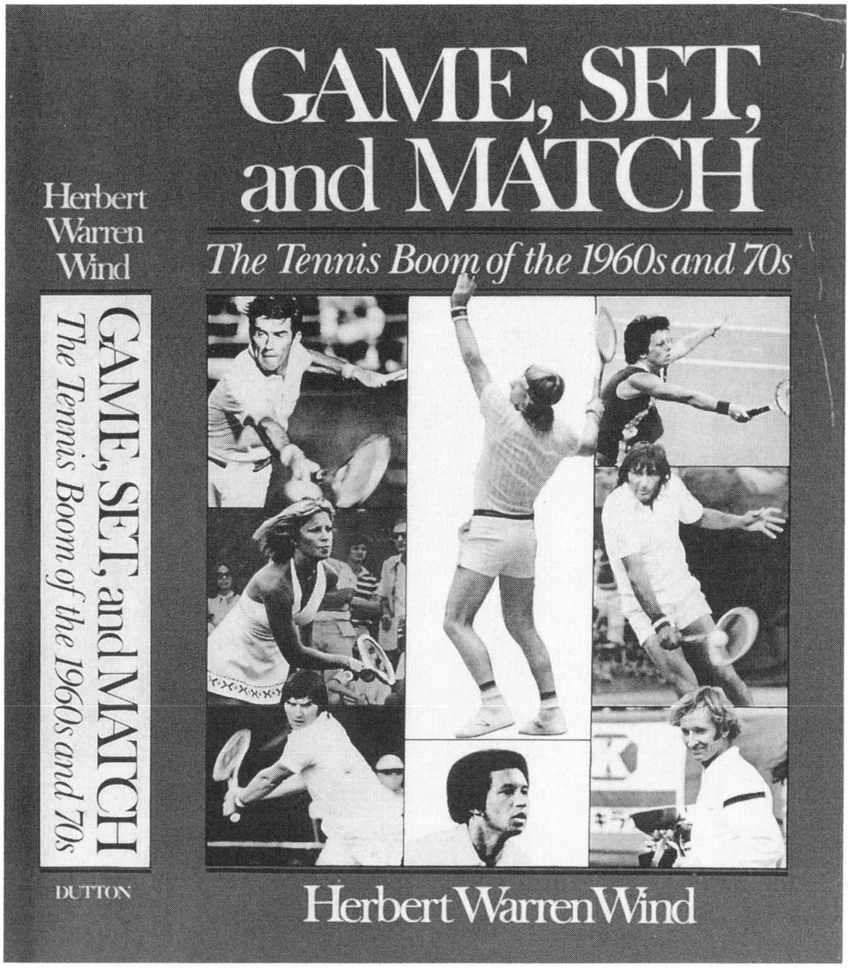


*Dust jacket for a collection of Herbert Warren Wind's* New Yorker *tennis articles*

Wind enrolled at Yale University, where he wrote his freshman thesis on golf.

He played golf whenever he could at Yale, from which he received his B.A. in 1937. As a graduate student at Cambridge University he often played on the Royal Worlington nine-hole course in Newmarket. His firsthand acquaintance with major tournament golf included the U.S. Open at Baltusrol in 1936 and Britain's only Walker Cup triumph in 1938. It is clear, though, that no golf event during that period influenced his future as much as the Oxford-Cambridge match of 1939, where he met Darwin and ran errands for the older writer. Although Wind has written that Bobby Jones was the greatest man he has known, it was Darwin, who "saw the game much more completely than anyone else who has written about it," whom he emulated professionally.

Wind received his M.A. from Cambridge in 1939. Serving in the U.S. Army Air Corps from 1942 to 1946, primarily as a public information officer, Wind was stationed in the China-Burma-India theater (where he played at the Royal Calcutta Golf Club) and, after the war, with the American Occupation forces in Japan before being discharged with the rank of captain. He became a staff writer for *The New Yorker* in 1947.

When Wind was researching and writing *The Story of American Golf* in the late 1940s, the early history of his subject still lay within the living memory of Henry O. Tallmadge, one of the six charter members of the St. Andrews Golf Club of New York, the first such club in America. Tallmadge's recollections of the "Apple Tree Gang" hitting gutta-percha balls over a Yonkers orchard establish the familiar tone of the chapters on little groups of well-to-do Americans, along with Scottish professionals, laying out the courses where, within four decades, tournaments would be won by popular heroes who would rank with Babe Ruth and Jack Dempsey. Wind's belief in a fellowship of golfers pervades his account of sixty years' play, even though, as he notes, at least five million people had taken up the game by the time his book went to press.

Tallmadge's anecdotes and others from the first twenty-odd years of American golf lead up to "the shots heard 'round the world," Wind's term for Francis Ouimet's victory in the 1913 Open. While the author neglects no significant early figures, their feats are merely a prelude to the glorious upset at Brookline, where Ouimet, a "workingman's son," had grown up across the road from the future site of his triumph over British champions Harry Vardon and Ted Ray and had caddied there. Wind's crisp account of the match is the earliest appearance of his widely praised gift for re-creating golf matches of the past. His account allows the reader to see the options Ouimet and his rivals faced at the tee and their trials on the fairway and the green.

Wind devotes the central section of his history to Bobby Jones's rather brief tournament career, beginning with his emergence as a national golfing celebrity in his teens and detailing his rise through the 1920s to become the peer not only of Hagen and Gene Sarazen in golf but also of football's Red Grange and tennis's Bill Tilden. Jones's Grand Slam in 1930 – his sweep of the Open and Amateur tournaments in the United States and Britain – was, according to Wind, "very probably, the greatest exhibition of skill and character by any individual athlete, bar none, since the beginning of recorded sports history." Illustrating both the skill and the character, Wind writes: "He placed his putter in front of the ball, as he always did to square the blade, but in doing this he cut off the breeze that had been holding the ball on the grade, and the ball moved a scant unnoticeable inch. Jones promptly called a penalty stroke on himself." After the Grand Slam, Jones – having, in Wind's phrase, "no more worlds to conquer" – abruptly retired.

Wind's *New Yorker* "Profiles" include a study of golf-course designer Robert Trent Jones (4 August 1951) that foreshadows the detailed descriptions of golf-course architecture that appear throughout Wind's later work. As a freelancer he sold as many articles on football and baseball as on golf, and he published a short story about baseball in *Collier's* (21 July 1951). Most of Wind's published references to his own golf game have been self-demeaning comparisons with professionals or accounts of his being tested on Scottish links by his local hosts, but he reached a playing level beyond the dreams of most so-called crack golfers when he competed in the 1950 British Amateur, losing his first-round match to a Walker Cup team member by 2 and 1.

*Thirty Years of Championship Golf: The Life and Times of Gene Sarazen* (1950), the first of Wind's three collaborations with noted golfers, is the story of Sarazen's rise from origins humbler than Ouimet's to a twenty-year-old prodigy's triumphs at the 1922 Open and all the glory that followed. He flourished during Bobby Jones's prime, but, perhaps because their origins were so similar, Sarazen's rivalry with Hagen stimulates Wind's best narrative efforts.

In 1954 Wind edited the anthology *The Complete Golfer.* Its 307 double-column pages of fine print includes golf writing that goes back as far as Tobias Smollett's reference to the game in *Humphry Clinker* (1771), as well as drawings by such cartoonists as Clare Briggs and H. T. Webster and foldout maps of six famous courses that accompany Robert Trent Jones's long essay on golf-course architecture.

Also in 1954 Wind became one of the founding contributors to *Sports Illustrated.* For the next six years he was the magazine's golf authority, covering all major events, writing a regular golf column, editing a golf tips feature, and even venturing into fiction about golf. He also wrote about other sports, especially hockey. His byline appears in each of the first six issues of *Sports Illustrated,* 30 August through 4 October, his only sustained period of meeting the deadlines and space restrictions that faced most other sportswriters. Even so, his reports on the 1954 U.S. Amateur are full of digressions from the action into opinion, remarks on how given shots should be played over particular segments of the course, and noticeably more history than is found in most other *Sports Illustrated* stories of the period.

A revised edition of *The Story of American Golf* appeared in 1956. The first lines of the introduction place the reader at Augusta as "the Sarazens and the Souchaks . . . stride side by side down the green fairways, and for a moment it is hard to know whether it is 1927 or 1957." The revised edition includes a new chapter, "The Age of Hogan"; but in spite of the title, neither Ben Hogan nor any other golfer – not even Babe Didrikson among the women – was as dominant as Bobby Jones had been in the Golden Age. Unfortunately, with Jones's Grand Slam, the high point of Wind's narrative, coming a little more than halfway through the book, the author still has to work through two hundred pages with "The Changing of the Guard" and "The Advent of the Modern Era," before "The Age of

Hogan" ends with the champion losing the 1955 U.S. Open to Jack Fleck.

Two short books are adaptations of material that first appeared in *Sports Illustrated.* The first, *Five Lessons: The Modern Fundamentals of Golf* (1957), a collaboration with Hogan, is an instructional treatise embellished with illustrations by Anthony Ravielli that precisely match the no-nonsense teaching. The whimsical *On the Tour with Harry Sprague* (1958) is, like Ring Lardner's *You Know Me Al* (1916), a series of fictional letters sent home by a marginal professional athlete – a "busher" in Lardner's classic, a hustling assistant club pro in Wind's book. Sprague functions in a setting that is positively genteel compared to Lardner's version of the White Sox in spring training fifty years earlier. Although Wind had noted toward the end of *The Story of American Golf* that venality was becoming "an occupational disease" among American tournament golfers, his Harry Sprague is not as crudely greedy as Lardner's Jack Keefe. Wind also lacks the ear for American speech that underlies Lardner's comic misspellings and antic syntax. The result in Sprague's letters is gentler satire than in *You Know Me Al* and a milder treatment of golf than in Lardner's "A Caddy's Diary" (1922) and "Mr. Frisbie" (1928).

Wind left *Sports Illustrated* in 1960 to become consulting writer and associate producer of the television series *Shell's Wonderful World of Golf* on NBC. He returned to *The New Yorker* in 1962 when the magazine's sports coverage was expanded from "Audax Minor's" racing columns and cursory notes on Ivy League football to become "The Sporting Scene." The great occasions for Wind in his twenty-seven-year second tenure with the magazine were the Masters each April, the U.S. Open each June, and, less frequently, the British Open, but the outcome of these events is often less significant than the digressions the play may evoke. Reading Wind on a given tournament is to be reminded not merely of an earlier Masters or Open but of Bobby Jones designing Augusta National, of courses that are "penal" and others that are "strategic," of "linksland" and the sport's Scottish endowment; and of Wind's scrambling with other "pilgrims" in a Master's gallery for vantage points learned over the years. Wind's study of tournament winners in the 1960s leads him to see that the eras of Hogan and Arnold Palmer are being superseded by the relentless progress of Jack Nicklaus, and he finds connections between Bobby Jones and Nicklaus that go back before the younger master was born: when Jones won a U.S. Open at Scioto, Nicklaus's father, Charles, was in the gallery at what would become his son's home course.

"The Call of the Masters" (28 April 1962) was Wind's first "Sporting Scene" article. A few lines deal with Palmer's victory in an essay that is really a celebration of Wind's being free to expatiate on what it means to be in the gallery in Augusta for Bobby Jones's tournament. Wind had first attended the Masters in 1947 and had immediately begun to absorb "the feeling of extraordinary kinship" with his fellow pilgrims that suffuses almost every piece he writes about the tournament. He is at least as good at evoking this atmosphere as he is at describing the play. Twenty-four years after the 1962 piece Wind's readers would know precisely how Jack Nicklaus played the last round of his unbelievable win in the 1986 Masters and, just as vividly, what it was like for Wind to be there. Perhaps nothing else startled Wind's old readers more than to pick up the first *New Yorker* golf article after his departure in 1989 and read Charles McGrath's uneasy reaction to the "religiosity" he found at the Masters, and then reflect that someone could approach this glorious weekend as a dry-eyed outsider.

The annual Masters article was frequently the setting for one of Wind's sketches of noted golfers – not necessarily the winner of that year's tournament. "Austin and Augusta" (10 June 1985) begins by acknowledging Ben Crenshaw's poor play in defense of his Masters title before launching into an admiring portrait of the young professional, who turns out to be as avid a golf historian as Wind himself. Sarazen's double eagle that won the 1935 tournament is mentioned more than once in Wind's Masters accounts, where it serves as a pretext for the writer to say much more about Sarazen before getting around to the current year's play. Above all, Wind's year-to-year coverage of the tournament develops his characterization of Bobby Jones, who was present at the event despite his long and painful illness; finally, in "Mainly about Jones" (29 April 1972) he describes the atmosphere at the Masters played the April after Jones's death in 1971.

While the Masters is played in a constant setting, the venue of the U.S. Open changes each year. The champion must meet the tests of what Wind calls a "penal course" and the bewildering changes officials contrive to toughen familiar landscapes. Wind has always been a traditionalist in golf, and "Foreign Accent," his coverage of the 1965 U.S. Open (17 July), begins with his objections to two decisions made that winter by the United States Golf Association: the end of match play for the U.S. Amateur and the stretching of the Open to four days from the more heroic three-day format in which qualifiers played thirty-six holes on the last day. "Like most people," he wrote, "I prefer to consider myself a fairly progressive type, but in my heart I know that where golf is concerned at least, my instincts are fairly conservative." Wind, who was usually generous with his praise, wrote of the "very indifferent collection of long-iron players" at Winged Foot in 1974, and at Atlanta two years later he reported a general view that "today's golfers demand too much of the courses they play and not enough of themselves."

All Wind's essays on the U.S. Open discuss the special features of the course, the earlier Opens that were played there, and the outcome of that year's tournament. Many exhibit his considerable skill with the controlled digression, such as the long sketch of Lee Trevino in the article about the 1980 Open at Baltusrol in which Trevino finished twelfth (14 July 1980). Wind's recollection of first noticing Trevino at Baltusrol in 1967 leads to his memory of Trevino's Open win the following year and his being one of the many favorites who fell by the wayside in Nicklaus's rush to the 1980 championship. Wind may have employed the device even more artfully in his article the year before (9 July 1979), in which changes in the course at the Inverness Country Club in Toledo, Ohio, include the vanishing of "old number seven," where Ray sealed his 1920 triumph over Vardon, leading Wind to remind his readers that Vardon was "the first modern golfer" because he revolutionized the golf swing with the steady head position and the overlapping grip. In due course, the reader learns that Hale Irwin, a deeply intelligent man, hung on for four rounds to win a rather dull tournament.

A superb example of Wind's work on the Open is his report from Pebble Beach (26 July 1982). As if recalling his own first round at the magnificent site, he begins by saying that "there are few experiences in sport that can compare with the thrill the average golfer or the skilled golfer receives when he plays a famous course for the first time." He describes the Old Course at St. Andrews as "a sea of erratically rolling duneland" that "must be memorized," while at Muirfield and other "frank, exceptionally straightforward" courses, "the individual holes clearly disclose the perils" to be faced. The grandeur of Pebble Beach, laid out by Jack Neville, who had never before built a golf course, confronts Wind's imaginary first-time golfer as he comes upon its eight holes that lie seaside to cliffside along Carmel Bay; the most daunting one requires a second shot to carry across an inlet, leaving the golfer with the choice of "a conservative layup" or "a career shot. He should give it a whirl nonetheless. After all, if he manages to get home in two he will have something to talk about for the rest of his life." After this Open he was moved to write: "The key to the endless fascination of golf is that it is the only game played on natural terrain or, at least, on land that, however unpromising it was originally, has been fashioned to resemble attractive and interesting natural terrain."

This is perhaps the most dramatic instance of Wind's familiarizing the reader with the course before getting to the details of a match; fortunately, the account of the 1982 action, once it is reached after a lengthy bit of historical exposition, is suited to the extraordinary setting. A deliberate Nicklaus and a higher-strung Tom Watson are tied for the lead as Nicklaus finishes his round with Watson on the tee of the long par-three seventeenth. From the rough just off the fringe of the green, Watson hits an "exquisite shot" with his sand wedge for a birdie and calms down sufficiently for a less dramatic birdie on the finishing hole. In Wind's account, this action almost seems to be a prelude to the display of sportsmanship that follows when Nicklaus ("no other loser in sports more gracious and warm") congratulates Watson, who, "apart from his superior personal qualities . . . is one of the purest shotmakers since Harry Vardon. . . . It was," Wind concludes, "just about as good as golf can get."

Of his other golf articles in "The Sporting Scene," those on his excursions to Scottish and Irish courses stand out. "North to the Links of Dornoch" (6 June 1964) has no tournament for its narrative line, only Wind's itinerary, with old friends, up out of "the rampant grime" of Glasgow into the Highlands, past Inverness to Dornoch. Many literary and historical allusions are made as Wind describes the landscape, before the first sight of the northern courses. At Nairn, Wind and his traveling companion are on the links with a local merchant, and the writer struggles with the wind ("I am not a proficient enough player . . . to control the ball when playing into a heavy wind") and the gorse ("no place to drive a golf ball into. You can seldom find it and I am beginning to think that if you can't you are undoubtedly better off"). At Dornoch, Wind plays a course whose records go back to 1616. It combines a primeval setting – "Everything *does* look as if it had been there, untouched, for centuries" – with the "touchy subtleties" of the mounds and slopes that make up its greens. Although Wind took a few more strokes than his hosts, he "played an exceedingly solid round, for me, and that was important to me. Dornoch was one course I wanted very much to like, and, say what you will, you cannot like a course unless you play it reasonably well." This candor is typical of Wind's ability to reach whatever golfing background a reader may bring to the article and give that reader an intense vicarious experience.

Wind wrote fifty-two articles on sports other than golf for *The New Yorker,* including thirty-one on tennis. On football, he wrote about placekickers rather than running backs and about Montana intrastate college rivalries rather than Notre Dame and the University of Southern California. Hockey took him to Canada and back to his childhood, when he picked up French-language broadcasts from Quebec on a crystal set. His basketball articles generally focused on the Boston Celtics. He also wrote on sports that were out of the mainstream, such as squash racquets and ski jumping.

Wind's work on tennis established him as the magazine's authority on the sport. The tone of his tennis essays oscillates between admiration for individual talent and disgust with the sport's organization. In "Trouble in Paradise" (4 August 1962) he discusses the decline in the popularity of tennis in the midst of a postwar boom in other sports; he proposes open competition at Wimbledon and Forest Hills to correct a situation in which the professional enterprise chokes off sustained, high-level amateur competition. Open competition quickly came into being, but without the results Wind had hoped for; "Tennis Troubles" (2 October 1971) and "Gold Diggers of 1974" (7 October 1974) record his disappointment. In his tennis columns Wind is a witness to a spectacle rather than the perennial pilgrim he is at Augusta. He had a wide acquaintance with people in tennis – Adrian Quist, Davis Cup competitor in the 1930s, wrote him a "long, chatty letter" from Australia (23 February 1963) – but it lacked the intimacy of his relationships with golfers such as Jones, Nicklaus, and Crenshaw. He describes tennis play precisely enough – some matches are rendered as vividly as a final round at the Masters – and, as in golf, he summons history and whatever atmosphere can be discerned in a given occasion. In "September Song" (28 September 1968), about Richard Gonzales's return to Forest Hills on the occasion of the venue's first Open championship, Wind characterizes Gonzales as deftly as he might Palmer or Nicklaus: "one of the strongest personalities in all sport . . . saturnine and disputatious, an unrepentant loner." The aging Gonzales is shown at the limits of his game, in one match recalling his dominance a decade and a half before but in the next worn down by the brilliance of the younger Ken Rosewall. Wind refers easily to Tilden and Suzanne Lenglen, but he knew Jones and Sarazen personally and could not have met any American tennis equivalent of Tallmadge or any other charter member of "the Apple Tree Gang." Though he noted instances of greed and too-rapid growth in golf, the sport continually surprised and reassured him, while tennis generally did not. He was, however, heartened by the emergence of Arthur Ashe, whom he describes in "September Song" as "intelligent, calm, honest, and natural . . . the antithesis of the spoiled American tennis hero." Many of Wind's tennis articles for *The New Yorker* were collected as *Game, Set, and Match: The Tennis Boom of the 1960s and 70s* (1979).

Nineteen years after his collaboration with Sarazen, Wind coauthored a book with Nicklaus, then the best golfer in the world. *The Greatest Game of All: My Life in Golf* (1969) is a revealing look at a twenty-eight-year-old athlete who, for all his prodigious achievement, was still apparently on the cusp of much more. The authors mix a relatively brief autobiography with the young champion's reflections on playing technique, course architecture, and the contemporary golfing scene. Businesslike and mature for his years, Nicklaus appears to have approached his work with Wind in the same thoroughgoing manner that he has shown in every other phase of his career. The book was written after the 1968 season; the poorest Nicklaus had experienced as a professional, it had forced him to realize that he was no longer "the young Jack Nicklaus." His age was, in fact, the same as that at which Jones had retired in 1930.

Wind's favorite humorist is not any of his fellow *New Yorker* writers but P. G. Wodehouse, of whom he wrote a "Profile" in 1971 that was republished the following year as a short book, *The World of P. G. Wodehouse.* According to Wind, Wodehouse wrote with "consistently high quality . . . exceptional in the marshy field of humor," and he includes himself among "the great many people in and out of the literary world" who thought Wodehouse "the outstanding humorist since Mark Twain." Wind notes of Wodehouse that, "Unlike most humorists, he is a basically cheerful man"; insofar as Wind emulated Wodehouse in his own humorous writing, this characterization may suggest why Wind's Harry Sprague is so little like Ring Lardner's Busher.

Perhaps the oddest example of Wind's devotion to golf is *Herbert Warren Wind's Golf Quiz* (1980), which gathers six years of the quizzes he contributed to *Golf Digest* and must be among the most arcane collections of trivia ever devised. Wind's humor helps the moderately well-informed reader pick the correct answer by deducing, for example, that the saying "Golf is a good walk spoiled" could not be attributed to "Max Beerbohm, the famous head pro at the Baden-Baden Racquet and Tennis Club."

*Following Through* (1985) is Wind's selection from his *New Yorker* golf articles; some are reprinted almost intact, while others are pared down to whatever he thought worth preserving in the mid 1980s. No more than a paragraph is saved from his account of the 1963 British Open, at which Bob Charles was the first left-hander to win one of the major tournaments. His headnotes are candid about himself ("The result was a much longer article than I had intended to write," he says of his forty pages on the fiftieth anniversary of the Masters), revealing about his subjects ("Gene [Sarazen] still plays the game seriously – very seriously – and still smiles like the Cheshire Cat when he comes off the course after a solid round"), and, in some cases, prophetic (as when he remarks that Nicklaus, then forty-five, "might just win his twentieth major some day").

After leaving *The New Yorker* in 1989 Wind moved to Weston, in his native Massachusetts. In retirement he has remained devoted to golf. He was the only golf historian to be interviewed for a promotional film sponsored by the U.S. Golf Association and shown prior to the broadcast of the 1994 U.S. Open.

Wind has been steadily praised for his work, beginning with *The Story of American Golf,* which drew objections only to its price. In his introduction to *The Complete Golfer* Bobby Jones cited Wind as "a fine, sensitive writer on the game whose works range from essays of the most accurately appreciative kind to some of the finest golf reporting I have ever read." Reviewing *The Greatest Game of All* in *Book World* (22 June 1969), George Plimpton noted Wind's "uncanny gift to describe past tournaments and give them immediacy and excitement." John Updike, whose own golf writings have been collected in *Golf Dreams* (1996), says in a dust-jacket blurb for *Following Through* that "Golf has attracted many fine writers, but none extols the game with more authority than Herb Wind, or more successfully conveys its gracious, fickle, generous spirit to the printed page."

# Stanley Woodward

*(5 June 1895 – 29 November 1965)*

Harold Rosenthal

MAJOR POSITIONS HELD: Reporter and city editor, *Worcester Evening Gazette* (1919–1923); copy-boy and sportswriter, *Boston Herald* (1923–1930); sportswriter (1930–1938), sports editor (1938–1948), war correspondent (1944–1945) *New York Herald Tribune;* editor, *Sports Illustrated* (1948–1949); sportswriter, *New York Daily Compass* (1949–1950); sportswriter, *Miami Daily News* (1952–1954); sportswriter, *Newark Star-Ledger* (1954–1959); sports editor, *New York Herald Tribune* (1959–1962).

*Stanley Woodward in 1962*

BOOKS: *Sports Page* (New York: Simon & Schuster, 1949);

*Paper Tiger* (New York: Atheneum, 1964);

*Sportswriter,* by Woodward and Frank Graham Jr. (Garden City, N.Y.: Doubleday, 1967).

OTHER: "Intelligentsia De La Boxe," in *Sports Extra,* edited by Stanley Frank (New York: Barnes, 1944);

"Objective Gained!" in *Best Sports Stories of 1944,* edited by Irving T. Marsh and Edward Ehre (New York: Dutton, 1945), pp. 76–80;

"Letter for Overseas" in *Best Sports Stories of 1944,* edited by Marsh and Ehre (New York: Dutton, 1945), pp. 266–268;

"The Emperor's White Horse," in *Best Sports Stories of 1945,* edited by Marsh and Ehre (New York: Dutton, 1946), pp. 252–254;

"Flare-Up on Franklin Field," in *Best Sports Stories 1947 Edition,* edited by Marsh and Ehre (New York: Dutton, 1947), pp. 105–109;

"One Strike is Out," in *Best Sports Stories 1948 Edition,* edited by Marsh and Ehre (New York: Dutton, 1948), pp. 21–25;

"Horsey at High Noon," in *Best Sports Stories 1950 Edition,* edited by Marsh and Ehre (New York: Dutton, 1950), pp. 249–251;

"The Pro Game Isn't Football," in *Best Sports Stories 1951 Edition,* edited by Marsh and Ehre (New York: Dutton, 1951), pp. 32–42; reprinted in *Collier's Greatest Sports Stories,* edited by Tom Meany (New York: Barnes, 1955), pp. 244–253;

"There Never Was a Football Upset," in *Best Sports Stories 1952 Edition,* edited by Marsh and Ehre (New York, Dutton, 1952), pp. 175–188;

"Please, No Peach-Cakes," in *Best Sports Stories 1953 Edition,* edited by Marsh and Ehre (New York: Dutton, 1953), pp. 249–250;

"Tony on the Waterfront," in *Best Sports Stories 1955 Edition,* edited by Marsh and Ehre (New York: Dutton, 1955), pp. 267–268;

"The Sea-Arm is Bowl-Bound," in *Best Sports Stories 1958 Edition,* edited by Marsh and Ehre (New York: Dutton, 1958), pp. 125–129;

"It's the *Lonely* End," in *Best Sports Stories 1959 Edition,* edited by Marsh and Ehre (New York: Dutton, 1959), pp. 125–126;

"The Big Three," in *Best Sports Stories 1960 Edition,* edited by Marsh and Ehre (New York: Dutton, 1960), pp. 110–113;

"The 1960 All-America," in *Best Sports Stories 1961 Edition,* edited by Marsh and Ehre (New York: Dutton, 1961), pp. 270–271.

Several eye operations left young Stanley Woodward with tunnel vision, a limited view of objects to his right and left sides, but clear sight of what is directly ahead. In that manner Woodward, one of the best-known American sports editors, was to pursue his livelihood, with his bottle-thick lenses locked on his target – whether in belaboring his journalistic rivals or in taking on wartime enemies.

Blind for a time before the surgeries, Woodward never became the major league catcher that his father had hoped the boy would become. But he did pitch ("fast and wild," as his friend and colleague Red Smith said), and he played tackle as a football player at Amherst College, from which he graduated in 1917. As a sports editor of the *New York Herald Tribune* Woodward left a legacy of having assembled the greatest sportswriting staff in the history of the American newspaper by recruiting writers such as Smith, Joe H. Palmer, Al Laney, and others.

Rufus Stanley Woodward was essentially an eastern newsman with national and international views. Born in Worcester, Massachusetts, on 5 June 1895, he worked on papers in New England, New York, New Jersey, and Florida. He could sense a story better than most reporters, largely because he associated with other newsmen and read all the papers, including his own.

Neither his background nor his attitudes as a journalist were particularly unusual. After college and service in the merchant marine during World War I, he got a job in 1919 on his local paper, the *Worcester Evening Gazette,* and by 1922 he had become its city editor. At the *Evening Gazette* he developed his stubbornly held notion that all who work in the news department of a paper – sports and entertainment writers included – should serve a stint as police reporters. He was convinced that such experience somehow taught one to focus the attention and become meticulous about detail. In his two books on sportswriting – *Sports Page* (1949) and the posthumously published *Sportswriter* (1967), the latter coauthored with Frank Graham Jr. – Woodward added that apprenticing on a small-town paper was a valuable source of experience.

In 1923 Woodward took a position as a copyboy with the *Boston Herald,* where he was eventually to become a sportswriter. After some years there he left for New York City, where his writing gained wider circulation: football coverage of a dozen big college teams was showcased in large Sunday sports sections of the *New York Herald Tribune,* which he joined in 1930. During his early years in New York he lived in Port Washington, Long Island, and enjoyed yachting on Long Island Sound. In 1932 he married Esther Rice of New Haven, Connecticut; the couple later had two daughters, Ellen and Mary.

Following the death of George H. Daley, Woodward took over as sports editor in 1938. For the Fresh Air Fund he devised the annual Eastern All-Star game, a promotion that featured the professional New York Giants against the pick of the graduating college football stars from eastern schools. The idea of holding a game between college all-stars and professional players was not originally Woodward's, but he refined it for regional football fans. The college players whom he selected beat the Giants only once, but proceeds that the *Herald Tribune* received from the game subsidized summer vacations in the outdoors for many city children.

His height, bulk, and the slightly sinister appearance that his thick lenses created made him stand out at any sports gathering. His reputation as a football authority, on a level with most of the important coaches, gained instant respect. "Actually, no other sportswriter (with the exception of Arthur Sampson in Boston)," wrote Quentin Reynolds, "comes close to knowing the game and the bewildering variations of offensive and defensive play as does the man whose colleagues . . . gave him the nickname of 'The Coach.' " This was an "honorary title," according to *The New York Times,* "traced specifically to his preoccupations with football" as well as to "the generally high esteem in which he was held." The quotes he drew from the college coaches he knew applied to him as well, for he wrote about sports as Frank Cavanaugh once told him football ought to be played – "to the uttermost limits of respectability." As he wrote in *Sports Page,* "Leaving symbolism and nature completely out of my work, I became unique as a football writer. Like the others, I didn't know what I was talking about, but I showed it in an entirely different way."

After World War II began and military demands decimated his staff, Woodward turned over the sports department in 1944 to Al Laney, one of his writers, and left to serve as a war correspondent – first with the 101st Airborne in Europe and then with Task Force 58 in the Pacific. When he returned to the *Herald Tribune,* his character remained unchanged: as Joe H. Palmer memorably characterized him, Woodward "is often contemptuous of superiors, barely tolerant of equals, and unfailingly kind and considerate to subordinates." Coworkers admired and respected him, but many were justifiably fearful about failing to perform up to his standards. His 3 December 1944 "Objective Gained!," about Army's first victory over Navy after five consecutive defeats, and his 26 December "Letter for Overseas" both appeared in *Best Sports Stories of 1944* (1945), the first volume in the long-running series published by Dutton.

Woodward returned from military service, for instance, to find that his staff had an opening for a baseball writer, a highly prestigious position. One of Woodward's reporters had been drafted but was hospitalized and then discharged because he could not ride on trains. When this reporter applied for the opening as a baseball writer, Woodward looked at him incredulously. "How are you going to get from here to St. Louis when the Dodgers go on the road?," he asked. "Walk?" The young man stammered an answer, but without further rejoinder Woodward picked up the phone and asked the accounting office to draw up a final paycheck for the reporter: Woodward was firing him.

Woodward's columns near the time of his military service often addressed the unusual. One concerned an art exhibit by Mickey Walker, a former prizefighter who had turned to painting. On 3 February 1944 Woodward evaluated the show by noting that "the drinks were good; the editors of 'Click,' which fostered the show, were hospitable; the Klieg lights were bright; and the painting was just about what you'd think it would be." Following his return to the paper, he lampooned in a 23 July 1945 column the dubious breeding of the Japanese emperor's white racehorse, which was "so ancient that he may not survive, even though preserved from the B-29's, until Beikoku No Kaigun Taisho ['big American Admiral' William] Halsey gets around to riding him through the streets of Tokyo." As Woodward wrote in *Sports Page,* a good columnist "is read not so much for his material as for himself. He is a stylist and in a sense an artist, and he has a following which reads him day after day no matter what he writes about."

On 9 May 1947 Woodward and the *Herald Tribune* broke one of the most important stories of the time, that of the Saint Louis Cardinals' threatened strike against the presence of Jackie Robinson when the Brooklyn Dodgers were to make their initial appearance of the season in Saint Louis. One of Woodward's writers got the story during a drinking bout with a Saint Louis official. Because the reporter did not want to reveal his identity to baseball officials in Saint Louis and elsewhere, he turned over the story to Woodward, his editor. The front office of the *Herald Tribune* suggested that the story, "Anti-Negro Action," be moved to the front page of the paper, but Woodward, who believed in the importance of the sports section, retorted, "Why bury a good story on Page One?"

The response to this story was strong and immediate. Woodward had quoted the National League president, who had threatened to banish anyone who participated in the strike, which defused the situation. Red Smith wrote of Woodward's general handling of the story that no one could have written "with more smashing impact. He could deflate the pompous, expose the ridiculous or explode a fallacy with one devastating line." In a follow-up story the next day Woodward continued:

> The blast of publicity which followed the New York *Herald Tribune*'s revelation that the St. Louis Cardinals were promoting a players' strike against the presence of Jackie Robinson, Brooklyn's Negro first baseman, in the National League, probably will serve to quash further strolls down Tobacco Road. In other words, it can now be honestly doubted that the boys from the Hookworm belt will have the nerve to foist their quaint sectional folklore on the rest of the country.

"Woodward's story," Reynolds noted, "read by every baseball owner, had the effect of stiffening their backbones to the point where they themselves laid down the law to their players – and from that point on everything was relatively sweetness and light." Franklin P. Adams added, "It demonstrated the no-discrimination proposition before either *Gentleman's Agreement* or *Crossfire* had been shown in the motion-picture palaces." The public reaction to the story added force to the wave that swept aside the last racial barriers in baseball, and Woodward's story won the Dutton Best Sports Stories prize as the best news story of 1947.

Woodward was at the peak of his own journalistic strength for several postwar years, and he demanded from his staff as much loyalty as he gave to it. Able to handle day-to-day prose with the best in the business, Woodward had the additional strengths of columnist Smith and horse racing ex-

**With the College Athletes In Their Diversified Affairs**

By Stanley Woodward

FRITZ CRISLER, the gent who snatched the Princeton brand from the burning, wishes to know whence comes all the conclusive information to the effect that Old Nassau has a guaranteed champion in readiness for the 1933 football season. He himself feels that there will be some improvement in offensive strength this season, but beyond that he will not go. Viewing the situation extramurally, he thinks that Columbia will be as strong as last year, that Yale will be stronger, and that Navy will place in the lists a team which will surprise every one.

Mr. Crisler attended a luncheon of the Downtown Harvard Lunch Club last week and made a speech in which he revealed that he once had been offered a job as end coach at Harvard and in which he said nothing about the renewal of football relations between the Codfish and the Tiger. "I knew nothing about the things that brought about the break, so why should I talk about the thing at all?" he said.

**Must Include Tiger Crew**

If the Los Angeles Chamber of Commerce presists in overlooking Princeton in the selection of the field for its "National Intercollegiate Regatta," it would seem necessary for the sake of accuracy to drop the first word in the event's title. Princeton, on its record, certainly qualifies for admittance to any national intercollegiate event. It administered severe beatings to two of the Los Angeles entries, Cornell and Harvard.

College, of Columbia University, and students who attend may earn two points credit toward a master's degree in physical education. Lou and Tuss plan to make the course both intensive and extensive. They plot detailed delineation of the football they themselves teach and an exposition of the other more successful methods. Several other coaches will aid in the instruction.

The eight-oared crew of Rollins College, Winter Park, Fla., is scheduled to depart today for Boston, where it is to participate in several races with school eights. Rowing is a new sport at Rollins and has developed slowly because of lack of competition. This spring, however, the Rollins oarsmen have had practice races with Asheville School, of Asheville, N. C., and Indian River School, of New Smyrna, Fla. Only one of the Rollins oarsmen is a Floridian. The rest are Yankees. The roster follows: Edward N. Jenks, Haverford, Pa., stroke; Edward H. Bonelli jr., West Newton, Mass.; Norman B. Weston, Wilton, Conn.; George W. Edwards, captain, Orlando, Fla.; Horace P. Abbott jr., Melrose, Mass.; Jay

*Column by Woodward from the 28 May 1933* New York Herald Tribune

pert Joe Palmer to provide the finest inside-stuff journalism and writing styles.

Woodward had acquired Smith from the *Philadelphia Record,* and Palmer, an instructor of English at the University of Kentucky, had arrived a short time later. Both had required little persuasion. Smith had previously come from the Midwest to work on one of the cheaper Philadelphia papers, where he found he needed to augment his paycheck by writing pieces for children. Palmer had found the job of a college instructor not particularly attractive in the 1940s. This talented pair was installed in the fifth-floor editorial room of the *Herald Tribune* – and shortly thereafter in the adjoining downstairs building, in a former speakeasy called Bleeck's (pronounced "Blake's"), a storied staff hangout that also attracted many nonjournalists. Woodward, also an imbiber, also used Bleeck's as a strategy room for coaches whom he brought in to tutor his college all-stars. Important newspaper personalities from other New York papers would also converge on this watering spot to drink and to enjoy a game of betting how many matches were in the grasp of a rival. The loser was to pay the bar bill.

Woodward's self-appraisal demonstrates the leanness in prose that he valued. "I want to be remembered," he said, "as someone who tried to use good English and tried to get others to do it." He felt about writers and the quality of their work as he did about athletes: "amateurs," such as college football players, were supposed to be trained by professionals – those who performed with unmistakable expertise for weekly paychecks – to become professionals. He wanted others to train amateur writers to be the professionals needed to put out his or any other newspaper section. His resultant rage when naive, job-hunting sportswriters told Woodward about their love for sports being so great that they would "write for nothing" shocked prospective applicants.

In 1948 Woodward was fired as sports editor of the *Herald Tribune* by the recently widowed Helen Rogers Reid, who had assumed the duties of president of the paper after the death of her husband the previous year. In response to a management memo requesting each department head to provide two names of staff members whose positions might be eliminated in possible staff reductions, Woodward had peremptorily listed his name above that of Smith on the form. When Woodward soon afterward refused Mrs. Reid's request to assign a reporter to cover an event he regarded as trivial, he was finished with his first term as sports editor of the paper.

When Woodward left the *Herald Tribune* in spring 1948, his first book, *Sports Page,* had been completed and set in type, but the galleys required much correction. Nothing had altered his professional opinion, and he praised his old paper highly, but he had to change his presentation of everything in the narrative to the past tense. *Sports Page* was to become the textbook on how to run a sports department, and it argued for proper recognition of the sports page and sportswriters. "The sports department," Woodward wrote, "should be organized on the same principle as the city department. Some one must judge the news, both in advance and after it happens. And some one must sit on the copy desk to write the heads and put the paper together." The book garnered rave reviews from Rex Lardner, Lloyd Lewis, and Smith, Woodward's former employee. The *Springfield Republican* found *Sports Page* to be "the cleanest exposition yet written as to what actually happens in the sporting page that you read every morning."

Soon after Woodward had left the *Herald Tribune* he joined Dell Publishing, where Albert Delacourt had plans for a new sports magazine, *Sports Illustrated,* that Woodward was to edit. The magazine was able to appear for only a few issues, however, before the publisher, attempting to cut its financial losses, sold the title to Time, the publisher for whom the new journal was to become one of the big successes among postwar sports magazines.

In 1950 *Collier's,* aware of Woodward's love for college football, approached him about the possibility of writing an article comparing the college football game to the financially struggling professional game. For such an article Woodward and any fellow contributor were promised $2,000 – a big fee for a single magazine article in the early 1950s. Could Woodward get a recognized name in the college coaching fraternity to agree with him that a good college team could beat a good professional team? Recalled Earl "Red" Blaik, the distinguished Army football coach, "Stanley told me of the money in it and I told him I wouldn't dare send my best army team out against the Cleveland Browns, but he gave me that hard look, and the next thing we had a story claiming just that. Stanley said he could use the money. I gave him my end of it as well."

"The Pro Game Isn't Football" appeared in *Collier's* on 28 October 1950 and won the 1951 Dutton Best Sport Stories prize for best magazine story. Woodward had caught Blaik in a most voluble mood. According to the article, professional teams are essentially "show business" and lack the aggressiveness of the college game, with its "youthful fire and emotional life." Professional teams had fallen in love with the forward pass, according to the story; college teams had frequent scrimmages, kept the running game alive, and encouraged players to pursue receivers and learn downfield blocking. As author-interviewer, Woodward had cast himself in the role of the doubting Thomas. Thinking of the professional game, he admitted, as did Blaik, that he found it entertaining. "But was it football?," he wondered.

Woodward carried the point further in another football piece for *Collier's* on 1 December 1951. Focusing on the Navy team's 1950 upset of an Army team that had been unbeaten in twenty-eight previous games, Woodward's conclusion is borne by the title of the article: "There Never Was a Football Upset." Through interviews with Navy coach Eddie Erdelatz and others, Woodward showed that "Navy was prepared for that game with a technical thoroughness that matched its spirit. Its scouts had put the microscope on Army throughout every game on the schedule. Its coaching staff had studied movies of previous Army-Navy games a hundred times." Finding fundamental weaknesses in the Army team, Navy had been able to defeat Army handily. Woodward examines other "upsets" only to find that

> The results which we call upsets are painstakingly worked out on the drawing board in advance. Back of each upset is a scouting report and a plan of campaign which often have taken months to devise. If the team's spirit is high, the scouting keen and intelligent, the plan of campaign clever, the underdog becomes the better team and wins the game.

The coaches know this, Woodward concludes, but "the coaches don't always say what they know."

After his abortive *Sports Illustrated* start in magazine journalism, Woodward moved to the *New York Daily Compass,* a tabloid that sold for three times the price of the other New York papers. His most memorable piece for the *Daily Compass* was a 2 November 1949 story, "Horsey at High Noon," about the sixty-first National Horse Show at Madison Square Garden – from which, Woodward writes, "all forms of sport will be barred" for more than a week after this event in the future. Noting that "at horse shows one encounters at least 10 rear ends for every front end of the horse," Woodward unleashes his contempt for such events. The three main functions of a horse, he insists, "are, in the order of their importance: a) 'Running for the money of my clients'; b) Plowing on hillsides; c) Helping out in war-time by acting as an ingredient in K rations." Woodward finds horses additionally useful as hillside ornaments or as creatures "perform[ing] the valuable function of kicking and biting horse-lovers."

When the *Daily Compass* folded, James Middleton Cox and James McMahon Cox hired Woodward in the hope of infusing some life into their *Miami Daily News,* which was steadily losing ground to the *Miami Herald.* Having watched plenty of southern football amidst plenty of Florida heat and humidity, Woodward later referred to his brief stay with the Miami paper as existing in a "green hell." While there he derisively tagged the football program of the University of Miami as "Ol' Sun Tan U." In a slightly facetious story dated 18 July 1952, "Please, No Peach-Cakes," he expressed views that would have offended many readers in later decades: "Even when a female attains horsiness of Calumet calibre, her track and field performances do not rise much above the status of the intermediate men's Sunday School picnic," he wrote, and he requested that not one penny of his gift to the Olympic Fund be used to "equip or maintain a single peach-cake in Helsinki, Finland," the site of the 1952 Summer

Olympics, where she could be "butchered by the Muscle Molls of Moscow." Woodward concluded that women could excel at a few sports (especially golf and tennis), but he preferred "our own women unathleticized. They're tough enough as it is."

By 1954 Woodward had returned to the North, where he had run his wartime farm in Princeton, New Jersey, in addition to doing his newspaper work. Publisher Samuel I. Newhouse soon sought out Woodward for the sports department of the profitable *Newark Star-Ledger*. Woodward appeared to like the role and the Newhouse newspaper family, which ultimately became the largest publishing empire in the United States.

The quality of his prose was best during his years on the Newark paper. Even with his faulty eyesight, Woodward saw more of what was happening on the field than others with better vision saw, and his analyses of games were written with pungent leads, as in this 1 December 1957 opening:

> The dynamic running of Captain Ned Oldham, the pitching and play direction of Tom Forrestal and an intelligent ferocity which reached the uttermost limits of respectability carried Navy to victory over Army, 14–0, in their 58th annual game today. A crowd of 102,000 in Municipal Stadium sopped up tons of rains as the Midshipmen took all the sting out of an Army team which had been rated unstoppable.

He continued to write his witty, salty essays on various sports-related subjects, such as a trip to the movies to see Elia Kazan's *On the Waterfront* (1954) with boxer "Two-Ton" Tony Galento, who had acted in the film, and a complaint about how other newspapers in 1958 were mangling the name he had given to the new Army offensive alignment, "the 'lonely end' formation." Their name for it, the "lonesome end" formation, "may be just as good if no better," he insisted, "but still brands the corrupters as finale-hopping second-guessers."

About the time that Woodward was about to become sports editor for the entire family of Newhouse papers in 1959, the *New York Herald Tribune* decided to change its sports editor. Red Smith, still the paper's sports columnist, was consulted. Smith asked, "Why not Woodward?" Eleven years later, in returning to replace the man who had replaced him, Woodward began his first column: "As I was saying before I was so rudely interrupted, . . . ." Lacking the vitality he had shown in his earlier years, however, Woodward was returning to a situation beyond his control.

John Hay Whitney had bought the paper from Helen Rogers Reid in 1958, and the *Herald Tribune* was in a losing battle for advertising revenue with *The New York Times,* which had ballooned to three times the size of Whitney's paper. Former magazine publishers controlled the editorial direction of the *Herald Tribune,* and one of them –John Denson, formerly of *Newsweek* – began wryly ridiculing Woodward and his columns. Woodward wrote, for instance, one tongue-in-cheek history of Princeton-Harvard-Yale: "There is one other Harvard academic course we forgot to mention. It is called 'The Archeology of Middle America.' . . . Probably a study of why the Western Conference gets all the tackles." He also compiled a parody of the yearly college all-star teams. Woodward's 1960 squad listed punning names of players and college teams such as Yurinda, Army; DeBosifromme, Syracuse; Proto, Col.; and Tregrozin, Brooklyn – despite the fact that "Brooklyn has no football team." Whatever Denson may have thought of pieces like these, the editors of the Dutton Best Sport Stories series included both of them in their volumes for 1960 and 1961, respectively. Woodward won his third and final Dutton Best Sports Stories prize in 1962 for a 1961 *New York Herald Tribune* news-coverage story about the Princeton football team's coming from behind to defeat Columbia.

Woodward resigned as sports editor in 1962 and retired to Connecticut, where he did some radio broadcasting and finished his second book, *Paper Tiger* (1964). The title of this autobiographical work was not original, but it described him well. Woodward's narrative offered what *The New York Times* book review called "a blow-by-blow exposition of the rewards and perils of a sportswriter's lot," some tales from the press box and Bleeck's, and some dismal predictions for the future of American journalism. Critics generally approved of the book, but the *Saturday Review* found it "flawed by its quixotic organization," probably because that literary journal might have preferred a straightforward sports memoir rather than the casual, anecdotal approach that the book takes.

Having suffered from the long-term effects of pulmonary and circulatory disorders, Woodward died on 29 November 1965. *The New York Times* called him "A purist in the use of language [who] possessed a rare ability to make words serve his command. While whimsy was not beyond him, the mace rather than the scalpel and the broadsword rather than the rapier were his favorite verbal weapons. And he struck out against incompetence, pomposity, hypocrisy and duplicity." To Smith, Woodward had been "the finest of craftsmen, a leader who commanded loyalties almost as fierce as the loyalty he gave, a giant of integrity, dedication and honor, a brooder, a worrier, a fighter, and the dearest of friends."

In his history of the *Herald Tribune,* Richard Kluger adds that "during the twenty years he graced them, the sports pages of the *Herald Tribune* achieved an unprecedented level of pungent literacy." Roger Kahn, who served Woodward as a copyboy and later became his friend, recalled Woodward's demanding editorial eye:

> Whoever the author, Woodward would not tolerate horrendous clashes of fearsome Tigers and snarling Wolverines, concluded in purple sunsets. He wanted, he said, to present his department "in such a manner that it would be intelligible to the nonsports reader, should one happen to fold back the paper at the wrong place." He liked verbs of action and exact adjectives.

Woodward's daily job provided metaphors for his own summary of his goals, immediate and long-term: "I don't care who wins or loses. All I care about is that they play a reasonably close game so that I'm not writing about some farce or a mismatch which should never have been played."

**References:**

Roger Kahn, *The Boys of Summer* (New York: Harper & Row, 1972);

Richard Kluger, *The Paper: The Life and Death of the* New York Herald Tribune (New York: Knopf, 1986).

# Dick Young

*(1917 or 1918 - 31 August 1987)*

Jack Ziegler
*Cochise College*

MAJOR POSITIONS HELD: Sportswriter, columnist, sports editor, *New York Daily News* (1942–1982); sportswriter and editor, *New York Post* (1982–1987).

BOOK: *Roy Campanella* (New York: Barnes, 1952).

OTHER: "Battle on the Plains," in *Best Sports Stories of 1944,* edited by Irving T. Marsh and Edward Ehre (New York: Dutton, 1945);

"How Gamblers Move In," in *Best Sports Stories: 1945,* edited by Marsh and Ehre (New York: Dutton, 1946);

"Dodgers on the Rocks," in *Best Sports Stories: 1955,* edited by Marsh and Ehre (New York: Dutton, 1955), pp. 75–76;

"The Outlawed Spitball," in *Best Sports Stories: 1956,* edited by Marsh and Ehre (New York: Dutton, 1956);

"From Bottom to Top," in *Best Sports Stories: 1960,* edited by Marsh and Ehre (New York: Dutton, 1960), pp. 23–26;

"It Isn't Over Yet," in *Best Sports Stories: 1961,* edited by Marsh and Ehre (New York: Dutton, 1961), pp. 19–22;

"The Big Leagues' Iron Curtain," in *Best Sports Stories: 1964,* edited by Marsh and Ehre (New York: Dutton, 1964);

"And What a Battler He Was," in *Best Sports Stories: 1966,* edited by Marsh and Ehre (New York: Dutton, 1966);

"Prince Charming Takes Cinderella to the Ball Game," in *Best Sports Stories: 1968,* edited by Marsh and Ehre (New York: Dutton, 1968);

"The Joe Namath System," in *Best Sports Stories: 1971,* edited by Marsh and Ehre (New York: Dutton, 1971), pp. 245–247;

"Roberto . . . 3,000 Wasn't Easy," in *Best Sports Stories: 1973,* edited by Marsh and Ehre (New York: Dutton, 1973), pp. 81–83;

*Dick Young (UPI/Corbis-Bettmann)*

"The Barbie Doll Soap Opera," in *Best Sports Stories: 1977,* edited by Marsh and Ehre (New York: Dutton, 1977);

"1947: Brooklyn Dodgers 3, New York Yankees 2," in *The Baseball Reader,* edited by Charles Einstein (New York: Bonanza Books, 1989).

SELECTED PERIODICAL PUBLICATIONS – UNCOLLECTED: "He Walked With the Stars," *Coronet,* 46 (July 1959): 25–28;

"Those Weak-Kneed Jets," *Look,* 32 (29 October 1968): 72–74+;

"It's Religion, Baby, Not Show Biz," *Sports Illustrated,* 38 (9 April 1973): 54–56;

"Men with a Yen for the Fences," *Sports Illustrated,* 41 (11 November 1974): 84, 89, 92;

"Can a Starting Pitcher Be an MVP?," *Sport,* 77 (April 1986): 52;

"Muhammad Ali," *Sport,* 77 (December 1986): 18–19, 136.

For millions of sports fans – and for more than forty years – Dick Young was the premier sportswriter of the *New York Daily News,* the newspaper that traditionally has had the largest circulation in New York City. In the 1940s and 1950s the *Daily News* arguably employed the best sports staff in New York – Joe Trimble, Jim McCulley, Jimmy Powers, Gene Ward, and cartoonist Bill Gallo, whose caricatures adorned Young's column. But Young was the star of the group. Ironically, when he first joined the *Daily News* in 1937, he was not interested in writing about sports: "I wanted to be a hotshot newspaperman like Walter Winchell," he told Ross Wetzsteon. "I wanted to be a stop-the-presses guy, competing with the other paper for the scoop and for the girl." In "Dick Young's America," his article in *Best Sports Stories: 1986,* Wetzsteon estimated that Young wrote four thousand words a week and more than ten million in his career. Typically he wrote five to seven "Young Ideas" columns per week and covered a baseball team six days a week. In his tribute "Dick Young, In His Time," (*The New York Times,* 2 September 1987) fellow reporter George Vecsey recalled that in the early 1960s he could hear Young working all morning down the hall. First Vecsey would listen to Young pounding out his "Young Ideas" column on a typewriter, then hear him phone it back East, then Young would leave to cover an afternoon or evening Mets game. To Vecsey's knowledge, Young was the only reporter he knew to perform continually such grueling double duty.

Wetzsteon identifies Young as a key transitional figure who steered sportswriting away from the gentlemanly pomposity of Grantland Rice, Arthur Daley, and their imitators at *The New York Times* and the *New York Herald Tribune.* As Wetzsteon notes, Young "singlehandedly replaced the pompous poetry of the press box with the cynical poetry of the streets." Randall Poe, in his seminal article on sportswriting in America, "The Writing of Sports" (*Esquire,* October 1974), types Rice and his ilk as the developers of a "Hero Sandwich" school of sportswriting, turning athletes into Homeric figures. Young approached sports from another angle. Trusting in his public's expertise, he wrote authentic, accurate accounts of games and players. His style was breezy, realistic, and resonant with the human voice – the New York human voice. As the *National Review* put it,

> Young was combative, direct, and sometimes disrespectful. Howard Cosell called him a "right wing, cultural illiterate." Young for his part despised Cosell as "Howie the Shill. A fraud. An ass." Young's youthful hard times often showed up in the personal values to which he held. While he supported Jackie Robinson's breaking of the baseball color barrier in 1947, he eventually believed that Robinson, like many black athletes, became the victim of a persecution complex. As Young said to Harry Waters (in *Newsweek,* 21 May 1973): "Some of these black guys believe that everything bad that is happening to them is happening because of their blackness. It's a terrible crutch."

Young never completely outgrew the attitudes he developed in the 1930s and 1940s. He was severely critical of the new generation of athletes that included Tom Seaver, Dwight Gooden, Muhammad Ali (whom Young persistently referred to as Cassius Clay years after Ali had legally changed his name), and Joe Namath. He considered them whiny, self-centered children. With other great satirists like Jonathan Swift, Samuel Johnson, and Mark Twain, Young always looked back to a golden age filled with heroic men like Rocky Marciano, Babe Ruth, and Roy Campanella. Occasionally he found a modern throwback like Joe Frazier, Kenny Norton, or especially Gil Hodges (then managing the Mets), whose strong work ethic, modesty, and poise set him off from his self-indulgent peers. Young was above all a moralist who applied high standards to athletes and management in both amateur and professional sports. As his colleague and friend Phil Pepe said in the *New York Daily News* (1 September 1987), Young was the "conscience of sports, the protector of the fan." Millions of Young's readers would have agreed.

Young was born in either 1917 or 1918 in the Bronx section of New York. His parents – his father was a Russian Jew and his mother an American Jew – divorced when he was three, and from ages six to twelve he boarded with an Italian Catholic family. He grew up in the lower-middle-class Washington Heights neighborhood of Upper Manhattan, graduating from George Washington High School. He then migrated to California to live with his father, a cameraman in Hollywood, and to attend Los Angeles Junior College. Unable to raise the seventy-five-dollar nonresident tuition fee ("All the money in the world," he noted), he moved back East and worked with the Civilian Conservation Corps, building state parks in upstate New York for thirty dollars a month.

In 1937 Young made his own big break. Hitchhiking to New York City, he talked himself into an eighteen-dollar-a-week messenger-boy job at the *New York Daily News* by promising to attend New York University at night. He remained with the *News* for forty-five years, rising from messenger boy in the publicity department to tabulator in the sports department, then beat reporter (with the Giants in 1943, the Dodgers in 1946, and the Mets in 1962), star columnist (his "Young Ideas" beginning in 1963), and sports editor. In 1982, in a move regarded as treasonous by many of his readers, he broke his contract with the strike-threatened *Daily News,* which could not guarantee his $131,000 annual salary, and jumped to a rival metropolitan tabloid, the *New York Post.* Ironically Young's move occurred just five years after he had lambasted New York Mets pitcher Tom Seaver for trying to renegotiate his contract. Young also wrote a column for *The Sporting News* from the late 1950s to 1985.

Despite a $1.5-million lawsuit brought by the *Daily News* for breach of contract, which was ultimately dismissed, Young remained at the *Post* until his death in 1987. He garnered many honors during his lifetime, including the presidency of the Baseball Writers' Association of America, the James J. Walker Award (from the Boxing Writers Association of America) for a lifetime's meritorious service to boxing, and election in 1978 to the writer's wing of the Baseball Hall of Fame. At his death at age sixty-nine, he was still writing, in semiretirement, two columns a week for the *Post.* At his peak he earned $150,000 annually, which made him one of the highest paid sportswriters in America. Young and his wife, Jay, had one son and seven daughters.

Young began writing sports in 1941 and by 1942 was earning a byline. Early in his career his biggest thrill was scooping New York in his account of Freddie Fitzsimmons's replacing Bucky Harris as manager of the Philadelphia Phillies in 1943. By 1944 Young was well on his way to journalistic stardom, making the first of at least seventeen appearances in *Best Sport Stories of the Year* for his 5 March 1944 *Daily News* story, "Battle on the Plains." His story centers on Army's 47–40 basketball win over Navy. Typical of a Young story, the piece does not concentrate on the obvious, the contributions of the two major scorers, Dale Hall and Doug Kenna. Rather, with his gift for seeing what others might miss, Young singles out the role of John Hennessey, who scored only three points. Young's practiced eye could see that Hennessey's hustle and spirit fueled the élan of the Army squad. With Hennessey's absence because of foul trouble, Army morale sagged and Navy made a runaway game a close one. Even in spot reporting, Young reveals a superb sense of narration. He does not provide a mere rehash of scoring and big plays. Instead, he identifies what made the contest unique – the intangible but very real contribution of Hennessey.

In 1945 Young was again included in *Best Sports Stories,* this time for a freelance piece he wrote for the April 1945 issue of *True.* Cowritten with another youthful reporter, Leonard Lewis of the *New York Daily Mirror,* "How Gamblers Move In" examines the subtleties of illegal betting. The article is notable for Young's knowledge and powers of analysis as it traces the evolution of the "middle-line" (a 13–15 point spread) to a "half-point line" (a team would be a 13½-point favorite) familiar to current bettors. Throughout the piece the reader senses another salient Young quality – his disdain for hypocrisy. In many ways "How Gamblers Move In" is a plea for legalized betting. Young sees friendly wagering as part and parcel of organized sports despite the disapproval of the more straitlaced. Young further notes that honest bookies – by far the majority – make only a modest profit of 2½ to 15 percent, a considerably smaller profit margin than most "legitimate" businessman.

One of the most interesting aspects of the article is Young's discussion of "Ben Hecht," the biggest bookmaker operating in America. In 1945 "Hecht" ran a highly organized operation from Minneapolis, coordinating a nationwide network of informants and overseeing a talented staff that perused out-of-town newspapers for injury reports and shifts in team morale. Young notes how "Hecht" laid great emphasis on home-field or home-court advantage. As a result, the Great Lakes Naval Station football team and the Penn State basketball team became prohibitive favorites at home.

In 1947 Young was entering his second year as beat reporter for the Dodgers. In April of that year he covered three big related stories – Jackie Robinson's making the Dodgers, Happy Chandler's suspension of Dodger manager Leo Durocher, and Dodger general manager Branch Rickey's handling of both crises. Much like Rickey, Young handled Robinson's chance of making the Dodgers in a low-key, matter-of-fact way. In a 6 April 1947 *Daily News* story, Young shows a tentative Robinson, unsure whether he will wind up playing second base for the Royals or playing somewhere in Brooklyn's infield. As Robinson told Young, "Like they sing in that song, 'I'm just a sittin' and a waitin' And a hopin'." Yet in the same article Young astutely wonders if Rickey hadn't already made up his mind (he had) and was letting the rest of the world catch up. By 17 April 1949 Young's *Daily News* article "Robinson

*Young (right) with Yogi Berra, manager of the New York Mets, 17 October 1973 (UPI/Corbis-Bettmann)*

Hunts Home, Looks Like He'll Stick" reveals that a landmark decision had just quietly happened. He also perceptively notes that two Dodgers southerners, Pee Wee Reese and Hugh Casey, volunteered a lot of their time to help Robinson hone his skills to a professional level.

Brooklyn manager Leo Durocher had been at odds with baseball commissioner Happy Chandler for some time because of Durocher's association with gamblers and his frequently abrasive personality. Young therefore reacted skeptically to the Lip's newfound conversion to sweetness and light. This story provides one of the first examples of Young's characteristically vitriolic reaction to hypocrisy: "Leo Durocher, who has surrendered himself to goodness like some sin-shunning revivalist, has taken the initiative in baseball's clean-up crusade."

Durocher's conversion happened too late. Young's 10 April 1947 story, "Chandler Bans Durocher for Season," is one of Young's finest spot reporting pieces. Besides providing the facts of the suspension, Young concentrates on the obvious hypocrisy of Chandler's decision. Despite Young's personal dislike of Durocher – ironically a man much like himself – he believes that Durocher had been singled out for unfair treatment by the commissioner. Young notes that Rickey and Larry MacPhail, the two leading protagonists in the Dodger-Yankee feud, were fined $2,000 each, while Durocher's yearlong suspension cost him $60,000. Durocher's firing looked to Young like a put-up job, likely rooted in Chandler's personal dislike of Durocher for his adulterous affair with film star Laraine Day (whom Durocher later married). From Chandler's one-thousand-word statement Young strategically extracted the single phrase "conduct detrimental to baseball," and his story was accompanied by a *Daily News* photo of a glowering Chandler.

Young followed the Durocher suspension with a 15 April 1947 article that details Joe McCarthy's disinterest in managing Brooklyn and notes how one reporter (doubtlessly himself) stayed behind to quiz Rickey after the interview formally concluded. Once again, Young provided the key question: Would Durocher be back in 1948? Rickey hemmed and hawed, even when pressed, causing Young again to drop a payload final sentence: "And that from the man whom Durocher referred to as a father." Young then started pointedly referring to Durocher's successor, the elderly, mild-mannered Burt Shotton, who wore a suit and rarely left the Dodger dugout, as "Kindly Old Burt Shotton" – eventually shortening that to KOBS.

Young sometimes used his newspaper background to write an occasional magazine article. For example, his familiarity with Preacher Roe's "special pitch" led him to write "The Outlawed Spitball" for the 4 July 1955 issue of *Sports Illustrated* (later included in *Best Sports Stories: 1956*). This piece is largely one long quotation from Roe. Young only intervenes with such third-person phrases as "Roe lit another cigar" or "Roe was handed a ball. He gripped it on the slick leather." "The Outlawed Spitball" is an excellent example of the minimalist interview, where the interviewer leads and guides the interviewee, allowing him to speak for himself. Young does provide some historical background, relating how in 1947 Roe felt washed up, having gone 4–15 with the Pirates in 1946. Roe then discovered his money pitch, which he used only in emergencies and which was an open secret around the league. Roe was not the least apologetic, and Young captures his defiance in the closing quote. Roe mentions how the spitter lengthened his career, enabling him to get three World Series cuts, a nice home, and a successful business. As Roe says, "Not bad for a little ol' country boy. I'm for spitballs. I like 'em."

After the Giants and Dodgers left New York, Young campaigned relentlessly in print for a new National League franchise, which materialized in 1962 when the New York Mets brought their own brand of athletic ineptitude to the aging Polo Grounds. Young suffered with a dozing Casey Stengel and millions of New York fans until 10 September 1969, when the Mets pushed past the Cubs into first place in the National League East. Young celebrated this redemptive act in a 11 September 1969 *Daily News* piece, "Day to Remember . . . Years We Can't Forget." In this generally light-hearted article, Young offers prayers of thanksgiving to all who suffered and worked to make the Mets respectable. In a serious vein he thanks Bill Shea, whose badgering of the "Lords of Baseball" (Young's abusive term for the sport's owners, movers, and shakers) resulted in a National League New York franchise. Young further pays tribute to the patience and stoicism of Stengel. He then remembers with grace and humor the early years – the years of Marvelous Marv Throneberry, Hot Rod Kanehl, Don Bosch, Choo Choo Coleman, and the worst pitching staff in baseball. It is no wonder, Young believes, that banners advised "Cheat," or that an amorous couple was escorted from the Polo Grounds for providing more interesting action in the stands than there was on the diamond.

Yet in his column the following day, Young castigated Mets fans for waving white handkerchiefs at the Montreal Expos – "the Mets of 7 years ago." He remembers that in the early years, when the Mets were struggling to win 25 percent of their games, other teams laughed but did not taunt. As Young puts it, "They didn't pull the wings off a fly." His stand is not surprising – he always showed respect for competitive losers and magnanimous winners. Young trusts that the Mets fans' outburst may have been thoughtlessly spontaneous rather than malicious. He concludes, "I would like to think that the New Breed, in their triumphal emergence, will show a little class; that they will be worthy of the Amazin' Mets, the No. 1 Mets, because the Met fan was considered No. 1 long before the ballclub was."

Young's collective standards for correct behavior extended to individual athletes. He could be merciless on modern athletes, whom he often considered spoiled brats. One of his best columns in this vein is "The Joe Namath System," a 1970 *Daily News* column that won the general reporting award in *Best Sports Stories: 1971*. Young begins the piece with what George Vecsey would call a "knee-in-the-groin opening": "Hello, stupid. You reading this while hanging by one arm in a D Train? You going to work, stupid?"

Young makes the point that working stiffs obviously are incapable of understanding the "Joe Namath System." Blue-collar slobs from Brooklyn and Queens do not have the luxury of whining to Jets boss Phil Iselin about how difficult their lives are and how they need time off (or more money) to solve their personal problems. (Young further notes that Iselin is far less receptive to similar complaints in his dressmaking business.) In a column heavy with Swiftian righteous indignation, Young remarks that the Bowery is full of people on the Joe Namath System: "They refined the system to the ultimate. They blamed the whole world and got off. They moped about their problems for a few days, a few weeks, a few years, and eventually the problems went away." Young, the self-made man, is quick to dismiss Namath's depression or personal upsets. In a knockout conclusion Young compares Broadway Joe's plight to that of Roy Campanella's, the wheelchair-bound former Dodger catcher: "The other day I saw Roy Campanella. He was going to work. I guess he doesn't have any problems."

Young became an apologist for older players, frequently dead ones. While he loathed most drugs, considering them a sign of moral degeneracy, Young was more understanding of alcohol abuse. Consider his response to a disillusioned fan writing to Young about Babe Ruth's alcoholism: "On and off the field Babe Ruth played hard. He sometimes drank hard, but was no drunkard. He was something of a super-being, given to excess." Ruth is classified a Herculean hero and excused for an alcoholic flaw. Namath, on the other hand, is pilloried, and Dwight Gooden is ripped by Young ("Stand Up and Boo") when the Mets pitcher returned from drug rehabilitation in 1987.

Probably no modern athlete fascinated and repelled Young as much as Muhammad Ali, who outraged many of the sportswriter's personal beliefs. In a 27 April 1967 "Young Ideas" column, Young comments on Ali's refusal to be inducted into the army. In a most politically incorrect way Young sees Ali as a sincere but gullible young man, being manipulated by the Black Muslims, "a sad little sect." In 1967 Young was fifty and very much stuck in the 1930s and 1940s. Like many old New Dealers in the 1960s, Young saw the American involvement in Vietnam in the shadow of Munich: "the commitment has been made. We are there and we must see it through, with the Cassius Clays, white and black." No doubt Young is sincere, but his patriotism sounds rather strident. He had no understanding that Ali, and many others, did not want to fight a war they did not believe in.

Young made further comments about Ali after his suspension from boxing and his return to the ring in 1970. In his "Young Ideas" column for 4 March 1971, "No White Hopes, So Ali Colors Frazier White," Young notes that Ali resembles Congressman Adam Clayton Powell in his cultivation of a "professional blackness." Young argues that Ali's blend of "commercial blackness, narcissism, [and] flamboyancy" makes him anathema to many middle-aged whites who served in World War II or Korea. Such men see Ali's avoidance of military service as cowardly and are drawn to the more traditional values of Frazier. Ali gleefully endorses such a distinction, painting Frazier as a black "white hope." Young, while not doubting Ali's devotion to his Muslim faith, believes that Ali is a cheerful cynic who serves as a questionable role model for black youths. Not only has Ali told Whitey to stick it, but "he does it laughing, all the way to the bank."

Young's eye for boxing detail made him question whether Ali's three-year layoff had not fatally weakened him for a bout with Joe Frazier. "Mr. Ali Regrets He Won't Be Talking Tonight" (9 March 1971) offers a sympathetic, even moving, picture of a game Ali who fought courageously against a stronger, remorseless Frazier. Once again Young uses a handful of words to cut to the heart of the matter. Young finds that Ali's former "mystique is gone. . . . The layoff had rusted his dancing slippers, sucked him of his stamina." Yet Young points out that even Ali's harshest critics admit that he showed plenty of heart.

In columns later in March 1971 Young examines the two men's reactions to the fight. His 21 March column is notable both for its precision of language as well as its capturing the megalomania of a sports hero. Young compares the chastened Ali to Napoleon on the island of Elba, brooding over past defeats and yearning for a triumphal return: "Now he thinks he is Napoleon in defeat, beating the drums, saying he will return from Elba, sending word to his people, the dreamers, that he shall be back, wait for me, die for me." Young pictures the triumphant Frazier behaving in a much more realistic, low-key way. "You Don't Take That Kind of Punishment Twice" (10 March 1971) shows us a plain-as-water champion who fights, not as a symbol of a religious or political group, but as the proud practitioner of a demanding vocation. Out of his blue-collar ethic, Young feels a great affinity for Frazier's pride: "It is the pride of realism, the sweaty pride, the kind that sticks to the ribs."

Frazier was always a hero in Young's athletic pantheon. Other representative figures included Marciano and Hodges, as well as Baseball Commissioner Ford Frick. In all three cases, Young seemed to be looking back to a simpler, more beneficent time. In a column of 3 September 1969 he remembers the stoic Marciano as a man in charge of himself who did not need to be prodded to train. Young recalls Marciano as a man of modest talent but tremendous desire, a fighter who would make every sacrifice to win. Thus, Marciano has a serious commitment to preparation: "Condition is everything to me. . . . If I abuse my body, the other guy has an edge." How soothing this must have sounded to Young, who felt himself drowning in a sea of athletic "me-first" egotism – the partying life of Namath or the drug-filled world of Gooden. Young recalls that the last time he saw Marciano, Rocky was lamenting the lack of boxers worthy of their profession, commenting sadly, "There aren't any good young heavyweights around now." Young provides a bittersweet reply for the boxer who died on 31 August 1969, "Not as many great old ones, now, either."

Young always admired control and grace, two characteristics of Hodges, first as a Brooklyn Dodgers first baseman, then as Mets manager. In a 13 October 1969 story, Young pays tribute to a self-discipline as remarkable as Marciano's. Involved in a tense World Series game with the Orioles, Hodges watches seemingly impassively as pitcher Jerry Koosman struggles to maintain a thin lead. When Hodges does go to the mound, he is a relaxed and controlled presence, imparting confidence to all

*Young in rain gear at ringside for the 20 June 1980 Roberto Duran–Sugar Ray Leonard fight in Montreal (UPI/Corbis-Bettmann)*

around him. Young's admiration is profound: "Gil Hodges had a heart attack a year ago, and now he sits there watching all this, and my heart is pounding like crazy, and 50,000 other people's in the seats here. I don't know how he does it." For Young, Hodges epitomizes grace, bearing, control. When the Mets finally win, Hodges walks to the clubhouse, "his head tilted high, like some pedigreed pointer."

Young also admired moral courage off the field, and he has warm praise for Ford Frick's memoirs in his "Young Ideas" column of 5 April 1973. At times Young gushes over Frick's *Games, Asterisks and People* (1973), hailing him as a "Walt Disney of the written word." Young rushes on to wholesomeness, praising Frick's book "wherein the only four letter words are Play Ball, mere ballplayers are swapped, and the pictures show Connie Mack with his clothes on." Young's enthusiasm almost approaches a Grantland Rice innocence that was perhaps inappropriate in a man whom Roger Kahn in *The Era* (1994) described as a notorious philanderer.

One achievement Young praises Frick for is his help in desegregating baseball. Not only does Young endorse Frick's demythologizing of baseball "saint" Kenesaw Mountain Landis as a diehard segregationist, he praises Frick's handling of the threatened 1947 Saint Louis Cardinals players' strike against Jackie Robinson. When Cardinal owner Sam Breadon asked what he should tell his recalcitrant players, Frick snapped, "Tell them this is America, and baseball is America's game. Tell them they will be barred from baseball for life." This firm moral stand resonated with Young, who admired heroes who cut through flummery and evasiveness to deliver clear messages like a firebell in the night: "Tell them this is America."

A certain elegiac tone to Young's writing could be heard as he grew older. In one article, for example, pitcher Sandy Koufax and announcer Rex Barney come together at the 1969 World Series between the Orioles and Mets to reminisce about the old days in Brooklyn. At one point a frowning Barney remarks, "Do you realize . . . we've lost three of our guys in the last week," naming Danny O'Connell, Willie Ramsdell, and Don Hoak, three former Dodgers who died in 1969. The message here seems to be that, while the Mets play entertaining, and now winning baseball, they are not the old Brooklyn Dodgers who played real baseball in cotton uniforms on real grass on sunny afternoons.

Perhaps one of the saddest themes in Young's columns of the 1960s and 1970s is that of the old heroes from an early era who cannot let go of the past. One such player was Robin Roberts, former Phillie whiz kid, who was toiling away for Reading in the Eastern League in 1967. At age forty Roberts, with 286 major league wins, wanted one more crack at the big leagues to win 300 (he did not get it). Young asks Curt Simmons, the interviewee and Roberts's former teammate, why Roberts does not retire. Simmons replies: "He could get a job. He could get into radio or sell insurance. He doesn't want that. Baseball is his life." Similarly, Willie Mays, playing for the Mets at age forty-two, seems wedded to a dying past. In a 15 April 1973 story Young relates how Mays pulled himself from the game in the sixth inning because of shoulder and knee pain. Young's commentary captures the pathos of a championship athlete's fading powers: Mays "has been trying to minimize it, hide it, so that runners won't take advantage. It can no longer be hidden."

Having worked with Branch Rickey, Jackie Robinson, Roy Campanella, and Mel Ott and having covered ball games in Ebbetts Field and the Polo Grounds, Young found Tom Seaver and Shea Stadium a disappointment. Young had walked with his own heroes on the plains of windy Troy – the present was a pale shadow of those glory days. Seaver was no Erskine, Cleon Jones no Willie Mays, the Ali-Frazier match no Louis-Conn match. It was galling to Young to see once-great athletes like Roberts and Mays continue to play when he believed that they should have retired from the game.

Young resented no symbol of modern life more than television. His feuds with technical people in locker rooms were legendary, with the five-foot-seven-inch Young offering to take on any technician in the house. One award-winning article in particular, "The Barbie Doll Soap Opera," reprinted in *Best Sports Stories: 1977,* demonstrated Young's scorn for what modern sports had become. Young's thinking put him at odds with the dean of New York sportswriters at the time, Red Smith of *The New York Times.* Smith, seventy-two years old in 1977, had a more gentlemanly, protective view of women that made him think of locker rooms as off-limits. Young typed Smith as a hopelessly outdated sentimentalist. According to Ira Berkow in *Red: A Biography of Red Smith* (1987), Young cruelly referred to Smith as "a pitiable shell of a once great writer, reduced to lingering on Connie Mack and Ed Barrow." Also according to Berkow, Smith was not especially perturbed. When asked to respond to Young's charge, Smith replied, "No comment. Nothing Dick Young writes merits comment."

A further surprise to many Young watchers was his reconciliation with Muhammad Ali, an event chronicled in Young's "Muhammad Ali" in December 1986 *Sport,* published just eight months before Young's death. In a surprisingly mellow article Young surveys his history with Ali, reminding his readers that in the 1960s he was wary of the heavyweight champion. Eventually Young realized that beneath his slick surface, Ali was a decent man, devoted to his family and possessed of courage and athletic skill. The final breakthrough occurred in 1981, following Trevor Berbick's pounding of Ali. With Ali holding his daughter on his lap in his makeshift dressing room, Young realized that the battered fighter possessed all the qualities he had admired all his life – ability, courage, and most important, class. A magical transformation takes place: "Ali's daughter slid from his lap. I was crouched in front of him, taking notes. I put down the pad. Ali and I embraced. We kissed on the lips."

Young excelled at "just-the-facts-ma'am" journalism and found a unique way to tell each story. As Randall Poe noted in *Esquire,* October 1974: "Young's work is coarse and simpleminded, like cave painting. But it is superbly crafted." Young took his profession seriously and achieved great depth, even profundity, in his analysis of sports.

Roger Kahn describes Young in *The Boys of Summer* (1987) as "spiky, self-educated and New York" with a penchant for extramarital sex and heavy drinking. Wearing a signature bright bow tie, waving a cigar in one hand and a drink in the other, Young was a familiar figure at New York sporting events for more than forty years. Aggressive, belligerent, and knowledgeable, he was intolerant of evasion and subterfuge – his caustic "Whaddya mean" could cut like a knife through deceitful rhetoric. His fanatical defense of the First Amendment often led to confrontation. When a much bigger Duke Snider got pushy with Young because of his rough handling of a Snider slump, Young advised him to hit baseballs, not writers. Likewise, when a towering Larry Holmes ordered him out of his dressing room because he did not like "what you say about me," Young shot back, "You've got a right to say things and I don't?"

While Young's high standards applied to everyone else, they did not always apply to his own personal life. As one anonymous colleague observed, "Nobody could live by the rules Dick set down for other people . . . so he didn't try to." Michael Katz, a younger writer on the *Daily News,* had pointed things to say about Young, whom he judged the quintessential New Yorker: "Dick Young was New York. Maybe he was the New York you always got in an argument with, but he was ours, and I wouldn't trade him for all the Carl Sandburgs in

Chicago." Katz further commented on the heavy streak of moralizing that runs deeply through Young's work: "Everything was black or white. Young was not very strong on gray." (*New York Daily News,* 2 September 1987).

In both *The Boys of Summer* and *The Era,* Kahn portrays Young as intelligent, worldly, knowledgeable, crude, and abrasive and concluded that Young's unsettled early years may have taken a heavy toll later. For all Young's exuberance, Kahn found the extroverted older man desperately scarred, "a man . . . who has been hurt by life and who lives behind rings of fortification." Young, of course, would dismiss such an evaluation corrosively, as he did when Wetzsteon quizzed him about the effects of growing up without his father: "Sure, my parents were divorced when I was 3, but it pisses me off when I hear about some guy who sobs his way to the electric chair because he came from 'a broken home.' I came from a broken home, and I always felt I was one of the luckiest guys alive."

Kahn also acknowledges that Young taught him much about writing sports. First, Young practiced an extraordinary work ethic, accumulating sports wisdom through diligent, if not relentless, application. Second, he taught Kahn the importance of the angle, of finding out the key ingredient in any sports contest. Speaking in *The Boys of Summer* about covering baseball games, Young urged Kahn to approach each story in an original way: "Now you're gonna write the game most of the time. Nothing you can do about that and it ain't bad. But anytime you . . . can get your story off the game you got to do it. Because that's unusual and people read unusual things. Fights. Bean Balls. Whatever. Write them, not the games." Finally, Young presented Kahn with the example of the well-crafted style that crackled across the pages of the *Daily News* and *Post* for forty-five years. Always provocative, Young held the attention of millions of readers for half the twentieth century. Katz, in his *New York Daily News* article, probably best expressed the firm hold Young exercised over his readers: "But as angry as Dick would make me, he was still the first thing I read in the morning."

Dave Sloan, in the *TSN Official 1988 Baseball Guide,* offered this assessment: "Young was perhaps the most colorful, innovative, controversial and disliked sportswriter of the past 50 years." Young himself wrote two epitaphs. The first is professional: "I'm not a writer. I'm a newspaperman." The second is personal: "I have a job that lets me laugh while I work. What more can a man ask?" But the simplest, and most telling, of all tributes comes from Vecsey, who, like Kahn, Pepe, and Katz, was a Young protégé: "Dick Young covered the Dodgers and the Mets, and in his time he was a giant."

**References:**

Roger Kahn, *The Boys of Summer* (New York: Harper & Row, 1987);

Kahn, *The Era* (New York: Ticknor & Fields, 1994);

Randall Poe, "The Writing of Sports," *Esquire,* 82 (October 1974): 173–176, 373–374, 376–379;

David Quentin Voight, "From Chadwick to the Chipmunks," *Journal of American Culture,* 7 (Fall 1984): 31–37;

Harry F. Waters, "Voice of the People," *Newsweek,* 81 (21 May 1973): 60, 65;

Ross Wetzsteon, "Dick Young's America," in *Best Sports Stories of 1986* (Saint Louis: Sports Network, 1986).

# Checklist of Further Readings


Adams, Caswell. *Fifty Famous Sports Stories.* New York: Stamford House, 1946.

Betts, John R. "Sporting Journalism in Nineteenth Century America," *American Quarterly*, 5 (Spring 1953): 39–56.

Cooke, Bob. *Wake Up the Echoes: From the Sports Pages of the New York Herald Tribune.* New York: Hanover House, 1956.

Danzig, Allison, and Peter Brandwein, eds. *Greatest Sports Stories from the New York Times.* New York: Barnes, 1951.

Danzig and Brandwein, eds. *Sport's Golden Age: A Close-Up of the Fabulous Twenties.* New York: Barnes, 1948.

Frank, Stanley. *Sports Extra.* New York: Barnes, 1944.

Holtzman, Jerome. *No Cheering in the Press Box,* revised and enlarged edition. New York: Holt, 1995.

Kluger, Richard. *The Paper: The Life and Death of the* New York Herald Tribune. New York: Knopf, 1986.

Krim, Seymour. "Sportswriting: Square and Avante-garde," in his *Views of a Nearsighted Cannoneer,* enlarged edition. New York: Dutton, 1968, pp. 198–212.

Lipsyte, Robert. *Sportsworld: An American Dreamland.* New York: Quadrangle, 1975.

Marsh, Irving T., and Edward Ehre, eds. *Thirty Years of Best Sports Stories.* New York: Dutton, 1974.

Oriard, Michael. *Dreaming of Heroes: American Sports Fiction, 1868–1980.* Chicago: Nelson-Hall, 1982.

Oriard. *Reading Football: How the Popular Press Created an American Spectacle.* Chapel Hill: University of North Carolina Press, 1993.

Plimpton, George, ed. *The Norton Book of Sports.* New York: Norton, 1993.

Poe, Randall. "The Writing of Sports," *Esquire*, 82 (October 1974): 173–176, 373–379.

Reisler, Jim. *Black Writers/Black Baseball: An Anthology from Black Sportswriters Who Covered the Negro Leagues.* Jefferson, N.C.: McFarland, 1994.

Rice, Grantland, and Harford Powel, eds. *The Omnibus of Sport.* New York: Harper, 1932.

Russell, Fred. *Bury Me in an Old Press Box.* New York: Barnes, 1957.

Silverman, Al, ed. *The Best of Sport, 1946–1971.* New York: Viking, 1972.

Silverman and Brian Silverman, eds. *The Twentieth Century Treasury of Sports.* New York: Viking, 1992.

Walker, Stanley. "Sports – Valhalla's Bull-Pen," in his *City Editor.* New York: Stokes, 1934, pp. 115–133.

Ward, Arch, ed. *The Greatest Sports Stories from the Chicago Tribune.* New York: Barnes, 1953.

Wheeler, John. *I've Got News For You.* New York: Dutton, 1961.

# Contributors

| | |
|---|---|
| Lawrence Baldassaro | *University of Wisconsin – Milwaukee* |
| R. C. Brignano | *Pennsylvania State University, Beaver Campus* |
| Pete Cava | *International Sports Associates* |
| Scott A. G. M. Crawford | *Eastern Illinois University* |
| William Curran | *Silverton, Oregon* |
| J. Douglas English | *Johns Hopkins University* |
| Tom Gallagher | *San Francisco* |
| Steven P. Gietschier | The Sporting News |
| Dennis Gildea | *Springfield College* |
| Adam R. Hornbuckle | *History Associates Incorporated* |
| Bill Hughes | *Brooklyn, NY* |
| James C. Kaufman | *Los Angeles* |
| Robert Klein | *Editorial Board, NINE* |
| Bill Knight | *Western Illinois University* |
| Mike Lackey | The Lima News |
| Terence Malley | *Long Island University* |
| William J. McGill | *Lebanon Valley College* |
| Andre Milner | *Upper Darby, Pennsylvania* |
| Tom Nawrocki | Worth *Magazine* |
| Maggie Nichols | Field & Stream |
| James E. Odenkirk | *Arizona State University* |
| Richard Orodenker | *Pennsylvania State University, Ogontz* |
| Sanford Pinsker | *Franklin and Marshall College* |
| William J. Plott | The Birmingham News |
| Josh Riley | *Marieta, Georgia* |
| Harold Rosenthal | *Bocca Raton, Florida* |
| Paul J. Sandin | *Indianapolis, Indiana* |
| David Sanders | *Harvey Mudd College* |
| Mark W. Schraf | Spitball *Magazine* |
| Leverett T. Smith Jr. | *North Carolina Wesleyan College* |
| Joel Sternberg | *Saint Xavier University, Chicago* |
| Edward J. Tassinari | *SUNY Maritime College* |
| James A. Vlasich | *Southern Utah University* |
| Timothy E. Walsh | *Butler County Community College* |
| Mark Royden Winchell | *Clemson University* |
| William H. Wood | *Community College of Philadelphia* |
| Jack Ziegler | *Cochise College* |

# Cumulative Index

*Dictionary of Literary Biography,* Volumes 1-171
*Dictionary of Literary Biography Yearbook,* 1980-1995
*Dictionary of Literary Biography Documentary Series,* Volumes 1-14

# Cumulative Index

**DLB** before number: *Dictionary of Literary Biography,* Volumes 1-171
**Y** before number: *Dictionary of Literary Biography Yearbook,* 1980-1995
**DS** before number: *Dictionary of Literary Biography Documentary Series,* Volumes 1-14

## A

## B

## C

## D

## E

## G

## H

# I

## J

## K

## L

## M

## N

## O

## P

## Q

## S

## T

## U

## V

## Y

## Z

ISBN 0-8103-9934-2
90000
9 780810 399341